P9-DUQ-045

Basic Physiology

Edited by P. D. Sturkie

With Contributions by
T. M. Casey, H. M. Frankel,
P. Griminger, R. L. Hazelwood,
C. H. Page, P. D. Sturkie,
E. J. Zambraski

With 286 Figures

Springer-Verlag
New York Heidelberg Berlin

Dr. PAUL D. STURKIE, Emeritus Professor of Physiology, Department of Physiology, Rutgers University, New Brunswick, N.J. 08903, U.S.A.

Library of Congress Cataloging in Publication Data

Basic physiology.

Bibliography: p.
Includes index.
1. Human physiology. 2. Vertebrates—Physiology.
I. Sturkie, Paul D. [DNLM: 1. Physiology. QT
4 B311]
QP34.5.B36 612 80-27740

Printed in the United States of America

9 8 7 6 5 4 3 2 1

ISBN 0-387-90485-9 Springer-Verlag New York Heidelberg Berlin
ISBN 3-540-90485-9 Springer-Verlag Berlin Heidelberg New York

Preface

Basic Physiology is an introduction to vertebrate physiology, stressing human physiology at the organ level, and including requisite anatomy integrated with function. One chapter deals solely with topographic anatomy in atlas form and microscopic anatomy of the principal tissues of the body. Additional chapters cover cellular and general physiology; nervous system, muscle; blood and tissue fluids, heart and circulation; respiration, digestion and absorption; intermediary metabolism; energy metabolism; temperature regulation; nutrition; kidney; endocrinology, including hypophysis, reproduction; thyroids, parathyroids, adrenals and pancreas. All concepts are emphasized and well illustrated, and controversial material is omitted. It is written at a level suited to undergraduate students who have had introductory courses in biology, chemistry, and mathematics, and to more advanced students who wish to review the basic concepts of physiology.

This volume should be especially useful as a text for departments of biology, zoology, nursing, health, and agricultural sciences that offer courses in vertebrate and human physiology.

Basic Physiology is written by seven subject matter specialists who have considerable experience in teaching their specialty to undergraduates studying physiology and biology.

Paul D. Sturkie

Table of Contents

List of Contributors

T. M. Casey	Department of Physiology, Rutgers University, New Brunswick, New Jersey 08903, U.S.A.
H. M. Frankel	Department of Physiology, Rutgers University, New Brunswick, New Jersey 08903, U.S.A.
P. Griminger	Department of Nutrition, Rutgers University, New Brunswick, New Jersey 08903, U.S.A.
R. L. Hazelwood	Department of Biology, University of Houston, Central Campus, Houston, Texas 77004, U.S.A.
C. H. Page	Department of Physiology, Rutgers University, New Brunswick, New Jersey 08903, U.S.A.
P. D. Sturkie	Department of Physiology, Rutgers University, New Brunswick, New Jersey 08903, U.S.A.
E. J. Zambraski	Department of Physiology, Rutgers University, New Brunswick, New Jersey 08903, U.S.A.

Chapter 1

Tissues, Organs, and Skeletal Organization

The body is made up of systems that comprise organs, tissues, and cells. These systems, which carry out all of the body's functions include the: 1) skeletal, 2) muscular, 3) nervous, 4) circulatory, 5) digestive, 6) respiratory, 7) urogenital (including reproductive and excretory systems), and 8) endocrine.

The unit of structure and function of systems, organs, and tissues is the cell, discussed in Chapter 2.

Anatomy is the study of the structure of an organ or system, both gross and microscopic. Gross or systematic anatomy is concerned with the appearance and characteristics of a system or organ as seen with the naked eye. Microscopic anatomy (histology and cytology) deals with structure at the level of tissues and cells, respectively.

This chapter will deal briefly with the histology of the principal tissues and organs systems of the body, and with gross and topographic anatomy. Further anatomic details will be considered and integrated with the physiology under the appropriate chapters and headings.

Terminology

The terms used to describe location and position of bodily parts are listed in Table 1.1 and Fig. 1.1

Planes or sections of the body and tisues are depicted in Fig. 1.2 and include: 1) sagittal, medial, or longitudinal; 2) transverse, cross, or horizontal; and 3) frontal or coronal.

The most commonly used histologic specimens are those that are cut in cross and longitudinal sections.

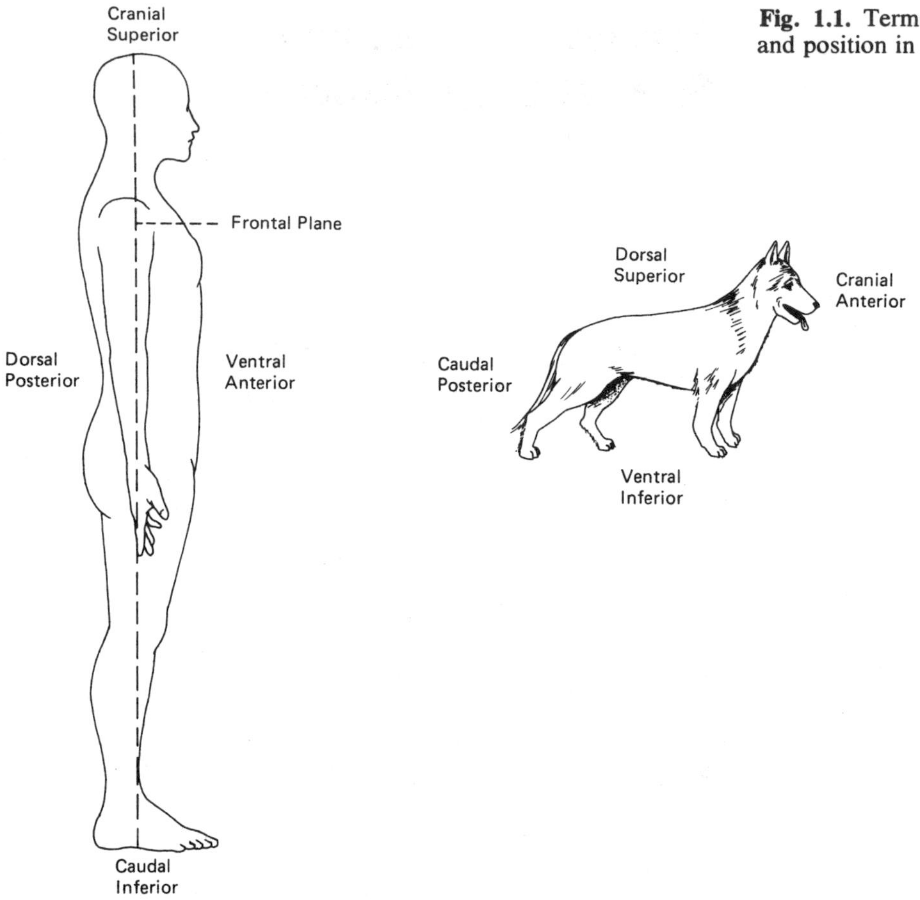

Fig. 1.1. Terms designating location and position in man and quadruped.

Table 1.1. Terms of location and position of body parts in man and quadruped

Man	Quadruped
Superior or cranial	*Anterior* or cranial
Anterior or ventral	Inferior or *ventral*
Posterior or dorsal	Superior or *dorsal*

Medial—near middle or median; nose is medial to eyes
Lateral—to the side (farther); eyes are lateral to nose
Proximal—nearest to; proximal joint of toe is nearest to foot
Distal—farther from; toenail is distal to foot
Peripheral—extension from central; spinal nerves are peripheral to the brain
Visceral—organs within a cavity; intestinal organs

* The commoner term is shown in italics.

Tissues

Organs and systems of the body are composed of one or more tissues with characteristic structures. They include: 1) epithelium, 2) skin, 3) membranes, 4) muscle, 5) nerve, 6)

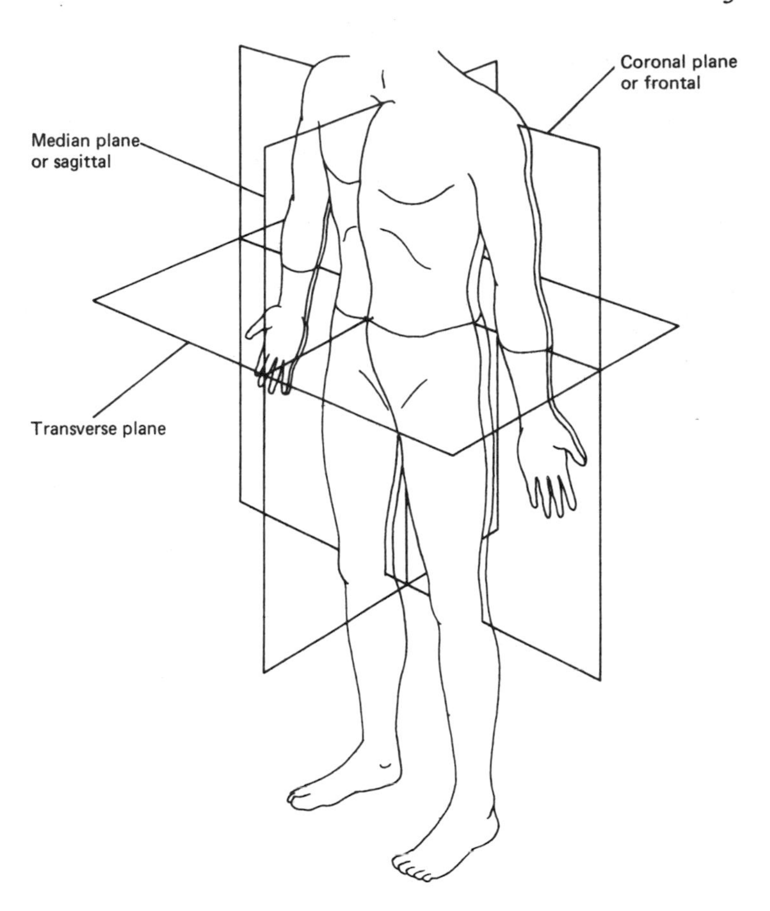

Fig. 1.2. The planes of the body.

connective tissue, 7) elastic tissue, 8) cartilage, and 9) bone. These tissues are derived embryologically from three germ layers, including: 1) ectoderm, 2) entoderm, and 3) mesoderm. Certain organs and tissues are derived from these germ layers as shown in Fig. 1.3.

Fig. 1.3. Derivatives of the three embryonic germ layers.

▼

Ectoderm

- Outer epithelium of body → Hair, skin, teeth, enamel, oral and anal epithelium, and sweat and mammary glands
- Lens of eyes, auditory vesicle, inner ear, and optic vesicle
- Nervous system → Neural tube, nerves, spinal cord, ganglia, and brain

Entoderm

→ Epithelium of: Digestive tube, liver, pancreas, urinary bladder, pharynx, trachea, bronchi, lungs, thyroids, parathyroids, and thymus

Mesoderm

→ Supporting tissue for organs: Bone, cartilage, muscle, connective and elastic tissue, and endothelium of blood vessels

Epithelial Tissues

Epithelial tissues are composed of adhering cells, arranged to form a covering membrane over the external surfaces and internal (lining) surfaces of the body. There are no blood vessels in these cells. Types and location of these cells include:

Simple squamous: One layer of flat cells found lining alveoli of lungs, lens of eye, and part of inner ear.
Stratified squamous: Flat and many layered cells found ir epidermis of skin (Fig. 1.4A). As skin thickens, flat squa mous cells are rubbed off and new cells are formed from lower layers (stratum germinativum).
Endothelial: One layer of cells lining interior surface of blood and lymph vessels.
Columnar: Cells are tall and cylindrical; nucleus is located near base of cell; cytoplasm contains numerous mitochondria and golgi apparatus (Fig. 1.4C); found in the lining of stomach and intestines; the cells may be plain or have cilia (hairlike motile projections); there may also be microvilli, small folds in the vascular projections of intestines that increase the absorption surface.
Cuboidal: Similar to columnar cells, but cuboidal and found in the liver.

Glands and Membranes

Exocrine glands have secretions which drain into a duct and are then carried to the bloodstream in contrast to a *ductless gland* (see Chap. 25). They have different shapes and numbers of ducts and are lined with columnar, cuboidal, or flat-squamous epithelium. The glands are classified according both to type and shape and to whether they are branched or unbranched (simple); their ducts may also be branched or unbranched (Fig. 1.5).
Epithelium and the underlying tissues form a membrane, a structure covering or enveloping another tissue (see Chap. 2). These membranes line the body cavities and organs therein and may secrete a thin serous fluid or a thick mucous one.
The lining of blood vessels is an endothelial membrane made up of endothelial cells and connective tissue. Other membranes include: *pericardium* (surrounding the heart), *peritoneum* (lining the abdominal cavity), and *pleura* (two in number, covering the lungs and chest cavity).
Synovial membranes are associated with bones and muscles and consist of a tough outer layer of fibrous tissue and an inner layer of areolar connective tissue containing elastic and collagenous fibers (Fig. 20.2). They secrete synovial fluid (tissue fluid and mucin), which serves as a lubricant to protect knee and elbow joints and to reduce friction therein.
Mucous membranes secrete mucus and line the alimentary

Fig. 1.4.A–C. Cellular morphology. **A** Flat or squamous epithelium. **B** Same type of epithelium in skin, showing three layers of epidermis and dermis, with nerve cells, tactile corpuscle and sweat glands. **C** Columnar (plain), ciliated, and with microvilli as seen in intestine, and connective tissue.

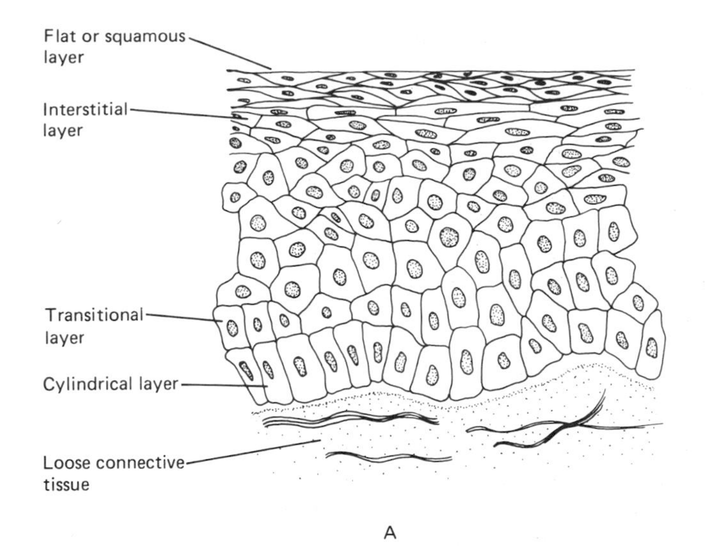

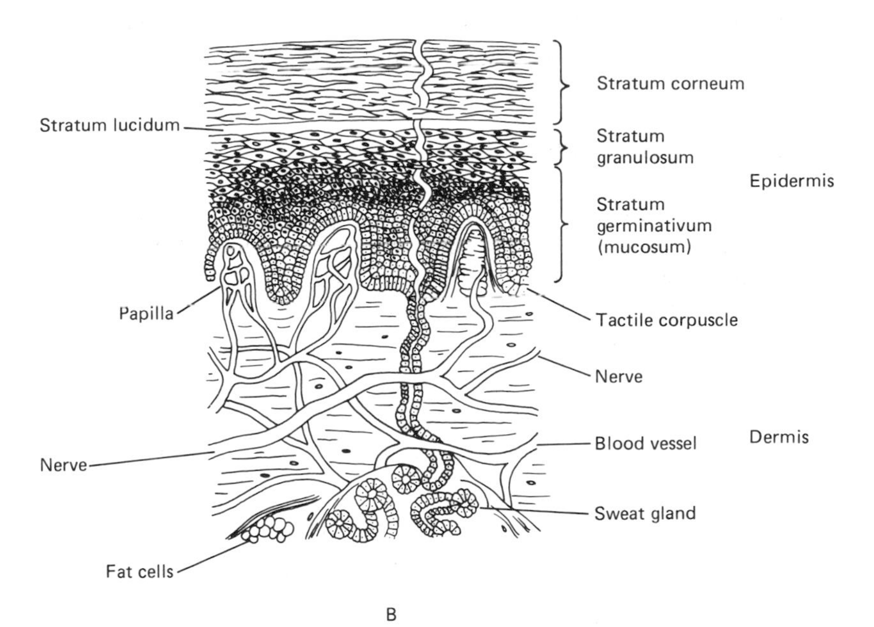

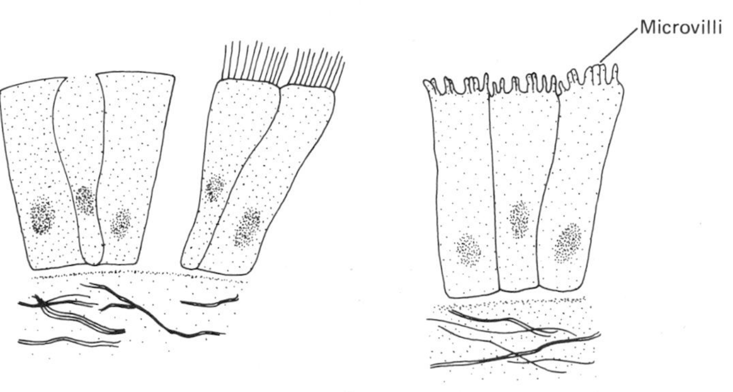

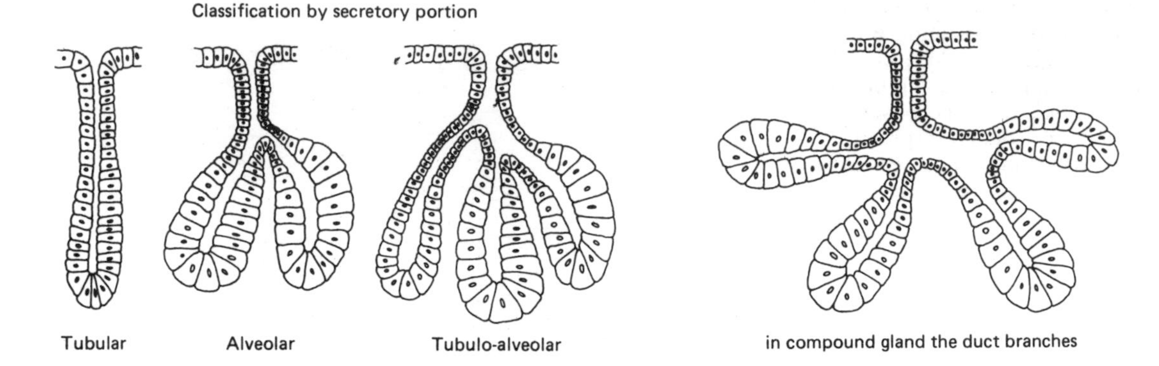

Fig. 1.5. Types of glands and branching of glands and ducts.

canal or digestive tube and trachea, bronchi and air sacs of lungs, and many other organs. A mucous membrane is made up of three layers including: 1) epithelium, 2) a supporting lamina propria, 3) a layer of smooth muscle (Fig. 20.2). The function of mucous membranes is 1) to support and protect blood vessels and lymphatics and 2) to provide for large absorptive surfaces such as occur in the intestines.

Connective Tissues

These tissues support and connect various structures. Although they contain few cells, the intercellular cement or fluid is abundant, and the tissue is highly vascular. The intercellular material and fibers vary and largely determine the type of classification.

Classification

The classification of connective tissue (CT) is as follows:

Table 1.2

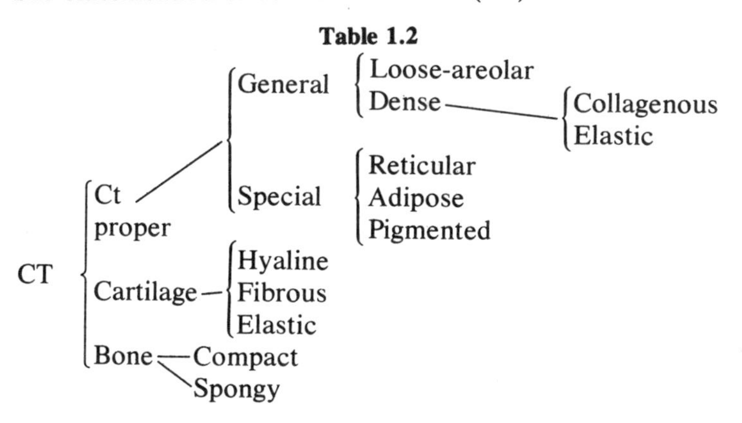

In CT proper the intercellular substance is soft, in cartilage it is firm but flexible, and in bone it is rigid because calcium salts have been deposited in the matrix.

Areolar CT is loose, widely distributed and it consists of cells and fibers in a fluid matrix forming a soft and displaceable and translucent tissue (Fig. 1.6A). The fibers are collagenous and elastic, forming a framework for most organs and an envelope or sheath for blood vessels and nerves. The

Fig. 1.6.A–C. Types of connective tissues (CT). **A** Showing elastic (*a*) and collagenous fibers (*b*), fibroblasts (*c*), plasma cell (*d*), histiocytes (*e*), mast cells (*f*), and lymphocytes (*g*). **B** Reticular CT showing reticular and collagenous fibers and cells. **C** Elastic CT seen in cross section of an artery in the tunica intima as black lines.

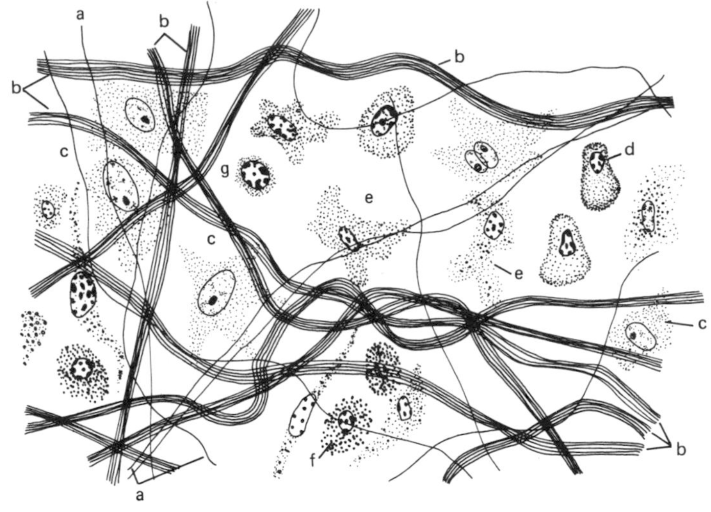

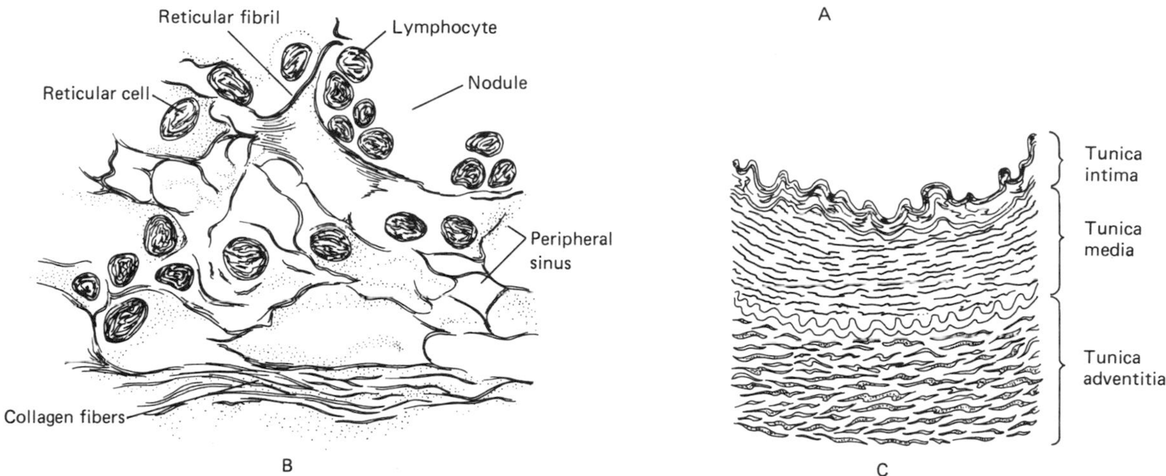

principal cell types are fibroblasts and histiocytes; the latter are phagocytic, i.e., they can physically engulf and digest cells.

Reticular CT is characterized by a cellular network (reticulum) of cells and fibrils, both reticular and collagenous (Fig. 1.6B). Reticular tissue forms the framework for lymphatic tissue (lymph nodes), spleen, and bone marrow; the last-mentioned also contains lymphocytes.

Adipose CT is characterized by large, ovoid fat cells, and the cytoplasm contains fat droplets. When the fat is dissolved out of the preparation (Fig. 1.7A), the cells appear as empty rings or ovals. Adipose CT is found in subcutaneous fatty deposits throughout the body.

Elastic CT is loose CT in which there is a preponderance of elastic fibers. The tissue is distensible and occurs in lung, some cartilage, blood vessels (Fig. 1.6B), bronchial tubes, and elastic ligaments.

Fibrous CT is loose CT in which collagenous (fibrous) fibers predominate. Large coarse whitish fibers, they are tough but not distensible. This tissue forms: 1) ligaments, which help to hold bones together at joints, and 2) tendons, which

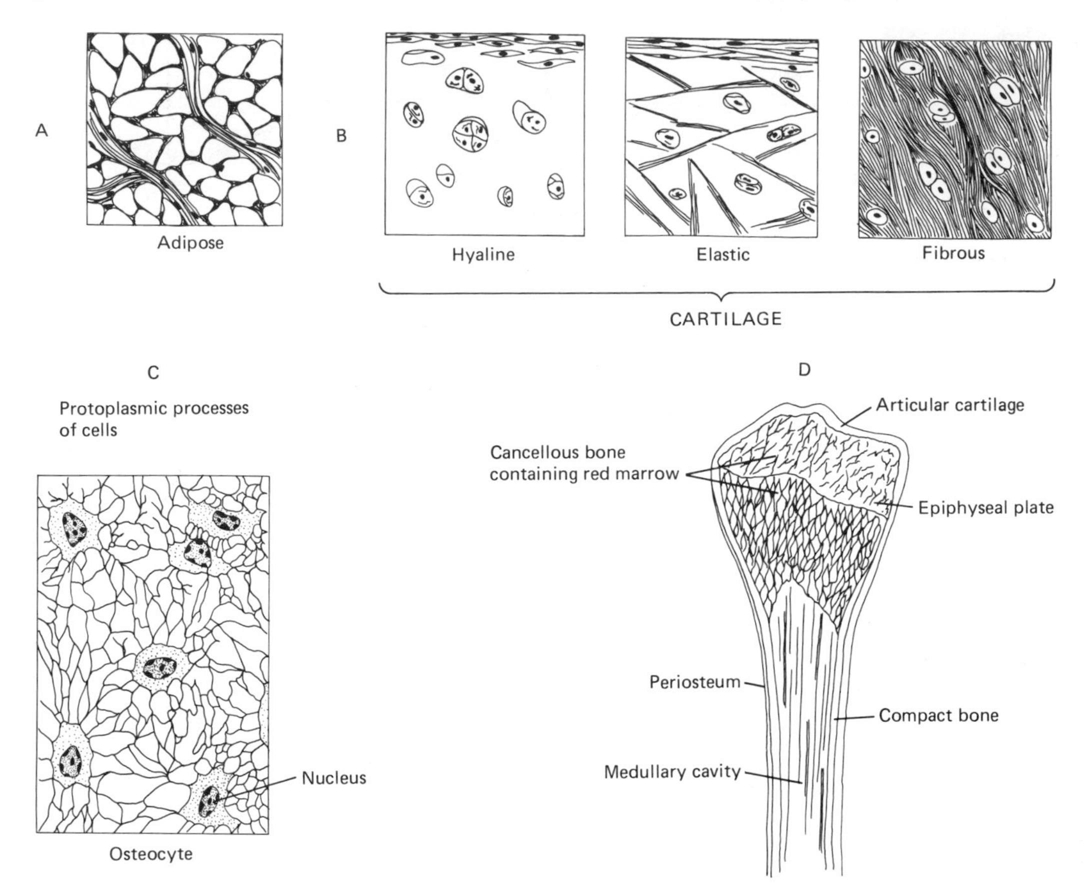

Fig. 1.7.A–C. Adipose tissue, cartilage, and bone. **A** Adipose tissue showing vacuoles (fat cell deposits absorbed) and fibers. **B** Hyaline, elastic, and fibrous cartilage. Cartilage cells in matrix containing elastic fibers and fibrous fibers. **C** Cross section of compact bone, showing cells (osteocytes) in matrix. **D** Longitudinal section of bone showing compact and cancellous (spongy) bone.

attach the muscles to bones and which are found in certain membranes that protect organs like heart and kidney.

Cartilage

Cartilage is firm but flexible and consists of cells suspended in a variable matrix, or intercellular cement and fibers (Fig. 1.7). *Hyaline* cartilage is a bluish-white, homogenous mass, containing cells and collagenous fibers. It covers the ends of bones at the joints and is found in the ventral ends of ribs of costal cartilage. It is also found in the nose, larynx, trachea, and bronchial tubes.

Elastic cartilage is CT in which elastic fibers predominate (Fig. 1.7). Found in the larynx, epiglottis, and external ear, it helps to maintain the shape of these organs but permits flexibility. *Fibrous* cartilage contains cells inbedded in fibrous connective tissue. Joining certain bones together, it forms a strong flexible connection where strength and rigidity are required.

Bone

Bone is connective tissue in which the intercellular substance (matrix) is made hard by salts of calcium and phosphate. This inorganic part comprises about two-thirds the weight of bone. The organic part consists of bone cells (osteocytes and osteoclasts), some cartilage and blood vessels, and nerves (see Chap. 28). Types of bone include: 1) spongy or cancellous (Fig. 1.7), and 2) compact.
The skeleton of an early embryo is preformed in cartilage; later ossification (change of cartilage to bone) begins and continues after birth. In the long bones (Fig. 1.7) the diaphysis (compact bone) is the center of ossification, which proceeds toward the epiphyses, or the ends of bones. Thin layers of cartilage extend between diaphysis and epiphysis.
Normal bone is constantly being resorbed by osteoclasts and reformed by osteocytes (see Chap. 28).

Muscle

Muscle tissue is made up of cells and cellular processes, or fibers. Three types of muscle fibers are delineated, based upon appearance under the light microscope and function. They include: 1) skeletal, 2) cardiac, and 3) smooth or visceral.
Skeletal muscles are under voluntary control by the somatic nervous system, but smooth and cardiac muscle are under involuntary control of the sympathetic nervous system. Skeletal muscles as the name implies are attached to bones of skeleton (see Figs. 1.13 and 1.14); cardiac muscle covers the heart, but smooth muscle forms the muscular portion of visceral organs (digestive organs) and blood vessels.
Skeletal muscle is made up of bundles, or fasciculi, containing many muscle fibers (Fig. 1.8A). Each bundle is separated by connective tissue that supplies blood vessels and nerves to the bundles. The individual fibers are cross striated (Fig. 1.8B) and enclosed in a tubular sheath (sarcolemma). Lengths range from 1 to 40 mm, and diameters from 0.01 to 0.15 mm; the fibers are made up of fibrils (myofibrils) that are close packed and which run lengthwise of the muscle fiber. For additional details on structure, see Chap. 11. The color of the striations alternate from light (isotropic, I band) to dark (anisotropic, A band). The combination of an A and I band is called a *sarcomere*. Each band is bisected by a thin dark line called the Z band.
The structure of *cardiac muscle* is similar to that of skeletal in that it is also striated, but unlike the latter there are points at the Z lines where individual fibers abut (merge) or *interdigitate* to form *intercalated discs* (Fig. 1.8C). They produce a strong union between fibers so that contraction in one unit or fiber can be easily transmitted to another unit. Cardiac fibers branch and interdigitate, but each fiber is enclosed by a separate membrane. Cardiac muscle receives

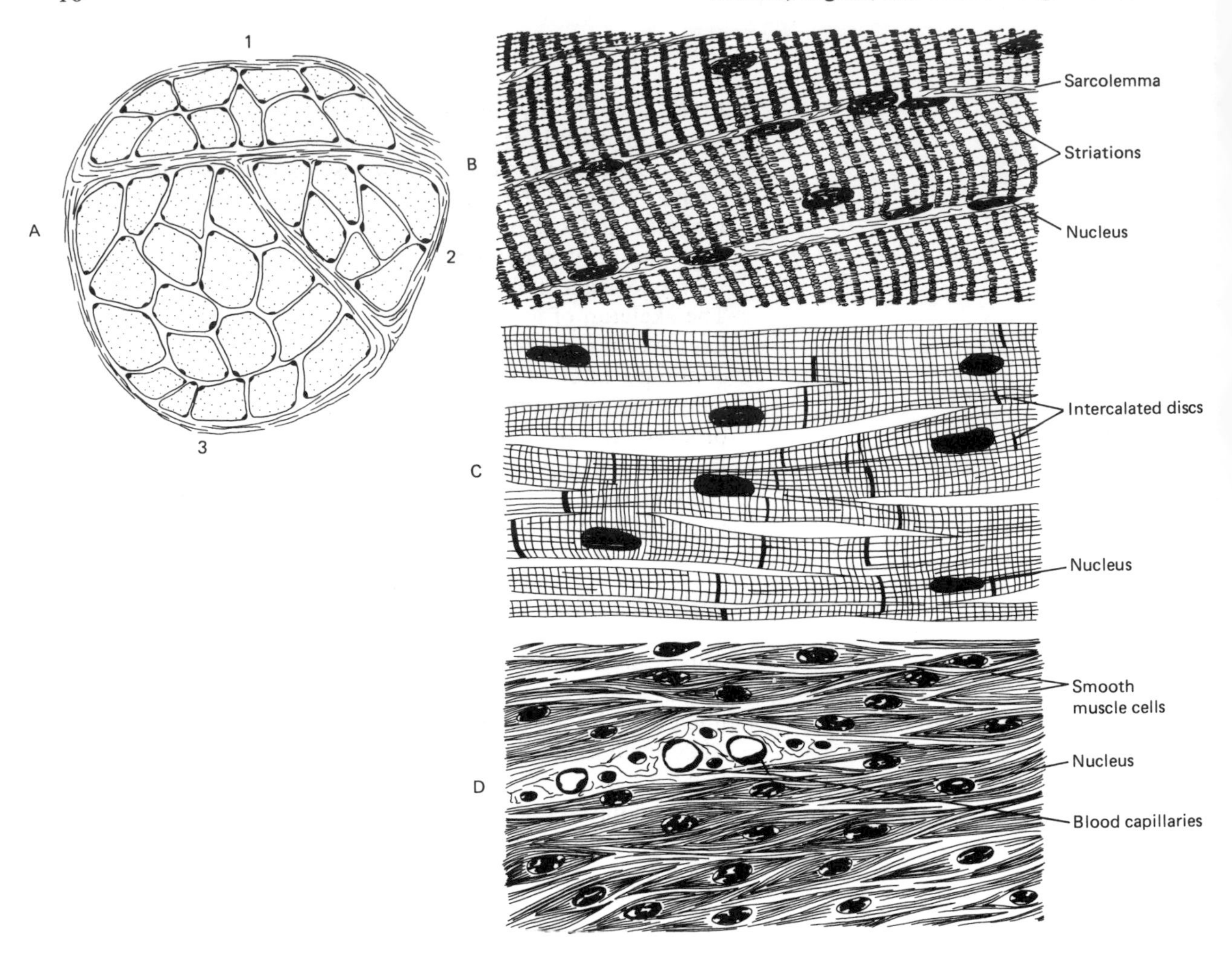

Fig. 1.8.A–D. Types of muscle. **A** Cross section showing three (1,2,3) muscle bundles (fasciculi), containing fibers and the connective tissue between the fibers and the bundles. **B** Longitudinal section of skeletal muscle showing individual fibers, sarcolemma, and cross-striations (dark and light bands). **C** Cardiac muscle showing striations and merging of fibers (intercalated discs). **D** Smooth muscle showing individual cells and fibers but no striations.

blood from the coronary arteries. See Chaps. 11 & 15 for further details on cardiac muscle.

Smooth muscle fibers are not striated (Fig. 1.8D) but are rather long and narrow; each fiber, however, is much shorter than skeletal fibers. Each fiber has only one nucleus, unlike skeletal muscle which has many. Each spindle-shaped cell is about 0.015 to 0.5 mm long and 0.002 to 0.02 mm in diameter. Smooth muscle in the visceral organs and blood vessels is usually arranged in two layers including an inner thick *circular* one, and an outer thin *longitudinal* one. The blood vessels run parallel to the smooth muscle fibers and between the muscle bundles (Fig. 1.8D).

Nervous Tissue

Nerve tissue is composed of cells call *neurons;* they differ somewhat in structure and function from other cells. Neurons display the unique functions of 1) irritability, 2) excitability, and 3) conductivity. In other words, neurons receive impulses and conduct or transmit responses to these impulses. Most of the details on the anatomy and histology of

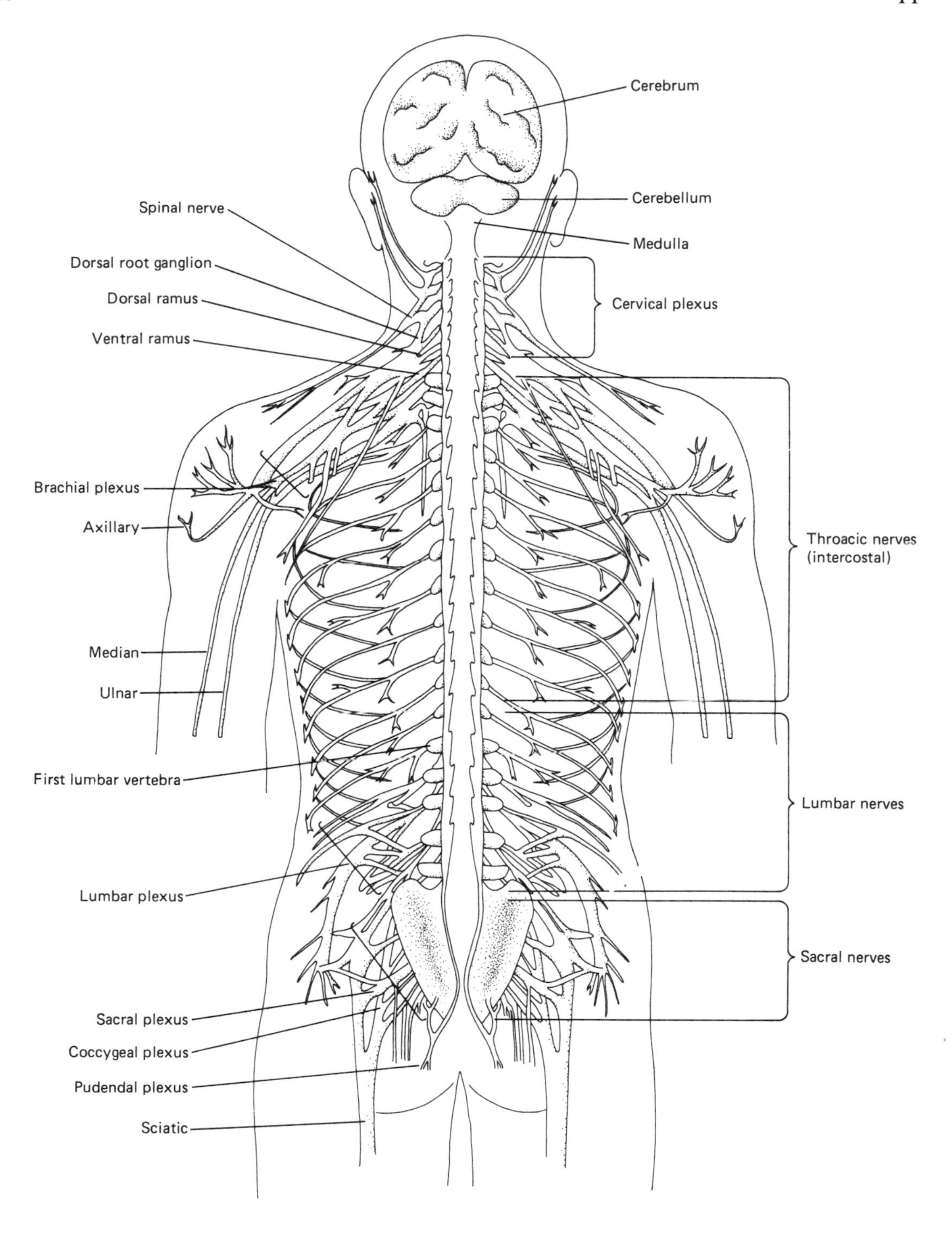

Fig. 1.9. The spinal nerves and plexuses.

the nervous system is discussed in the chapters on the nervous system. Distribution of the peripheral nerves, their emergence from the spinal cord, and their relationship to regional vertebral ganglia and nerve plexuses are shown in Fig. 1.9.

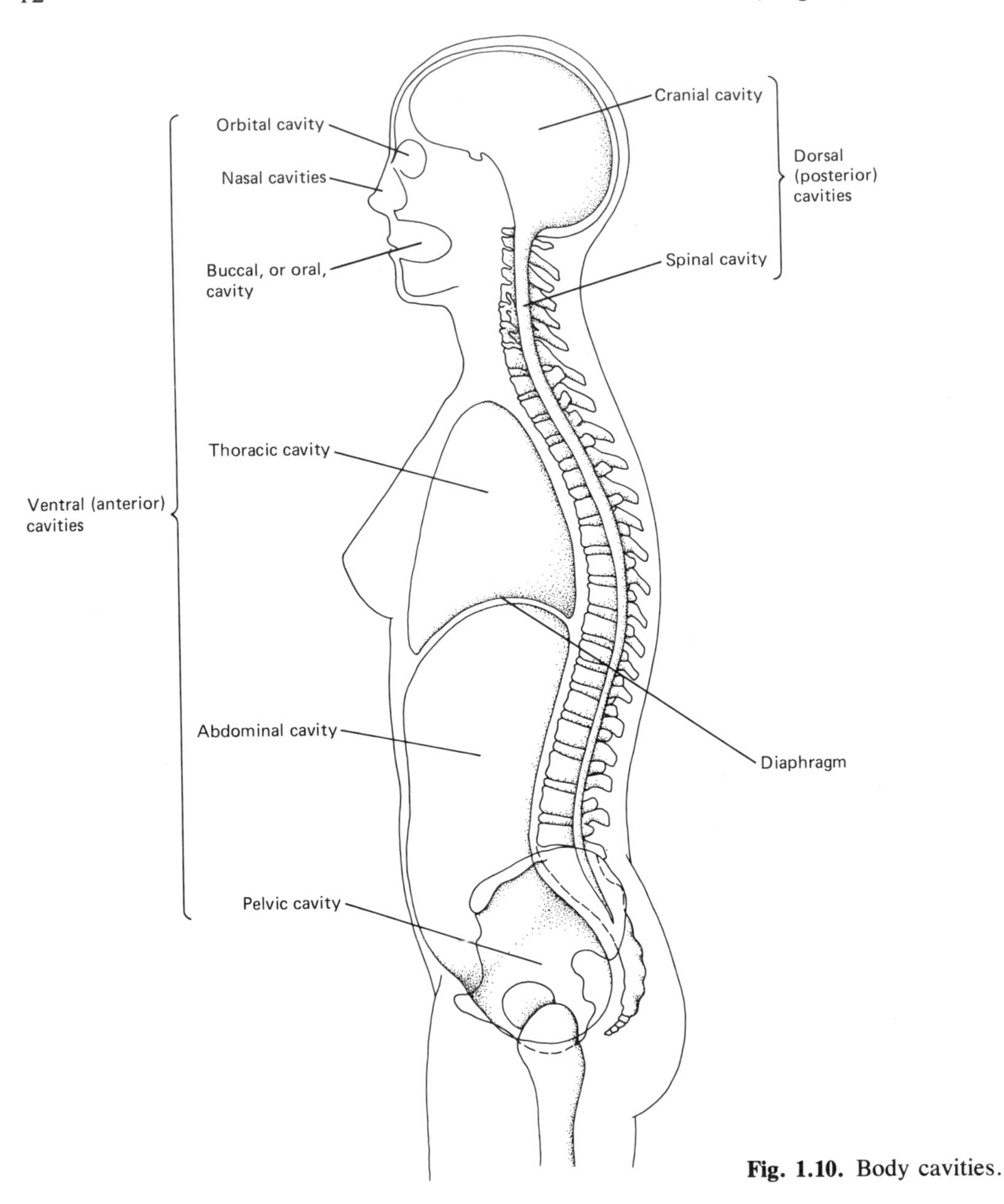

Fig. 1.10. Body cavities.

Anatomy

The body is supported by the skeletal and muscular systems, which enclose several body cavities containing organs (Figs. 1.10 and 1.11). The thoracic cavity contains the lungs, heart, trachea, and esophagus and is bounded and enclosed caudally by the diaphragm. The abdominal cavity comprises the liver, spleen, gall bladder, stomach, small intestines, and most of the large intestines. The pelvic cavity encloses the lower part of the large intestine, colon, rectum, urinary bladder, and some of the reproductive organs.

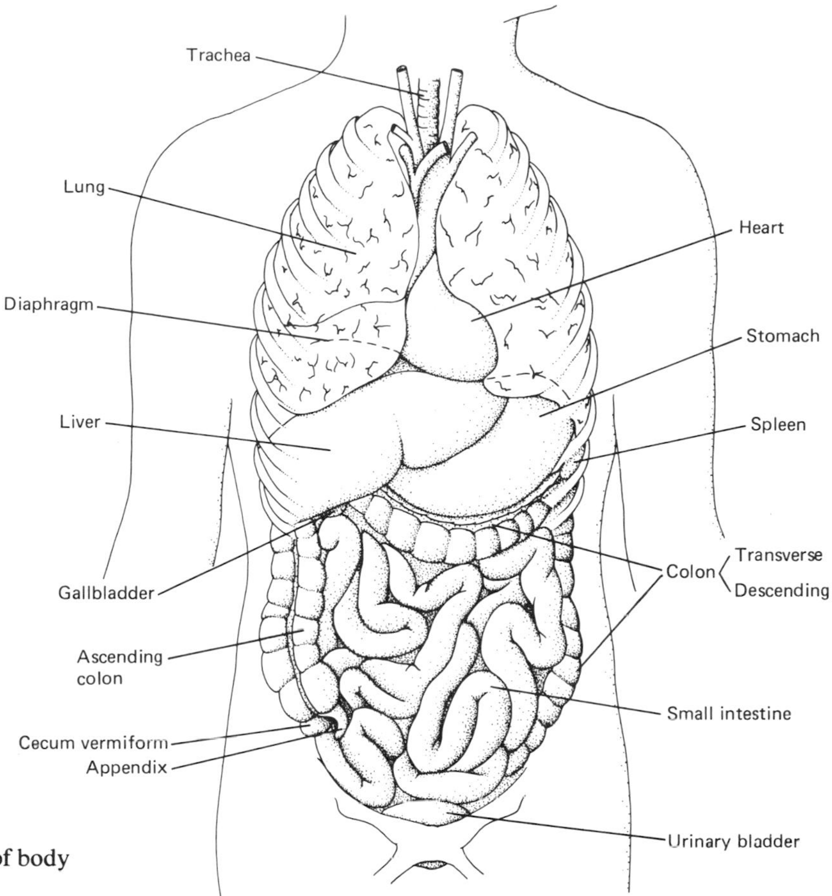

Fig. 1.11. Anterior projection of body showing principal organs.

Skeletal System

The skeleton of the human adult is composed of about 200 bones classified as follows:

Spine and vertebral column	26
Cranium	8
Face	14
Hyoid bone, sternum, and ribs	26
Upper extremities	64
Lower extremities	62
	200

The names and locations of the principal bones of the body are shown in Figs. 1.12 and 1.13. Bones are classified as long, short, flat, and irregular. *Long bones* are found in the limbs, clavicle, humerus, tibia, fibula, metacarpals, metatarsals, and phalanges. Most long bones are built for strength and support. *Short bones,* designed for some strength and extensive motion, consist of a number of small pieces, such as the bones of carpus, tarsus, and patella. *Flat bones* provide extensive protection and broad surfaces for attachment of muscles; bones of the skull and shoulder blades are exam-

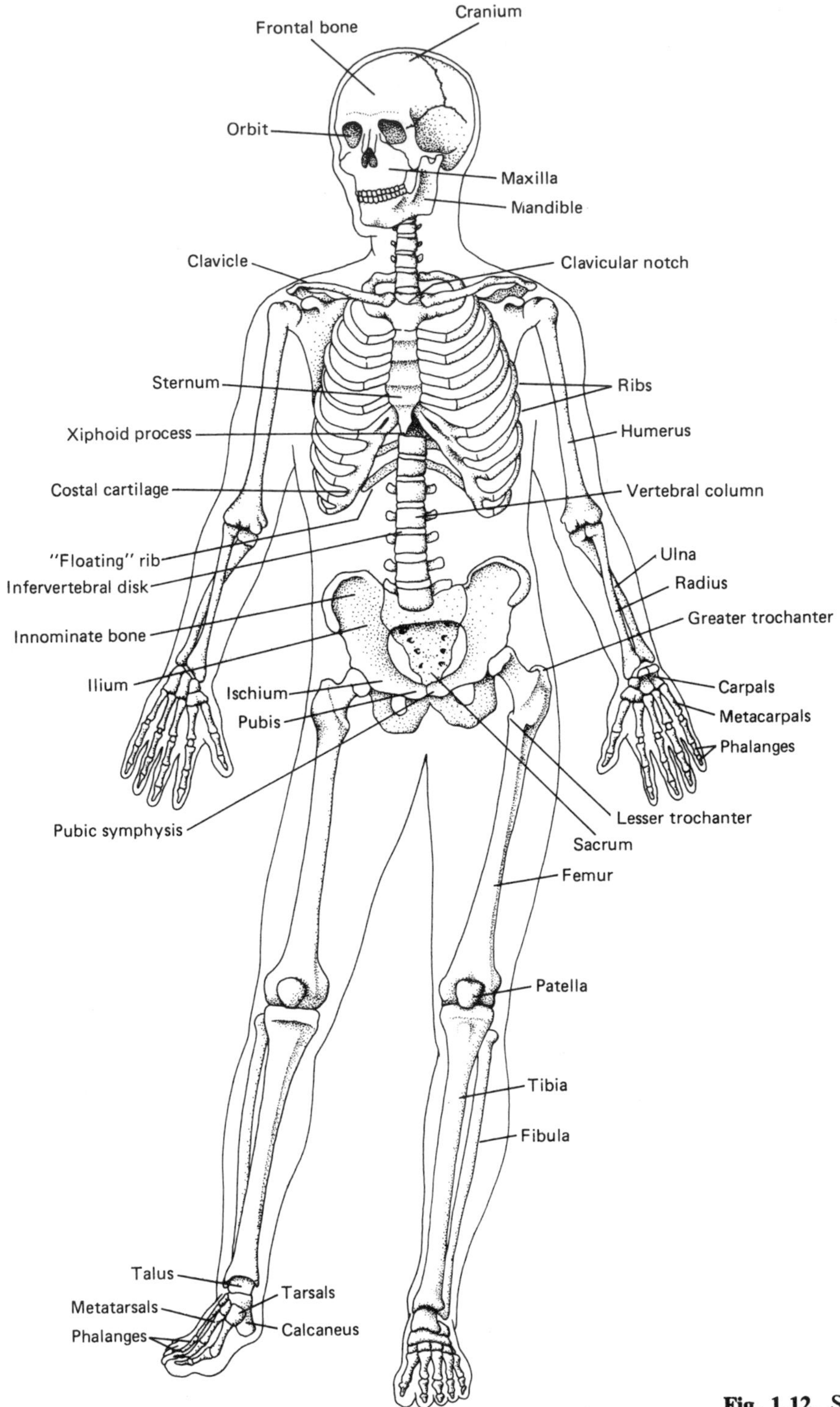

Fig. 1.12. Skeleton, anterior view.

ples. *Irregular,* or mixed bones include vertebral, sacral, coccygeal, temporal, sphenoid, maxillary, and others. Their structures are similar to flat and long bones and incorporate features of both. They likewise contain both compact and spongy bone.

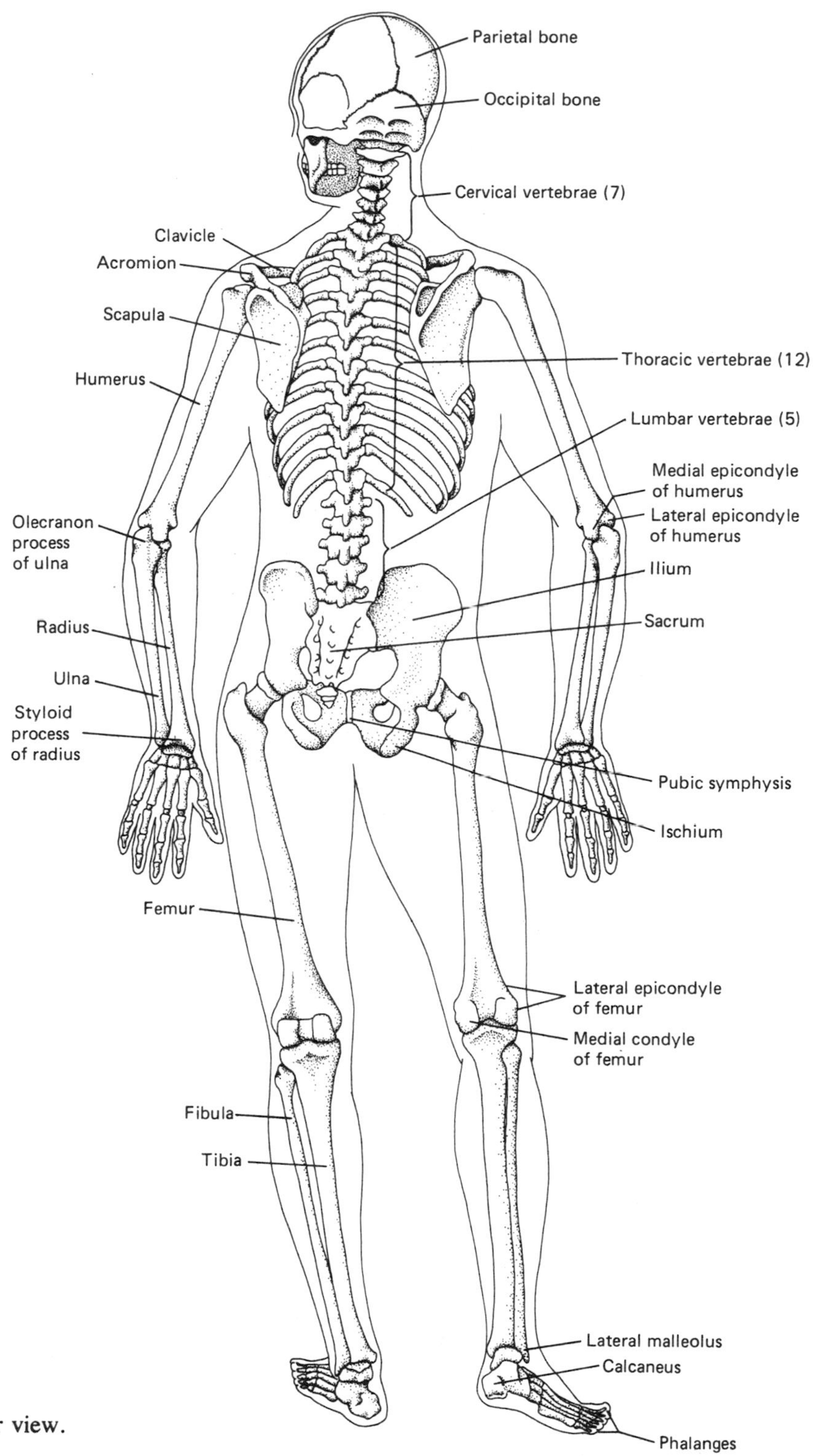

Fig. 1.13. Skeleton, posterior view.

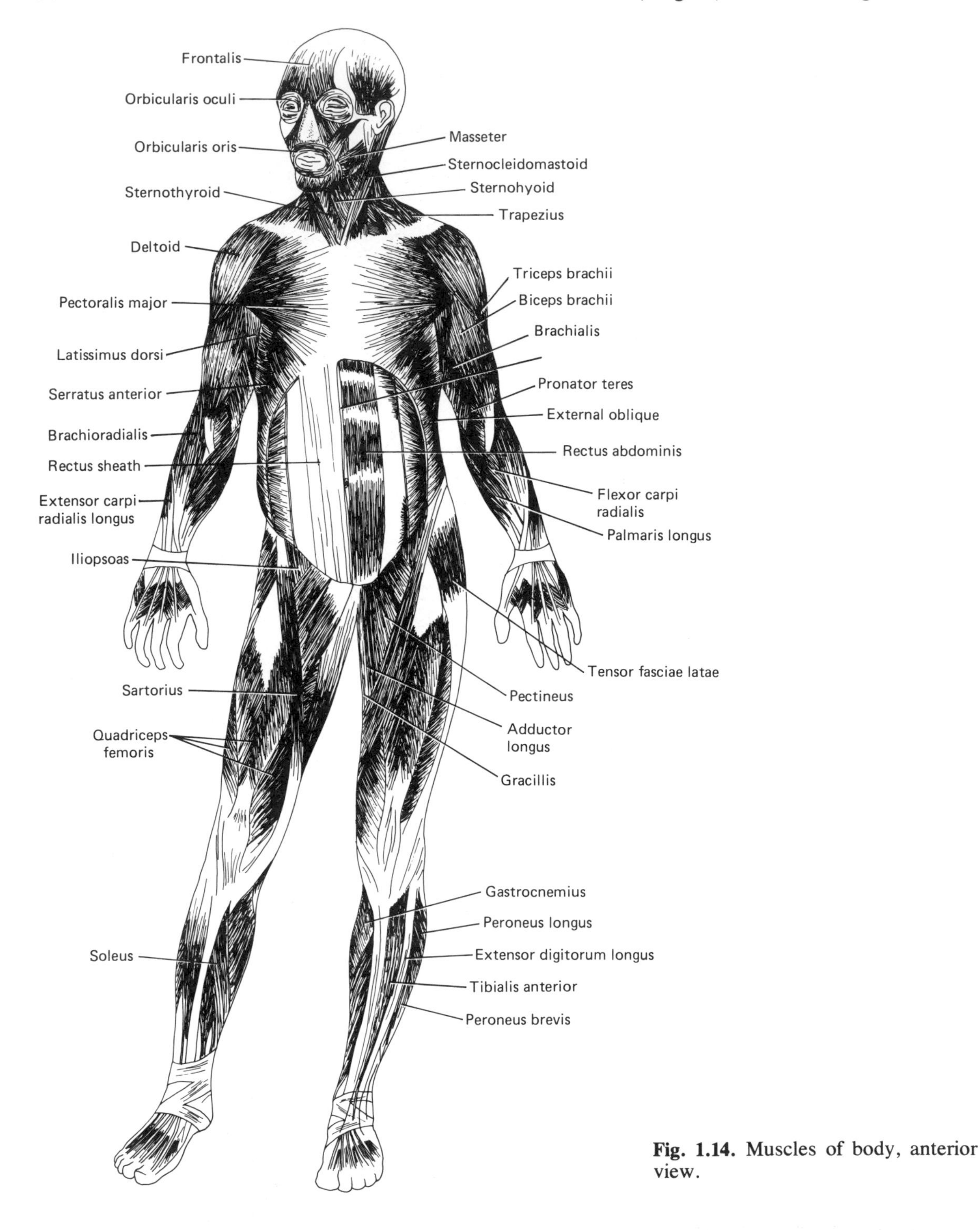

Fig. 1.14. Muscles of body, anterior view.

Muscular System

The skeletal muscles are connected to the bones and to other structures either directly or by the intervention of fibrous tendons (aponeuroses). The size and shape of muscles vary considerably, depending on the bone that is covered. They are long in the legs and broad and flattened in the

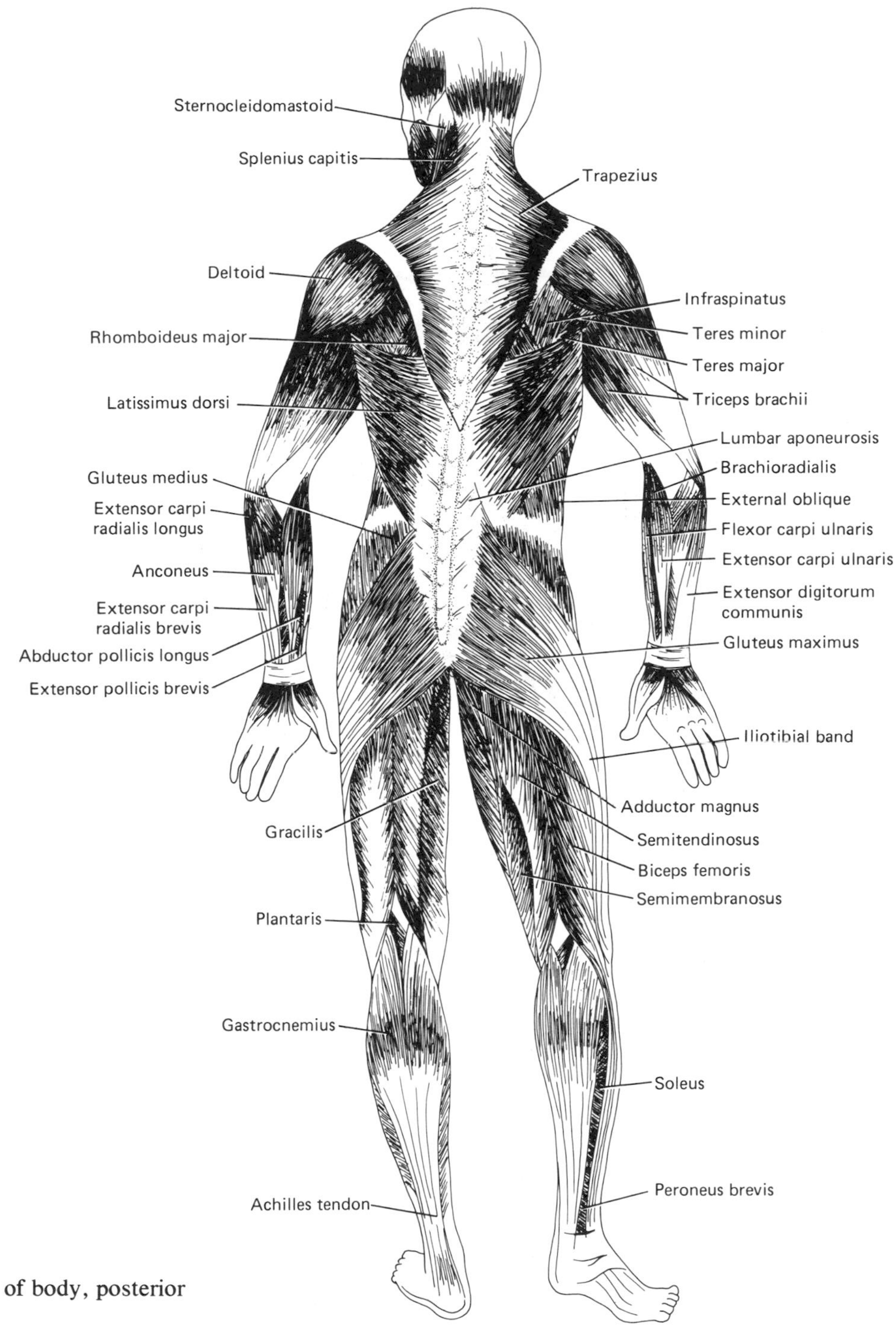

Fig. 1.15. Muscles of body, posterior view.

trunk. The gastrocnemius muscle is very large, and the fibers of the sartorius muscle are nearly two feet long.

The *arrangement* of the muscle fibers also varies, depending on the specific functions required (see Chap. 11). The names and location of the principal external muscles of the body are shown in Fig. 1.14 and 1.15.

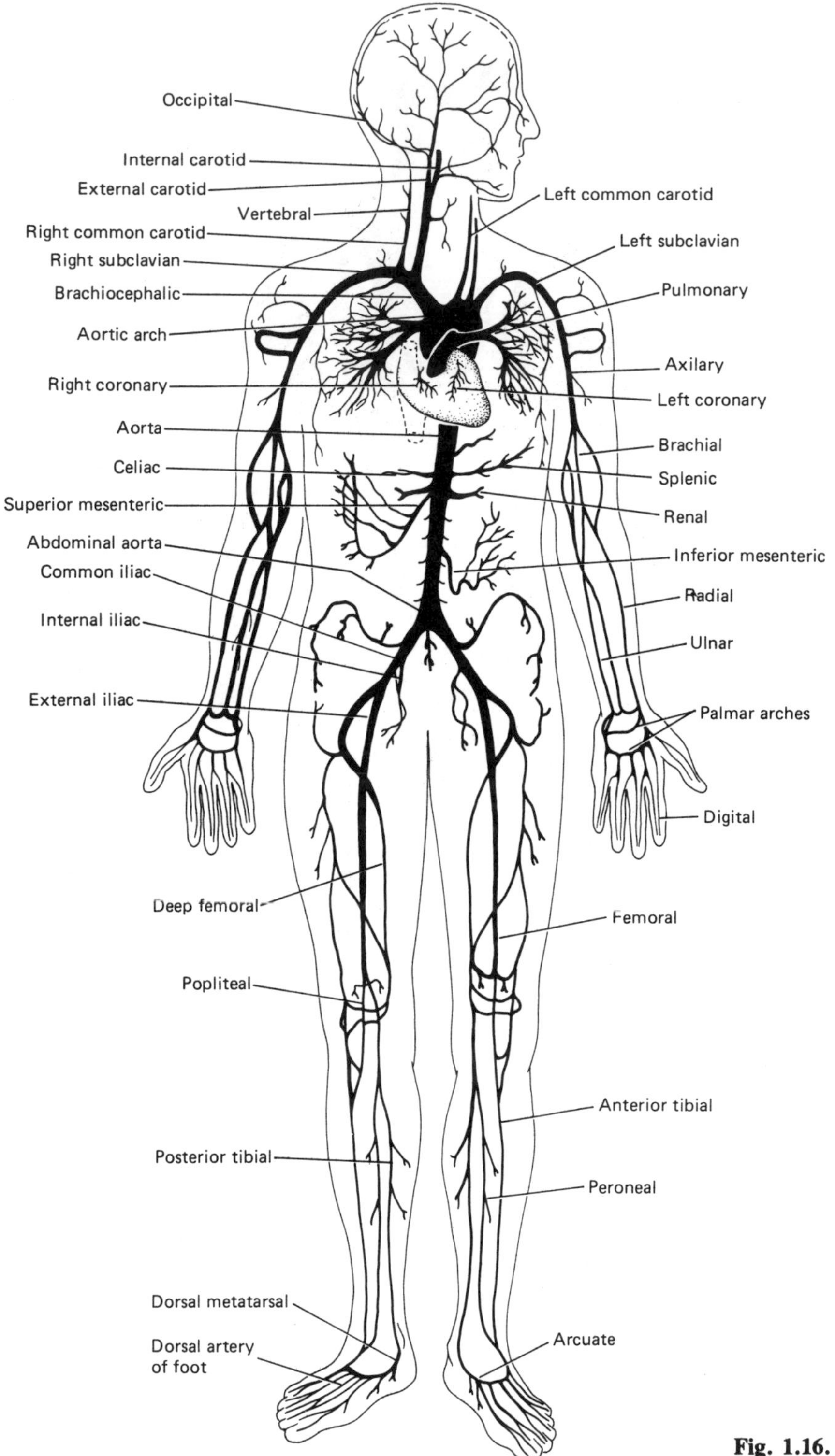

Fig. 1.16. Major arteries of body.

Circulatory System

This system comprises the heart, blood, and blood and lymph vessels. For additional details on anatomy and histology, see the chapters dealing with the heart and circulation.

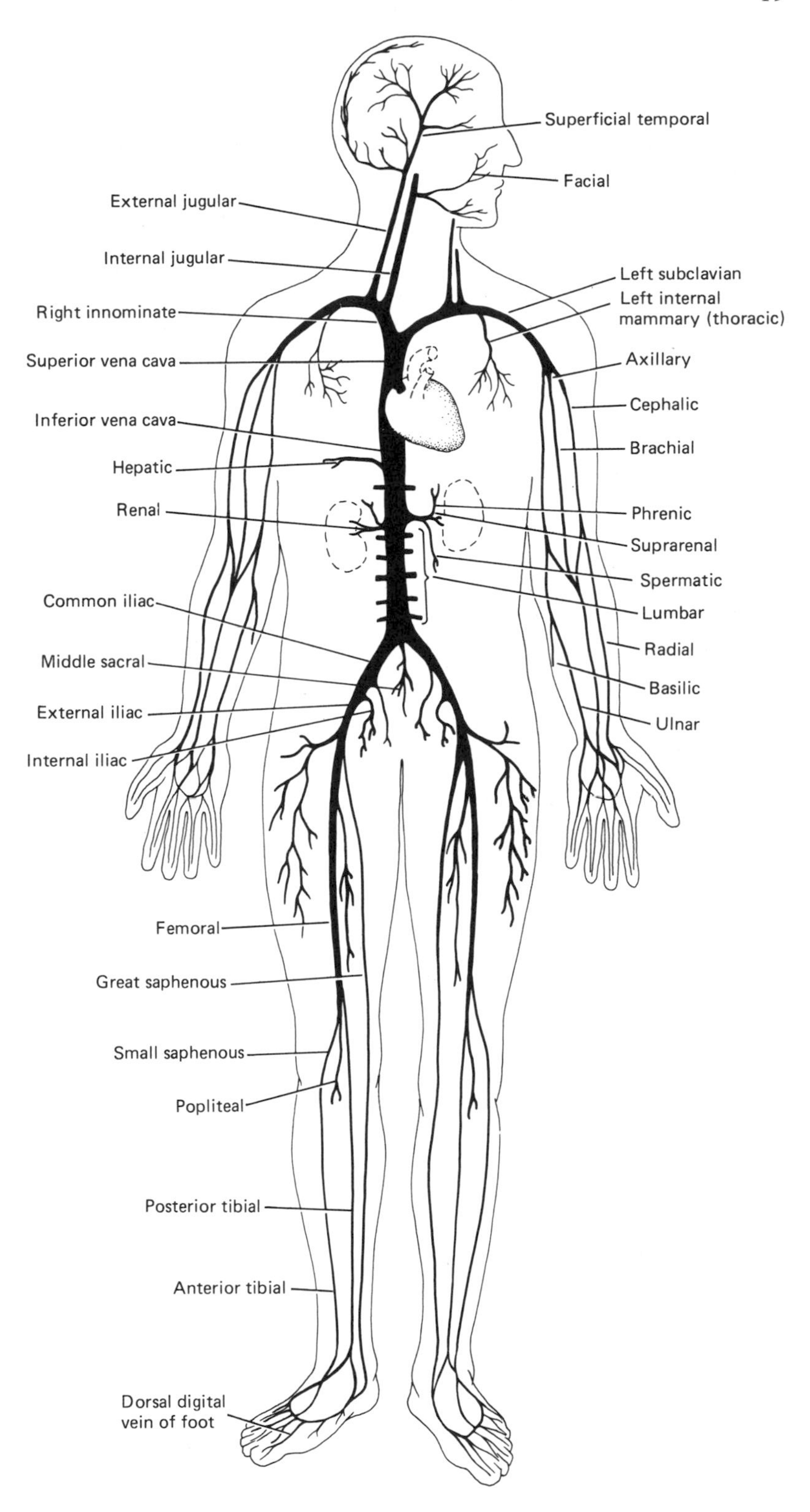

Fig. 1.17. Major veins of body.

The topographic anatomy of the principal arteries and veins are shown in Figs. 1.16 and 1.17 and Chap. 15, and lymphatic vessels and nodes are discussed in Chap. 14.

Selected Readings

Ebe T, Kobayashi S (1973) Fine Structure of Human Cells and Tissues. John Wiley and Sons, New York

Miller MA, Drakontides AB, Leavell LC (1979) Kimber-Gray-Stackpoles-Anatomy and Physiology (17th ed). MacMillan, New York

Pick TP, Howden R (1977) Gray's Anatomy (revised from 15th English ed). Bounty Books, New York

Weiss L (1977) The cell *in* Weiss L, Greep RO (eds): Histology (4th ed) McGraw-Hill, New York

Review Questions

1. Name the systems of the body.
2. Name the principal tissues of the body.
3. Define gross anatomy and histology.
4. Name at least two organs or tissues derived from a) ectoderm, b) entoderm, and c) mesoderm.
5. Define and characterize a) squamous epithelium, b) branched tubular gland with branched ducts (show diagram), and c) exocrine and endocrine glands.
6. Make a diagram of a membrane.
7. What is areolar connective tissue and where is it found?
8. What is elastic connective tissue and where is it found?
9. What is the difference between hyaline and elastic cartilage?
10. What is bone composed of?
11. What are the differences between skeletal and smooth muscle and cardiac muscle?
12. Name and locate the principal muscles of leg and arms.
13. Locate the carpal and metacarpal bones.

Chapter 2 General and Cellular Physiology

As early as 1665 Robert Hooke discovered that cork was made up of cells, but it was not until 1839 that Schleiden and Schwann developed and enunciated the cell doctrine, which stated that the cell is the unit of structure and function in all organisms. Their studies were based on the use of the light microscope. With the advent of electron microscopy—which has 400 to 2000 times the magnification of light microscopy—much more has been learned about the cell. Many of the particles (organelles) of the cell, particularly of the cytoplasm, were not known or seen before electron microscopy. Many of these organelles have been separated, isolated, and identified by various techniques including ultracentrifugation whereby particles are separated by virtue of size and specific gravity. Some of these organelles have been subjected to critical physicochemical analysis and are discovered to have important metabolic and biochemical functions.

Tissues and organs are made up of an aggregation of cells, the size, shape, and number of which vary according to the organ and its specialized function. These cell types include those making up (1) epithelium, (2) muscle (striated and smooth), (3) nerves and brain, (4) blood, (5) bone, and (6) connective tissue. These tissues and cell types are discussed in detail in Chap. 1.

Thus the collective function of cells, tissues, and organs—and therefore the function of the whole organism (activity and motility, or organ physiology)—are governed by the exchange and transport of biochemical substances and reactions between the interior and exterior of cells, or intra- and extracellular activities. These activities are regulated in part by the structure and function of the cell boundary, namely, the *cell membrane*.

Structure of Cells

Cells are composed of protoplasm (living substance) found in the nucleus, and cytoplasm (Fig. 2.1). Size is greatly variable, ranging from 10 μg for a human ovum to 0.7 μg for human smooth muscle cell.

Nucleus

The nucleus occupies the central portion of the cell and is the main regulator of its activity. Its surgical removal (enucleation) disorganizes cytoplasmic functions. The nucleus is bounded by a membrane and contains a *nucleolus* (one), a small spherical body, or nucleoli (more than one), and numerous chromatin granules, which are made up of chromosomes—the structural units that carry the hereditary agents (the genes). The chromosomes consist of a very large molecule of deoxyribonucleic acid (DNA) and a supporting protein; each gene occupies a part of the DNA molecule.
During *mitosis*, or normal cell division, the cells and chromosomes divide equally to give a full complement of such bodies (diploid number). Later these cells undergo reduction division (*meiosis*) to reduce the chromosome number to one-half in the sperm and ova (haploid number). Union of the sperm and ova (gametes) produces the fertilized egg, or zygote which becomes first the embryo and then the fetus, and both have diploid number of chromosomes (see Chap. 26). The *nucleoli* have no membranes and they consist mainly of nucleic acids (RNA) and proteins. Nucleoli are most numerous in growing cells, and are probably the site of the synthesis of RNA found in the ribosomes. Further details on the synthesis of RNA and DNA are presented in Chap. 22.
The *nuclear membrane* is double walled with a space between. It is highly permeable even to molecules as large as RNA, which pass freely from the nucleus to the cytoplasm.

Cytoplasm

The cytoplasm includes all of the cell except the nucleus and the outer plasma membrane. The more important of the many different types of bodies and particles (organelles) that it contains include: 1) endoplasmic reticulum, 2) ribo-

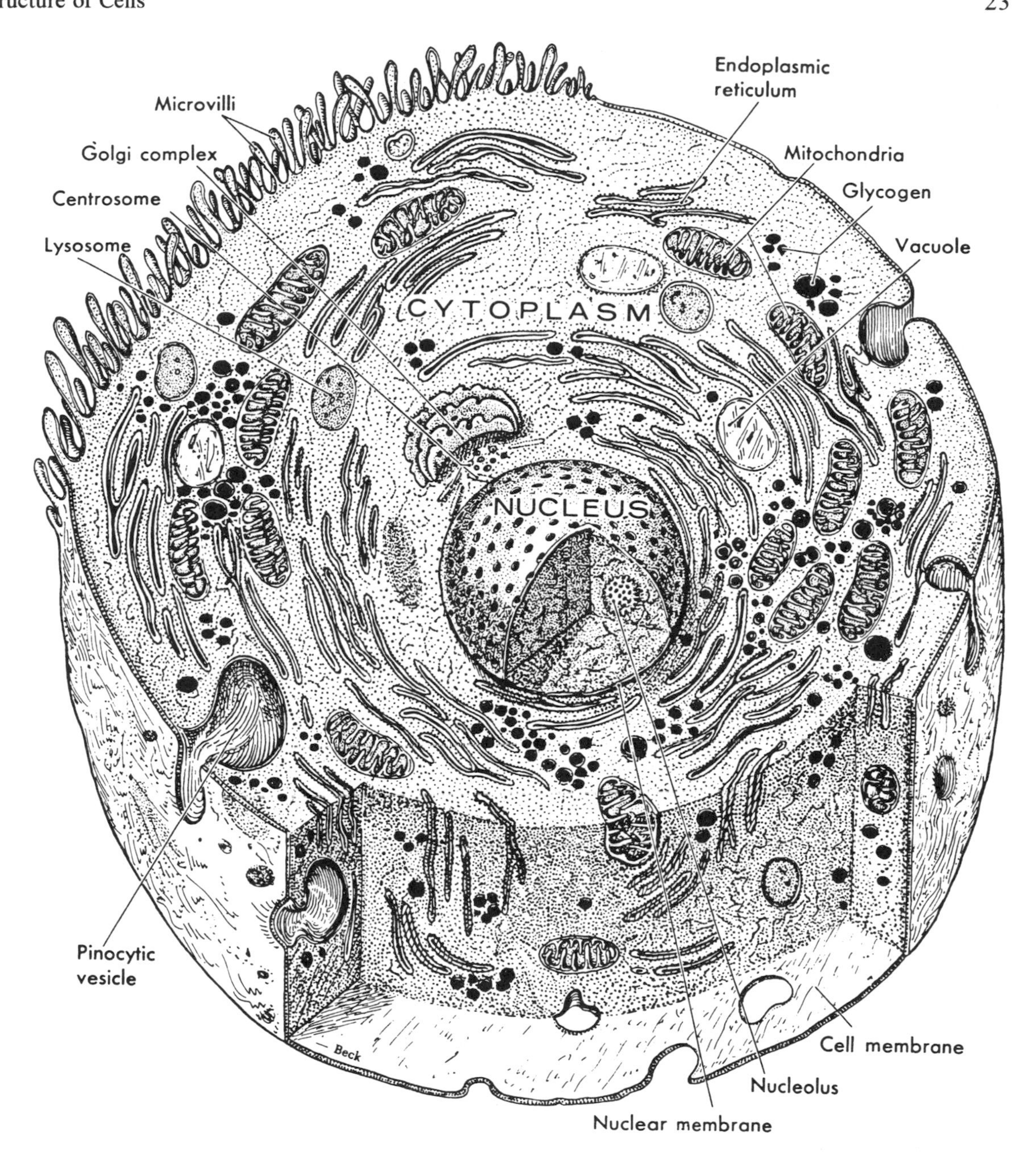

Fig. 2.1. Typical cell (composite) as might be seen under an electron microscope (more than one view). (Reproduced with permission from Anthony, C.P., and Kolthoff, N.J.: Anatomy and Physiology. Copyright © 1975 by C.V. Mosby, St. Louis.)

somes, 3) mitochondria, 4) Golgi apparatus, 5) centrosome, and 6) lysosomes.

Endoplasmic Reticulum

A network prominent in the cytoplasm of all cells, it is made up of thin membrane-bound cavities that vary considerably in size and shape under different physiologic conditions; indeed it may be completely absent in some cells such as erythrocytes. At one time it was thought to be confined to the endoplasm of the cell; it may, however, extend into the peripheral part of the cell (ectoplasm) and connect with the cell membrane.

The endoplasmic reticulum (ER) is a system of interconnected canals and cavities extending throughout the cytoplasm. There are two types of ER, rough and smooth. The enclosing membranes of ER are double walled and contain proteins, phospholipids, and enzymes involved in the synthesis of lipids.

Ribosomes

These electron-opaque particles containing RNA and protein, are 10 to 25 nm in size (1×10^{-6} mm) and usually attached to the rough ER; some of them, however, are unattached in the cytoplasm. Those attached to ER synthesize proteins that migrate to other parts of the cell; the protein made by the unattached or free ribosomes is used at the site produced.

Mitochondria

Mitochondria (Mi) are usually 0.5 to 1.0 μm in diameter and up to 7 μm long. Their shape and structure vary, depending on the physiologic state.

Mitochondria may be motile in certain cells, particularly those that are dividing, and may aggregate to form distinct arrangements in certain tissues, especially those exhibiting secretory activity. They have outer and inner membranes; the inner membranes, by folding, form a sac with extensions called cristae, each of which contain many small knobs. These knobs are storehouses of enzymes, mainly *adenosine triphosphate* (*ATP*), which is the major source of energy for many of the body's biochemical reactions.

Mitochondria are called the power plants of the cell because of the energy released from enzymatic reactions in the inner membrane. The number of Mi appears to be highest in the cells exhibiting the most physiologic activity. The active liver cell may contain as many as 2500; a relatively inactive cell as few as 25, Mi.

Golgi Apparatus

The Golgi apparatus comprises a system of canals with sacs or cisternae. Although variously shaped, these canals usually consist of stacks of flattened sacs associated with small vesicles and vacuoles. The organization of the Golgi apparatus, or complex, varies. It is small in some cells (muscle), but large and well developed in others (secretory cells and nerves). Granules produced in the Golgi apparatus (GA) contain hormones and enzymes, mainly proteins.

Recent evidence suggests that the GA synthesizes carbohydrates, and combines them with proteins to form glycoproteins.

Lysosomes

Lysosomes are organelles enclosed in a membrane, the size of which ranges from 0.25 to 0.8 μm and which appear dense and granular under the electron microscope (Fig. 2.1). They contain many enzymes that digest proteins, nucleic acids, polysaccharides, lipids, and even cells and bacteria. Certain diseases and disorders may be caused by lack of activity or destruction of lysosomes. Lysosomes may originate from either the ER or the vesicles of the Golgi apparatus.

Centrosome

As its name implies, the centrosome, is located near the center of the cell (Fig. 2.1), near the nucleus. It contains two dots or centrioles when viewed under the light microscope, but under the electron microscope they appear as small cylinders. Although their function is unknown, they may be involved in spindle formation of the cell during mitosis.

Cell Membrane

The cell membrane is often called the plasma membrane and it is about 75 Å thick (7.5×10^{-6} mm) and it folds inward to form an internal membrane. It contains lipids and proteins, the molecular arrangement of which is uncertain; however, certain arrangements of them, have been inferred from indirect evidence. In the past, the Davson-Danielli model has been a widely accepted model for the structural arrangement of the lipids and proteins in the membrane. This model envisions two layers of lipid molecules arranged radially, their polar groups extending outwardly and their hydrocarbon chains (nonpolar) extending toward each other inwardly. A globular protein layer is attached to each polar lipid layer.

This model views the proteins as being extended sheets, but recent spectrographic analysis indicates that the proteins are globular and arranged in a helical fashion, suggesting the fluid mosaic model illustrated in Fig. 2.2. There is a bilayer of phospholipid molecules (·) attached to fatty acid chains (wavy lines in the figure). The proteins (heavy lines) are shown partly folded and also arranged helically with

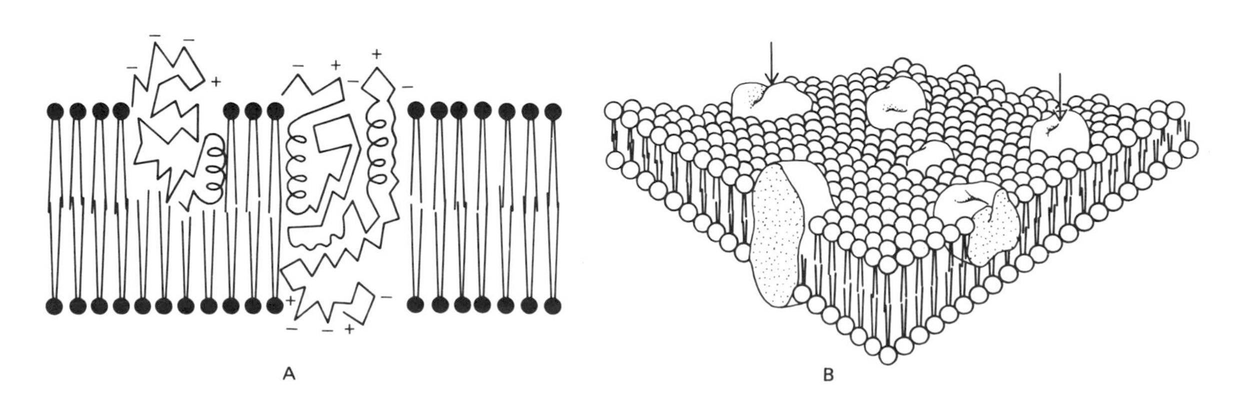

Fig. 2.2A and B. Cell membrane represented as a fluid mosaic of lipids and protein (globular). **A** In cross section. **B** In three dimensions. (See text.) (After Singer, S.J., and Nicolson, G.L. Science 175:720, 1972.)

charged (+ and −) ends on the outside or inside of the cell membrane. In the three-dimensional figure, the proteins appear as large globules irregularly placed; recent evidence indicates that some of them lie external and some internal to the phospholipid bilayers.

Some cells have *microvilli,* finger-shaped projections of the outer membrane, which increase the absorptive surface of the cell.

Intercellular connections are links between cells in tissues. Among them is the "tight junction," wherein the membranes of the two cells become apposed and fused. In a second type of interconnection, the two membranes are separated by a relatively large space of 15 to 35 nm. In a third type there is a junction or opening (gap junction) about 2 nm between the membrane of cells, permitting the passage of ions and small molecules between cells, but not through the extracellular spaces.

The existence of small openings, or *pores* (about 0.7 nm in diameter), in the cell membrane has been postulated, based on the rate of penetration of water salts into cells (physiologic pores). However, such pores have not been observed with the electron microscope.

The movement of substances through the cell membrane into and out of the cell is the main *function of the membrane*. There are several mechanisms by which this is accomplished, and they are discussed in a later section of this chapter and in Chap. 12. The plasma membrane serves as a barrier to, and a gateway for certain substances. It is highly selective, excluding some substances and permitting passage of others (diffusion); it also actively transports or pumps certain agents across the membranes.

The *permeability* of the membrane is influenced by the size of the molecules traversing it and the structure of the membrane. The membrane is virtually impermeable to large proteins and organic anions (A^-) that make up most of the intracellular anions, and is fairly permeable to Na^+ and freely so to K^+ and Cl^-. It is permeable also to lipid-soluble agents.

Movement Through Cell Membranes

Substances move back and forth through membranes by 1) filtration, 2) passive diffusion and osmosis, 3) facilitated or carrier-mediated diffusion, 4) dialysis, 5) active transport, and 6) pinocytosis.

Filtration

This process is governed by the pressure difference (ΔP) between inside and outside of the blood vessel and the permeability of the vessel to plasma fluid and substances with small molecules. Substances with large molecules do not pass through the pores of the membrane of capillary walls. (see Chap. 12 for the role of filtration in the formation of tissue fluid).

Diffusion

Passive diffusion is the spreading or expanding of molecules from areas of high concentration to areas of low concentration (net flux). This motion is random, and the molecules frequently collide in concentrated solutions. The rate of diffusion is governed by: 1) the *concentration gradient* of the solute (chemical gradient) or concentration; 2) *electrical gradient,* i.e., the electrical charge of ions (positive or negative; positive charges move to the negative areas and vice versa); 3) *permeability of membrane,* or the size of pores of the membrane in relation to size; and 4) *size of molecule,* in which diffusion rate varies inversely with size; water is much more diffusible than most substances.

Donnan Equilibrium

The exchange of fluid and certain diffusible ions (K^+, Cl^-, Na^+) from blood to tissues and from intra- to extracellular compartments is brought about by diffusion, whereby the ions cross freely through the membrane down their concentration gradients (downhill). Large plasma proteins are not diffusible through the membrane, and their charged ions (−) hinder the transfer and distribution of the diffusible ions in a definite and predictable manner as follows:

A	M	B
K^+		K^+
Cl^-		Cl^-
$Protein^-$		

K^+ and Cl^- are freely permeable to the membrane (M), but the protein is not, and its negative charge (−) hinders the diffusion of the cations (+). The end result is that more osmotically active particles are on the A side of the membrane so the: [K + A side] + [Cl^- A] + [$Prot^-$ A] is greater than (>) [K + B] + [Cl^- B]. However, the concentration ratios of the diffusible ions (K^+ and Cl^-) distribute themselves equally on each side of membrane as:

$$M\frac{[K + A] = [Cl^- B]}{[K + B] = [Cl^- B]}.$$ This is the *Donnan Equilibrium.*

Proteins, an important constituent of plasma, rarely diffuse out of the capillaries to form tissue fluid. Therefore the Donnan effect is an important one in the body. The quantitative distribution of these ions on either side of the membrane and their role in initiating the electrical discharge of the cell (the action potential) and changes in membrane potential are discussed in Chaps. 4 and 5.

Calculating Rate of Diffusion

Fick's Equation. The rate of diffusion of a solute, $\frac{(ds)}{(dt)}$ represents the amount passing from the area of higher concentration (S_1) to that of lower concentrations (S_2) in a given time; thus, $\frac{ds}{dt} = -D\,A\frac{(S_1 - S_2)}{dx \text{ or } T}$ where D = diffusion coefficient (constant for a given solute) and varies inversely with molecular size (see Chap. 1).

$S_1 - S_2$ = difference in concentration of solute (net flux)
A = cross sectional area
dX = diffusion distance or thickness (T) of membrane area
− = downhill diffusion (from area of high concentration).

In summary, the rate of diffusion is: 1) directly proportional to D and A, and concentration difference ($S_1 - S_2$) and 2) inversely proportional to thickness, or diffusion area. An example of diffusion of a more concentrated solution into another solution of lesser concentration is shown in Fig. 2.3. Here the membrane is diffusible to both the solute (NaCl) and solvent (H_2O).

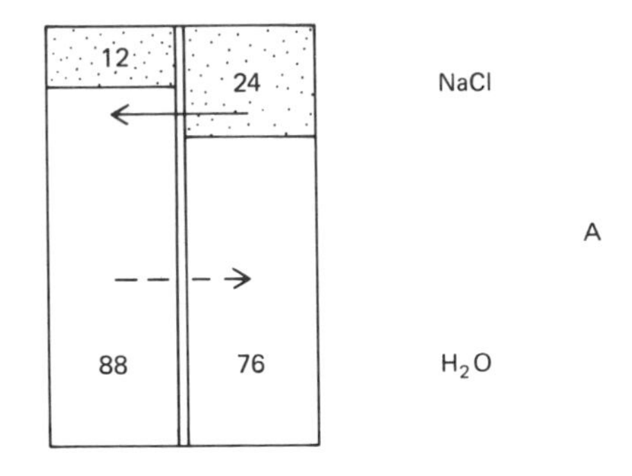

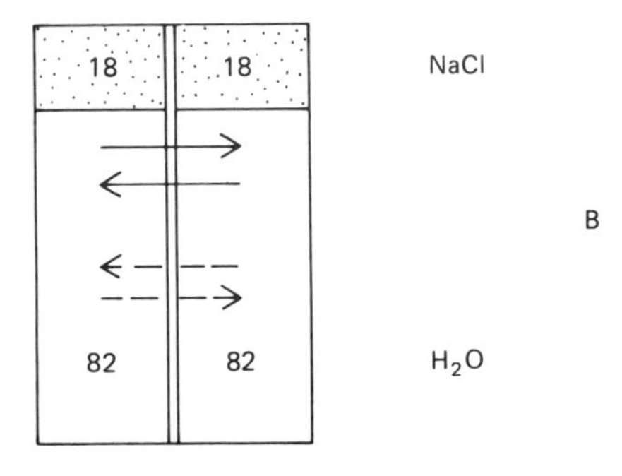

Fig. 2.3. Diffusion. (A) Membrane separating two NaCl solutions (12 and 24 percent) and H_2O solutions (88 and 76 percent). There is a concentration gradient between NaCl and the water. Net diffusion is from higher to lower concentration (downhill diffusion). Main direction of diffusion of NaCl is indicated by → and by --→ for water (in the opposite directions). Downhill diffusion is completed later (B), and solutions are in equilibrium, i.e., concentration of NaCl and H_2O are the same on both sides of membranes, and diffusion continues now in both directions (double arrows).

Facilitated Diffusion

Facilitated diffusion is represented by a solute that normally is diffusible but which is carried or helped by another substance, or molecule; gradient of the solute in question, however, is downhill. The transfer or diffusion of glucose from the blood (where concentration is high) to the tissues (where the concentration is low) is facilitated or mediated by a carrier molecule. Transport of glucose against its concentration gradient (absorption from the intestinal tract) is active and requires energy (Fig. 2.4).

Dialysis

Dialysis is a type of diffusion occurring when a solution containing small molecules (crystalloids) and large molecules

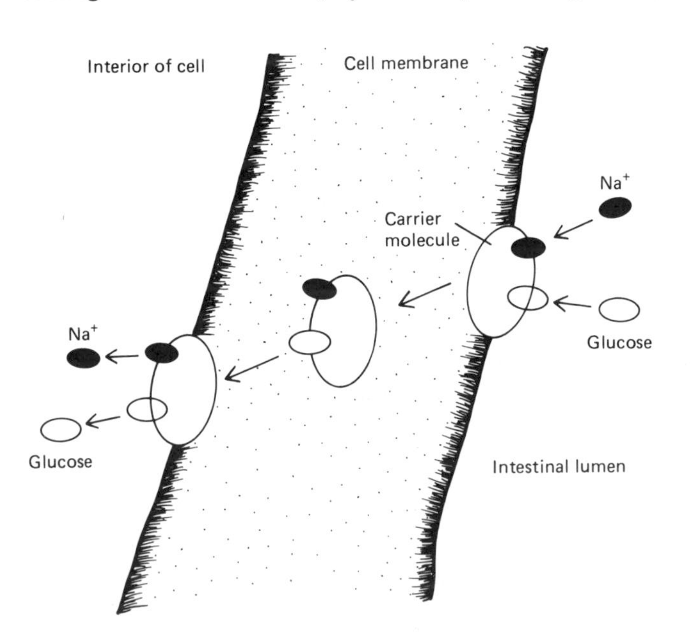

Fig. 2.4. Model of glucose absorption from intestine. An example of carrier-mediated (active) transport. Na^+ facilitates absorption of glucose. Energy required for maintenance of NA pump is derived from ATP. (From Crane, R.K. Fed. Proc. 24:1000, 1965.)

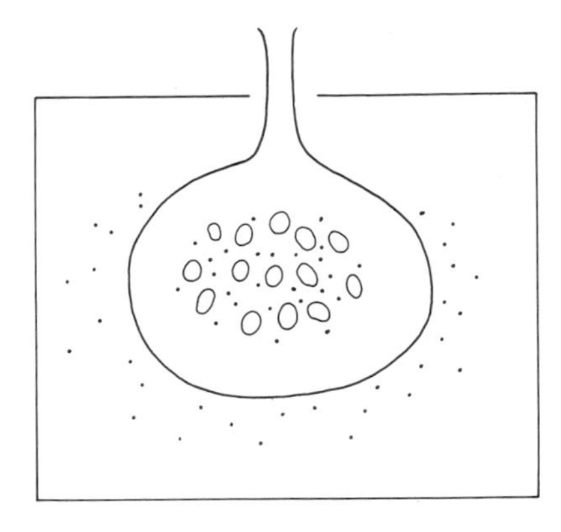

Fig. 2.5. Dialysis. Parchment bag containing solution of crystalloids (glucose) and colloids (protein ⌘) is placed in container of water. Glucose diffuses through semipermeable membrane into water, but protein molecules will not pass through membrane and remain in the bag.

(colloids) is separated from a water solution by a semipermeable membrane, i.e., permeable to small molecules but not to large ones (Fig. 2.5).

Osmosis

Osmosis refers to the diffusion of a solvent (water) through a semipermeable membrane, or one that is impermeable to certain solutes. Plasma proteins, e.g., are relatively impermeable to the membrane of blood vessels and do not diffuse through or filter out of the vessels because the protein molecules are too large. As a result, plasma proteins exert an osmotic pressure (see Chap. 12) which tends to counteract the hydrostatic pressure or filtration pressure and movement of water and certain solutes through the walls of blood vessel. This osmotic pressure tends to hold fluid within the blood vessels and significantly influences the absorption of tissue fluid. An example of osmosis is shown in Fig. 2.6.

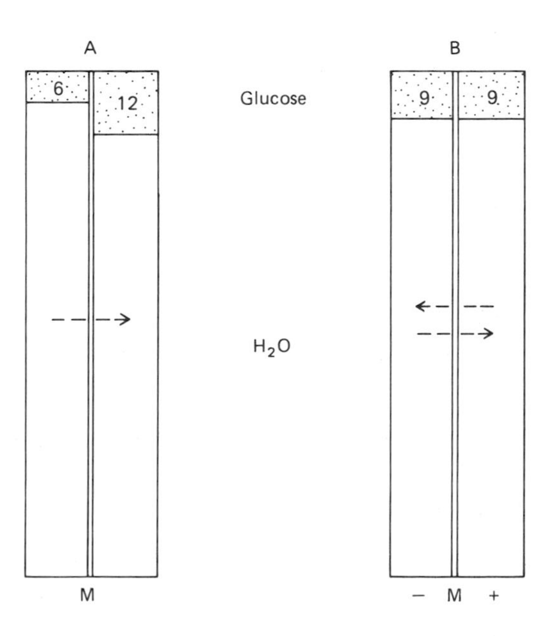

Fig. 2.6A and B. Osmosis. Membrane separating two solutions (6 and 12 percent glucose) in **A** is permeable to H_2O but assumed to be impermeable to glucose. Therefore only H_2O can diffuse through membrane and from solution with higher H_2O concentration (6 percent glucose in **A**) indicated by --→). The only way for glucose to reach same concentration on both sides of membrane is to add H_2O to more concentrated glucose side. This causes decrease (−) in volume of H_2O on less concentrated glucose side and an increase (+) in volume on more concentrated side in **B**. At the same time there is increased pressure on increased volume side and a decrease on decreased volume side. This assumes that compartment receiving the added water is expandable. In **B,** H_2O can diffuse both ways (double arrows).

Osmotic Pressure

Osmotic pressure depends on the number of solute particles in the solvent, and van't Hoff's law relates the osmotic pressure of solutions to the same for gases (Avogadro's law). Thus 1 mole of any substance dissolved in 1 liter of solution (1 molar solution) contains the same number of molecules and exerts a potential pressure of 22.4 atmospheres or 17,024 mmHg at 0° C or 19,300 mmHg at 37° C. Thus,

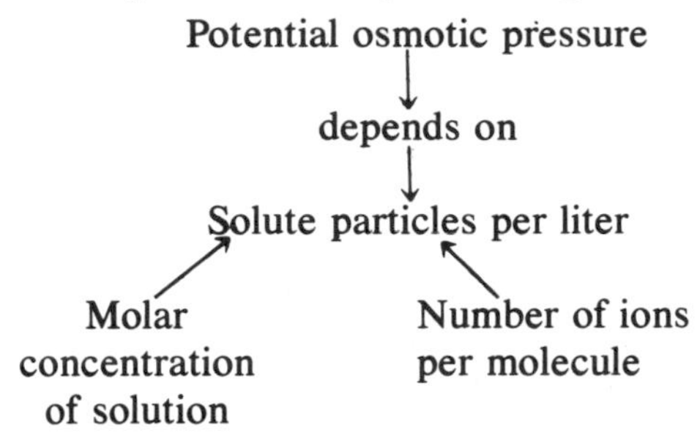

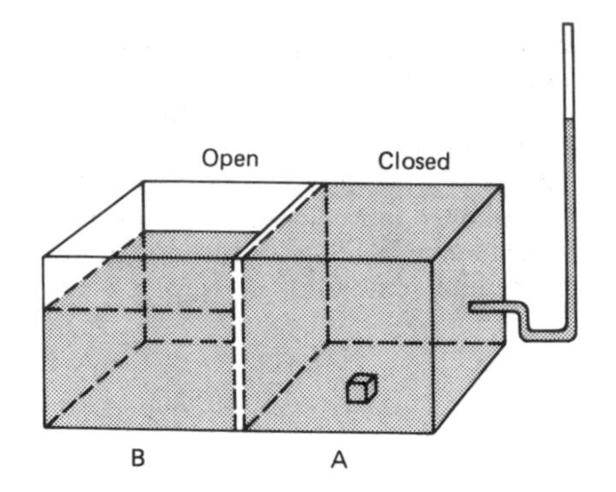

Fig. 2.7. Osmotic pressure. Tank of water is divided into two compartments by a semipermeable membrane, permeable to water but not to sucrose. Compartment A is closed and filled with water and cube of sugar. Compartment B is open and contains water. Water now moves from B to A until pressure of water head in manometer tube (open) is equal to osmotic pressure of sugar solution. (From Dowben, R.M. In: Goldstein, L. (ed.) Comparative Physiology, Holt, Rinehart and Winston, New York, 1977.)

Potential pressure represents the pressure that might develop in a solution if it were separated from a water solution by a selectively permeable membrane. The pressure required to prevent solvent migration under these conditions is the effective osmotic pressure (Fig. 2.7). In this figure it is observed that the compartment (A) containing the sucrose is fixed in volume and filled but is open to a manometer. Water moves from compartment B into A until the pressure in the manometer equals the osmotic pressure of the sucrose solution.

Normally, when water flows or diffuses to a solute that is impermeable to a membrane, there is an increase in volume and pressure on the side of the membrane containing the more concentrated solution.

Calculation of Osmotic Pressure at 37° C

Osmotic Pressure of nonelectrolyte in mmHg = Molar concentration × 19,3000 of solution

$$\frac{\text{Glucose solution, 10\%}}{\text{10\% solution}} = 100 \text{ g/liter}$$

Molecular weight of glucose = 180

$$\text{Thus, } \frac{100}{180} = 0.555 \times 19{,}3000 = 10{,}711 \text{ mmHg}$$

Electrolyte solution:

4% NaCl = 40 g/liter
Molecular weight of NaCl = 58
Ions per molecule = 2

$$\frac{40}{58} = 0.69 \times 2 \times 19{,}300 = 26{,}634 \text{ mmHg}$$

Isosmotic, Hyposmotic and Hyperosmotic Solutions

Isosmotic solutions have the same osmotic pressures; no net osmosis occurs between them, and they are considered to be isotonic.

Hyposmotic, like hypotonic solutions have osmotic pressure less than that of a reference solution. Hyperosmotic, like hypertonic, solutions have osmotic pressures higher than a reference solution. For further details concerning osmotic and filtration pressures of blood, see Chaps. 12 and 14.

Active Transport

In active transport the substance or solute moves uphill against a concentration or electrochemical gradient. The energy to move the substance uphill is derived from a metabolic reaction, usually from the hydrolysis of ATP (adenosine triphosphate). There is a specific enzyme that hydrolyzes ATP, called ATPase, which is believed to mediate the active transport of Na^+ and may be considered the sodium pump. Following depolarization of the cell, the Na^+ is pumped outward through the cell membrane against a higher concentration of Na^+ outside the cell. For every mole of ATP hydrolyzed, roughly 2 to 3 moles of Na^+ are pumped out of the cell, and the entry of 2 occurs when 3 Na^+ are extruded (sodium-potassium pump). Wherever more than 1 Na^+ is extruded for each K^+ entering the cell, there is a net migration (flux) of positive charges out of the cell, and the cell becomes hyperpolarized. The exact nature of the operation of the sodium pump has received much research attention recently (see also Chaps. 3 and 4).
The operation of the sodium pump in facilitating the carrier-mediated transport of glucose is shown in Fig. 2.4.

Pinocytosis, Exocytosis, Phagocytosis

These types of active transport are brought about by the activity of the cell itself.
Phagocytosis describes the ability of leukocytes to engulf bacteria (see Chap. 13). In *pinocytosis* the plasma membrane of a cell may form a small pocket around a bit of material outside the cell and pinches it off from the enveloped material; it then migrates inward as a closed vescicle, as shown:

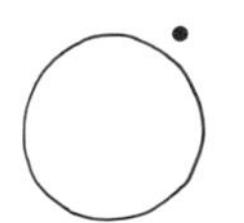

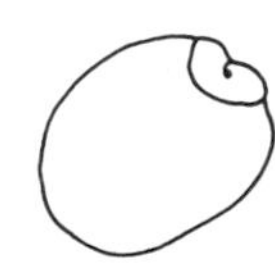
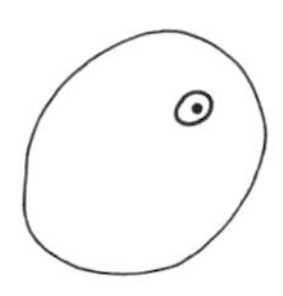
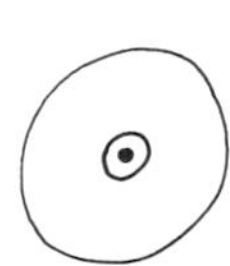

Exocytosis is the opposite of pinocytosis. A granule within the cell migrates toward the cell membrane, and its membrane fuses with that of the cell; finally the granule is extruded from the cell, as shown:

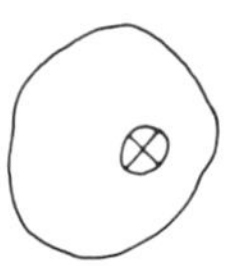
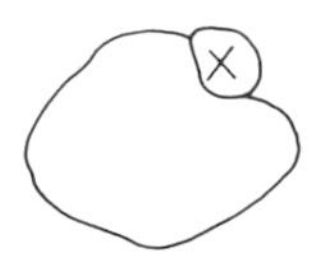

Membrane Potentials

Cells exhibit an electrochemical gradient, based on the difference in the concentration of intra- and extracellular ions. This is the membrane potential. It varies in different cells and tissues, and under different physiologic conditions. When the cell is resting or inactive, it is polarized or balanced, the interior of the cell being relatively negative to the outside, which is positive.
The potential is measured in millivolts (mv). The resting potential ranges from -10 to -100 mv, and for myocardial cells (heart) it is about -80 to -90 mv. When the cell becomes depolarized (as when muscle or nerve is stimulated), ions move across the membrane and an electrical impulse is generated; this causes the spread of the excitation and the action potential and a change in membrane potential to one of lesser negativity (see Chap. 4 for details of ions and action potentials).

Acid-Base Balance

Acid-base balance refers to the maintenance of the constancy (homeostasis) of hydrogen ion concentration $[H^+]$ or acidity and alkalinity of body fluids. Acidity is usually expressed in terms of hydrogen ion concentration or pH.
pH is a measure of the $[H^+]$ that are dissociated or unbound (ionic acidity). *Titratable acidity* is based on potential acidity with bound and free ions and is expressed in terms of milliequivalents/liter (mEq/liter) or grams/liter of replaceable H^+. The equivalent acid (one replaceable H^+) neutralizes one equivalent of a base solution.
Dissociation of an acid is expressed by Ka, a dissociation or equilibrium constant:

$$pKa = -\log Ka = \log \frac{1}{Ka}$$

Ka is directly proportional to the degree of ionization and the strength of the acid, but pKa is inversely proportional to the strength of the acid and magnitude of Ka. Thus a strong acid like 0.1 N HCl ionizes or dissociates more readily (91%) than a weak acid.

		Ka	*pKa*
Strong acid (sulfuric)	H_2SO_4	1.2×10^{-2}	1.9
Weak acid (carbonic)	H_2CO_3	7.9×10^{-7}	6.1

$$pH = -\log [H^+] \text{ or } \frac{1}{[H^+]}$$

As acidity increases, pH decreases and as it increases to 7 at 25° C, there are equal concentrations of OH^- and H^+ and the solution is neutral.

$$\text{Thus a pH of } 1 = 10^{-1} \text{ or } \frac{1}{10} [H^+]$$

$$7 = 10^{-7} \text{ or } \frac{1}{10{,}000{,}000} [H^+]$$

Above pH7, the solution is alkaline.

Buffers

Many of the reactions of the body occur at or near the neutral pH, and maintenance of the near-neutral pH is effected by the operation of a system of buffers or substances that prevents marked changes in pH of a solution when strong acids or bases (alkaline) are added.
Buffers occur in pairs of a weak acid and a base or salt. The principal ones in the blood and tissues are:

1. Hemoglobin $\frac{\text{K-Hb}}{\text{Hb}}$ and $\frac{\text{K-HbO}_2}{\text{HbO}_2}$ $\frac{\text{Salt}}{\text{Acid}}$

   HbO (reduced)
   HbO_2 (oxygenated)
2. Plasma proteins $\frac{\text{Na-proteinate}}{\text{Proteins-weak acid}}$
3. Phosphate $\frac{Na_2HPO_4 \text{ (base phosphate)}}{Na_2H_2PO_4 \text{ (acid phosphate)}}$
4. Bicarbonate $\frac{NaH_2\,CO_3}{H_2CO_3}$ or $\frac{KHCO_3}{H_2CO_3}$ $\frac{\text{Salt}}{\text{Acid}}$

Buffers replace strong with weak acids, which on dissociation yield fewer H^+ and have less effect in lowering pH than has a strong acid (see Chap. 17). The body forms more carbonic acid than any other acid, but it is a weaker acid than lactic acid, which is buffered or neutralized by bicarbonate (HCO_3^-) and is replaced by carbonic acid (H_2CO_3).
The permeability, absorption, transport, and excretion of various substances in the body depend on the degree of ionization and dissociation of H^+; this in turn depends on thc pH and temperature of the surrounding medium (see Chap. 24).

Henderson-Hasselbalch (H.H.) Equation

The pKa of an acid is the pH at which the concentrations of ionized and nonionized forms are equal. The pK's of different substances and buffer systems are known and have been tabulated for easy reference. In the H.H. equation,

$$pH = pKa + \log \frac{\text{salt}}{\text{acid}}$$

The equation may be used to calculate 1) the buffer ratio of salt and acid if pH and pKa are known and 2) the pH if the

buffer ratio and pKa are known. The concentrations of HCO_3^- and H_2CO_3 in blood are normally about 26 and 1.3 mEq/l, respectively.
Thus the pH of the medium at which dissociation occurs is:

$$pH = pK + \log \frac{HCO_3^-}{H_2CO_3} = \frac{26 \text{ mEq}}{1.3 \text{ mEq}}$$

$$pH = 6.1 + \log \frac{26}{1.3} = \log 20$$

$$= 6.1 + \log 20\ (1.3)$$
$$= 6.1 + 1.3 = 7.4$$

For further details on the role of respiration and the kidneys in regulation of acid-base see Chaps. 17 and 24.

Selected Readings

Davson H (1970) A textbook of general physiology (4th ed). Williams and Wilkins, Baltimore
Dowben R (1977) Membrane physiology *in* Goldstein L: Introduction to Comparative Physiology. Holt, Rinehart and Winston, New York
Ganong WF (1977) Review of Medical Physiology (8th ed). Lange Medical, Los Altos, Cal.
Giese AC (1973) Cell Physiology (4th ed). Saunders, Philadelphia
Singer SJ, Nicolson JL (1972) The fluid mosaic model of the structure of cell membranes. Science 175: 720

Review Questions

1. Diagram a cell and show the principal particles (organelles) in nucleus and cytoplasm.
2. What are ribosomes and what do they do?
3. What are mitochondria and their function?
4. What are lysosomes and their function?
5. What is the plasma membrane and why is it so important to the cell?
6. Name the mechanisms or ways by which substances are transported across the plasma membrane.
7. What are the differences between passive diffusion and active transport?
8. What is the Donnan Equilibrium?
9. Define and describe a) isosmotic, b) hyposmotic, and c) hyperosmotic.
10. What is the osmotic pressure of a 10% solution of glucose? Show your calculations.
11. Where is the energy derived from to bring about active transport?
12. Define dialysis.
13. Define pH.
14. What are buffers and their principal functions?
15. What constitutes the bicarbonate buffers (pairs)?
16. What is meant by pK of an acid?

Chapter 3

Organization of the Nervous System

The human nervous system is the most complex structure known to man. It contains 50 billion nerve cells, which are interconnected to form a network of unimaginable complexity. The brain contains the sensory centers, which analyze changes in both the external and the internal environments. It is the principal controller of body function, including the contractions of muscle fibers and the secretions of gland cells.

Neuron

The neuron, or nerve cell, is the functional unit of the nervous system. Specialized both morphologically and physiologically for the transmission and integration of information. Each neuron is divided into four different regions, including: soma, axon, axon terminals, and dendrites (Fig. 3.1). Each region is specialized to perform a specific function. The nerve cell body, or *soma,* contains the nucleus, ribosomes, endoplasmic reticulum, and associated organelles

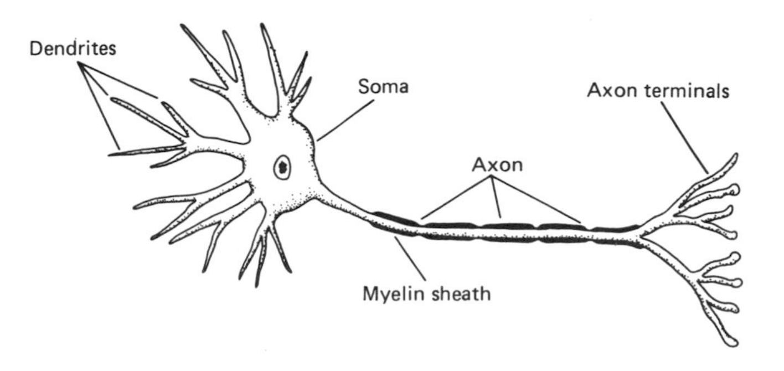

Fig. 3.1. The neuron.

and is the major synthetic region of the neuron. Many neurotransmitters, cell proteins, and other critical components are synthesized in the soma. The soma is essential to the survival and integrity of the neuron. Destruction of the neuronal cell body results in the degeneration of the entire nerve cell, including: dendrites, axon, and axon terminals.

The primary function of the *axon* is to provide a pathway for the transmission of nerve impulses to other cells: nerve, muscle, or secretory. Most axons are long fiber-like processes arising from the base of the nerve-cell soma and running from several millimeters to several meters through the brain before terminating on the receptive processes of other neurons. Many axons provide connections between the periphery and the central nervous system (brain and spinal cord). The axons of sensory neurons transmit information from sensory receptors in the periphery to the central nervous system (CNS). Motor neuron axons provide pathways for the conduction of nerve impulses from the CNS to the muscles of the trunk and limbs. Other axons provide pathways connecting the CNS with the sensory receptors, muscle cells, and secretory cells located in the viscera.

The nerve axon is specialized for the *transmission of nerve impulses*. Each nerve impulse is produced by a small change in the permeability of the axon membrane. This change produces an electrical potential, which is transmitted as a wave along the entire length of the axon from the soma to the axon terminals.

As an axon approaches its termination, it undergoes a series of divisions to form a fine spray of terminal branches (*axon terminals*). Each terminal branch ends by forming a specialized contact, or synapse, with a postsynaptic cell. The postsynaptic cell may be either a nerve, muscle, or secretory cell. Within the CNS the vast majority of synapses are formed by axon terminals ending on the dendrites of other neurons.

The *synapse* is specialized for the intercellular transmission of information. When a nerve impulse arrives at an axon terminal, the latter secretes a small amount of a specific chemical—the *neurotransmitter*. Upon its release from the axon terminal, the neurotransmitter binds to the dendritic membrane of the postsynaptic neuron where it produces a change in the permeability of the dendritic membrane. This change in permeability causes a shift in the electrical potential of the dendritic membrane. The resulting *synaptic potential* may be either excitatory or inhibitory. If excitatory, it increases the probability that the postsynaptic neuron will discharge a nerve impulse. In contrast, an inhibitory synaptic potential prevents the discharge of an impulse in the postsynaptic neuron.

Dendrites are formed by the tree-like branching of neural processes extending from the soma and are specialized to receive the synaptic input. The axon terminals of hundreds or thousands of neurons terminate on the dendrites of a typical nerve cell. The surface of the dendrite is covered with these synaptic terminals. Each axon terminal, when active, releases a neurotransmitter that produces a localized change in the permeability of the dendritic membrane. As a result of this change, the electrical potential of the dendritic membrane is altered. This change (synaptic potential) spreads from the dendrites to the initial segment of the axon. If the synaptic potential is excitatory, the rate of nerve impulse discharge increases; if inhibitory, the discharge of nerve impulses is suppressed.

Glial Cells

Despite the fact that nerve cells are the functional units that process neuronal information, they make up only 10% of the total cells in the nervous system. Most of the cells in the nervous system are glial or support cells. They fill the spaces between the neurons. There are four major kinds of glial cells (Fig. 3.2), including: *astrocytes, oligodendrocytes,* and *microglia*—all of which are found in the brain and spinal cord—and *Schwann cells,* located in the peripheral nerves. Many glial cells—oligodendrocytes in the CNS and Schwann cells in the peripheral nerves—are closely associated with the long fiber tracts that are formed by bundles of nerve axons.

Most large axons are enveloped by sheet-like extensions of glial cell membrane. These membranes produce a *myelin sheath* (Fig. 3.1) that insulates the axon membrane. This insulation is very important because it increases the conduction velocity of the nerve impulse. Other glial cells, the astrocytes, are interposed between blood vessels and neuron

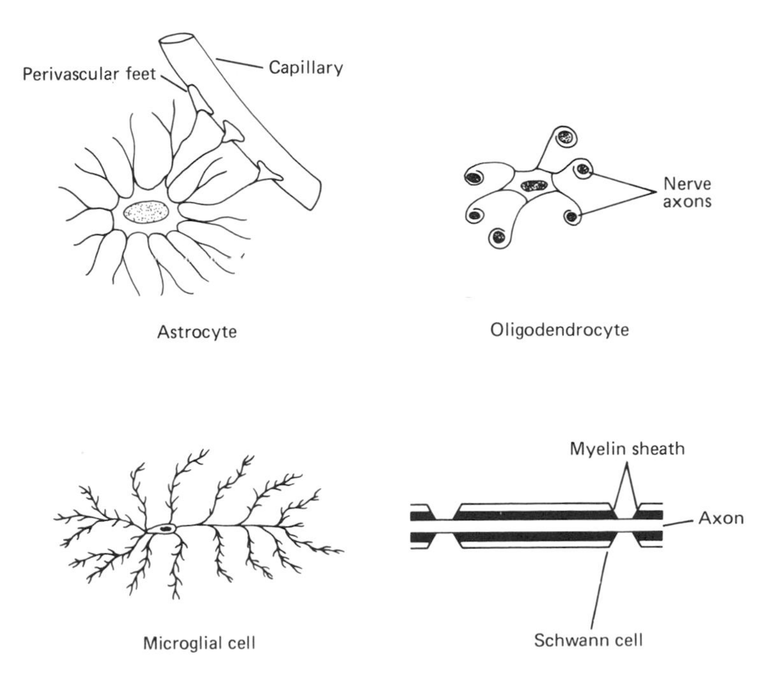

Fig. 3.2. The four major types of glial cells.

cell bodies. Some processes of astrocytes are in contact with the capillary wall. These perivascular feet form part of the blood–brain barrier. Many neurobiologists believe that these glial cells regulate the transport of nutrients from the capillaries to the neurons. It is suggested that glia and their associated neurons exchange proteins, nucleic acids, and other important molecules. Some evidence suggests that neuronal activity can affect the membrane potential of glial cells by increasing the concentration of K^+ in the extraneuronal space.

Microglia are the scavenger cells, or phagocytes, in the brain and form part of the reticuloendothelial system (see Chap. 13). Although rarely observed in the undamaged brain, large numbers of microglia are always found in the vicinity of damaged brain tissue.

Organizational Plan

The neurons composing the nervous system can be divided into three classes: sensory, interneuronal, and effector (Fig. 3.3). Both *sensory* and *effector* neurons provide pathways linking peripheral structures—sensory receptors, muscles, and glands—with the CNS (brain and spinal cord). The sensory neurons provide afferent pathways for the transmission of impulses from the sensory receptors to the CNS, whereas effector neurons provide the efferent pathways that conduct impulses from the CNS to the effector organs (muscles and glands). Effector neurons include motor neurons, which innervate the skeletal muscles, and autonomic neurons, which provide pathways for the regulation of the visceral muscles and glands by the CNS.

Interneurons are those neurons whose processes are confined to the CNS. They include all neurons except sensory and effector neurons. Almost all of the neurons in the nervous system are interneurons. They form the circuits in the

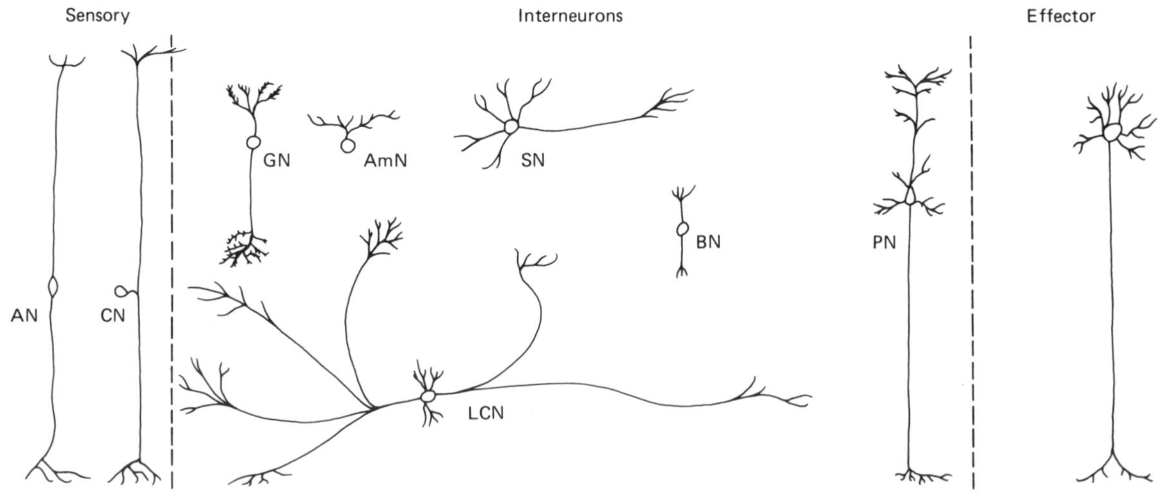

Fig. 3.3. The three classes of neurons. Note the wide range of interneuronal structures. The sensory neurons are *AN*, an auditory fiber in the vestibulocochlear nerve; *CN*, a neuron which is sensitive to cutaneous stimulation. Interneurons include *AmN*, amacrine neuron of retina; *BN*, bipolar neuron of retina; *GN*, granule neuron of olfactory bulb; *LCN*, neuron of locus cerulus; *PN*, pyramidal neuron of cerebral cortex; and *SN*, stellate neuron of cerebellum. The effector neuron is a spinal motor neuron.

CNS that permit the analysis of sensory inputs, the storage of neural experiences as memories, and the generation of appropriate neural outputs.

Interneurons occur in an almost infinite variety of structural forms (Fig. 3.3). The amacrine neurons in the retina or the granule neurons of the olfactory bulb are small neurons lacking axons, whereas others, such as the large pyramidal neurons of the motor region in the cerebral cortex, have axons that can be more than a meter long. The processes of some interneurons are confined to a very small neural region (bipolar neurons of the retina, or stellate neurons in the cerebellar cortex), whereas others extend throughout much of the brain (neurons of the locus ceruleus in the brain stem).

The nervous system is organized so that interneurons with similar functions—receiving similar inputs and having similar outputs—are grouped together to form structures called *nuclei* (Fig. 3.4). The brain is subdivided into hundreds of different nuclei, each containing thousands of neurons all of which are involved in the integration of closely related sets of neuronal activities. Examples of nuclei considered later in the discussion include the nuclei of basal ganglia, cerebellum, hypothalamus, and thalamus.

The *cortex* is the most complex form of neuronal organization. Each cortex—neo, cerebellar, and hippocampus—is

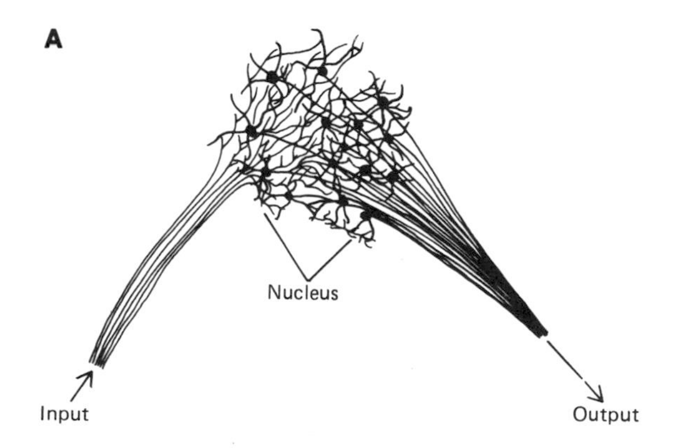

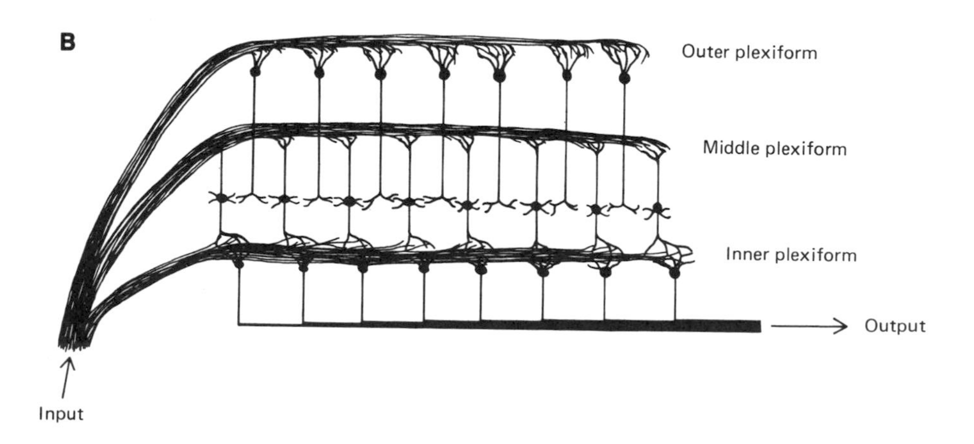

Fig. 3.4A and B. Complex units of neuronal organization. **A** Nucleus—a cluster of neurons which share similar inputs and outputs. The neurons in any one nucleus have a common function. **B** Cortex—a multilayered neural structure composed of alternate nuclear and plexiform layers. Neurons situated in the same region of the cortex share similar inputs.

composed of several layers of neurons and their processes (Fig. 3.4). Usually the cortex is formed by alternating nuclear layers (containing cell bodies) and plexiform layers (dendrites and synaptic terminals). Different classes of neurons are confined to specific layers of the cortex. Inputs to the cortex usually terminate in one or more of the plexiform layers. Axons that transmit the cortical output to other areas of the CNS usually originate from the base of one of the nuclear layers. For a detailed discussion of both cerebral and cerebellar cortices, see Chaps. 6 and 8.

Neuroanatomic Organization

The nervous system is divided into two major divisions—the *central nervous system* and the *peripheral nervous system*. The former is composed of all the neurons, neuronal processes, and glia in the brain and spinal cord. In contrast the peripheral nervous system includes all neurons, neuronal processes, and glia outside the CNS. This includes all of the neuronal processes running in the peripheral nerves—cranial, spinal, and autonomic nerves—as well as all of the clusters of nerve cells (ganglia) located in the periphery, i.e., the sensory and autonomic ganglia.

Anatomically the CNS is divided into four divisions, which are established during the third week of human embryonic development (Fig. 3.5). They are the *forebrain* (prosencephalon), *midbrain* (mesencephalon), *hindbrain* (rhombencephalon), and *spinal cord*. Continued embryonic growth leads in the seventh week to further subdivision of forebrain and hindbrain to form the five major divisions of the adult brain: *telencephalon, diencephalon, mesencephalon, metencephalon,* and *medulla*. The telencephalon and diencephalon are subdivisions of the forebrain, whereas the metencephalon and medulla are derived from the hindbrain.

Each of these divisions of the brain surrounds a fluid-filled space, or *ventricle* (Fig. 3.5), of which there are four. The lateral ventricles (first and second) are within the two lobes of the telencephalon, the third ventricle is in the diencepha-

Fig. 3.5A and B The developing nervous system. **A** at three weeks; **B** at seven weeks.

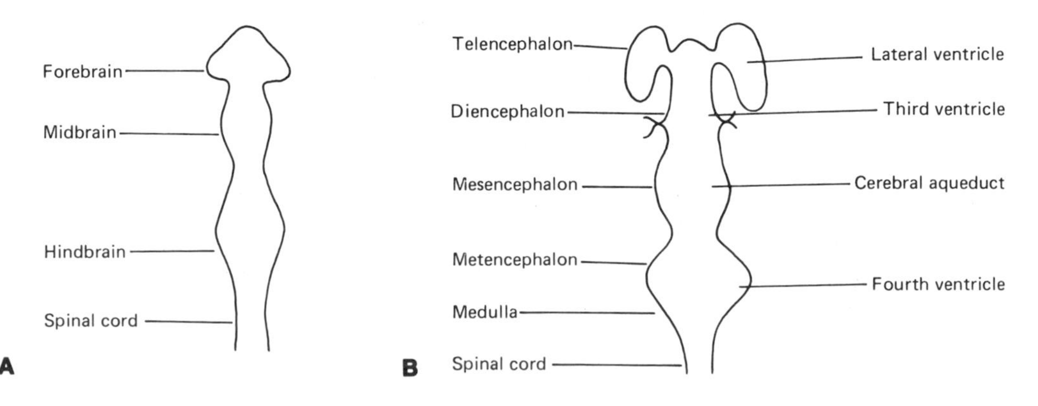

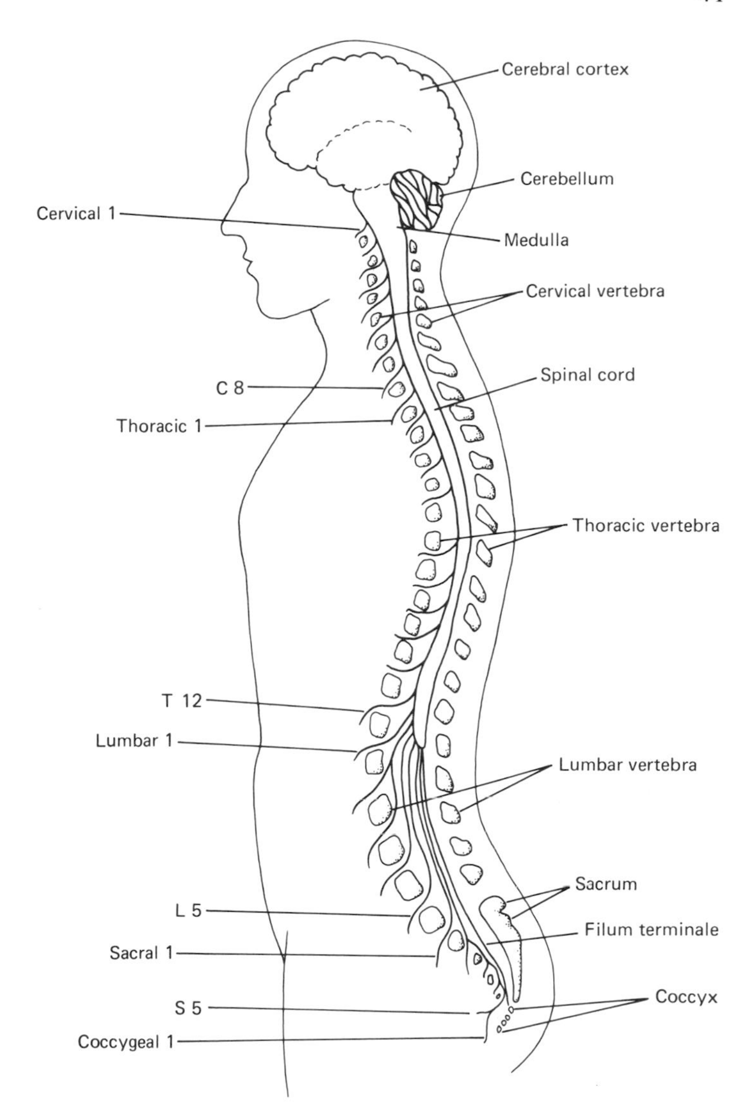

Fig. 3.6. Spinal cord and its relationship to the 31 pairs of spinal nerves.

lon, and the fourth ventricle extends through the metencephalon and medulla. The third and fourth ventricles are interconnected by the cerebral aqueduct, which forms a channel through the mesencephalon.

Spinal Cord and Spinal Nerves

The spinal cord lies within the vertebral canal of the vertebrae. In man it is about 45 cm long and extends from the medulla to the second lumbar vertebra (Fig. 3.6). At its anterior end it is continuous with the medulla. A non-neural filament, the filum terminale, projects from the posterior end of the spinal cord to attach to the first segment of the coccyx. The spinal cord gives rise to *31 pairs of spinal nerves,* one pair for each spinal segment, including: 8 cervical, 12 thoracic, 5 lumbar, 5 sacral, and 1 coccygeal (Fig. 3.6). Each spinal nerve is formed by the fusion of two spinal roots—

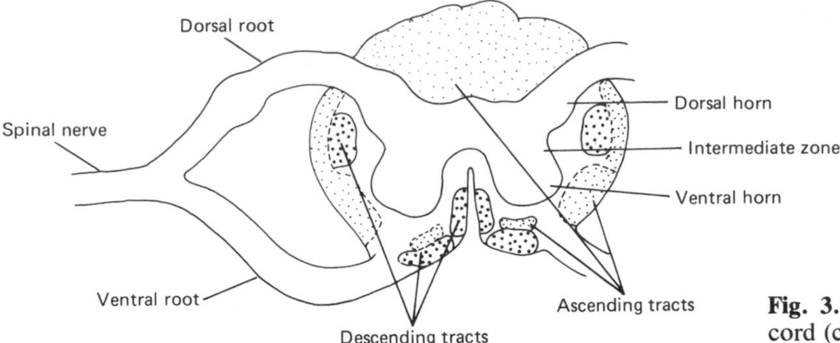

Fig. 3.7. Internal structure of spinal cord (cross section).

the *dorsal* (sensory) *root* and the *ventral* (motor) *root* (Fig. 3.7). Each spinal nerve exits through an intervertebral foramen in the appropriate vertebra and runs to the peripheral structures, sensory receptors and muscles, which it innervates.

The 31 pair of spinal nerves innervate all cutaneous sensory receptors except those located in the face and anterior scalp (Fig. 3.8). Each nerve contains sensory fibers, which innervate the sensory rceptors in its body segment. The distribution of spinal nerve innervation to body segments has been mapped out by measuring the size and extent of the regions of the body surface (*dermatomes*) innervated by each spinal nerve. The dermatomes have a segmental distribution over the body surface. The cervical dermatomes include the back of the head, the neck, the shoulders, and the anterior part of the forearm. Thoracic sensory neurons innervate the remainder of the forearm, the thorax, and most of the abdomen. Sensory fibers in the lumbar, sacral, and coccygeal spinal nerves innervate the remainder of the abdomen and the legs.

Internal Organization

The spinal cord is divided into two divisions: *spinal gray matter* and *spinal white matter* (Fig. 3.7). The gray matter, occupying the central core of the spinal cord, is concerned with the integration of spinal sensory and motor activities. It contains the cell bodies, dendrites, and axon terminals of the spinal neurons. The white matter is formed by the large tracts of myelinated nerve fibers that provide pathways for the transmission of neural activity both between different segments of the spinal cord and between the spinal cord and the brain.

The major divisions of the gray matter: dorsal horn, intermediate zone, and ventral horn (Fig. 3.7) are associated with three of the major functions of the spinal cord. They include: (1) receipt of sensory input (dorsal horn), (2) control of efferent activity to the viscera (intemediate zone in the thoracic and lumber segments), and (3) control of motor neuron activity to the skeletal muscle (ventral horn).

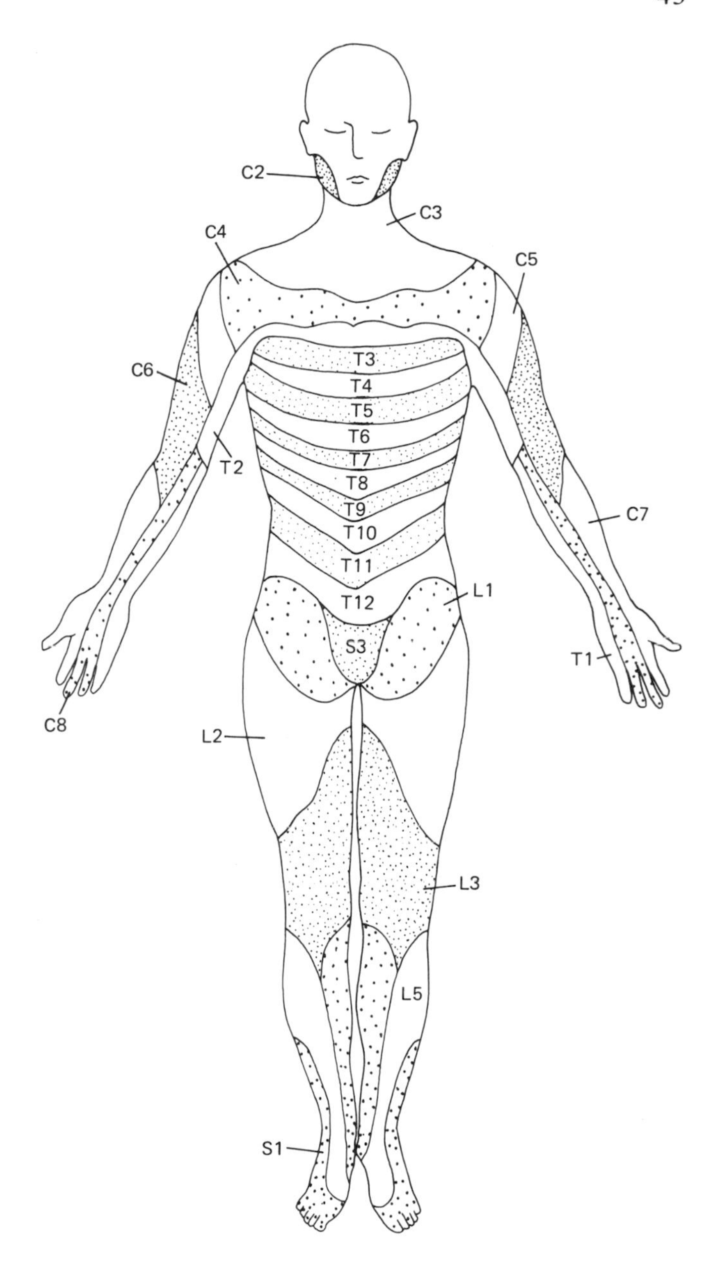

Fig. 3.8. Spinal dermatomes. The innervation of body surface by the 31 pairs of spinal nerves. (*C*, cervical; *T*, thoracic; *L*, lumbar; and *S*, sacral.

The *spinal tracts,* which compose the the white matter, can be divided into those transmitting information either toward the brain (*ascending tracts*) or in a posterior direction (*descending tracts*) (Fig. 3.7). The ascending tracts are formed by the axons of both sensory and spinal interneurons. They furnish pathways for transmission of sensory information—*touch, pressure, temperature,* and *pain*—to different regions of the brain, including: medulla, cerebellum, and thalamus. Axons originating from both interneurons in the cerebral cortex and the motor nuclei of the brain stem form the descending spinal tracts. They provide pathways over which centers in the brain can *regulate the sensorimotor activities* of the spinal cord.

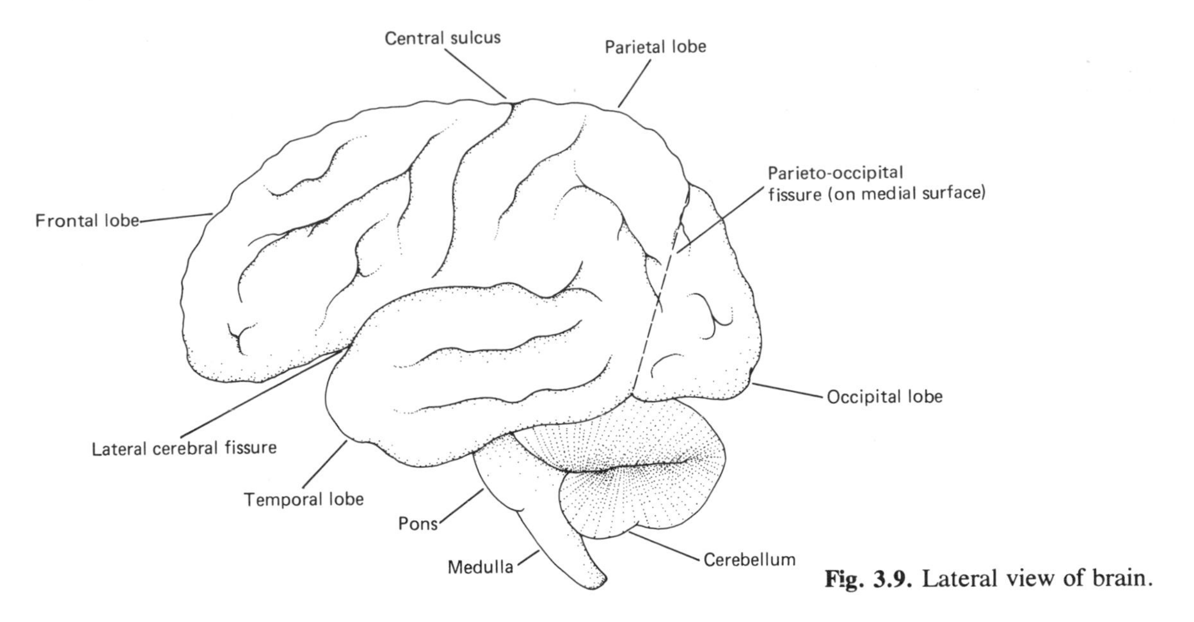

Fig. 3.9. Lateral view of brain.

Medulla

The medulla extends from the spinal cord to the pons and cerebellum (Fig. 3.9). As its posterior end the medulla resembles the cervical spinal cord with which it is continuous. The anterior medulla is a pyramid-shaped structure whose dorsal aspect is bounded by the fourth ventricle (Fig. 3.10). Most of the tracts between the spinal cord and the higher levels of the brain pass without interruption through the medulla. The fibers of the *corticospinal tract,* decending from the motor areas of the cerebral cortex to the spinal cord, form two longitudinal ridges (the medullary pyramids) on the ventral surface of the medulla (Fig. 3.11). Other charac-

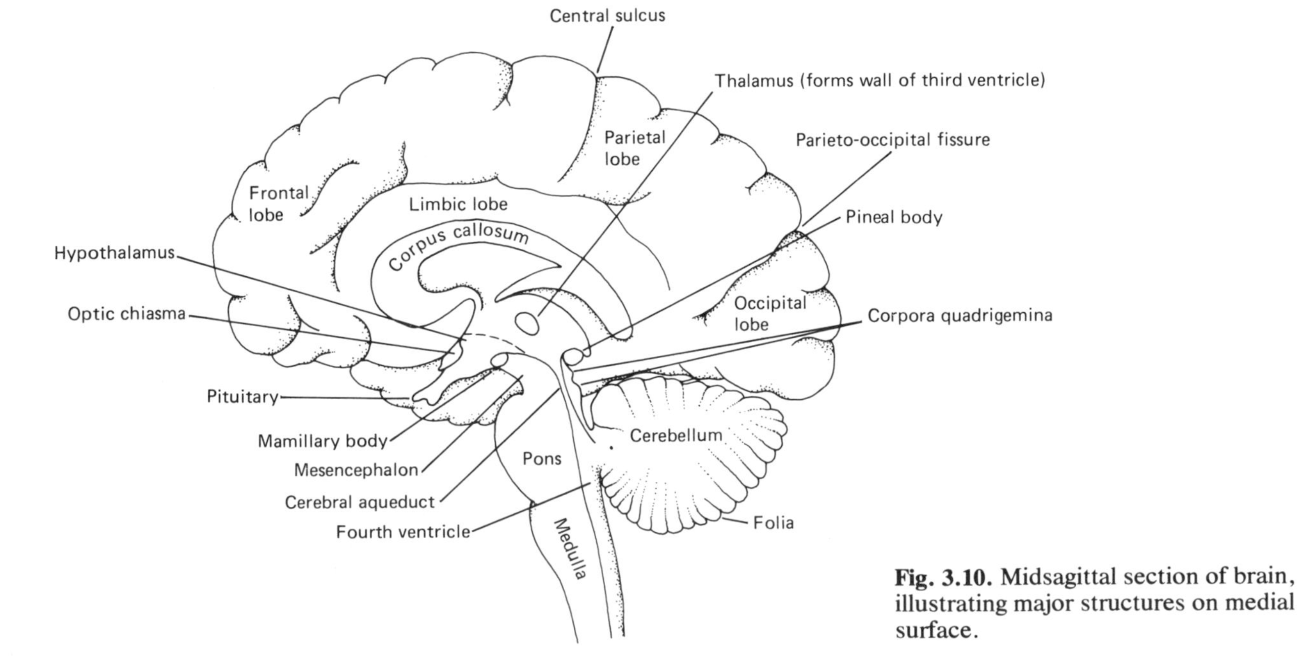

Fig. 3.10. Midsagittal section of brain, illustrating major structures on medial surface.

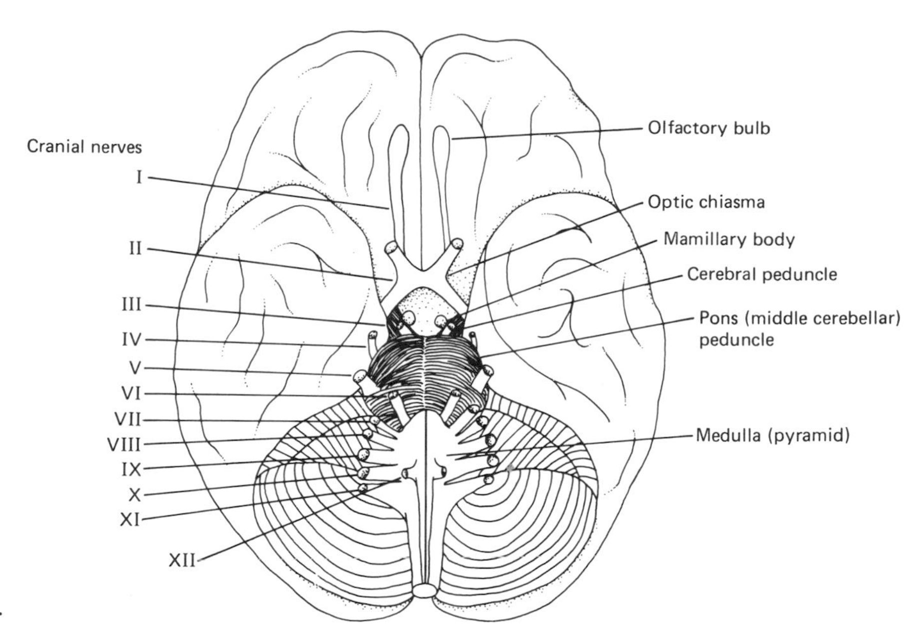

Fig. 3.11. Ventral view of brain.

teristic external landmarks include the roots of the *four cranial nerves,* which exit from the medulla (Fig. 3.11), including: *glossopharyngeal* (IX), *vagus* (X), *accessory* (XI), and *hypoglossal* (XII). These cranial nerves innervate tongue (IX and XII), parotid gland (IX), pharynx (IX and X), larynx (X and XI), carotid sinus (IX), and visceral organs of thorax and abdomen (X) (Table 3.1). The glossopharyngeal and vagus nerves contain both sensory and motor fibers; only motor fibers run in the accessory and hypoglossal nerves.

Table 3.1. Functions of the cranial nerves.

Cranial	Organ(s) Innervated	Function
Olfactory (I)	Olfactory epithelium	Sensory: olfaction (smell)
Optic (II)	Retina	Sensory: vision
Oculomotor (III)	Eye muscles	Motor: eye movement and constriction of pupil and lens shape (accommodation)
Trochlear (IV)	Eye muscle	Motor: eye movement
Trigeminal (V)	Head and jaw	Motor: chewing and swallowing Sensory: somatosensory (touch, pressure, and temperature) from anterior half of head and face
Abducens (VI)	Eye muscle	Motor: eye movement
Facial (VII)	Muscles of anterior half of head, salivary glands, lacrimal glands, and taste buds on tongue	Motor: facial movement and secretion of saliva and tears Sensory: taste
Vestibulocochlear (VIII)	Auditory and vestibular receptors	Sensory: auditory (hearing) and equilibrium
Glossopharyngeal (IX)	Pharynx, salivary glands, taste buds on tongue, and carotid sinus	Motor: swallowing and salivation Sensory: taste, blood pressure and gas content
Vagus (X)	Pharynx, larynx and viscera of thorax and upper abdomen	Motor: swallowing, movement of larynx, and parasympathetic control of viscera Sensory: visceral sensation
Spinal accessory (XI)	Muscles of neck, shoulder, larynx, and pharynx	Motor: movements of head, shoulder, larynx, and pharynx
Hypoglossal (XII)	Muscles of tongue	Motor: movement of tongue

Internal Structure

The core of the medulla contains the most posterior part of the *reticular formation*. This diffuse network of interneurons extends through the entire brain stem from medulla to diencephalon. It plays important roles in both the integration of sensory information and the control of the activity of all efferent neurons (both motor and autonomic neurons) and is also critical to the maintenance of arousal or consciousness in the cerebral cortex.

The medulla also contains many discrete clusters of neurons or nuclei that surround the diffuse reticular core. Examples of these nuclei include: the *nuclei of the four cranial nerves* (glossopharyngeal, vagus, accessory, and hypoglossal) exiting from the medulla; the *nuclei gracilis and cuneatus* forming part of the *dorsal column pathway* that transmits information from the spinal cord to the somatosensory region of the cerebral cortex; the *cochlear nuclei* of the auditory system; the *inferior olivary nucleus,* an important source of input to the cerebelllum; and "vital centers" controlling the *respiratory* and *cardiovascular systems*.

Metencephalon

The metencephalon is divided into two divisions: *pons* and *cerebellum* (Fig. 3.9). The pons is the section of the brain stem between the medulla and the mesencephalon. The cerebellum is formed by a large outgrowth of the roof of the anterior metencephalon, which extends over the dorsal surface of the pons and medulla (Fig. 3.10). The pons is interconnected with the cerebellum by the *middle cerebellar peduncle,* a large broad band of fibers, that is responsible for the characteristic swollen appearance of the ventral surface of the pons (Fig. 3.11).

Pons

The posterior end of the pons is directly continous with the medulla. Many medullary structures *ascending* and *descending fiber tracts* and *reticular formation* among others, extend uninterruptedly through the pons. The junction between the pons and medulla is marked by the exit of the *vestibulocochlear cranial nerve* (VIII) (Fig. 3.11). Other cranial nerves that exit from the pons include the *trigeminal* (V), the *abducens* (VI), and the *facial* (VII). They innervate the face, mouth, and scalp (V and VII), the tongue (VII), the auditory and vestibular receptors in the inner ear (VIII), and the lateral rectus muscles of the eyes (VI) (Table 3.1). Whereas the vestibulocochlear nerve contains only sensory fibers, sensorimotor fibers run in the trigeminal and facial nerves. The abducens is a purely motor nerve.

Interspersed throughout the pons are some important sensory and motor nuclei. In addition to the *nuclei* of the trigeminal, abducens, facial, and vestibulocochlear nerves, the pons contains the pontine nuclei which give rise to the middle cerebellar peduncle, which interconnects the pons and the cerebellum. Other important nuclei located in the pons include the *superior olivary nucleus* of the auditory system and *visceral centers* which regulate the *respiratory* and *cardiovascular systems*.

Cerebellum

The cerebellum is the dominant feature of the hindbrain. It is situated over the dorsal surface of the pons and medulla immediately behind the cerebral hemispheres of the telencephalon (Fig. 3.9). The cerebellum is attached to the brain stem by three sets of peduncles (inferior, middle, and superior) which interconnect the cerebellum with the medulla and spinal cord (*inferior peduncle*), the pons (*middle peduncle*), and the mesencephalon and thalamus (*superior peduncle*).

The cerebellum is divided into two divisions, the *superficial cortex* and the *deep cerebellar nuclei*. The surface of the cerebellum is covered by numerous ridges, or folia, derived from the complex folding of the cerebellar cortex (Fig. 3.10).

The complete output of the cerebellar cortex is transmitted to the cerebellar nuclei located deep within the cerebellar peduncles. Fibers arising from these cerebellar nuclei innervate both motor nuclei in the brain stem and (via the thalamus) motor regions of the cerebral cortex.

Mesencephalon

The mesencephalon, or midbrain, interconnects the forebrain with the hindbrain. Characteristic external landmarks of the mesencephalon include the rounded *corpora quadrigemina* (the *superior* and *inferior colliculi*) (Fig. 3.10), the roots of the *oculomotor* (*III*) and *trochlear* (*IV*) *cranial nerves* (Fig. 3.11), and the two large bundles of nerve fibers, the *cerebral peduncles,* which provide direct descending pathways for the fibers from the motor regions of the cerebral cortex (Fig. 3.11).

The mesencephalon forms the anterior end of the brain stem. Consequently many structures found in the posterior brain stem (pons and medulla), including: *ascending* and *descending fiber tracts,* and the *reticular formation* continue uninterruptedly through the mesencephalon.

The mesencephalon contains a number of important sensory and motor centers. The corpora quadrigemina in the dorsal mesencephalon form important integrative centers for the visual (the superior colliculi) and the auditory (inferior colliculi) systems. Important motor centers include both the *substantia nigra,* which forms part of the basal ganglia complex, and the *red nucleus*. Other important nuclei are those of the oculomotor and trochlear cranial nerves, which innervate the eye muscles.

Diencephalon

The diencephalon, located at the anterior end of the brain stem, is surrounded on three sides (anterior, dorsal, and lateral) by the fiber tracts (internal capsule) and lobes of the cerebral cortex (Figs. 3.10 and 3.12). Its anteroposterior boundaries are associated with two prominent landmarks on the ventral surface of the brain (Fig. 3.11). The anterior boundary is marked by the *optic chiasma* (the crossing of the optic nerves); the posterior boundary is located at the posterior edge of the prominent *mamillary bodies*.
The diencephalon is divided into: (1) *thalamus,* (2) *hypothalamus,* and (3) *epithalmus,* both thalamus and hypothalamus are major neural centers. The epithalamus contains two structures: the pineal body, a secretory organ (Fig. 3.10), and the habenula, an olfactory center associated with the limbic system.

Thalamus

The thalamus is located in the central medial region of the forebrain (Fig. 3.10) and is subdivided into a large number of nuclei, each of which supplies neural input to a specific region of the cerebral cortex. The neurons in these thalamic nuclei form the final stage in the pathways that transmit neural activity to the cortex. All neural inputs to the cortex (except the olfactory) must pass through the thalamic integrative and relay centers. Most of the fibers of the optic nerve terminate in the lateral geniculate nucleus of the thalamus. The *thalamic nuclei* can be classified according to the general function—sensory, motor, or associative—of the region of the cerebral cortex that they innervate. The major sensory nuclei are the *ventrobasal complex,* which processes input to the somatosensory and gustatory regions of the cortex; the *medial geniculate,* which innervates the auditory cortex; and the *lateral geniculate,* which transmits in-

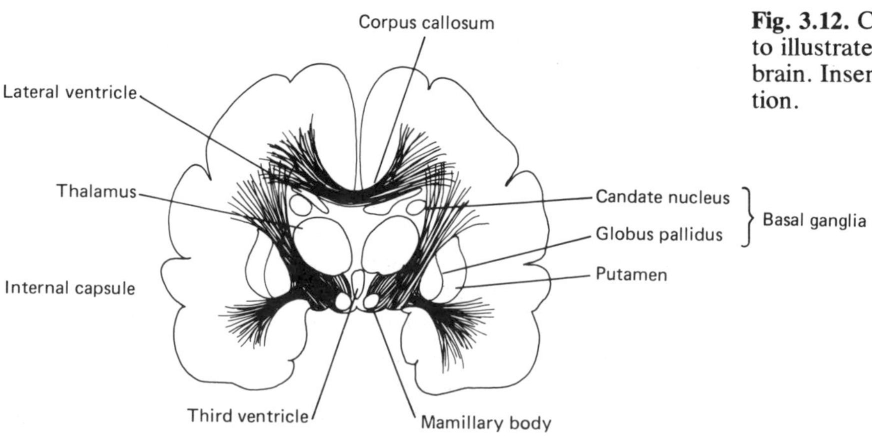

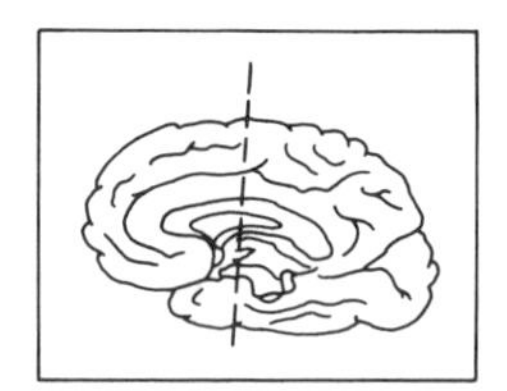

Fig. 3.12. Cross section through brain to illustrate internal structure of forebrain. Insert shows level of cross section.

formation to the visual cortex. The *anterior ventral* and *ventrolateral nuclei* furnish relay stations for the transmission of information from the basal ganglia and cerebellum to the cortical motor regions. Other important nuclei include the *anterior nuclei,* which transmit inputs from the limbic system to the cortex, and the interlaminar nuclei, which are relay stations for the transmission of crude pain sensations to the cortex.

Hypothalamus

The hypothalamus is located in the ventral part of the diencephalon (Fig. 3.10) and lies immediately above the pituitary gland whose function it regulates.
It contains eight small nuclei that control most of the visceral activities. In addition to control of *pituitary secretion,* the hypothalamic nuclei regulate *body temperature, water balance,* and *food intake*. The hypothalamus also plays a critical role in the regulation of *emotional and sexual behavior*.

Telencephalon

The telencephalon is divided into two anatomic parts, namely, the *cerebral cortex* and the *basal ganglia*. The former is the largest division of the brain and is divided into two hemispheres interconnected by huge tracts of nerve fibers forming the *corpus callosum* (Fig. 3.12). Each hemisphere extends over the anterior dorsal surface of the brain stem (Fig. 3.9). The basal ganglia form a series of interconnected nuclei located deep within each cerebral hemisphere (Fig. 3.12).

Cerebral Cortex

Although the cerebral cortex is a highly folded structure (Fig. 3.9), the locations of the major folds (gyri) and furrows (fissures, or sulci) are consistent between different brains. As a result, the major gyri and sulci are used as morphologic landmarks for the demarcation of the cortical regions. The external surface of the cerebral cortex is divided into four lobes, including: *frontal, parietal, occipital,* and *temporal* by three major furrows (Figs. 3.9 and 3.10). The furrows include: (1) the central sulcus, separating frontal and parietal lobes; (2) the parieto-occipital fissure, forming a boundary between parietal, occipital and temporal lobes; and (3) the lateral cerebral fissure, separating temporal from frontal and parietal lobes.
Each of these lobes receives input fibers from one or more of the thalamic nuclei. *Sensory regions* in each lobe (parietal for somatosensory and gustatory, occipital for visual, and temporal for auditory) receive their inputs from the ventrobasal complex, lateral geniculate, and medial geniculate in the thalamus, respectively. The *motor areas* of the frontal

lobe are innervated by the anterior ventral and ventrolateral thalamic nuclei. Neurons in other regions of the cortex supply most of the neural input to the association regions of the cortex.

The *olfactory nerve* (I) provides a direct neural input to the cerebral cortex. The nerve fibers terminate in the olfactory bulb, which extends from the ventromedial surface of the cortex (Fig. 3.11). The input is closely associated with the ring of *limbic cortex* that lies inside the four external lobes (Fig. 3.10). The limbic cortex is interconnected with several subcortical and hypothalamic structures to form the *limbic system,* which contains the highest centers of visceral control.

The cortex is interconnected with the lower regions of the brain (basal ganglia, thalamus, mesencephalon, and pons) by the large fiber tracts forming the *internal capsule* (Fig. 3.12). These broad sheets of white matter contain millions of nerve fibers, some of which (the axons of thalamic neurons) transmit neural activity to the cortex, whereas others (the axons of cortical neurons) transmit the neural output of the cortex to lower neural centers.

Basal Ganglia

The basal ganglia form a chain of subcortical nuclei—the globus pallidus, putamen, caudate nucleus, and amygdala—located deep beneath the cortex surrounded by the fiber tracts of the internal capsule (Fig. 3.12). Three of these nuclei—globus pallidus, putamen, and caudate—are interconnected with several motor nuclei in the brain stem to form the basal ganglia system, an important center for the coordination of movement. Although the amygdala is morphologically part of the basal ganglia, functionally it forms part of the visceral control centers of the limbic system.

Selected Readings

Bodian D (1967) Neurons, circuits and neuroglia. In: Quarton GC, Melnechuk T, Schmitt FO (eds) The neurosciences. Rockefeller University Press, New York

Eyzaguirre C, Fidone SJ (1975) Physiology of the nervous system, 2nd edn. Yearbook Medical, Chicago

Heimer L (1971) Pathways in the brain. Sci Amer 225: 48

Kuffler SW, Nicolls JG (1976) From neuron to brain: a cellular approach to the function of the nervous system. Sinauer, Sunderland, Mass

Noback C, Demarst RJ (1975) The human nervous sytem, 2nd edn. McGraw-Hill, New York

Palay SI, Chan-Palay V (1977) Morphology of neurons and neuroglia. In: Kandel ER (ed) Handbook of physiology: Cellular biology of neurons. Amer Physiol Soc, Bethesda, Maryland

Truex RC, Carpenter MB (1975) Human neuroanatomy. Williams and Wilkins, Baltimore

Review Questions

1. What are the four functional regions of the neuron? Briefly describe their functions.
2. What is a glial cell? How does it differ from a neuron?
3. Briefly describe the two higher levels of neuronal organization —the nucleus and the cortex.
4. Draw a cross section of the spinal cord. Label its principal features.
5. Make a table summarizing the anatomical and functional characteristics of the cranial nerves. Include information concerning their principal functions, the region of the brain from which the nerve exits, and the major organs innervated.
6. Draw a midsagittal section of the brain to show the locations of the medulla, pons, cerebellum, mesencephalon, hypothalamus, thalamus, corpus callosum, and cerebral cortex.
7. Construct a diagram showing the anatomic relationship between the cerebral cortex, corpus callosum, internal capsule, thalamus, and basal ganglia.

Neuronal Activity

Chapter 4

The most important function of the nervous system is the rapid and accurate *transmission of information*. Communications from sensory receptors to central sensory centers, between central sensory and central motor centers, and from central motor centers to the effector organs (muscles and glands) must be fast and precise. The survival of the organism may depend on its ability to detect and respond rapidly to a threatening change in its environment.

Nerve cells are specialized, morphologically and physiologically, for information transmission (see Chap. 3). The major process of most neurons is the axon. It provides an intracellular pathway for the rapid conduction of nerve excitation between two points in the nervous system. Axons are specialized for the intraneuronal conduction of nerve impulses (*action potentials*).

Transmission of neuronal activity from one neuron to the next occurs at the specialized contacts (*synapses*) that axons form with the dendrites and somata of other neurons. The synapse is the site of both neuronal integration and the interneuronal transfer of information. The release of a specific chemical transmitter by the presynaptic axon transfers neuronal activity across the synapse.

The transmitter interacts with the postsynaptic dendrites and soma, producing brief changes in their membrane potentials (*synaptic potentials*).

Both the conduction of action potentials and the generation of synaptic potentials depend on the intrinsic capacity of nerve cell membranes to undergo brief changes in electrical potential. Although an electrical potential exists across the membranes of all cells, only the membranes of nerve, sensory and muscle cells can sustain action and synaptic potentials. Because of their ability to support transient changes in their membrane potentials, nerve, sensory, and muscle cells are "excitable." All other cells of the body are unexcitable. The unique capacity of nerve cells to transmit and process information is a consequence of this characteristic.

Membrane Potential

Transmembrane Ion Distribution. The intracellular concentrations of the principal monovalent ions—chloride (Cl^-), potassium (K^+), and sodium (Na^+)—differ markedly from their levels in the extracellular fluids bathing the cell. As indicated in Fig. 4.1 the major cation (positively charged ion) within the cytoplasm is K^+; most of the intracellular anions (negatively charged ions) are amino acids and other organic molecules. The major cation in the extracellular fluids is Na^+, and Cl^- is the major anion.

These differences in ionic distribution are the result of two factors: (1) intracellular *negatively charged organic molecules* and (2) *active transport systems* within cell membranes that "pump" Na^+ from the cell and K^+ into the cell (see Chap. 2). Although small ions, such as Cl^-, K^+, and Na^+ can penetrate the cell membrane, organic anions like amino and organic acids in the cytoplasm are too large to pass across the membrane. Therefore a large pool of negative charges (organic anions) is trapped within the cell. These

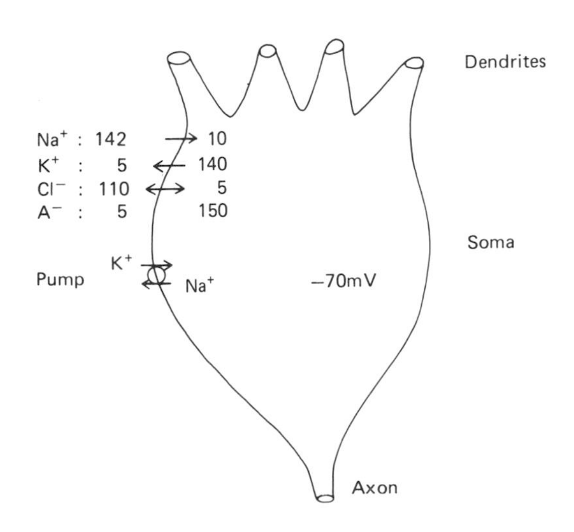

Fig. 4.1. Distribution of ions across neuron membrane. Typical values for intra- and extraneuronal concentrations of Na^+, K^+, Cl^-, and organics (A^-) are listed. Low level of intraneuronal Na^+ is maintained by the Na^+/K^+ transport system, which "pumps" Na^+ from cell in exchange for extraneuronal K^+. Low concentration of Cl^- within neuron is result of repulsive effects of negatively charged organic molecules trapped within membrane. Concentrations shown are all in mEq/liter.

trapped negative charges repel negative ions, e.g., Cl^-, that attempt to enter the cell while attracting positively charged cations into the cell (K^+ and Na^+). Most Na^+ that enter the cell, however, are immediately removed by the *Na^+/K^+ transport system*. Because of the rapid removal of Na^+ from the cell by the Na^+/K^+ pump, only K^+ can accumulate within the cell, attracted by the negative charge of the organic anions and transported into the cell by the Na^+/K^+ pump.

Measurement of Cell Potentials

A difference in electrical potentials exists between the inside and outside of all cells. Depending on the characteristics of the particular cell, the *resting membrane potential* may be −40 to −95 mV. Typical values for nerve cells are −60 to −80 mV.

The membrane potential is usually determined by inserting a probe or electrode into the cell and measuring the potential difference between that electrode and a similar one located in the fluid medium around the cell (Fig. 4.2). The electrodes are connected to an amplifier that magnifies the potential. The magnitude of the potential is measured with a voltage-sensitive device, such as an oscilloscope (Fig. 4.3).

Microelectrodes are usually used to measure membrane potential. The smallness of the electrode tip permits penetration of the cell membrane without damaging any except the smallest cells.

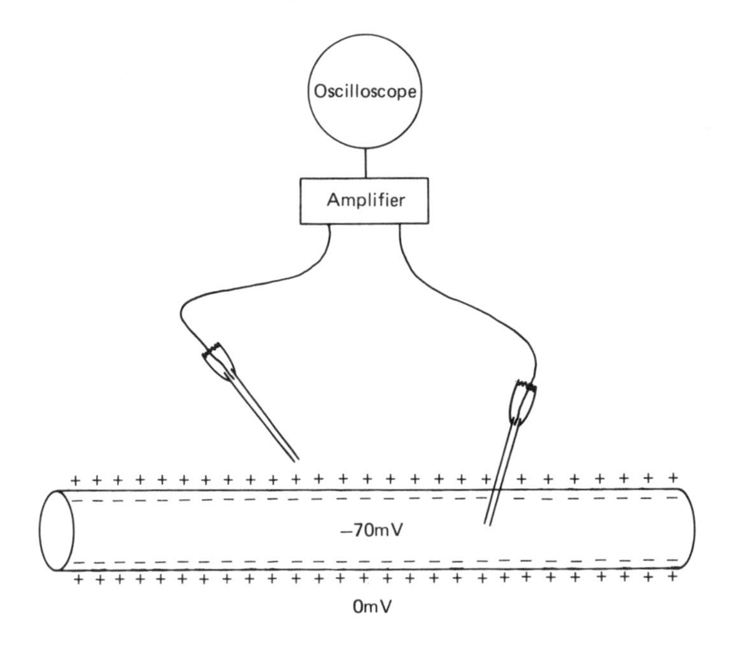

Fig. 4.2. Intracellular measurement of membrane potential in an axon. The oscilloscope determines the magnitude of the amplified potential.

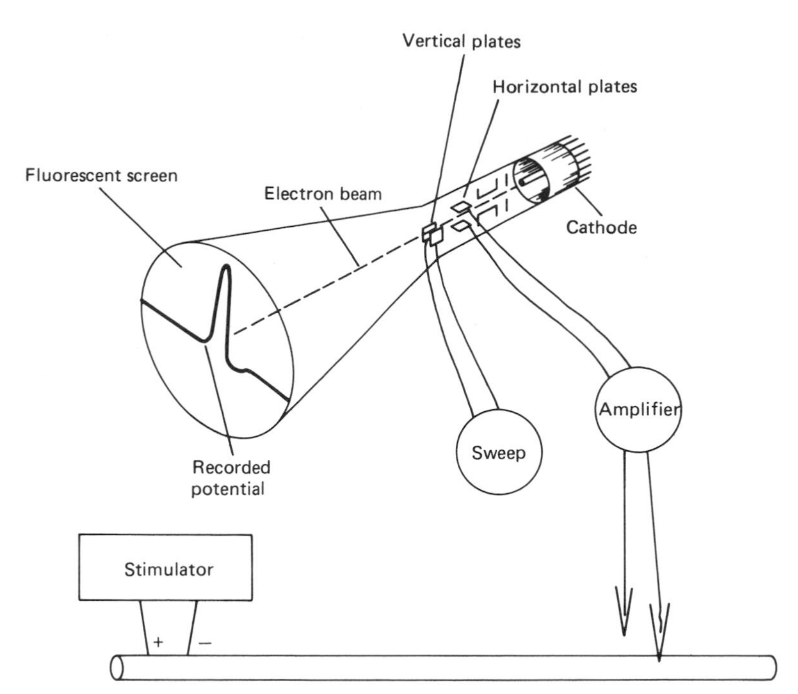

Fig. 4.3. Cathode ray oscilloscope. Electron beam produces a bright spot when it strikes fluorescent screen. Movement of beam across screen is controlled by potentials of vertical and horizontal plates. The sweep circuit generates a potential which moves beam at constant rate along horizontal axis. The sweep of beam provides time base for voltage-time plot, which is displayed on screen. Vertical deflection of beam is produced by potentials applied to the vertical plates. Since recording electrodes are usually connected to these plates (through an amplifier), the vertical deflection indicates magnitude of the potential. Potential displayed on screen is voltage-time plot of nerve impulse initiated by stimulator and recorded with intra-axonal microelectrode.

Ionic Basis of Membrane Potential

The resting membrane potential is primarily the result of the asymmetric distribution of K^+ across the cell membrane. Since intracellular K^+ is about 30 times more concentrated than extracellular K^+, a chemical concentration gradient exists across the membrane, which favors the diffusion of K^+ out of the cell (Fig. 4.1). Each positively charged K^+ diffusing from the cell leaves an unbalanced negative charge behind (organic anions). These negative charges within the cell form a negative potential (see also Chap. 1).

Nernst Equation

Potassium movement reaches an equilibrium when the chemical concentration gradient driving K^+ from the cell is just counter-balanced by the negative intracellular potential drawing K^+ into the cell. At equilibrium the net movement of K^+ is zero. The diffusion potential at equilibrium is the *equilibrium potential* (E_{K^+}). It (at 25°C) may be calculated from the Nernst equation as follows:

$$E_{K^+} = -59 \log \frac{K^+ \text{ in}}{K^+ \text{ out}}.$$

The Nernst equation can also be used to calculate equilibrium potentials for both Cl^- and Na^+. Typical values for neurons are: $E_{K^+} = -86$ mV; $E_{Cl^-} = -78$ mV; and $E_{Na^+} = +66$ mV.

The resting membrane potential of most cells is nearly identical to E_{K^+}. Actually if the membrane were only permeable to K^+, i.e., if it were impermeable to Cl^- and Na^+, the membrane potential would equal E_{K^+}. Although their membrane permeabilities are much lower than the K^+ permeability, both Cl^- and Na^+ can penetrate the membrane and contribute to the membrane potential. The relative contribution of each ion's equilibrium potential to the membrane potential is proportional to its membrane permeability. The greater the relative permeability of an ion, the larger its contribution to the membrane potential. Since the permeability of K^+ in neuronal membranes is about seven times greater than the permeability of Cl^- and 25 times greater than the permeability of Na^+, the resting membrane potential is primarily determined by E_{K^+}.

Action Potential

The nerve axon is specialized for the transmission of nerve impulses. The action potential is produced by a brief reversal in the resting membrane potential, which is propagated as a wave over the axon membrane. Normally an action potential begins in the initial segment of the axon near the soma and is conducted down the axon to the axon terminals. An action potential can be initiated in an isolated axon by applying a brief pulse of electrical current to the axon membrane (Figs. 4.3 and 4.4). Because of its small size (100–125 mv) and brief duration (1–2 ms), the action potential that is recorded must be amplified and displayed on an oscilloscope to obtain a voltage-time plot of its wave form.

The wave form of an action potential has two phases (Fig. 4.4), including: (1) the *rising phase,* which is produced by an abrupt positive shift or depolarization of the membrane potential of about 110 mV. The membrane is depolarized from its resting level (about −70 mV) to a point near E_{Na^+} (about 40 mV); and (2) the *falling phase,* during which the membrane potential drops or repolarizes to the resting potential and then hyperpolarizes to a level about 10 mV more negative than the resting potential (about −80 mV).

The rising phase of the action potential is the result of a transient *increase* Na^+ *permeability* of the axon membrane (Fig. 4.4). Membrane channels specific to Na^+ open, permitting Na^+ to rush into the axon. This influx of positive ions depolarizes the membrane.

The closing of the Na^+ channels and the opening of K^+ channels in the membrane produce the falling phase of the action potential (Fig. 4.4). Sodium movement into the axon decreases because of the *decreased* Na^+ *permeability,* whereas the increased K^+ permeability increases K^+ efflux from the axon. Since the increased K^+ efflux removes positive charges from the axon, the membrane repolarizes. The hyperpolarization of the membrane below the resting potential (more negative) is the result of the very high K^+ permeability during the falling phase of the action potential. The membrane is

Fig. 4.4 A Voltage wave form of action potential. **B** Changes in membrane permeability during action potential.
Depolarization to threshold elicits increase in Na^+ permeability (g Na^+) which generates rising phase of action potential. Falling phase of action potential is result of decreased Na^+ permeability coupled with increased K^+ permeability (g K^+). Resting membrane potential is −70 mV; threshold for action potential initiation is −50 mV. Permeability is measured in mmhos/cm^2 (the unit of conductance—the mho—is the reciprocal of the unit of resistance, the ohm).

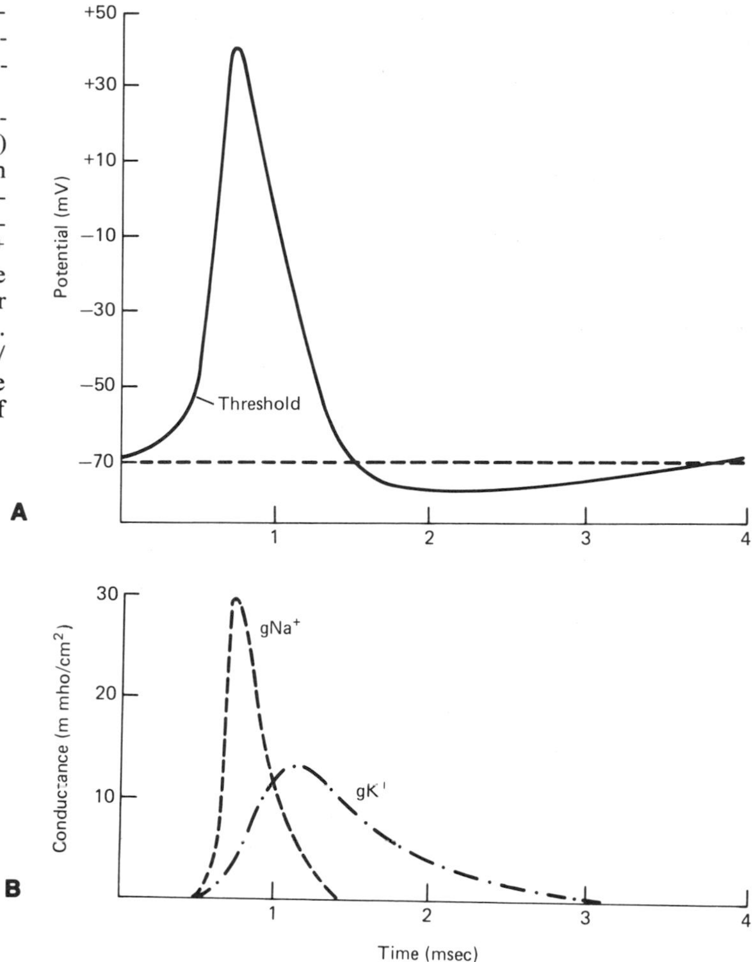

restored to its resting level by the closing of the K^+ channels; the K^+ and Na^+ permeabilities of the membrane return to their resting levels.

Initiation of Nerve Impulse

All excitable cells have a *threshold* membrane potential for the initiation of an action potential. An action potential is initiated whenever the membrane is depolarized to the threshold level (about −50 mV in neurons). The threshold is the membrane potential at which the sodium channels begin to open. The opening of the sodium channels rapidly depolarizes the neuron membrane toward E_{Na^+} (about 40 mV), thereby generating an action potential.
To initiate an action potential, the stimulus must depolarize the membrane to its threshold level. The initiation of an action potential is an "all-or-none" event, i.e., either the stimulus intensity is sufficient to depolarize the neuron membrane to the threshold or it cannot initiate an action potential. Although the usual stimulus employed in a laboratory is a brief pulse of depolarizing current, actually any

stimulus that depolarizes the cell to its threshold, e.g., mechanical distortion of the membrane or a brief flash of a laser, will initiate an action potential.

Nerve Refractoriness

The axon membrane supporting the action potential is in a refractory state—its excitability is depressed. The interval of this state can be examined by injecting twin pulses of stimulating current into an axon. If the interval that separates the twin pulses is large enough (more than 10 ms), each stimulus pulse evokes an action potential.
As the interval between the pulses is reduced (less than 10 ms) the second pulse no longer elicits an action potential. This period of reduced excitability is the *relative refractory period* because an action potential can be generated if the intensity of the second pulse is increased (Fig. 4.5). During the relative refractory period the action potential threshold is raised; therefore a greater current is required to depolarize the membrane to threshold. The relative refractory period includes most of the falling phase and after-hyperpolarization of the action potential; during this time the K^+ permeability is very high and the Na^+ channels are recovering their excitability.
If the interval between the twin pulses is very short (less than 3 ms), the second pulse occurs during the *absolute refractory period* when the axon is completely unexcitable

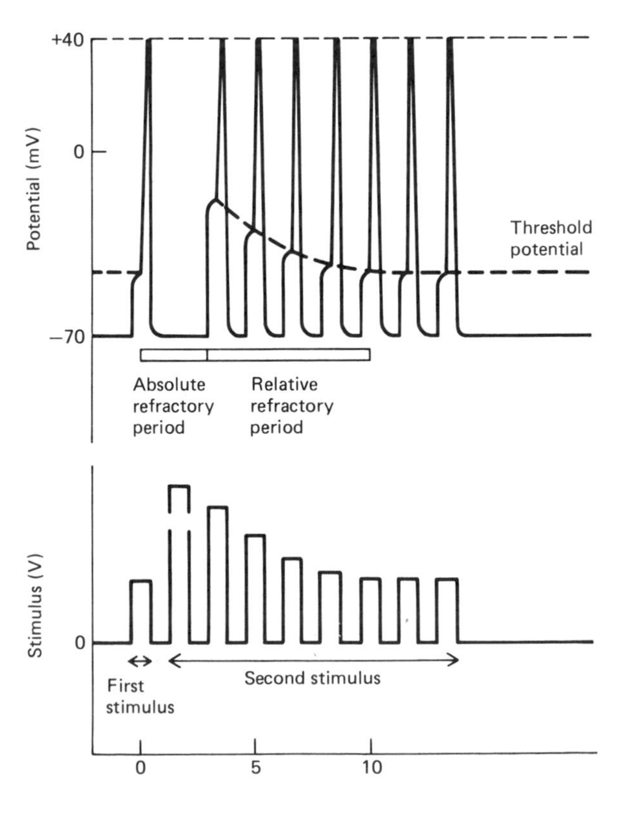

Fig. 4.5. Absolute and relative refractory periods. To measure time course of refractory periods, twin pulses of stimulating current are injected into axon. First stimulus is sufficient to initiate an action potential. Effect of second stimulus depends on length of interval between first and second pulses. If second pulse occurs during absolute refractory period, membrane cannot generate a second action potential. Although the threshold for action potential initiation is raised during the relative refractory period, an action potential can be elicited if stimulus strength is increased.

(Fig. 4.5). This is the period of the rising phase of the action potential when Na^+ permeability is high.
Nerve refractoriness is critical to axon function because it limits the minimum interval between successive action potentials. As a result the maximum rate at which most nerve fibers can transmit nerve impulses is about 200 impulses/s.

Recovery of Ionic Gradients

Although only about 1×10^{-12} equivalent (1 p. mole) of Na^+ enters the axon during an action potential (with about the same amount of K^+ leaving the axon), it can lead to a significant change in internal Na^+ (and external K^+) levels if the axon conducts a large number of impulses. A critical function of the Na^+/K^+ membrane pump is to prevent a long-term change in the Na^+ and K^+ concentration gradients across the membrane. The pump rapidly removes any Na^+ entering the axon and exchanges it for external K^+; however, it does not play a direct or immediate role in the generation of an action potential. Large axons whose Na^+/K^+ pumps are poisoned with the cardiac glycoside ouabain, can conduct thousands of nerve impulses before internal levels of Na^+ and external levels of K^+ build up to a point that they block conduction. In contrast, action potentials can be blocked by the specific poison tetrodotoxin (it blocks the opening of the Na^+ channels) without affecting either the Na^+/K^+ pump or the Na^+ and K^+ gradients across the membrane.

Nerve Impulse Conduction

The conduction of an action potential down an axon is analogous to the propagation of a flame down a fuse. Both of them are all-or-none phenomena with discrete thresholds; the axon must be depolarized to a threshold potential to open the sodium channels; the fuse must be heated to a critical threshold temperature to ignite. Propagation of the flame down the fuse results from the spread of heat from the segment of the fuse that is burning to the unburned fuse immediately in front of the flame. When the heat is sufficient to heat the unburned segment of fuse to the threshold temperature, the fuse ignites.
The nerve impulse travels down an axon in a similar manner. The action potential is propagated down the axon by small "local currents" generated by the inward flux of Na^+ during the rising phase of the action potential (Fig. 4.6). The influx of Na^+ produces a positive current, which spreads to the membrane just in front of the action potential wave. The positive current depolarizes this membrane from its resting potential to the threshold for action potential initiation, opening the Na^+ channels and thus advancing the action potential wave down the axon.

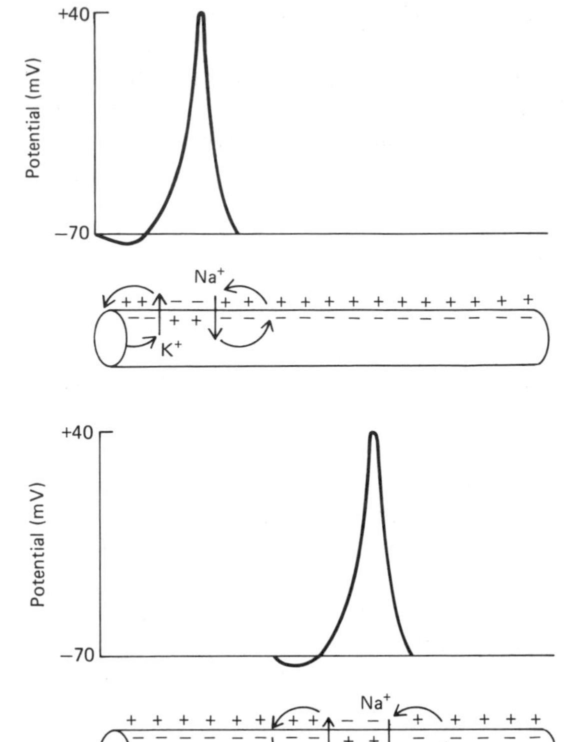

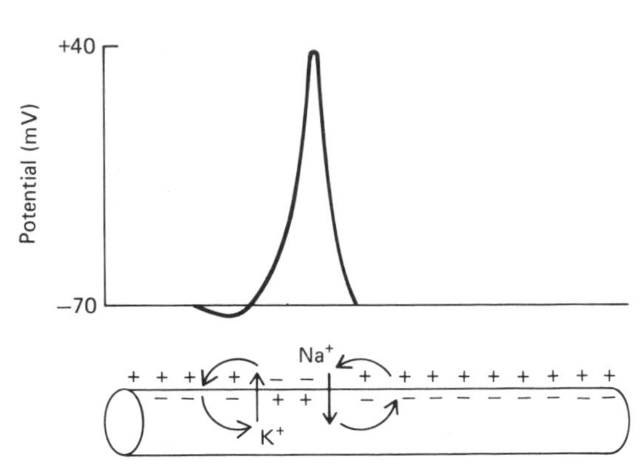

Fig. 4.6. Propagation of action potential. Each diagram shows distribution of potential and ionic currents at a successive instant in time as the nerve impulse propagates down the axon. Note that influx of Na^+ produces "local circuits" of current which depolarize adjacent membrane to threshold. Efflux of K^+ generates currents which hyperpolarize membrane below resting potential.

Saltatory Conduction

Most nerve axons are myelinated (see Chap. 2). A *myelin sheath* increases the rate at which an action potential is conducted over an axon. Myelinated axons conduct action potentials at rates of 30 to 100 m/s; the maximum rate of conduction for a mammalian unmyelinated fiber is about 3 m/s. The large difference in conduction velocity is because action potential propagation in unmyelinated axons requires the depolarization of each successive segment of the axon membrane by local circuits of current. In myelinated fibers the depolarizing currents do not affect those axon segments that are enveloped by the insulating myelin sheath (*internodes*). The action potential is conducted by a saltatory process. It "skips" down the myelinated axon involving only those small areas of exposed membrane (*nodes*) which separate

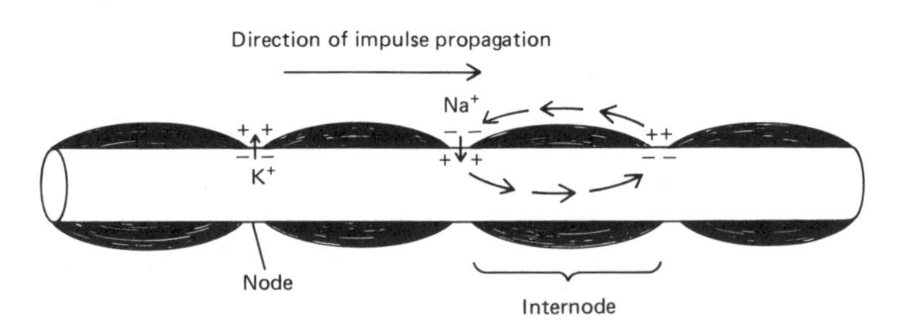

Fig. 4.7. Saltatory conduction. Each node in its turn is subjected to (1) depolarization to threshold, (2) influx of Na^+, (3) efflux of K^+, and (4) recovery. Influx of Na^+ at node generates depolarizing current which passes around internode to next node. When node is depolarized to its threshold, its Na^+ permeability increases, permitting Na^+ influx.

successive myelin sheaths (Fig. 4.7). Because only a small portion of the total axon membrane (nodes) ever supports the action potential, impulses are rapidly conducted down myelinated axons.

Saltatory conduction results from the influx of positive current flowing across the nodal membrane and passing forward to the next node. The current depolarizes the nodal membrane to threshold, and the Na^+ channels open; the action potential has advanced one node down the axon.

Mixed Nerves

Each peripheral nerve is composed of hundreds or thousands of axons. All except the smallest axons in a mixed nerve are myelinated; their diameters range from 0.1 μm for the smallest unmyelinated fibers to 20 μm for the largest fibers. The conduction velocity of an action potential in an axon is proportional to the diameter of the axon. *Conduction velocities* in a mixed nerve range from 120 m/s for the largest myelinated fibers to 0.5 m/s for the smallest unmyelinated fibers.

When a mixed nerve is stimulated with a brief electrical pulse, a compound action potential is propagated down the nerve (Fig. 4.8). The compound action potential has a biphasic wave form. The biphasic wave form is measured by a pair of recording electrodes. As it is propagated down the nerve, the action potential is recorded in succession by each electrode. As the impulse approaches, the first electrode records a negative potential relative to the second. When the impulse passes the first electrode and reaches the second electrode, the latter records the negative potential relative to the first electrode. The resulting wave form is biphasic.

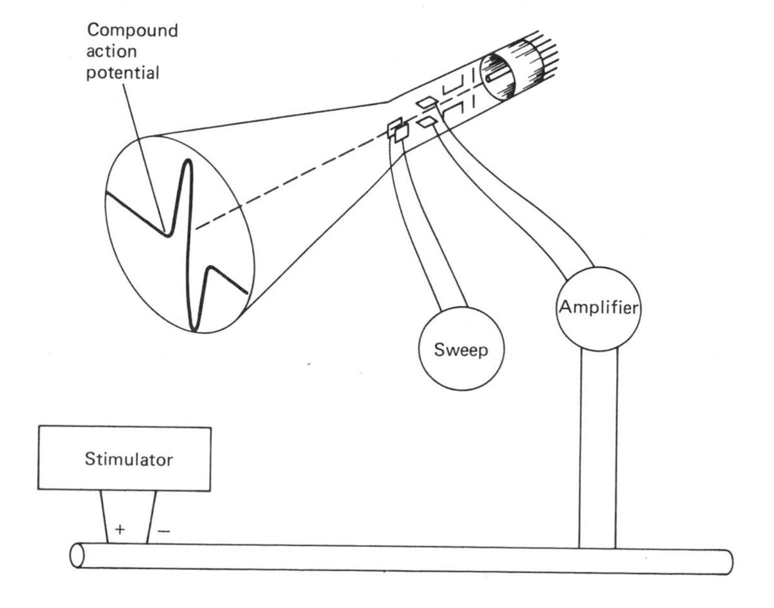

Fig. 4.8. Recording a compound action potential. Peripheral nerve is stimulated with brief voltage pulse. Resulting compound action potential is recorded by placing nerve over pair of wire electrodes. As impulse passes each electrode, it causes deflection of oscilloscope. Wave form is biphasic because electrodes are connected to opposite vertical plates.

The compound action potential is a summed potential—produced by the summation of the action potentials in each of the axons of the nerve (Fig. 4.9). The amplitude of the compound action potential is a function of the stimulus intensity. The largest fibers have the lowest threshold for action potential initiation. The minimum stimulus intensity required to evoke a response excites only the largest fibers. With each increase in stimulus intensity, the potential increases because fibers of successively smaller diameter are excited. The maximum stimulus strength is the stimulus intensity that is just sufficient to excite all of the axons in the nerve.

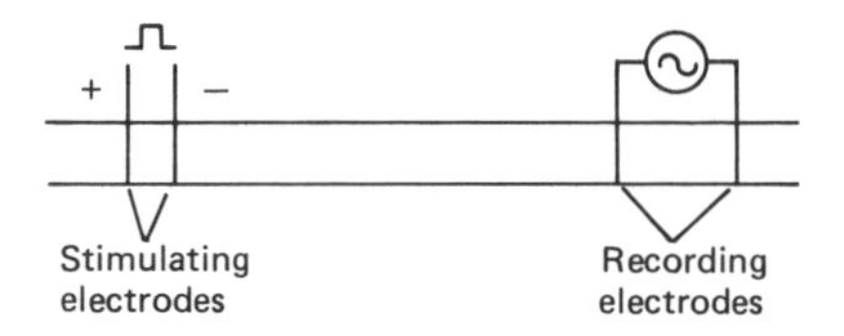

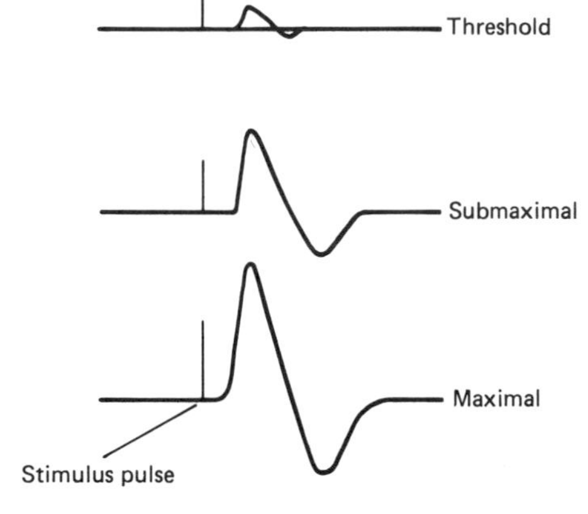

Fig. 4.9. Relationship between size of compound action potential and stimulus intensity. Peripheral nerve is stimulated by voltage pulses of increasing intensity. Maximal stimulus intensity occurs when action potential is initiated in every axon in nerve.

Synaptic Transmission

A synapse is a functional contact formed by the presynaptic terminals of a nerve axon and the membrane of a postsynaptic cell. All synapses provide sites at which nerve impulses can influence (excite or inhibit) the activity of the postsynaptic cell. The two kinds of synapses are: (1) *chemical,* in which the presynaptic terminal secretes a neurotransmitter, which reduces a synaptic potential in the postsynaptic cell and (2) *electrical,* in which current flows from the pre- to the postsynaptic neurons.

Although all synapses share a common plan, synapses formed by different neurons differ in (1) the morphology of their axon terminals, (2) the identity of their chemical transmitter and postsynaptic cell, and (3) the nature of their postsynaptic potential. The simplest synaptic contacts are those in which only one axon innervates the postsynaptic cell. An example is the synapse or neuromuscular junction at which a motor neuron innervates a muscle fiber. The most complex synaptic junctions are those in which thousands of axon terminals end on the postsynaptic neuron. These neurons usually have very elaborate dendritic branching patterns; the Purkinje cell of the cerebellum and the pyramidal cell in the cerebral cortex are examples.

Neuromuscular Junction

Each skeletal muscle fiber is innervated by a single motor neuron. The axon terminals of the motor neuron form synaptic contacts with a specialized region of the muscle membrane—the *motor end-plate* (Fig. 4.10). The membrane of the presynaptic axon terminal is separated by a 30-nm wide gap—the *synaptic cleft*—from the postsynaptic motor endplate.

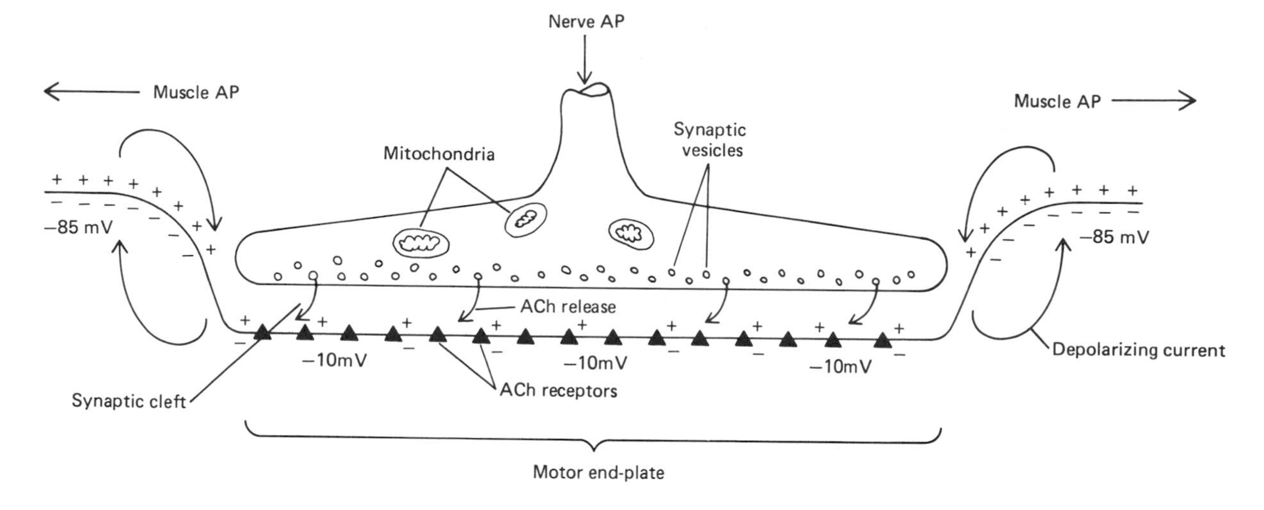

Fig. 4.10. Transmission across neuromuscular junction. (1) Motor neuron terminal is invaded by action potential (*AP*); (2) synaptic vesicles release acetylcholine into synaptic cleft; (3) acetylcholine (ACh) attaches to receptors embedded in postsynaptic membrane (motor end-plate); (4) motor end-plate to depolarized (from −85mV to −10mV), producing end-plate potential; (5) muscle AP is initiated by flow of current from depolarized end-plate to adjacent muscle membrane.

Neurotransmitter: Presynaptic Release and Postsynaptic Response

The neurotransmitter, *acetylcholine,* is stored in small round 50-nm-diameter vesicles in the presynaptic terminal (Fig. 4.10). These *synaptic vesicles* are clustered along the internal surface of the presynaptic membrane. The terminal also contains numerous mitochondria, which provide a source of adenosine triphosphate (ATP) for synthesis of the neurotransmitter and other essential components. When the axon terminal is depolarized by a nerve impulse, the synaptic vesicles fuse with the presynaptic membrane, discharging their contents into the synaptic cleft. The trigger for release of the transmitter is a rise in the calcium (Ca^{2+}) level within the terminal, which itself is triggered by the depolarization of the terminal.

Acetylcholine diffuses across the synaptic cleft and binds to receptor proteins embedded in the postsynaptic motor end-plate (Fig. 4.10). This binding causes an increase in the ionic permeability of the motor end-plate for K^+ and Na^+. The resulting influx of Na^+ and efflux of K^+ depolarizes the motor end-plate region to a potential of about −10 mV (about midway between E_{K^+} and E_{Na^+}; see "Ionic Basis of Membrane Potential" in this chapter. The depolarization of the motor end-plate forms an excitatory postsynaptic potential (*EPSP*) or end-plate potential (*EPP*).

Initiation of Muscle Impulse

The neuromuscular junction is a *repeater junction;* each motor neuron impulse always evokes a muscle action potential. In contrast to the EPSPs formed at a neural-neural junction, a single EPP will always initiate an action potential in the muscle fiber. The muscle action potential is elicited by the flow of depolarizing current from the depolarized motor end-plate to the surrounding membrane of the muscle fiber (Fig. 4.10). The current generated by an EPP is more than sufficient to depolarize the muscle membrane to its threshold for action-potential initiation. The resulting action potential is propagated over the surface of the muscle fiber, providing a trigger for the initiation of muscle contraction (see Chap. 11).

Recovery of Neuromuscular Junction

The external surface of the motor end-plate contains two protein complexes that bind with the acetylcholine transmitter. They are the *receptor protein,* which controls the ionic permeability of the membrane, and the enzyme *acetylcholinesterase,* which inactivates acetylcholine by breaking it down into its components, acetate and choline. Choline is transported into the presynaptic terminal for the resynthesis of acetylcholine. The activity of the acetylcholinesterase is very high. The half-life for an acetylcholine molecule is only several seconds. All of the acetylcholine released into the cleft is broken down within 20 s. The permeabilty of the end-plate decreases with the decline of the acetylcholine available for binding to the receptor protein. The EPP ends when the ionic permeability of the end-plate membrane returns to its resting level.

Vesicles are reformed in the axon terminal by the pinching off of membrane at the edges of the presynaptic terminal. The vesicles accumulate acetylcholine, which is synthesized in the terminal from its precursors—acetylcoenzyme A (produced in the mitochondria) and choline—by the enzyme *choline acetylase.*

Pharmacologic Inhibitors of Neuromuscular Junction

Numerous pharmacologic agents block synaptic transmission at the neuromuscular junction. The most widely known poison, *curare,* which was originally used by South American indians, blocks neuromuscular transmission by binding to the receptor proteins in the postsynaptic membrane. This action prevents acetylcholine from binding to the receptor protein. Depending on the concentration of curare, the EPP is either reduced or totally eliminated.

Several organophosphates—physostigmine and neostigmine—interfere with neuromuscular transmission by inactivating acetylcholinesterase. The blockade prolongs the EPP, producing multiple firing of muscle action potentials. Each motor nerve impulse elicits a barrage of muscle action potentials, which produces an intense muscular spasm.

One of the most potent poisons known is botulin, produced by the bacterium, *Clostridium botulinum.* The toxin inhibits neuromuscular transmission by blocking the release of acetylcholine from the presynaptic terminals. This toxin is responsible for a type of fatal food poisoning.

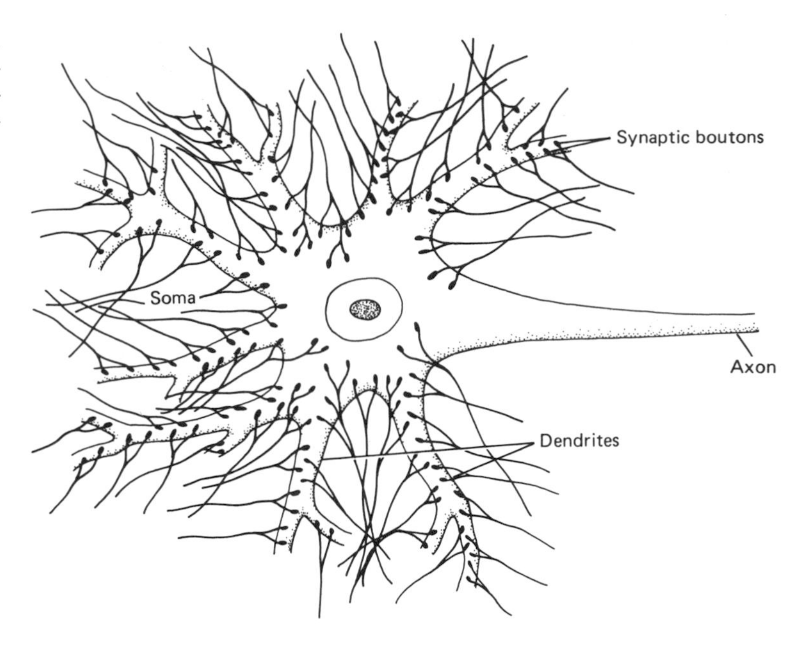

Fig. 4.11. Synaptic inputs to neuron. Terminals (synaptic boutons) of presynaptic axons form synaptic contacts on dendrites and soma of postsynaptic neuron.

Neuronal Synapse

The axon terminals of hundreds or thousands of different neurons may form synaptic contacts with a single neuron (Fig. 4.11). Most axons subdivide repeatedly before they terminate. Synaptic contacts are almost always formed on the dendrites of the neuron. Only about 15% of the axon terminals synapse with the soma. Each axon enlarges to form a small end-bulb, or *synaptic bouton* as it terminates (Fig. 4.12). This presynaptic terminal contains synaptic vesicles filled with neurotransmitter and is separated from the postsynaptic membrane of the neuron by a 30-nm-wide synaptic cleft.

The sequence of events producing a postsynaptic potential at a neuron synapse is identical to that already described for the generation of an EPP at the neuromuscular site. However, neuron synapses differ from neuromuscular junctions in two important characteristics: (1) Many neuron synapses are inhibitory, i.e., their synaptic boutons secrete a neuro-

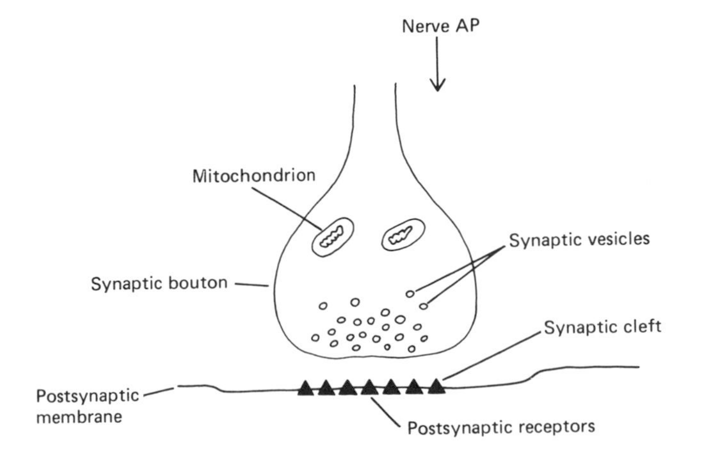

Fig. 4.12. Neuronal synapse. When presynaptic terminal (synaptic bouton) is depolarized by action potential (*AP*), synaptic vesicles release neurotransmitter into synaptic cleft. Binding of neurotransmitter to postsynaptic receptors produces either local depolarization (EPSP) or local hyperpolarization (IPSP) of postsynaptic membrane.

transmitter that produces an inhibitory postsynaptic potential (*IPSP*) in the postsynaptic neuron. (2) The EPSP which is generated by depolarization of a single synaptic bouton is small (1 to 2 mV); it is insufficient to elicit an action potential. Impulses in the postsynaptic neuron are evoked by the summation of many overlapping EPSPs.

Postsynaptic Responses

Depending on its chemical identity and that of the postsynaptic cell, a neurotransmitter either depolarizes or hyperpolarizes the postsynaptic membrane (Fig. 4.13). If the neurotransmitter elicits a general increase in membrane permeability to Na^+, K^+, and Cl^-, a depolarizing EPSP re-

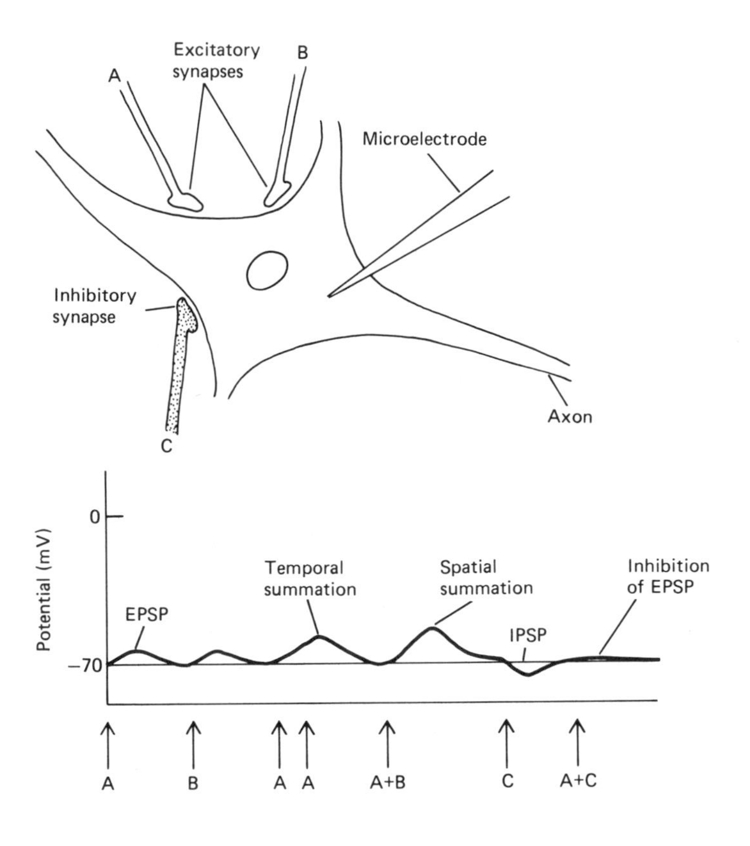

Fig. 4.13. Interaction of EPSPs and IPSPs at neuronal synapse. The EPSPs produced by activity of terminals A and B and IPSP generated by terminal C are recorded with microelectrode. Compound EPSP may be formed by either temporal (repetitive discharge of same terminal) or spatial summation (simultaneous activity at two or more terminals). An IPSP inhibits by reducing EPSP.

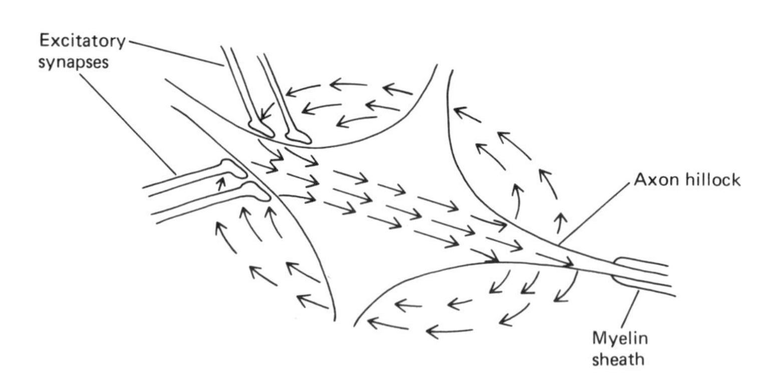

Fig. 4.14. Initiation of action potential in axon hillock. EPSPs on dendrites and soma produce depolarizing currents which flow toward axon hillock. If resulting depolarization of axon hillock exceeds its threshold, an action potential is initiated.

sults. If the permeability for only K^+ and Cl^- increases, the membrane is hyperpolarized, forming an IPSP.
The action potential is initiated in the *axon hillock region* of a neuron (Fig. 4.14). This is the initial segment of the axon, which arises from the soma. The axon hillock is the most excitable part of the neuron and has the lowest threshold for the initiation of an action potential. To initiate an impulse in the postsynaptic neuron, the membrane potential of the axon hillock must be depolarized by −10 to −25 mV.

Neuronal Integration

Because of its small size and remote location, the EPSP generated at a single excitatory synaptic junction produces only a small perturbation (1 mV or less) in the membrane potential of the axon hillock. Action potentials are initiated by either the repetitive discharge of single excitatory boutons (*temporal summation*) or the simultaneous discharge of many different terminals (*spatial summation*). The individual EPSPs produced by either repetitive or simultaneous discharge summate to form a compound EPSP (Fig. 4.13). If the local depolarizing currents produced by the compound EPSP are large enough to depolarize the axon hillock to its threshold, an action potential is initiated.
The IPSPs inhibit the firing of an action potential by reducing the size of the compound EPSP (Fig. 4.13). The initiation of the nerve impulse depends on whether the compound synaptic potential produced by the summation of all of the EPSPs and IPSPs is sufficient to depolarize the membrane at the axon hillock to its threshold for impulse initiation.

Presynaptic Inhibition

Two kinds of inhibitory interactions occur at neuron synapses: (1) postsynaptic inhibition, which produces hyperpolarizing IPSPs, and (2) presynaptic inhibition, which reduces the release of excitatory neurotransmitter. Presynaptic inhibition occurs whenever an axon terminal forms

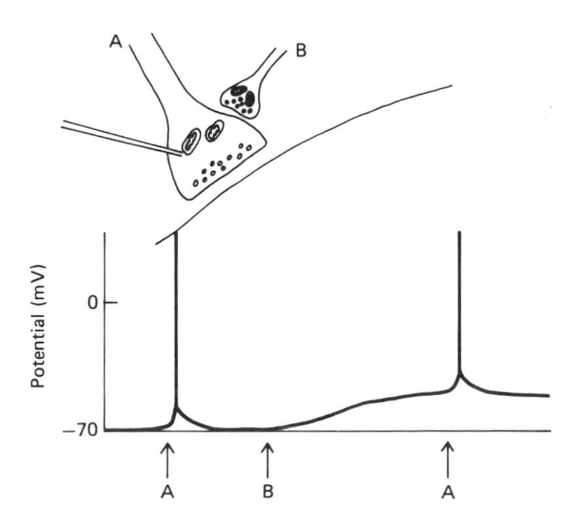

Fig. 4.15. Presynaptic inhibition. Release of neurotransmitter by terminal A is depressed when terminal B is active. Microelectrode recording from terminal A demonstrates that terminal B inhibits terminal A by depolarizing it and thus reducing the amplitude of action potential in terminal A.

a synaptic contact on another synaptic bouton (Fig. 4.15). When an impulse invades the presynaptic inhibitory terminal, it releases a neurotransmitter that depolarizes the "postsynaptic" bouton. This action reduces the amount of neurotransmitter that the "postsynaptic" bouton can release by reducing the size of the nerve impulse which invades the postsynaptic bouton. The amplitude of the invading impulse is reduced because the postsynaptic bouton is already partially depolarized by the presynaptic inhibitory terminal.

Neurotransmitters: Excitatory and Inhibitory

Several small organic molecules are believed to be neurotransmitters. *Acetylcholine* is released by branches of motor neuron axons within the spinal cord. It is also an excitatory transmitter within the autonomic nervous system (ANS) and is thought to be an excitatory transmitter in several areas of the brain as well. *Norepinepherine* is an excitatory neurotransmitter within both the ANS and several regions of the brain. *Dopamine* is an excitatory neurotransmitter in the basal ganglia. Other possible excitatory neurotransmitters are *aspartate, glutamate,* and *serotonin*. Evidence suggests that several peptides—*substance P*, hypothalamic releasing factors, and enkephalin—can modulate neuronal activity, perhaps functioning as neurotransmitters.

Within the CNS there are at least two inhibitory neurotransmitters: *gamma-aminobutyric acid* and *glycine*. The release of acetylcholine by parasympathetic neurons inhibits cardiac and some smooth muscles. The sympathetic neuron release of norepinepherine also inhibits several smooth muscles (see Chap. 11 for a discussion of these parasympathetic and sympathetic effects).

Electrical Synapses

In several regions of the mammalian brain synaptic contacts are observed between neurons that are not sites of chemical synaptic transmission. These are regions of close membrane apposition, which form *gap junctions*. Gap junctions are characterized by a 2-nm-break between the pre- and postsynaptic neurons, which is bridged by tiny (1 nm in diameter) cytoplasmic channels. These channels provide a path for current flow from the pre- to the postsynaptic neuron. Depolarizing or hyperpolarizing currents can spread from neuron to neuron across these junctions. Conduction across a gap junction is bidirectional. (Across a chemical synapse transmission is unidirectional, from presynaptic terminal to postsynaptic neuron.)

Selected Readings

Aidley DJ (1971) The physiology of excitable cells. Cambridge University Press, New York

Axelrod J (1974) Neurotransmitters. Sci Amer 230: 59

Baker PF (1966) The nerve axon. Sci Amer 214: 74

Eccles JC (1965) The synapse. Sci Amer 212: 56

Eccles JC (1977) The understanding of the brain. McGraw-Hill, New York

Hodgkin AL (1964) The ionic basis of nerve conduction. Science 145: 1148

Katz B (1968) Nerve, muscle and synapse. McGraw-Hill, New York

Katz B (1971) Quantal mechanisms of neural transmitter release. Science 172: 123

Krnjevic K (1974) Chemical nature of synaptic transmission in vertebrates. Physiol Rev 54: 418

Lester HA (1977) The response to acetylcholine. Sci Amer 236: 106

Mountcastle VB (1974) Medical physiology. Mosby, St. Louis

Shepherd GM (1974) The synaptic organization of the brain: An introduction. Oxford University Press, New York

Shepherd GM (1978) Microcircuits in the nervous system. Sci Amer 238: 92

Review Questions

1. What is an excitable cell? List several examples.
2. Prepare a table showing the distribution of the major cations and anions across the cell membrane. Explain how these distributions arise.
3. Explain the ionic basis of the resting membrane potential. Why is the Nernst equilibrium for potassium (E_{K^+}) not equal to the resting membrane potential?
4. Draw the wave form of an action potential. Label to indicate both the time course and magnitude (voltage) of the wave form.
5. Describe the sequence of changes in the ionic permeability of the nerve membrane which produce an action potential, i.e., rising phase, falling phase, and hyperpolarization.
6. What is an "all-or-none" event? Explain why the initiation of an action potential is an example of an "all-or-none" event.
7. Differentiate between the absolute and the relative refractory periods. How are they measured?
8. Why is the Na^+/K^+ membrane pump important for the maintenance of nerve excitability?What happens to nerve excitability if the pump is inhibited?
9. Explain how a nerve impulse propagates down an axon. What are "local circuits"?
10. What is saltatory conduction?
11. Draw a diagram of a motor end-plate.
12. List in order the events in the transmission of a nerve impulse across the motor end-plate. Begin with propagation of the nerve impulse into the motor neuron terminal.
13. What are the major differences—morphologic and physiologic—between the neuromuscular junction and the neuron-neuron synapses.
14. Define spatial and temporal summation.

15. Compare and contrast presynaptic and postsynaptic inhibition. Consider morphology, transmitter release, and the effects of each on neuron excitability.
16. Discuss how the neuron serves as an integrator. What is the function of the axon hillock in this information transfer process?
17. What are gap junctions? How is neural activity transmitted across the gap junction?

Chapter 5

Sensory Reception and Somatesthesia

Sensory receptors provide information to the central nervous system (CNS) about the status of the environment. Each sensory receptor responds to a particular kind of stimulus energy (chemical, electromagnetic, mechanical, or thermal). Sensory receptors are transducers that are specialized to transform the stimulus energy into an electrochemical potential. Information about the stimulus is encoded by the discharge of action potentials in the sensory nerve. These sensory discharges are then conducted to the sensory areas of the nervous system where they are decoded and analyzed.

Sensory Receptors

Each sensory receptor is morphologically and physiologically specialized to respond to only one kind of stimulus energy. This is the "adequate stimulus" for that receptor; it is the stimulus to which the receptor is most sensitive.

One convenient scheme for the classification of sensory receptors is to group them according to their adequate stimuli. They are usually arranged into five groups, including: (1) *electromagnetic receptors,* which respond to light; (2) *mechanoreceptors,* which respond to mechanical displacement (touch, pressure, and sound waves); (3) *thermoreceptors,* which respond to temperature (warmth or cold); (4) *chemoreceptors,* which respond to specific chemicals in the internal and external environment (taste, smell, blood levels of respiratory gases, and blood glucose levels); and (5) *nociceptors,* which respond to tissue damage (pain).

Receptors can also be classified according to the usual location of the stimulus to which they respond. In this classification receptors may be divided into four groups, including: (1) *teloreceptors,* responding to stimuli distant from the body (vision, smell, and hearing); (2) *exteroreceptors,* responding to stimuli impinging on the external body surface (taste, touch, pressure, and temperature); (3) *interoceptors,* responding to stimuli within the body (stimulation of the viscera and concentrations of specific chemicals in blood); and (4) *proprioceptors,* responding to the position of the body in space (joint position and muscle length).

Sensory Transduction

The primary response of all sensory receptors is the generation of a *receptor potential,* which is produced by an interaction between the stimulus energy and the receptor membrane (Fig. 5.1). Depending on the adequate stimulus, this interaction can take one of several forms, including: (1) mechanical distortion—mechanoreceptors; (2) photoexcitation of a membrane-bound photopigment—visual receptors; (3) thermal alteration of membrane permeability—thermoreceptors; (4) chemical binding to membrane receptors—chemoreceptors; (5) peptide (released by tissue damage) binding to membrane receptors—nociceptors. Each of these interactions causes a general increase in the ionic permeability of the receptor membrane, which results in an influx of Na^+ into the receptor terminal. This influx depolarizes the receptor terminal, producing a receptor potential (the visual receptors hyperpolarize, see Chap. 6).

Fig. 5.1. Generation of receptor potential in sensory nerve terminal. The stimulus energy interacts with receptor membrane to increase its ionic permeability. Resulting influx of Na^+ depolarizes receptor, producing a receptor potential. The receptor potential provides a source for local circuits of current which depolarize initial segment of sensory nerve. An action potential is initiated whenever initial segment is depolarized to its threshold.

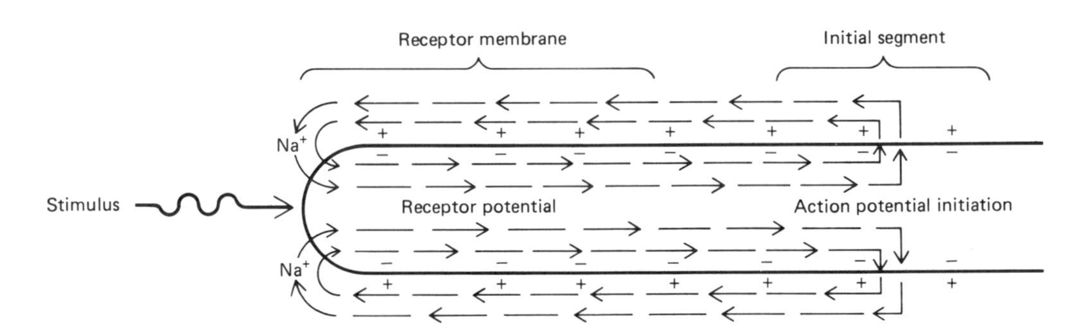

Initiation of Action Potential

Nerve impulses are initiated in the initial segment of the sensory nerve as a result of the excitatory influence of the receptor potential (Fig. 5.1). The actual sequence leading to the initiation of action potentials in the sensory nerve depends on the anatomic relationship between the sensory receptor which generates the receptor potential and the sensory nerve. A sensory receptor can be either the terminal of a sensory nerve specialized to function as a sensory trans-

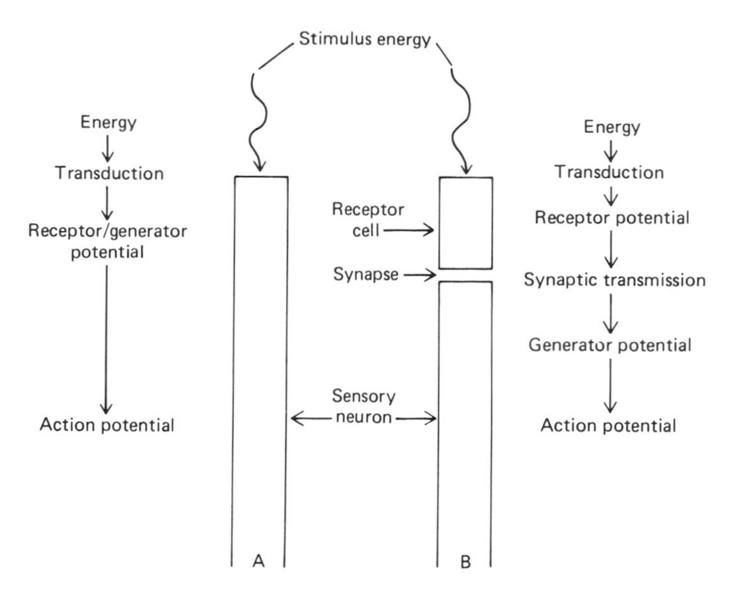

Fig. 5.2. Sensory nerve excitation showing sequence by which a sensory receptor transforms stimulus energy into discharge of sensory serve impulses. **A**, receptor is a terminal of the sensory nerve; **B**, receptor is a separate cell innervating sensory nerve terminal.

ducer (Fig. 5.2A) or a discrete cell forming a chemical synapse with the sensory nerve (Fig. 5.2B). In the former instance the receptor potential produced in the specialized sensory terminals of the sensory nerve serves as the generator potential (Fig. 5.2A); in the latter the receptor potential evokes the release of a chemical transmitter from the "presynaptic" receptor cell, producing a depolarizing generator potential in the "postsynaptic" sensory nerve terminals (Fig. 5.2B).

In both instances action potentials are initiated in the sensory nerve by depolarizing currents produced by the generator potential. These currents spread to the initial segment of the axon where the threshold for action potential initiation is lowest (Fig. 5.1). The sequence of events leading to the firing of impulses in the sensory nerve is similar to what is involved in the excitation of nonsensory neurons and muscle; an action potential is fired if the generator potential is large enough to depolarize the initial segment to its threshold for action potential initiation (Fig. 5.3) (see Chap. 2 for a more detailed description).

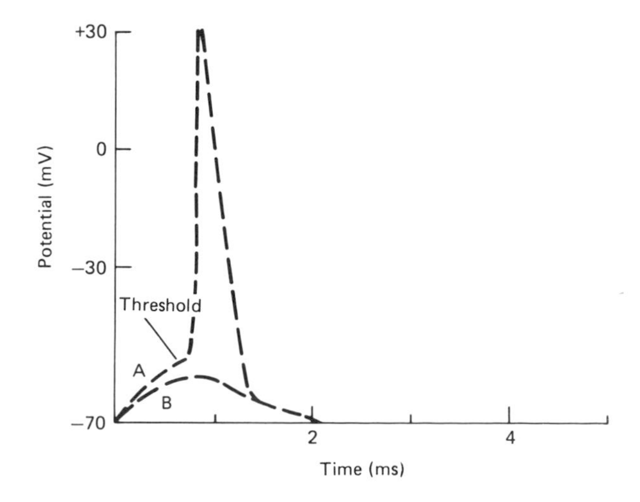

Fig. 5.3. Initiation of an action potential by receptor/generator potential. **A**, stimulus energy is sufficient to produce receptor/generator potential which exceeds threshold potential for nerve impulse firing at initial segment; **B**, stimulus energy is reduced from that in *A* so that resulting receptor/generator potential is too small to depolarize membrane to its threshold potential.

Sensory Coding and Adaptation

The first principle of sensory coding is that the sensation produced by excitation of a sensory nerve is determined by its site of termination in the CNS, not by the nature of the stimulus energy. The nervous system is wired so that excitation of a sensory receptor is always interpreted as if it resulted from stimulation by the receptor's adequate stimulus. This principle is sometimes called the *law of specific nerve energies*. When, e.g., the eye is struck by a mechanical blow stimulating the visual receptors, the person experiences an intense, visual sensation ("sees stars") rather than the sensation of mechanical force.

Information about stimulus intensity is encoded by the amplitude of the receptor potential. The receptor potential increases with the logarithm of the stimulus intensity. Since the rate of discharge of impulses in the sensory nerve is proportional to the size of the receptor potential, *the rate of action potential discharge is also proportional to the logarithm of the stimulus intensity*. It has recently been shown that the logarithmic relationship between stimulus intensity and sensory response is only approximate. A more accurate description is a power function: $R = KI^A$, where R is the sensory response, I is the stimulus intensity, and K and A are constants.

A decline is observed in the responses of all sensory receptors when they are subjected to continued stimulation by a constant stimulus (Fig. 5.4). This is *sensory adaptation*. Both measures of sensory excitation—the rate of discharge of sensory action potentials and the size of the receptor potential—decrease during sensory adaptation. The decline of the receptor potential is believed to result from a gradual de-

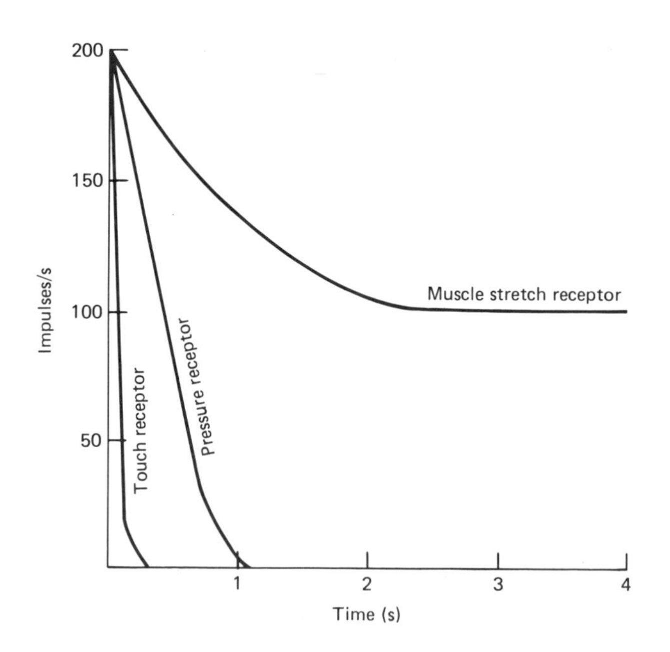

Fig. 5.4. Sensory adaptation. Decrease in discharge of action potentials during constant stimulation of receptor. Touch and pressure receptors are phasic—they rapidly adapt to continued stimulation. Muscle stretch receptors are tonic—they slowly adapt over many hours to maintained muscle stretch.

crease in the ionic permeability of the receptor membrane. Adaptation of sensory nerve discharge is, of course, a direct result of the adaptation of the receptor potential. The rate of discharge decreases as the amplitude of the receptor potential declines.

Although sensory adaptation is characteristic of all receptor systems, the *rate of adaptation* varies according to the identity of the receptor (Fig. 5.4). Some sensory receptors, such as touch or pressure, respond to the stimulus with only a few impulses, regardless of the duration of stimulus. Others, such as the muscle stretch receptors or carotid body chemoreceptors, respond to constant stimulation with a sensory discharge that slowly decreases over many hours. Receptors may be classified according to their rates of adaptation: *phasic* receptors adapt rapidly; *tonic* receptors adapt slowly.

Somatosensory System

The sense concerned with the state of the body is called somatesthesia. The *somatosensory receptors* monitoring this internal physical state include: *cutaneous receptors* sensitive to touch, pressure, temperature, and pain as well as *proprioceptors,* which respond to joint and muscle movement. Somatosensory receptors may be contrasted with the other major group of sensory receptors, the *special sensory receptors,* which include the visual, auditory, olfactory, gustatory, and vestibular receptors (see Chap. 4). All of the special sensory receptors are confined to the head where they are innervated by cranial nerves. Somatosensory receptors are located throughout the body in the limbs, trunk, and head. The vast majority of the somatosensory receptors —those in the limbs and trunk—are innervated by spinal nerves. Cranial nerves innervate the somatosensory receptors in the head.

Sensory Fibers and Spinal Nerves

The sensory fibers within a spinal nerve can be divided into four classes based on size and conduction velocity (Table 5.1). Fibers of different size innervate different populations of somatosensory receptors. The largest fibers (Class I) arise from the muscle proprioceptors, the annulospiral endings of the muscle spindle and the Golgi tendon organ. Class II fibers include cutaneous receptors that respond to touch and pressure as well as the flower spray endings of the muscle spindles. Crude touch and pressure, temperature, and pain receptors form Class III fibers. The smallest fibers (Class IV) transmit information from temperature and pain receptors.

Table 5.1. Sensory fibers in mammalian nerve.

Class	Origin	Diameter (μm)	Conduction Velocity (m/s)
I	Annulospiral endings of muscle spindle and Golgi tendon organ	12–20	75–120
II	Flower spray endings of muscle spindle, touch, and pressure	4–12	24–72
III	Pain, temperature, touch, and pressure	1–4	6–24
IV	Pain and temperature	0.5–1	<6

There are two kinds of pain, including: *sharp pricking pain* (experienced from a sharp cut or stab wound) and *burning pain* (experienced from a burn). Noxious stimulation of Class III pain receptors produces the sensation of sharp, pricking localized pain. Excitation of Class IV pain receptors evokes a diffuse, burning pain sensation.

Receptor Endings and Sensory-Receptive Fields

Many somatosensory receptors have specialized sensory terminals (Fig. 5.5). Some of the largest touch- and pressure-sensitive terminals are encased in a capsule of connective tissue; they include *Meissner's corpuscle, Krause's corpuscle,* and the *pacinian corpuscle*. In others the mechanosensitive nerve endings are enlarged—*Merkel's discs* and *Ruffini's end organ*—or entwined around the base of a

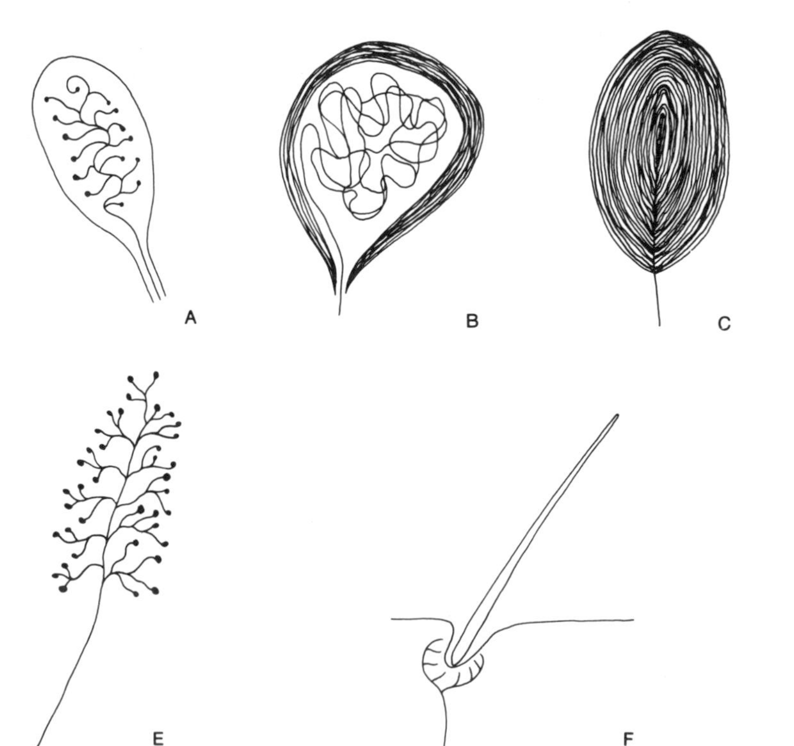

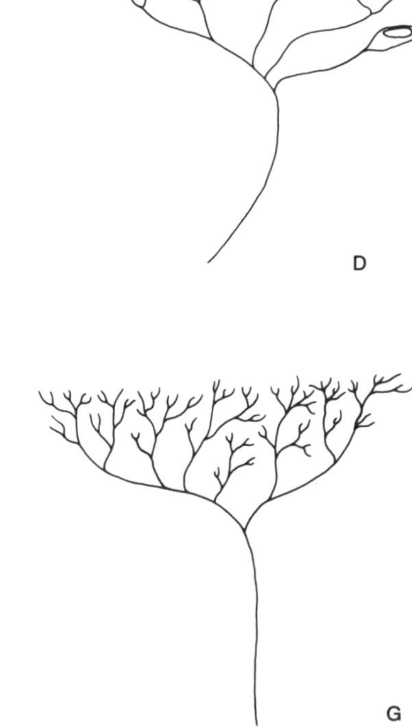

Fig. 5.5A–G. Types of sensory nerve terminals. Encapsulated endings include Meissner's corpuscle (**A**), Krause's corpuscle (**B**), and Pacinian corpuscles (**C**). Mechanosensitive endings of Merkel's discs (**D**) and Ruffini's end organ (**E**) are enlarged. Terminals of a sensory nerve may innervate a hair follicle (**F**) or form free nerve endings (**G**).

hair follicle. The *muscle spindles* and *Golgi tendon organs* are specialized to sense changes in muscle length and tension (see Chap. 5 for details). The smaller somatosensory fibers form free nerve endings throughout the body. They include all of the receptors for warmth, cold, and pain as well as those mechanosensitive receptors that innervate regions lacking more complex sensory endings. The cornea, e.g., which is sensitive to mechanical stimulation, contains only free nerve endings.

The *receptive field* of a sensory neuron is that area of the body surface which if stimulated excites the neuron. The size of the receptive field of a sensory neuron is roughly proportional to the size and spread of its sensory terminals. Class II sensory neurons arise from encapsulated receptors and tactile hairs. These receptors are affected only by stimuli applied to the tissues in their immediate vicinity. They have small, discrete receptive fields. In contrast, Class III and Class IV sensory neurons the sensory terminals of which extend over a large areas as free nerve endings have large, diffuse receptive fields.

Central Somatosensory Pathways

The nerve fibers arising from somatosensory receptors in the limbs and trunk run in the spinal nerves to the spinal cord. They enter the spinal cord through the dorsal root. Within the spinal cord these fibers either terminate on spinal-tract neurons in the gray matter (sensory fiber Classes I, III, and IV) or enter the white matter to form a spinal sensory tract (Class II fibers).

The four major spinal sensory tracts (Fig. 5.6) include: (1) the dorsal columns, (2) the spinocerebellar, (3) the lateral spinothalamic, and (4) the ventral spinothalamic. The *dorsal columns* are formed by the axons of Class II sensory nerves. Class I neurons synapse on the spinal interneurons the

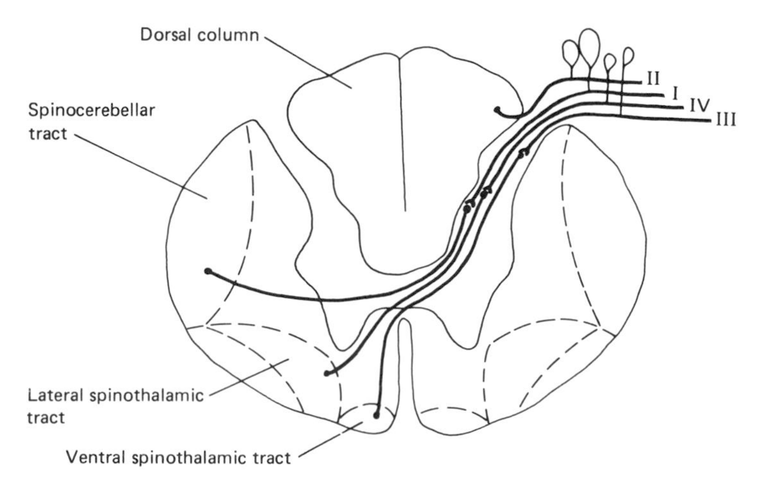

Fig. 5.6. Principal spinal sensory tracts (see text).

axons of which form the *spinocerebellar tract*. The *lateral spinothalamic tract* is formed by the axons of spinal interneurons receiving their inputs from Class IV neurons. Interneurons whose axons form the *ventral spinothalamic tract* are excited by Class III sensory fibers.
Somatosensory nerves innervating the face and head run in the trigeminal nerve to the brain and terminate in the sensory nucleus of the trigeminal nerve.

Dorsal Column System

The dorsal columns are formed by the axons of Class II sensory neurons, which run from the spinal cord to the medulla (Fig. 5.7). They synapse on interneurons in the *dorsal column nuclei of the medulla* (the nucleus cuneatus and nucleus gracilis). The medullary neurons cross to the contralateral side of the medulla and enter the medial lemniscus, which ascends to the *ventroposterior nucleus of the thalamus*. The medullary neurons synapse on thalamic neurons, which run to the *somatosensory area of the cerebral* cortex. Thus the dorsal column system consists of a three-neuron chain connecting the Class II somatosensory receptors to the somatosensory cortex, including: (1) Class II sensory neurons (2) medullary neurons, and (3) thalamic neurons. It

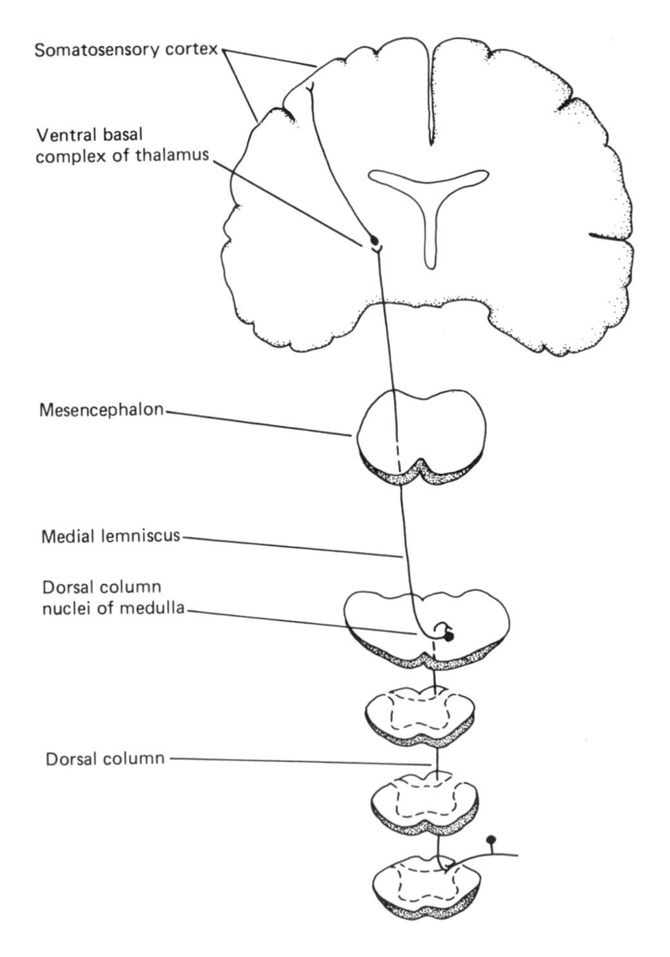

Fig. 5.7. Dorsal column system (see text). This three-neuron chain transmits information from touch and pressure receptors to somatosensory cortex.

provides a rapid conduction pathway for the transmission of tactile information from the periphery to the cortex.

Each sensory neuron in the dorsal columns is sensitive to stimulation of a small, discrete receptive field. Most are excited by light cutaneous stimulation (touch or pressure) of encapsulated receptors. Dorsal column neurons transmit precise information about the location of the stimulus on the body surface.

A correlation exists between the position that each neuron occupies in the dorsal column system and the location of its receptive field on the body surface. Neurons that are excited by stimulation of adjacent points on the body surface run together in the dorsal column system. The position of each fiber in the dorsal columns is correlated with its point of origin. First-order fibers in the medial portion of the dorsal columns innervate receptors in the legs and trunk. Fibers arising from receptors in the arms and neck run in the lateral portion of the dorsal columns. The more anterior the point of origin of the fiber, the more lateral its position in the dorsal columns.

Spinothalamic System

The spinothalamic system is composed of two major divisions; they are the ventral spinothalamic and the lateral spi-

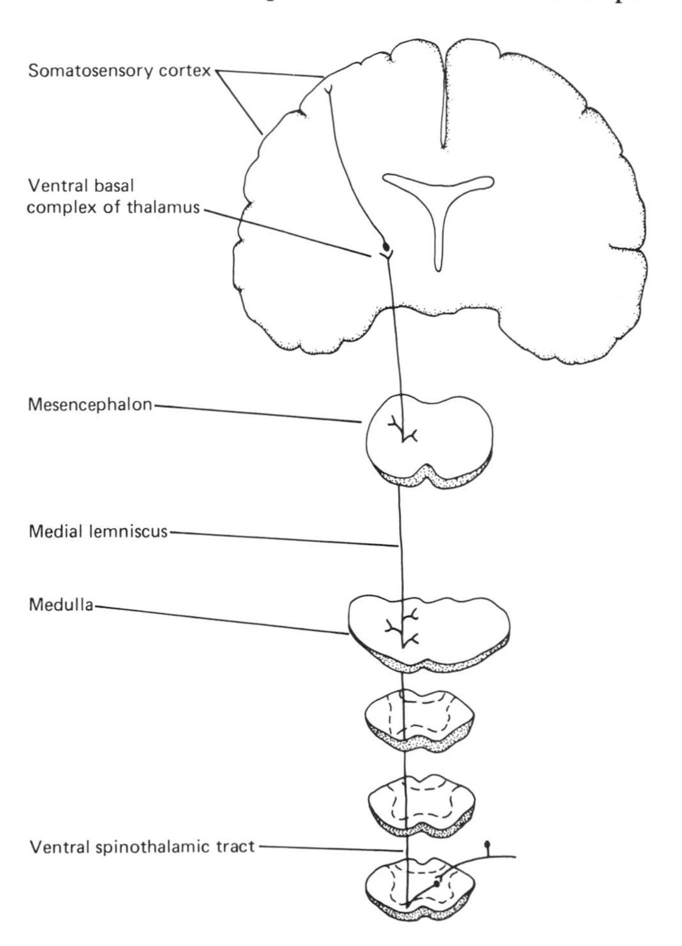

Fig. 5.8. Ventral spinothalamic system. This pathway transmits information from touch, pressure, temperature, and pain receptors to somatosensory cortex.

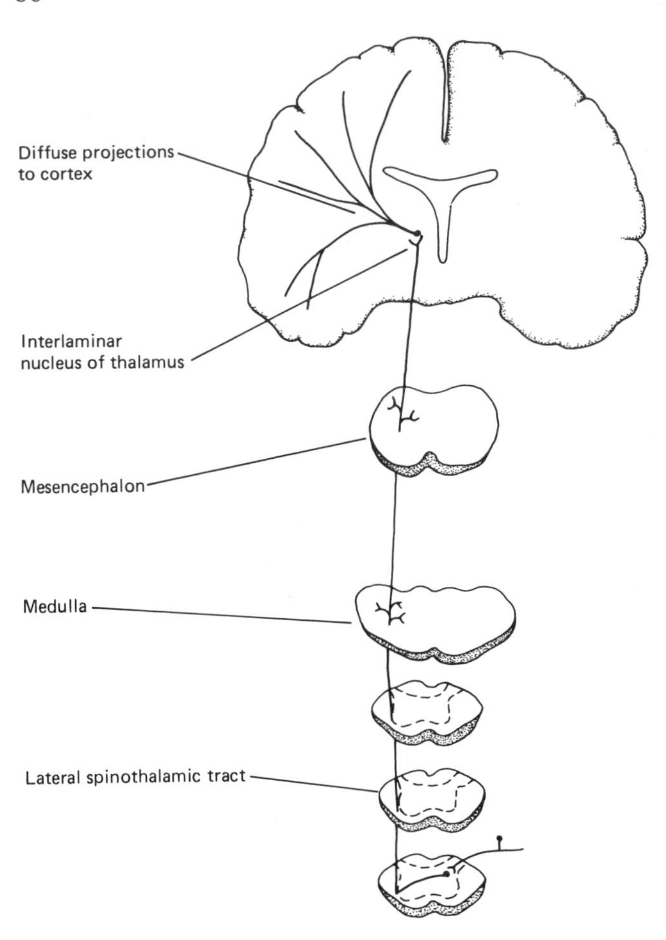

Fig. 5.9. Lateral spinothalamic system. This system transmits information from temperature and pain receptors to broad regions of cerebral cortex (see text).

nothalamic tracts. The *ventral spinothalamic tract* is formed by the axons of spinal interneurons in the dorsal horn that receive their inputs from Class III sensory neurons sensitive either to touch, pressure, temperature, or sharp pain (noxious) stimulation (Fig. 5.8). Class IV neurons sensitive to either thermal or noxious (burning pain) stimulation terminate on the spinal interneurons, which form the *lateral spinothalamic tract* (Fig. 5.9).

Both sets of spinal interneurons cross the spinal cord before entering either the ventral or lateral spinothalamic tracts and running up the spinal cord and through the brain stem to the thalamus. As they pass through the medulla, pons, and mesencephalon, the spinothalamic axons give off a large number of side branches. They are a source of excitatory input to the core areas of the brain stem, especially to the reticular formation—an important consideration in maintaining a state of arousal (see Chap. 7).

The sites of termination within the thalamus differ for second-order neurons running in the two subdivisions of the spinothalamic tract. Axons running in the ventral spinothalamic tract terminate in the *ventral posterior lateral nucleus of the thalamus*. Second-order neurons from the lateral spinothalamic tract terminate in the *posterior and interlaminar nuclei of the thalamus*. Third-order neurons arise from the thalamus and run to the *cerebral cortex*. Neurons sensitive to either tactile, thermal, or noxious (sharp pain) stimulation terminate in the somatosensory cortex; the endings of neurons excited by noxious (burning pain) stimuli have a widespread distribution, ending throughout the entire cerebral cortex.

The receptors of Class III and IV sensory nerves are formed by the branching of sensory nerve terminals. These receptors have large, diffuse receptive fields. As a result, spinothalamic neurons are not capable of fine spatial discriminations concerning stimulus location. Fibers in the spinothalamic system are excited whenever tactile, thermal, or noxious stimuli are applied to large areas of the body. In contrast each dorsal column fiber is sensitive only to tactile stimulation of a small localized area of the body. The spinothalamic system is concerned with the slow transmission of stimulus sensation to the cortex, whereas the dorsal columns provide rapid conducting pathways that inform the cortex of stimulus location.

Thalamic Ventrobasal Complex

The ventral posterior lateral thalamus receives input from two sets of second-order neurons: those from the dorsal column system and those forming the ventral spinothalamic tracts. Their points of termination overlap; their sites of termination are *somatotopically organized,* i.e., the point at which they terminate within the ventral posterior lateral thalamus is correlated with the location of their receptive fields on the body surface. Second-order fibers with adjacent receptive fields, i.e., that are excited by stimulation of adjacent points on the body surface, synapse on adjacent thalamic neurons.

Somatosensory fibers innervating the face run in the trigeminal nerve to the trigeminal nucleus in the mesencephalon where they synapse on second-order neurons. These latter fibers terminate in the thalamus in the *thalamic ventral posterior medial nucleus*. Together the ventral posterior lateral and medial nuclei form the *ventrobasal complex* of the thalamus.

The somatotopic organization of the ventral posterior lateral nucleus is extended into the ventral posterior medial nucleus. The more medial a neuron is in the ventrobasal complex, the more anterior is the location of its receptive field.

Somatosensory Cortex

Thalamic neurons arising in the ventrobasal complex terminate in the somatosensory cortex (Figs. 5.7 and 8). The *primary somatosensory cortex* is located in the postcentral gyrus immediately posterior to the central sulcus (Fig. 5.10). Posterolateral to the primary somatosensory cortex is a sec-

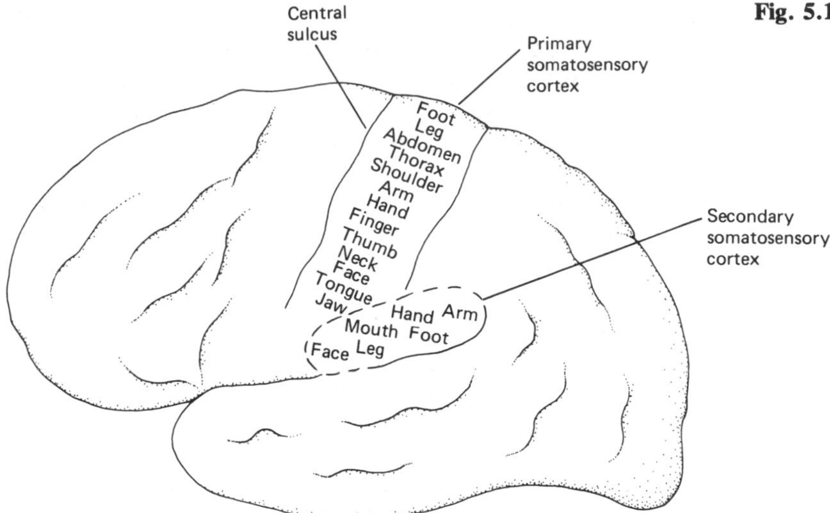

Fig. 5.10. Somatosensory cortex.

ond area of somatosensory integration, the *secondary somatosensory cortex.*

The primary somatosensory cortex has a *somatotopic organization*. The point at which each thalamic neuron terminates in the cortex is correlated with its point of origin in the thalamus. Fibers excited by stimulation of the foot terminate in the dorsomedial portion of the postcentral gyrus, whereas inputs driven by receptors in the head end on cortical neurons in the ventrolateral-most part of the somatosensory cortex (Fig. 5.10). There is a direct correlation between the position of a neuron in the somatosensory cortex and the location of its receptive field on the body. All receptive fields of postcentral gyrus neurons are on the contralateral side of the body. Neurons in the secondary somatosensory cortex receive inputs from both sides of the body, ipsi- and contralateral.

Cortical Structure

The cortex is composed of *six layers of neurons*. Inputs from the ventrobasal complex terminate on neurons in layer IV. The processes of these neurons extend in two directions: toward the surface (layers I and II) and downward to layers V and VI. The axons of neurons in layers V and VI form the output and connect the somatosensory cortex to other areas of the cerebral cortex, especially the somatosensory association and the motor cortex.

The primary orientation of the processes of the cortical neurons is vertical. As a result, interactions among neurons are confined to *vertical columns* exending perpendicularly from the cortical surface down through all six cortical layers. The cortex is organized into columns for processing somatosensory information. Each column is concerned with processing information about a specific kind of somatosensory stimulus at a specific location. Many columns are concerned, e.g., with the quality of tactile stimuli (sharpness, roughness, or intensity) that touch a specific region of the body; there is a different column for each point on the body. Other columns are concerned with movement and position of specific joints.

The primary somatosensory cortex is concerned with both localization of the stimulus to a specific point on the body and the evaluation of differences in location and intensity of two stimuli simultaneously applied to the body. It evaluates the nature of the stimulus (sharpness, roughness, or temperature) and correlates information—at the conscious level—about the body's position and movement through space.

The function of the secondary somatosensory cortex is poorly understood. Receiving inputs from both ipsi- and contralateral sides of the body, it is involved in the correlation of somatosensory activity over the entire body.

Selected Readings

Catton WT (1970) Mechanoreceptor function. Physiol Rev 50: 297
Eyzaguirre C, Fidone SJ (1975) Physiology of the nervous system, 2nd edn. Year Book Medical, Chicago
Lim RKS (1970) Pain. Ann Rev Physiol 32: 269
Lowenstein WR (1960) Biological transducers. Sci Amer 203: 98
Lynn B (1975) Somatosensory receptors and their CNS connections. Ann Rev Physiol 37: 105
Mountcastle VB (1974) Medical physiology, 13th edn. Mosby, St. Louis
Sensory receptors (1965) Cold Spring Harbor symposium on quantitative biology. Cold Spring Harbor Laboratory, Cold Spring Harbor, New York
Wall PD, Dubner R (1977) Somatosensory pathways. Ann Rev Physiol 34: 315

Review Questions

1. What is the "adequate stimulus" of a sensory receptor?
2. Define: (a) teloreceptor, (b) exteroreceptor, (c) interoceptor, and (d) proprioceptor.
3. Describe the sequence of events through which the stimulus excites a sensory nerve to generate a discharge of nerve impulses.
4. What is the law of specific nerve energies?
5. Construct a table showing the different classes of sensory fibers in a spinal nerve. Include information on stimulus sensitivity, associated spinal tracts, and relative size.
6. What is a sensory receptive field?
7. Name the two spinal tracts that conduct somatosensory information to the thalamic ventrobasal complex. Briefly discuss how these two tracts differ, in their neuroanatomic organization and in the kind of sensory information which they transmit.
8. Describe the neural pathways which transmit somatosensory information from the face to the somatosensory cortex.
9. The somatosensory cortex has both a columnar and a somatotopic organization. Briefly describe the nature of these two aspects of cortical organization.

Special Sensory Systems

Chapter 6

The special senses are those sensory systems confined to the head. They include the sensory systems which respond to visual, auditory, vestibular, gustatory, and olfactory stimulation, and are responsible for the five special senses of sight, hearing, equilibrium, taste, and smell.

Visual System

The eye is specialized to collect visual information from the environment and to transmit it to the sensory areas of the brain. The eye uses an elegant optical system to project the visual image onto the retinal visual receptors. The retina is composed of a complex network of receptors and associated neurons which extracts information from the visual stimulus related to, inter al., its intensity, color, size, curvature, and rate of movement. This information is then transmitted over the optic nerve to the visual areas of the brain for interpretation.

Fig. 6.1. Structure of eye. The eye is divided into two chambers. The anterior chamber is filled with the aqueous humor; the posterior chamber contains the vitreous humor. To strike the photoreceptors in the retina, light must follow a path which passes in succession through the cornea, aqueous humor, lens, and vitreous humor.

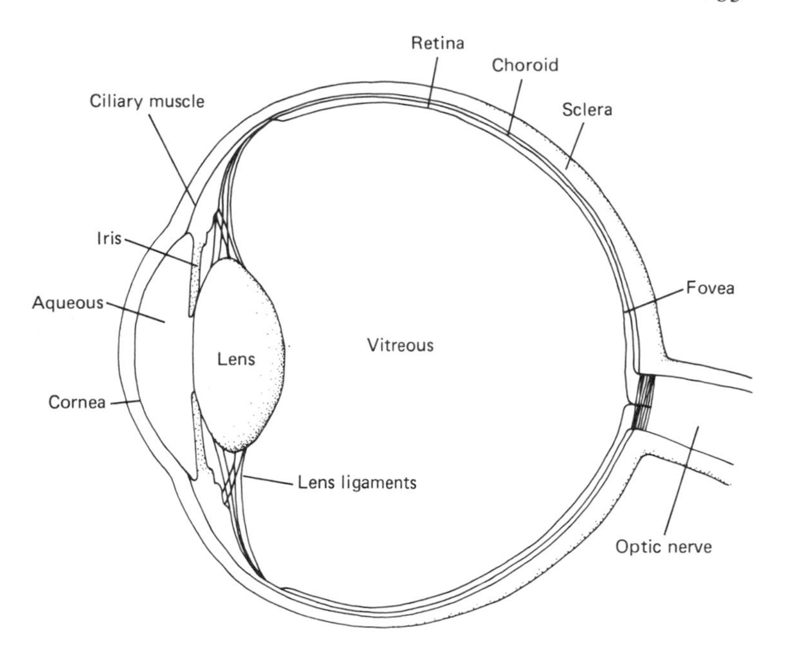

Structure of Eye

The eye is a ball-shaped organ enveloped by a tough fibrous sheet of connective tissue called the sclera (Fig. 6.1). The sclera is continuous anteriorly with the transparent *cornea*. The inner surface of the *sclera* is lined by two delicate membranes, the *choroid* and the *retina*. The former contains the blood vessels that supply the eye. This heavily vascularized membrane lies between the sclera and the retina. The retina is a neural membrane containing photoreceptors and interneurons. The optic nerve arises from the axons of the retinal ganglion cells.

The *crystalline lens* divides the eye into two fluid-filled chambers; the anterior one is filled with *aqueous humor;* the posterior contains a gelatinous fluid, the *vitreous humor*. Light entering the eye through the transparent cornea must cross the aqueous humor, pass through the pupil and lens, and cross the vitreous humor to strike the retina.

Pupillary Light Reflex

The iris regulates the amount of light entering the eye by controlling pupil size. It contains two sets of muscle fibers, *pupillary sphincter* and *pupillary dilator* (Fig. 6.2). The pupillary sphincter muscle consists of a circular band of muscle fibers surrounding the pupil. The pupillary dilator is formed by radially directed muscle fibers positioned like spokes in a wheel around the pupil. Parasympathetic nerves innervate the pupillary sphincter; sympathetic nerves innervate the pupillary dilator (see Chap. 10).

In bright light the iris contracts to reduce the amount of light entering the eye. If illumination increases, neurons in the

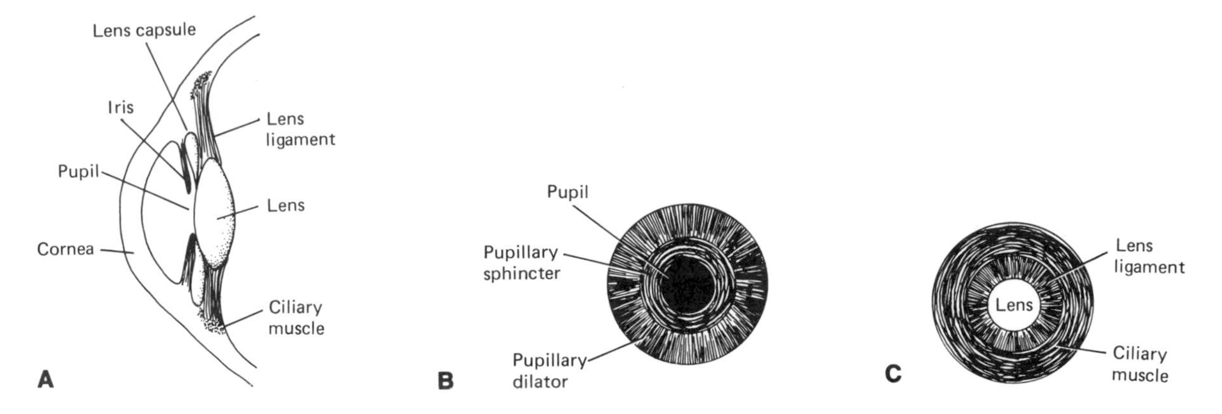

Fig. 6.2A–C. Mechanisms controlling reflexes of iris and lens. **A** arrangement of iris and lens. **B** pupillary light reflex. The iris is composed of two sets of muscle fibers. Contraction of the pupillary sphincter constricts the pupil. The pupil is dilated by contraction of the pupillary dilator. **C** accommodation. The lens is suspended from the lens ligaments. Contraction of the ciliary muscle fibers reduces the tension in the lens ligaments, permitting the elastic lens capsule to assume a more spherical form.

pretectal nucleus excite those parasympathetic fibers innervating the pupillary sphincter (see this chapter, "Central Visual Pathways"). The resulting contraction of the sphincter constricts the pupil to reduce the light entering the eye. If the ambient level of lighting decreases, the pretectal neurons inhibit the parasympathetic fibers. This inhibition of parasympathetic activity relaxes the sphincter, permitting the pupil to dilate. Active dilation of the pupil (contraction of the pupillary dilator muscle) results from the excitation of the sympathetic nervous system during emotional or physical stress (see Chap. 10).

Image Formation

The eye resembles a camera in its optical organization. The cornea and crystalline lens form a system that focuses light on the retina. The latter contains a layer of photoreceptors that records the visual image by alterations in its electrical activity. Light stimulation of these receptors is analogous to exposure of a film in a camera.

When light rays cross a curved surface separating two mediums of differing optical densities, they are *refracted,* or bent. Light rays from a distant source are refracted by a *biconvex lens* so that they focus at a point (*focal point*) behind the lens (Fig. 6.3). The degree of refraction depends on the angle formed when the light ray strikes the lens surface. The larger the angle of incidence, the smaller the refraction of the light ray. Those rays that strike the outer edges of the lens are strongly refracted toward its center. Those that strike the center of the lens perpendicularly are not refracted at all.

The total refractive power of the eye is about 66.7 diopters. *Diopters* are equal to the reciprocal of the focal length (diopter = 1/focal length). The *focal length* is the distance in meters from the lens to its focal point (Fig. 6.3).

Light rays are refracted at four interfaces as they pass through the eye, including the interfaces between (1) air and cornea, (2) cornea and aqueous humor, (3) aqueous humor and lens and (4) lens and vitreous humor. The principal re-

Fig. 6.3. Formation of an image by a biconvex lens. Light rays which originate from a distant source are focused at the focal point behind the lens. The distance from the lens to the focal point is the focal length.

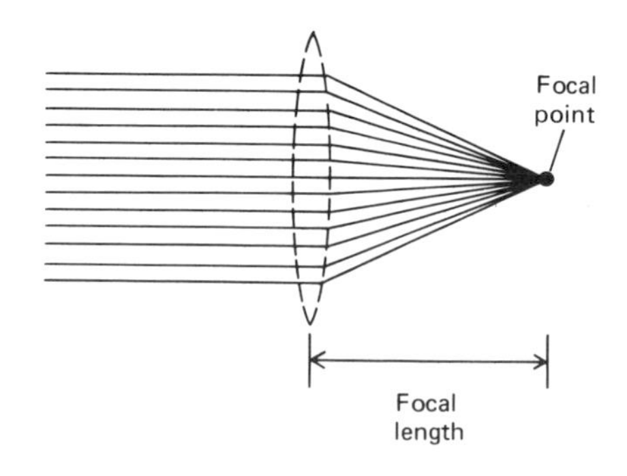

fractive interface is that between air and the cornea, which has a refractive power of about 45 diopters. This high refractive power reflects the curvature of the cornea and the large difference between the optical densities of air (1.0) and the cornea (1.38). The refractive power of the lens (about 20 diopters) reflects the relatively small difference between the optical densities of the aqueous humor (1.33) and the lens (1.39).

Accommodation

If an object located within 6 m of the eye is viewed by the resting eye, the visual image is blurred. The refractive power of the eye is insufficient to focus an image on the retina (Fig. 6.4). The accommodation reflex provides a mechanism for increasing the refractive power of the eye by as much as 14 diopters, thus enabling it to form focused images of nearby subjects on the retina.

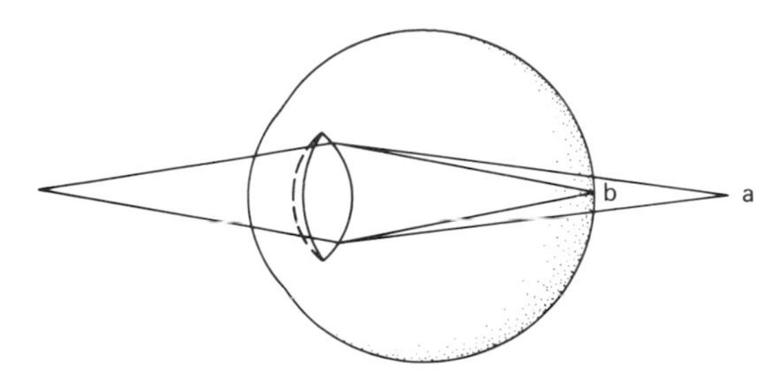

Fig. 6.4. Accommodation. When the resting eye views a nearby subject, it forms an image of that subject behind the retina (*point a*). The accommodation reflex increases the refractive power of the eye so that the focused image falls on the retina (*point b*).

The *accommodation reflex* increases the refractive index of the eye by increasing the curvature of the crystalline lens. The curvature is controlled by contraction of a ring of ciliary muscle fibers surrounding the lens (Fig. 6.2). The lens is suspended in the center of this ring by the radially orientated lens ligaments.

Contraction of the ciliary muscle is controlled by parasympathetic fibers that run in the oculomotor nerve. When the eye shifts its gaze from a distant to a near object, the resulting excitation of these fibers increases contraction of the ciliary muscle. This contraction in turn reduces the diameter of the muscle fiber ring. Since the lens ligaments insert on the muscle fiber ring, any decrease in the ring's diameter relaxes the lens ligaments. Reduction in the tension permits the elastic capsule of the lens to assume a more spherical shape, thus increasing the refractive index of the eye.

The parasympathetic excitation triggered by the eye's sighting a near object also activates the pupillary sphincter muscle, which constricts the pupil. The resulting narrowing of the pupillary aperture increases the depth of focus by preventing divergent light rays from striking the retina. It in effect increases the sharpness of the image formed on the retina. Optically, pupillary constriction during accommodation is equivalent to increasing the f-stop on a camera.

Visual Acuity

Visual acuity is a measurement of the ability of the eye to form a focused image on the retina. It is measured by determining the minimum distance by which two points must be separated to resolve them as discrete entities. The separation must be sufficient so that their focused images strike separate sets of retinal receptors. The *normal visual acuity* is 1′ (one minute) of arc; if two points are separated by 1′ of arc, they are resolvable. If, however, the two points are separated by less than 1′ of arc or if the retinal image is blurred (because of poor focusing), the retinal images of the two points will overlap, producing the visual sensation of a single point.

Visual acuity can be tested by a *Snellen chart*. It contains several lines of different-sized letters so constructed that the gaps distinguishing one letter from another e.g., a C from an O subtend arcs of 1′, 2′, 3′, 4′, and so on when viewed from 6 m (20 ft). Measurement of visual acuity depends on the subject's ability to discriminate a sequence of letters with similar size gaps. Ability to discriminate between letters with 1′ gaps reflects a visual acuity of 20/20, a normal reading. Visual acuities of 20/40 (ability to resolve letters with 2′ gaps) or 20/60 (letters with 3′ gaps) are below normal and require corrective lenses.

Optical Abnormalities

Many eyes cannot form sharply focused images on their retinas. Usually this deficiency is the result of either an abnormally shaped eyeball (Fig. 6.5) or an abnormal curvature of the surface of cornea or lens.

Farsightedness, or *hypermetropia*, results from a short eyeball. The distance from lens to retina is abnormally short so that the focal point of the eye falls behind the retina. The defect is correctable by using a *convex* lens to increase the eye's refractive power.

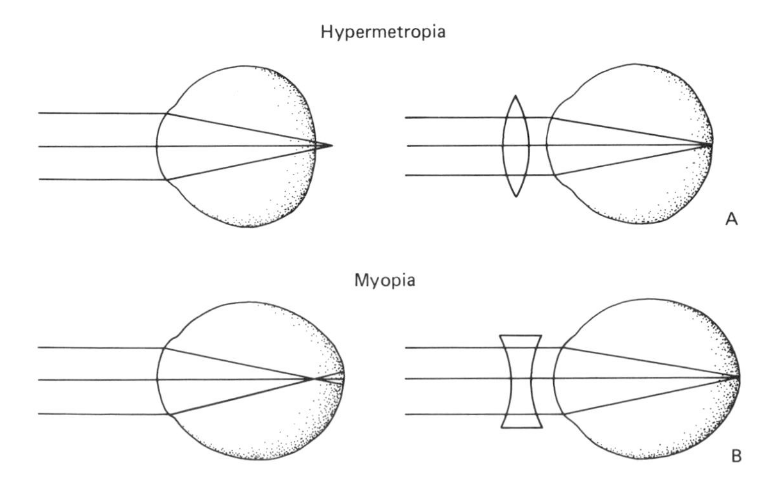

Fig. 6.5A and B. Optical abnormalities. **A** In hypermetropia the eyeball is abnormally short so that the image forms behind the retina. Correction requires the placement of a convex lens before the eye. **B** Myopia is a result of an abnormally long eyeball so that the visual image is formed before the retina. To correct myopia a concave lens is placed before the eye.

If the eyeball is too long, the eye is nearsighted. In nearsightedness (*myopia*) the visual image is focused in front of the retina. To correct, the refractive power of the eye is reduced by using a *concave* lens.

Astigmatism is usually the result of an abnormally curved cornea. The focus of the retinal image is distorted: some points are in focus, others are not. Correction requires lenses to compensate for the abnormal curvature.

Photoreception

The retina is formed by four layers of cells (Fig. 6.6), including a *pigment cell layer*, a *layer of photoreceptors*, and two *layers of retinal neurons*. The outermost layer (adjacent to the sclera) is formed by the pigment cells. The photoreceptor layer is sandwiched between the pigment cell layer and the two layers of retinal neurons. The axons forming the optic nerve arise from ganglion cell neurons located in the innermost layer of retinal neurons, which is adjacent to the vitreous humor.

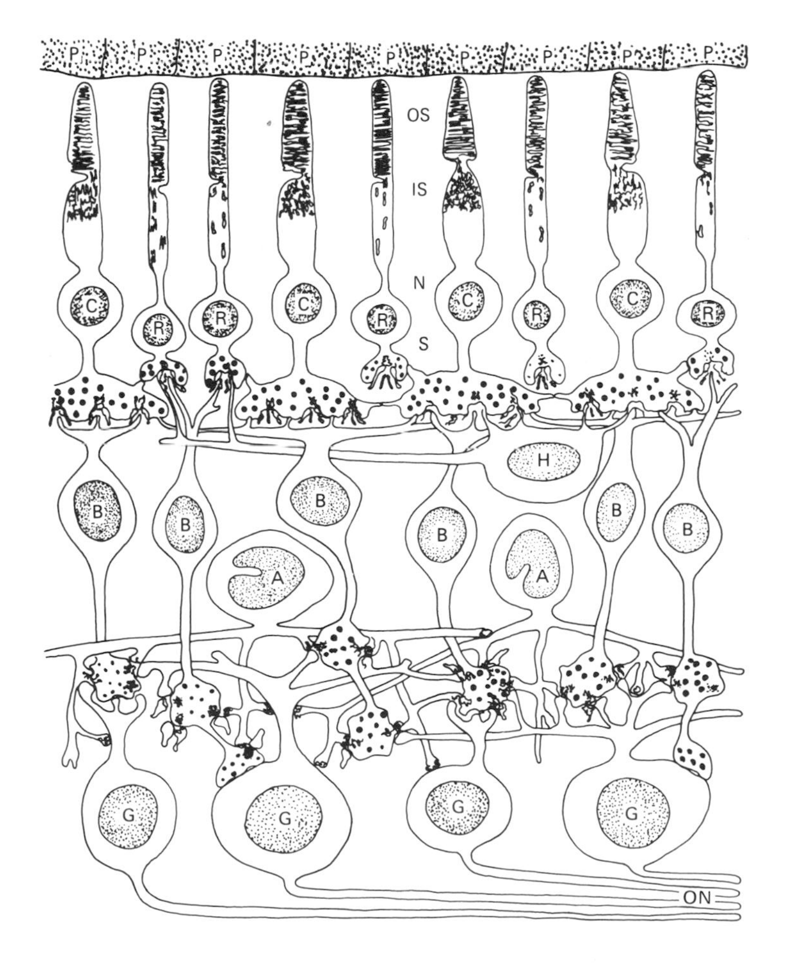

Fig. 6.6. Cellular organization of retina. Pigment cells (*P*) are located in the outermost layer. The receptor cells, rods (*R*) and cones (*C*), are divided into four morphologic zones: outer segment, inner segment, nuclear, and synaptic. The external layer of neurons contains the bipolar (*B*), horizontal (*H*), and amacrine (*A*) cells. Inner neural layer is composed of ganglion cells (*G*) whose axons form the optic nerve (*ON*).

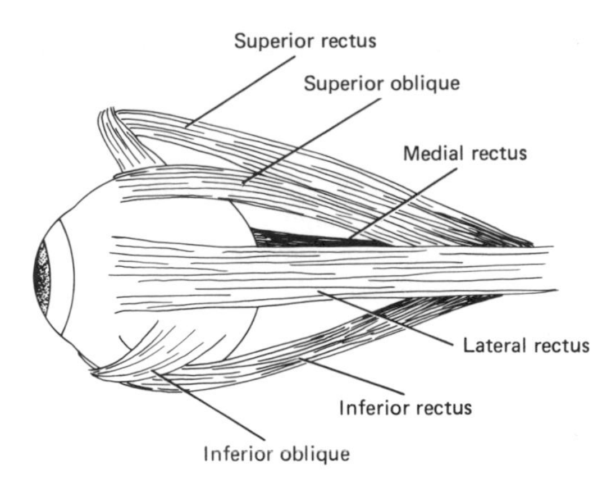

Fig. 6.7. Extraocular muscles.

As a result of the "inverted" structure of the vertebrate retina, light passing through the eye must cross both layers of retinal neurons before striking the photoreceptors. In their passage through these layers many rays are scattered as a result of striking a retinal neuron. This scattering decreases the quality of the visual image formed on the retina.

Only in one small area of the retina—the *fovea*—is a sharp visual image formed. The high visual acuity of the fovea is a result of a structural specialization in which the two layers of retinal neurons are drawn to one side to expose the photoreceptors to direct light stimulation. Light strikes the photoreceptors in the fovea directly without any interference from the retinal neuron layers. The fovea contains a dense concentration of small cone photoreceptors that further enhances its visual acuity (see this chapter, "Central Visual Pathways").

Since the fovea is the only area of the retina that can form a sharp visual image, it is important that the image of any object viewed by the eye fall on the fovea. One major function of the *three pairs of extraocular muscles* (Fig. 6.7 and Table 6.1) is to ensure that the image of an object looked at by the eye is formed on the fovea.

Table 6.1. Extraocular eye muscles.

Muscle	Innervation	Movement of Eyeball
Superior rectus	Oculomotor	Up and inward
Inferior rectus	Oculomotor	Down and inward
Medial rectus	Oculomotor	Inward
Lateral rectus	Abducens	Outward
Superior oblique	Trochlear	Down and outward
Inferior oblique	Oculomotor	Up and outward

Rods and Cones

The retina contains two kinds of photoreceptors: rods and cones. *Rods,* more sensitive than cones in dim light, are responsible for "night vision," whereas *cones,* which are color-sensitive, are associated with "day vision." There are more than 100 million rods and about 5 million cones in the retina. The cones are concentrated in the central portion of the retina, especially in the fovea. Most of the photoreceptors in the peripheral regions of the retina are rods.

Both cones and rods are elongated cells and can be divided into four morphologic zones, including (1) the outer segment, which contains the photopigment, (2) the inner segment with mitochondria and other organelles, (3) the nuclear zone, and (4) the synaptic zone (Fig. 6.6). The outer segment of the rod is narrow and elongated, whereas that of the cone is short and cone-shaped. The membrane of the outer segment is enfolded to produce a series of discs wherein the photopigment is located.

The electrical activity of rods and cones is unique among sensory receptors; their receptor potentials are hyperpolarizing, and they do not generate action potentials. Absorption of light by the photopigment molecules in the disc membranes produces a decrease in the sodium permeability of the outer segment membrane. This decrease in sodium permeability hyperpolarizes the photoreceptor. The production of a *hyperpolarizing receptor potential* is unique—all other sensory receptors respond to stimulation by producing a depolarizing receptor potential (see Chap. 5).

Photochemical Properties

All photopigments contain retinal (retinene) and opsin. *Retinal* is an aldehyde of vitamin A; *opsin* is a protein. The photopigments of the rods and cones differ in their opsins. *Rhodopsin,* the rod photopigment, is composed of retinal and *scotopsin.* It is most sensitive to blue-green light having its peak absorption of light at a wavelength of 505 nm.

The cone photopigment, *iodopsin,* is composed of retinal and cone opsin. There are three kinds of cone opsins, or photopsins—one for each kind of cone. These three opsins

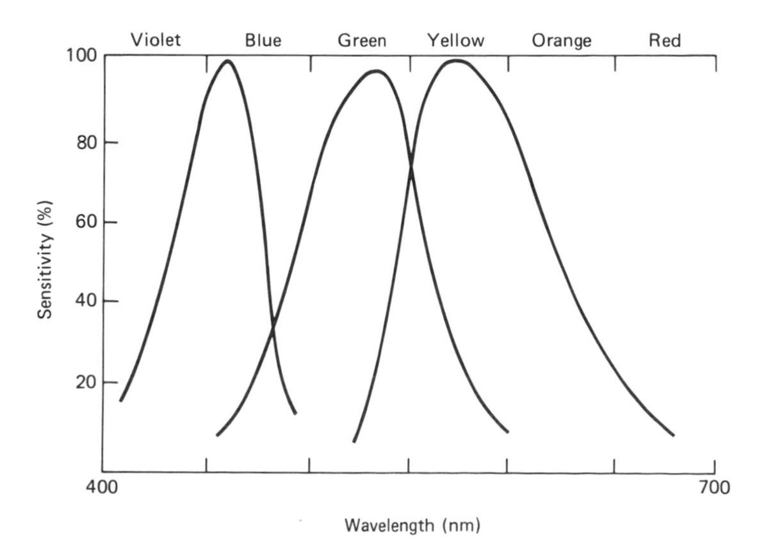

Fig. 6.8. The three cone pigments. Note that their photoabsorption curves are in the violet-blue (blue cone), blue-green-yellow (green cone), and the green-yellow-orange-red (red cone) ranges, respectively.

have their peak absorptions at wavelengths of 430 (blue cone), 535 (green cone), and 575 (red cone) nm (Fig. 6.8). It is these three different types of cones (blue-sensitive, green-sensitive, and red-sensitive) that enables the eye to produce the sensation of color. Since blue, green, and red are the primary colors, photoexcitation of various combinations of blue, green and red cones gives rise to all possible color sensations.

Rhodopsin contains retinal in the form of the 11-cis isomer. When a rhodopsin molecule is struck by light, 11-cis retinal is transformed to all-trans retinal (Fig. 6.9). This photoisomerization of retinal to the all-trans form converts rhodopsin to prelumirhodopsin, which is very unstable. It rapidly decomposes in stages to release all-trans retinal and scotopsin. During the decomposition two unstable intermediates are formed, lumirhodopsin and metarhodopsin. Being light-independent, the decomposition can proceed in light or darkness.

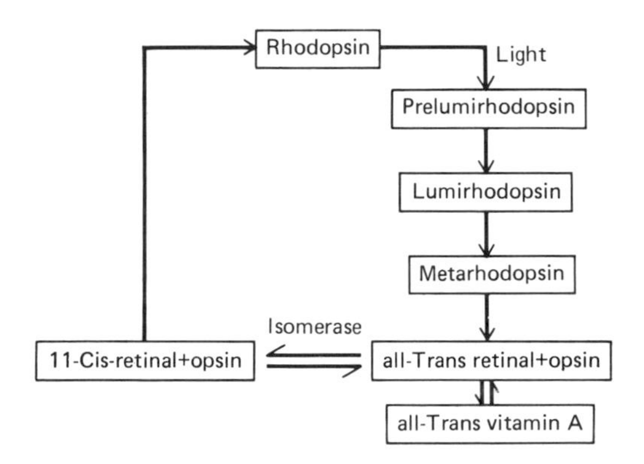

Fig. 6.9. Synthesis and photodegradation of rhodopsin.

To regenerate rhodopsin all-trans retinal must be converted to 11-cis retinal. This isomerization of all-trans retinal to the 11-cis form is catalyzed by the enzyme *retinal isomerase*. Rhodopsin is produced by the coupling of 11-cis retinal to opsin. Vitamin A provides an alternative source of 11-cis retinal. To be converted into 11-cis retinal, vitamin A must be oxidized to form all-trans retinal, which is then isomerized to 11-cis retinal.

Neural Processing

A network of retinal neurons connects the photoreceptors with the ganglion cells (Fig. 6.6). The axons of the latter form the optic nerve, which transmits visual information from the retina to the brain.

Three kinds of *retinal neurons* are in this network, namely, bipolar, horizontal, and amacrine cells. *Bipolar cells* form a direct link between photoreceptors and ganglion cells; they provide a pathway for the vertical transmission of information through the retina. In contrast the *horizontal* and *ama-*

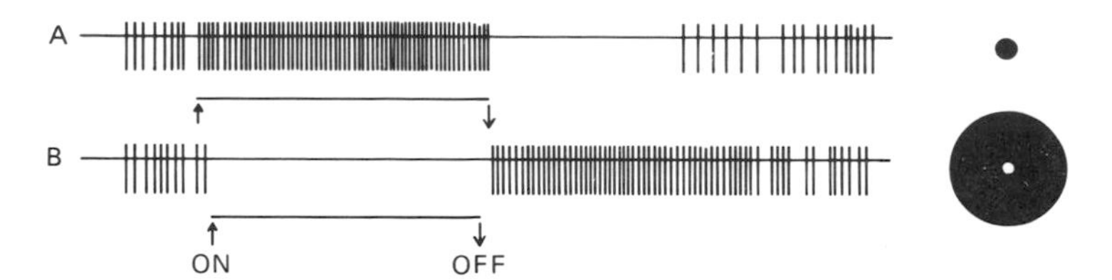

Fig. 6.10A and B. Response of ganglion cell excited by stimulation of its center and inhibited by stimulation of its surround. **A** stimulation of the center with a 0.5 mm spot of light excites the ganglion cell; **B** cell is inhibited when its surround is stimulated with a 3 mm annulus.

crine cells provide pathways for the horizontal diffusion of visual information through the retina by way of the interconnection of receptors and bipolars.

Ganglion cells have concentric receptive fields (Fig. 6.10). That is, for each ganglion cell there is a small circular region on the retina which when illuminated either excites or inhibits the discharge of that ganglion cell. The receptive field is divided into two antagonistic regions, a *center* and a *surround*. About one-half of all ganglion cells are excited whenever the center is stimulated with a spot of light. In contrast the ganglion cell is inhibited when the surround is stimulated by a ring, or annulus, of light. Other ganglion cells have receptive fields in which the center is inhibitory and the surround is excitatory.

The antagonistic organization of the receptive fields of ganglion cells is a result of the two retinal neural pathways (Fig. 6.6). (1) The center response is mediated by the *vertical pathway* over which information is transmitted from receptors to bipolars to ganglion cells. (2) The *horizontal transmission* of information is responsible for the surround effect. Excitation of receptors in the surround is transmitted to the ganglion cell through the lateral network of amacrine, bipolar, and horizontal cell processes.

The differences in visual acuity between the peripheral and foveal (central) areas of the retina can be correlated with the sizes of the receptive fields of the peripheral and foveal ganglion cells. Ganglion cells in the peripheral retina receive input from as many as 600 rods. Consequently the peripheral retina is characterized by low visual acuity. In the fovea the centers of many ganglion cell receptive fields contain only a single cone. The convergence ratio of one cone per ganglion cell is responsible for the high visual acuity of the fovea.

Central Visual Pathways

The optic nerve transmits visual information from the eye to the visual areas of the brain. The optic nerves from the two eyes meet at the base of the brain to form the *optic chiasma* (Fig. 6.11). At the optic chiasma one-half of the optic nerve fibers cross over (decussate) to run in the contralateral nerve. Fibers from the nasal half of the retina cross over to run in the contralateral nerve, whereas fibers originating in the temporal half of the retina remain in the ipsilateral nerve. As a result of the crossover at the optic chiasma,

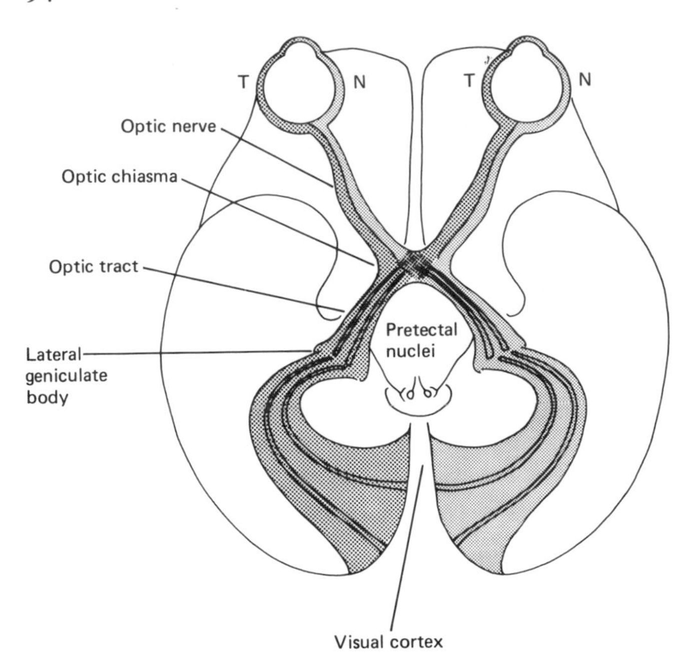

Fig. 6.11. Optic pathway from retina to visual cortex (ventral view). Retina is divided into nasal (*N*) and temporal (*T*) halves. Crossing of fibers at optic chiasma results in right cortex receiving inputs from right halves of two retinas while left cortex receives its input from two left retinal halves. Pretectal nucleus and superior colliculus (*not shown*) also receive terminations from optic nerve.

fibers from the right half of each eye terminate in the right thalamus, whereas those from the left half of each eye end in the left thalamus.

Most of the optic nerve fibers terminate in the *lateral geniculate body of the thalamus* (Fig. 6.11). Those fibers which do not terminate in the thalamus run posteriorly to the mesencephalon where they terminate in the *superior colliculus* and the *pretectal nucleus*. The superior colliculus also receives inputs from the visual cortex. Neurons in the superior colliculus and pretectal nucleus control eye movement through their inputs to the nuclei of the three cranial nerves (oculomotor, trochlear, and abducens), which innervate the extraocular muscles. The pretectal nucleus is also concerned with the regulation of visual accommodation reflexes.

Visual Cortex

The primary function of the lateral geniculate is to provide a relay station for the transmission of visual information to the visual cortex, which is located in the posterior half of the occipital lobe (Fig. 6.12). The visual cortex has a retinotopic

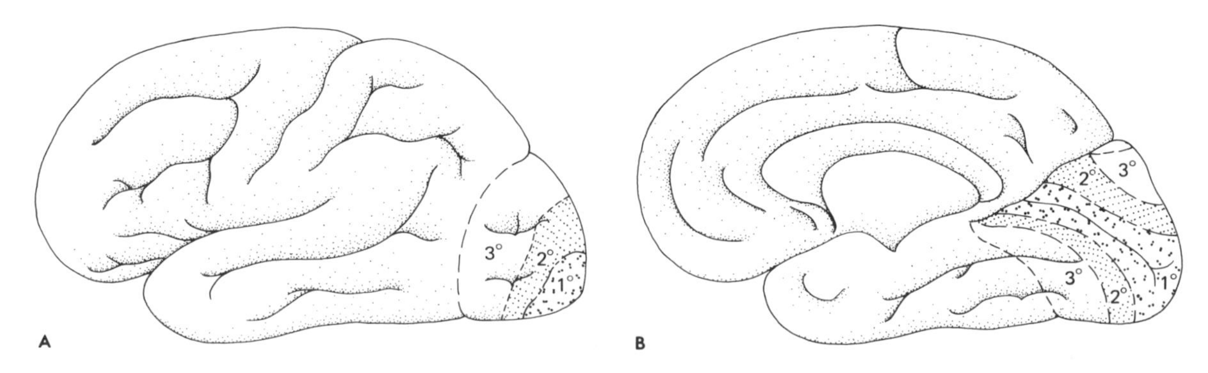

Fig. 6.12A and B. Location of primary, secondary, and tertiary areas of visual cortex. **A** lateral surface, **B** medial surface.

organization, i.e., there is a spatial correlation between the position of a neuron in the visual cortex and the location of its receptive field on the retina. The central region of the visual cortex receives inputs from the fovea, which is the retinal region of greatest visual acuity. Light stimulation of the dorsal retina excites neurons in the dorsal half of the visual cortex. Inputs from the ventral retina terminate in the ventral half of the visual cortex. There is a point-to-point correspondence between the locations of photoreceptors on the surface of the retina and the positions of the cortical neurons that they drive.

Cortical Activity

Neurons in the visual cortex are concerned with the extraction of visual information relating to stimulus form. Most visual cortical neurons are excited when a line, bar, or edge is projected on a particular small area of the retina (receptive field). Each neuron responds to a very limited range of visual stimuli. These cortical cells have been divided into three classes depending on the complexity of the stimulus that must be projected onto the retina to excite them.

Simple cells are excited by a black bar on a white background or a slit of light on a black background. To excite a particular cell, the bar must be precisely positioned in the visual field. Present evidence suggests that there is at least one simple cell for every possible orientation and position that a bar can occupy in the visual field.

Complex cells also respond best to bars with a precise orientation—vertical or horizontal. However, the position of the bar is much less critical than for simple cells. As a result complex cells can be excited by bar stimuli that are projected over broad areas of the retina. *Hypercomplex cells* are similar to complex cells except that they are sensitive to bar length in addition to orientation and position in the receptive field. If the bar is too long or too short, it will not be able to excite the hypercomplex cell.

All of the neurons in the visual cortex are organized into columns each of which forms an integrating unit (see Chap. 5 for discussion of columns in the somatosensory cortex). Each column contains neurons that respond to bar stimuli of one particular orientation which fall on one small region of the retina. A different column exists for each possible orientation of the stimulus, and a set of such columns is present for every receptive field on the surface of the retina.

The visual cortex is usually divided into primary, secondary, and tertiary regions (Fig. 6.12). The secondary and tertiary divisions form the *visual association areas* of the visual cortex. They receive inputs from both the lateral geniculate and primary visual cortex. They also contain almost all of the hypercomplex neurons and are concerned with more complex abstraction of visual information.

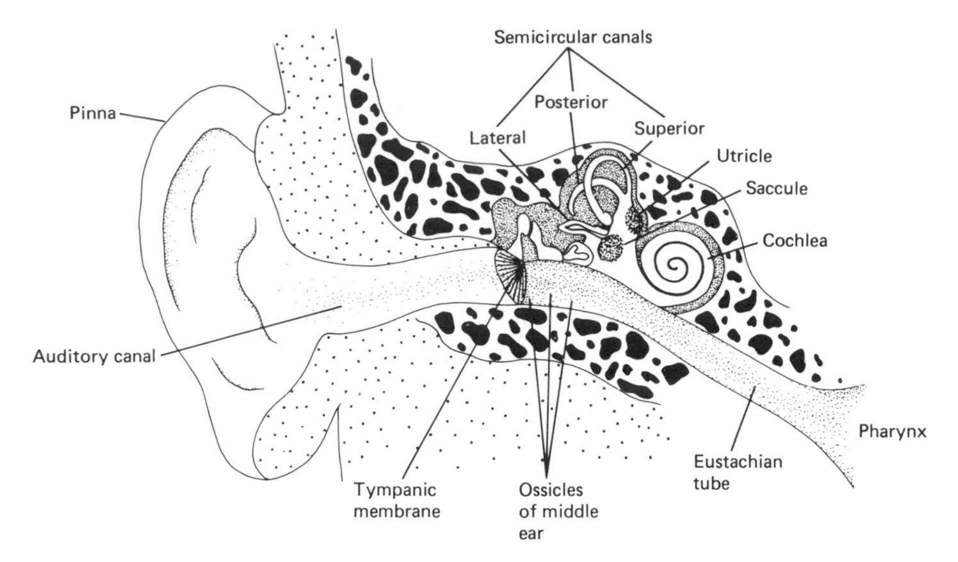

Fig. 6.13. Structure of the ear. The pinna and auditory canal direct sound waves at tympanic membrane. Cochlea and the vestibular apparatus make up the inner ear. Vestibular apparatus includes saccule, utricle, and three semicircular canals. Semicircular canals are orientated in the three planes of the body: lateral, posterior, and superior. The middle-ear cavity is connected to the pharynx by the eustachian tube.

Auditory System

The ear is divided into three compartments, the external, the middle, and the inner ear (Fig. 6.13). The external and middle ears are accessory sensory structures. Their function is to conduct sound to the auditory receptors in the cochlea of the inner ear. The inner ear contains sensory receptors belonging to two sensory systems, namely, the auditory receptors in the cochlea and the vestibular receptors in the vestibular apparatus (see this chapter, ''Vestibular System'').

Sound

Sounds are generated when *pressure waves* produced by the longitudinal vibration of air molecules strike the auditory apparatus of the ear. A sound source, such as a tuning fork or vibrating string, produces alternating waves of compression (high density) and rarefraction (low density) of air molecules that spread out from the source much like ripples on the surface of a pond. Each sound has two fundamental qualities, intensity and pitch.

The pitch of a sound is a function of its frequency or number of pressure waves per second. Frequency is measured in units of hertz (Hz). One Hz is equal to one cycle of vibration per second. The greater the frequency of a sound, the higher its pitch. The human ear is capable of detecting sounds over a frequency range of 20 to 20 000 Hz. Its maximum sensitivity is in the range of 1000 to 4000 Hz.

The intensity of a sound is proportional to the amplitude of the sound vibration. Intensity is measured in *decibels,* which are logarithmic units. A decibel is equal to 10 log I/I_s, where I_s is a reference sound intensity. The standard reference intensity is 0.0002 dyne/cm^2, which is very close to the threshold for human hearing.

External and Middle Ears

The *pinna* of the external ear forms a funnel that directs sound into the *auditory canal* (Fig. 6.13). Sound waves travel down the canal to strike the *tympanic membrane,* which separates outer and middle ears. Vibrations of the tympanic membrane are transmitted across the air-filled cavity of the middle ear by a chain of three small bones, or ossicles. The bones of the middle ear are the malleus (hammer), incus (anvil), and stapes (stirrup). The malleus is attached to the tympanic membrane; the stapes contacts the *oval window membrane* in the *cochlea* of the inner ear. Thus vibrations of the tympanic membrane are transmitted in succession by the malleus, incus, and stapes across the middle ear to the oval window.

An important function of the middle ear is to provide the impedance matching required to transmit sound from a medium of low density (air) to a medium of high density (inner-ear fluid). The amount of energy required to vibrate a membrane is a function of the density of the medium in which it is immersed. Vibration of a membrane in inner-ear fluid requires about 130 times more energy than vibration of the same membrane in air.

When sound waves are transmitted from the tympanic membrane to the oval window by the ossicles, sound pressure is increased 30-fold. This amplification is primarily the result of the large difference in size of the tympanic membrane (0.55 cm^2) and the oval window (0.032 cm^2). Vibrations of the large tympanic membrane are transmitted by the ossicles to the tiny oval window. The result is an increase in the sound pressure per unit area of membrane (when pressure at the oval window is compared to that at the tympanic membrane).

Vibration of the ossicles can be reduced or damped by contraction of two middle-ear muscles: the tensor tympani and the stapedius. They attach to the malleus (tensor tympani) and to the stapes (stapedius). Contraction of the middle-ear muscles increases the rigidity of the ossicular chain, thus reducing its ability to conduct vibrations to the cochlea. A loud sound elicits a reflex contraction of these middle-ear muscles. This tympanic reflex protects the auditory receptors in the cochlea from damage resulting from exposure to loud sounds.

Inner Ear

The cochlea is formed by the coiling of three fluid-filled tubes: the scala vestibuli, the scala media, and the scala tympani (Fig. 6.14). The scala vestibuli and the scala tympani are interconnected at the far end of the cochlear tube by the helicotrema. The scala media is sandwiched between the scala vestibuli and the scala tympani (Fig. 6.15). A thin

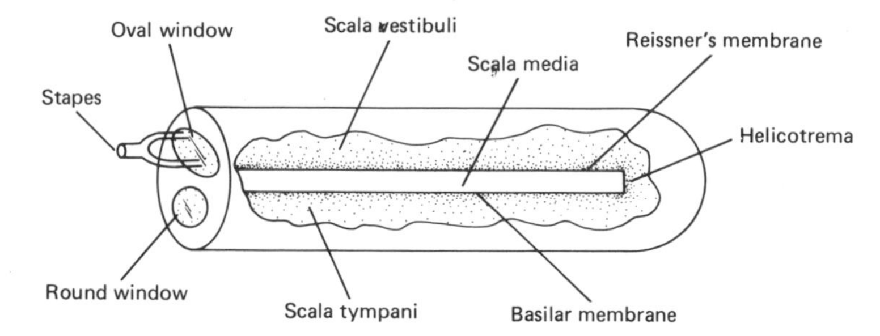

Fig. 6.14. The cochlea uncoiled.

membrane—Reissner's membrane—separates the scala media from the scala vestibuli. The basilar membrane separates the scala media from the scala tympani.

The cochlea contains two fluid systems; in one the scala tympani and scala vestibuli are filled with *perilymph;* in the second the scala media contains *endolymph*. These fluids differ in their composition: perilymph is high in sodium (150 mEq) and low in potassium (5 mEq); endolymph is sodium-poor (16 mEq) and potassium-rich (144 mEq). As a result of these differences in ionic composition, an endocochlear potential of about +80 mV is developed between the endolymph in the scala media and the perilymph-filled scala vestibuli and tympani. Since the resting potential of the hair cell is about −80 mV, there is a potential gradient of 160 mV (from −80 to +80 mV) between the endolymph and the receptor cells. This potential gradient is important for maintenance of hair-cell excitability.

The *oval window* is located at the proximal end of the scala vestibuli (Fig. 6.14). Low-frequency vibrations of the oval window produce pressure waves in the perilymph of the scala vestibuli. These fluid movements are conducted down the entire length of the scala vestibuli through the helicotrema and into the scala tympani. The *round window* is located at the proximal end of the scala tympani. Conduction of the pressure wave in the scala tympani results in fluctua-

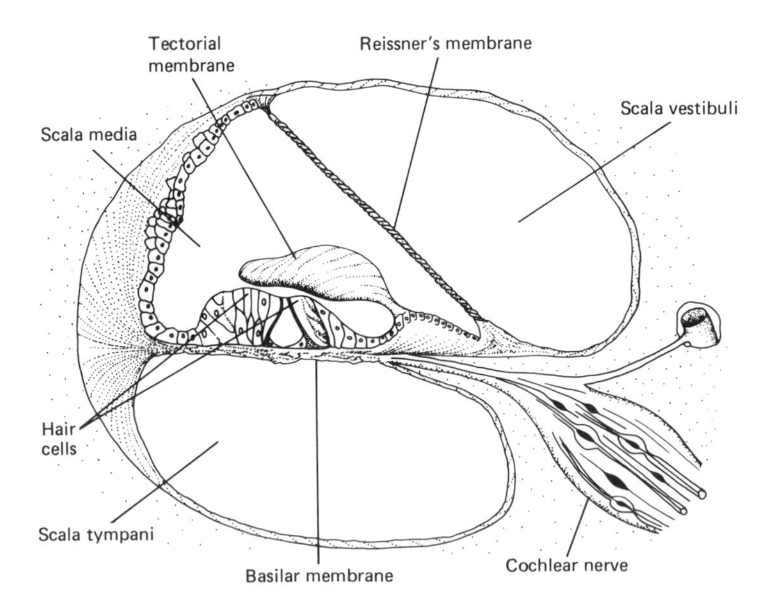

Fig. 6.15. Cross section of cochlear tube. Hair cells are embedded in basilar membrane. Tectorial membrane overhangs hair cells.

tions in the pressure which the perilymph exerts on the round window. Movement of the round window absorbs the energy of the pressure waves and acts as a damper.

Organ of Corti

The auditory receptors are hair cells (Fig. 6.16). There are about 20 000 of them in the human cochlea, embedded in the basilar membrane. Their apical surfaces contain hair-like cilia which project into the endolymph-filled scala media. A specialized flap of the basilar membrane, the tectorial membrane, overhangs the ciliated surface of the hair cells. The terminals of the cochlear nerve fibers form a synapse with the basal end of each hair cell. The cochlear nerve fibers combine with the fibers in the vestibular nerve to form the vestibulocochlear nerve (cranial nerve VIII). The auditory nerve is formed by the fibers of the cochlear nerve. The association of hair cells, cochlear nerve terminals, and tectorial and basilar membranes forms the *organ of Corti*.

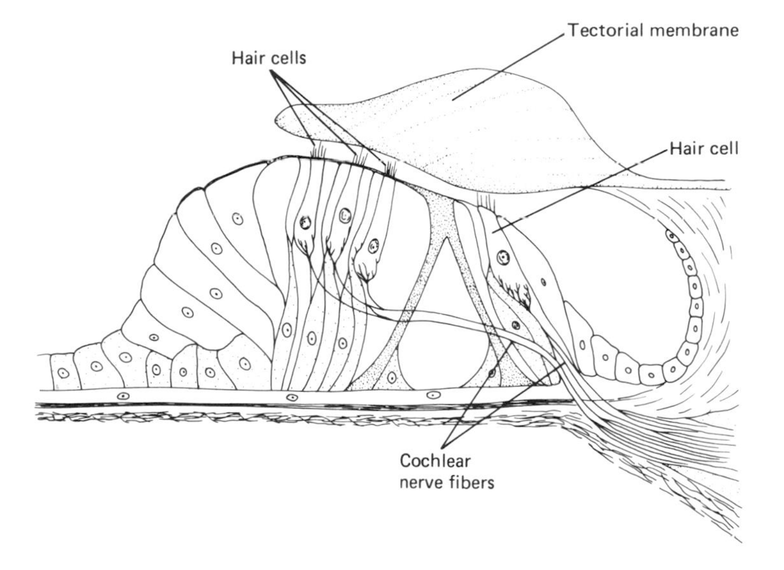

Fig. 6.16. The basilar membrane. Ciliated surfaces of hair cells project into endolymph of scala media. Tectorial membrane overhangs ciliated surfaces of hair cells. Cochlear nerve fibers innervate hair cells.

Receptor Excitation

Hair cells are excited by vibrations of the tectorial membrane, which is deflected by sound pressure waves traveling through the cochlea. Normally the tectorial membrane is in light contact with the ciliated surface of the hair cells. Vibrations of the tectorial membrane bend the cilia, thereby altering their ionic permeability and depolarizing the hair cells. The resulting receptor potential provides a source for the excitation of the cochlear nerve terminals. The exact mechanism for initiating action potentials in the cochlear nerve fibers is not known. There are two hypotheses: (1) The hair cells form chemical synapses with the cochlear nerve terminals. When the hair cell is depolarized, it releases a chemical transmitter which excites the postsynaptic terminals of the cochlear nerve. (2) Transmission is electrical; the terminals of the cochlear nerve are directly excited by a large receptor potential, which is produced when the hair cells are stimulated.

Pitch Discrimination

Vibration of the basilar membrane is frequency- or pitch-sensitive (Fig. 6.17). The elasticity of the basilar membrane gradually increases with increasing distance from the oval window. It is narrow (0.04 mm) and very stiff at the proximal (oval window) end of the cochlea. At the helicotrema the basilar membrane is both wider and more flexible. Consequently there is a gradual change in the vibratory characteristics of the basilar membrane along the length of the cochlea. The proximal basilar membrane responds best to high-frequency vibrations, whereas the distal basilar membrane is stimulated only by low-pitch sounds.
The *place theory of pitch discrimination* proposes that the basilar membrane serves as a pitch or frequency analyzer. The portion of the basilar membrane which vibrates maximally is a function of the stimulus pitch (Fig. 6.17). The lower the pitch of the stimulus, the greater the distance from the oval window to the point of maximum movement of the basilar membrane. As a result the frequency sensitivity of each hair cell is determined by its location. Hair cells that respond best to high-pitched tones are embedded in the narrow taut basilar membrane near the oval window. Those responding to low-frequency tones are on the wider, more relaxed basilar membrane in the distal cochlea.
Information concerning the pitch of a low-frequency tone is also encoded by the pattern of discharge produced in the cochlear nerve fibers. The *volley theory of frequency coding* proposes that the rate of action potential discharge is directly correlated with the sound frequency. Cochlear nerve fibers that respond to tones below 2000 Hz tend to discharge their action potentials at the stimulus frequency, i.e., a fiber responds to a 200-Hz tone by discharging impulses at a rate of 200/s.

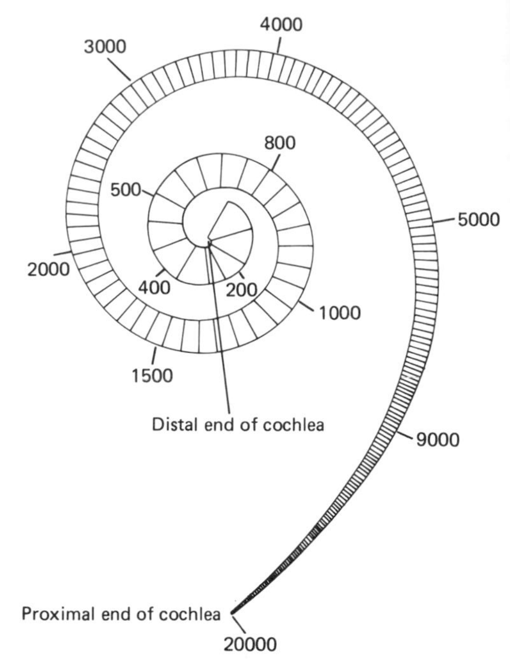

Fig. 6.17. Frequency sensitivity of cochlea. There is a gradual change in the vibratory characteristics of the basilar membrane along the length of the cochlea.

Central Auditory Pathways

The cochlear nerve fibers run in the vestibulocochlear nerve to the medulla where they terminate in the cochlear nucleus (Fig. 6.18). From the cochlear nucleus auditory information is transmitted to the auditory cortex through a chain of auditory interneurons located in the medulla (cochlear and superior olive nuclei), the mesencephalon (inferior colliculus), and the thalamus (medial geniculate). The final termination of the auditory input is in the dorsal lateral margin of the temporal lobe, forming the primary auditory cortex (Fig. 6.19). The auditory association cortex forms a band surrounding the primary auditory cortex.

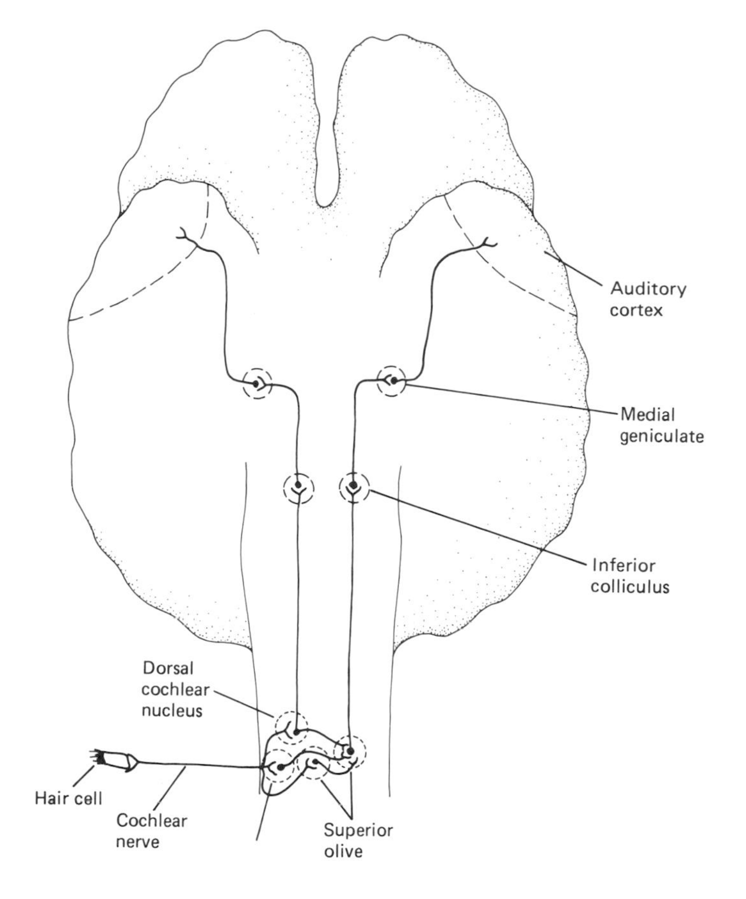

Fig. 6.18. Central auditory pathways. Cochlear nerve fibers terminate in several ipsilateral medullary nuclei—superior olive, dorsal cochlear, and ventral cochlear. Most fibers from these ipsilateral medullary nuclei terminate in the contralateral superior olive. Fibers from these medullary nuclei ascend to the inferior colliculus in the mesencephalon. Neurons from the inferior colliculus terminate in medial geniculate of the thalamus. Fibers from medial geniculate terminate in auditory cortex.

The auditory cortex is concerned with the recognition of complex sounds—both the frequency and intensity patterns of the sound are correlated. The auditory association cortex is concerned with interpreting the meaning of the sound stimulus. Information about sound location and pitch are extracted and processed by neurons in the lower auditory centers, namely, the inferior olive, inferior colliculus, and medial geniculate.

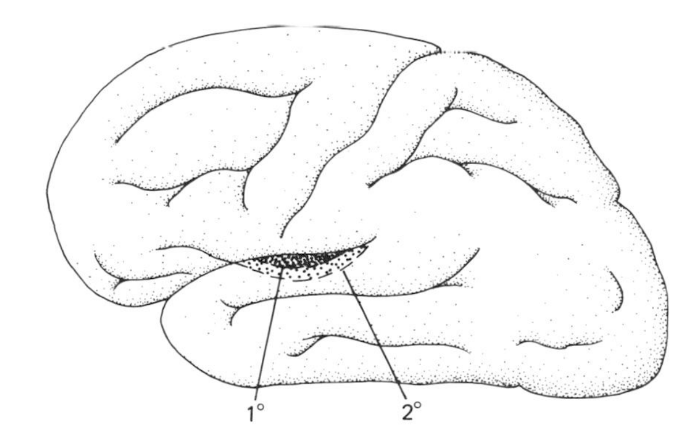

Fig. 6.19. Auditory cortex. Located on dorsolateral aspect of temporal lobe, is divided into primary and secondary areas.

Vestibular System

The labyrinth of the inner ear contains both auditory and equilibrium receptors (Fig. 6.20) and is formed by a series of fluid-filled, membrane-lined tubes embedded in the temporal bone of the skull. The labyrinth includes both the vestibular apparatus and the cochlea. The vestibular apparatus is subdivided into two functional units, including (1) the *utricle* and *saccule,* which respond to both position and linear acceleration; and (2) the three *semicircular canals,* which are excited by rotational acceleration. The cochlea contains the auditory receptors. The vestibular and auditory divisions of the labyrinth are connected by the cochlear duct, which runs between the scala media of the cochlea and the saccule. As a result of this interconnection, both the scala media and the vestibular apparatus are filled with a common endolymph.

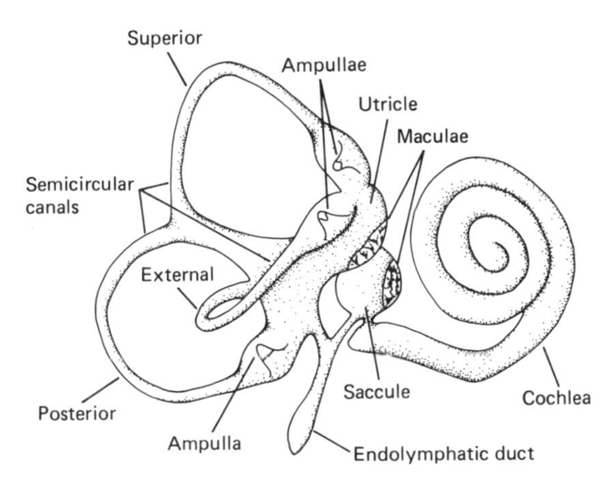

Fig. 6.20. The labyrinth. The vestibular receptors are located in the ampullae of the semicircular canals and the maculae of the saccule and utricle.

In both the auditory and vestibular systems the sensory receptors are ciliated hair cells. They are attached to the walls of the labyrinth so that their ciliated surfaces project into the endolymph. Apparently the endolymph provides the proper ionic environment for maintaining the excitability of the hair-cell receptors.

Utricle and Saccule

The portion of the labyrinth connecting the semicircular canals and the cochlea is enlarged to form two chambers—the utricle and the saccule (Fig. 6.20). A patch of sensory receptors, or macula is attached to the inner wall of each chamber (Fig. 6.21). Each macula contains several thousand hair cells. The hair-like cilia of the hair cells are embedded in a thick gelatinous layer. Tiny crystals of calcium carbonate (otoconia) are embedded in the gelatinous matrix. The hair cells are sensitive to displacement of their cilia, which are bent by the weight of the overlying gelatinous matrix.

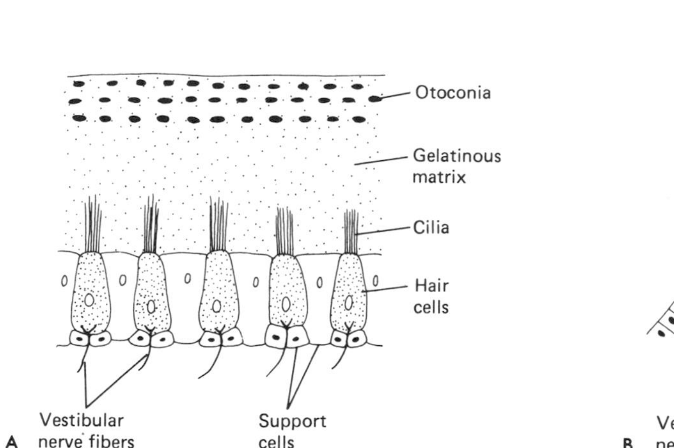

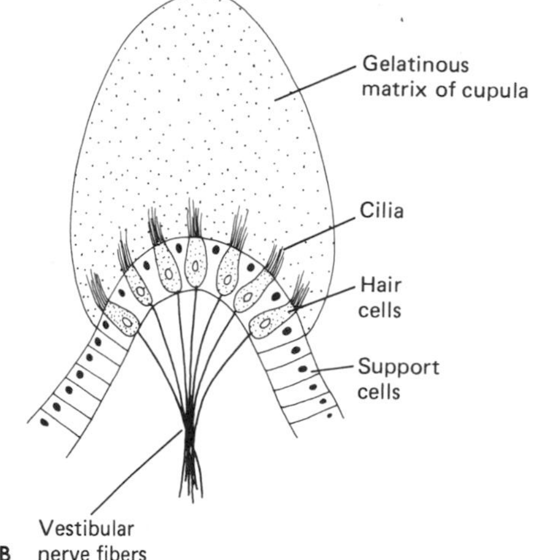

Fig. 6.21A and B. Vestibular receptors. **A** Maculae of utricle and saccule. The hair cells project cilia into an overlying gelatinous matrix. Outer layers of gelatinous matrix contain otoconia. **B** Semicircular canal receptors. Cilia project from apical surface of hair cell into gelatinous matrix of cupula.

Each hair is directionally sensitive. Directional sensitivity is correlated with the presence of two different kinds of cilia (Fig. 6.22). Each hair cell has a single large kinocilium flanked by many small stereocilia. Depending on the direction of the force bending the cilia, each hair cell may be either excited or inhibited. Deflection of the stereocilia toward the kinocilium excites the hair cell; if the stereocilia are bent away from the kinocilium, the hair cell is inhibited. Each macula contains several thousand hair cells oriented in all possible directions. Therefore the total pattern of hair-cell excitation and inhibition in the macula reflects the direction of the stimulation force.

The utricle and saccule are concerned with monitoring the *position* and *linear acceleration* of the head. Whenever the orientation or position of the head changes, the weight of the otoconia-containing gelatinous layer shifts, thereby stimulating a different set of hair cells. These vestibular organs are highly sensitive to change in the position of the head and will respond to as little as a one-half degree shift in the orientation of the head.

The utricle and saccule are also stimulated by linear acceleration of the head. Linear acceleration is produced by a sudden change in the rate of forward or backward movement. If

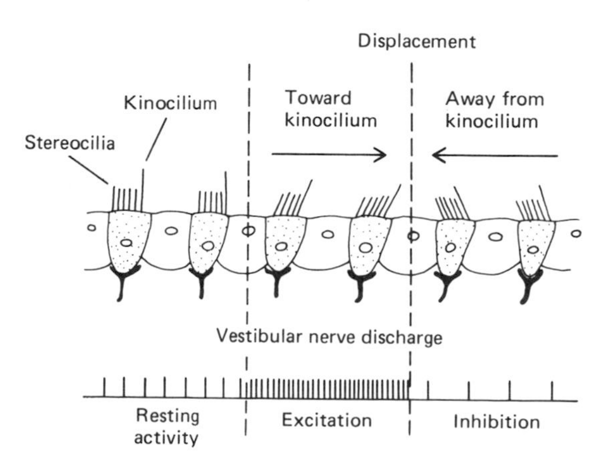

Fig. 6.22. Directional sensitivity of hair cells.

the head is suddenly moved, the resulting acceleration produces a change in the pressure which the otoconia exert on the hair cells. Information about linear acceleration is encoded by the different patterns of hair-cell excitation, which are elicited by the acceleration stimulus.

Semicircular Canals

Extending from the utricle, at right angles to each other, are the three semicircular canals (Fig. 6.20). They are so positioned that each is stimulated by rotational acceleration (sudden turning of the head) along one of the three planes—superior, posterior, or horizontal.

In each canal there is an enlarged region, or *ampulla,* containing the hair cell receptors (Fig. 6.21). A gelatinous sheet, the *cupula,* covers the ciliated surface of the receptor cells. This flap-like cap projects into the center of the canal and is readily deflected by movement of the canal-filled endolymphatic fluid. Deflection of the cupula excites the hair cells, which are embedded in its gelatinous matrix. Each hair cell contains a single large kinocilium and many small stereocilia. All hair cells in an ampulla have the same orientation so that deflection of the cupula in one direction excites, and in the other direction inhibits, the hair cells.

The hair cells are stimulated by rotational acceleration of the head, produced by the sudden turning of the head. The endolymph within the semicircular canal has an inertia so that it lags behind during movement of the canal. The fluid tends to remain stationary in space while the head, canal, and attached cupula are turning. The result is that the fluid presses on the cupula, deflecting it. Deflection bends the stereocilia toward the kinocilium, stimulating the hair cells. If the head suddenly stops turning, the semicircular canals experience *rotational deceleration*. Movement of the endolymph continues for a short time after the head, canal, and attached cupula have stopped moving. The fluid in the canal once again deflects the cupula. Since the cupula bends the stereocilia away from the kinocilium, the hair cells are inhibited by the deceleration stimulus.

Stimulation of the semicircular canals supplies information on the plane and rate of rotational acceleration (or deceleration). Since each semicircular canal is sensitive to acceleration in only one plane, the plane of the acceleration is encoded by the identity of the semicircular canal (superior, horizontal, or posterior) which is excited by the accelerating stimulus. The rate of acceleration determines the extent of deflection of the cupula and therefore the discharge rate in the vestibular nerve fibers, which innervate the hair cells.

Central Vestibular Pathways

The hair cells in the vestibular apparatus are innervated by vestibular nerve fibers which run in the vestibulocochlear nerve to the medulla where they terminate in the *vestibular nucleus* (Fig. 6.21). Fibers run from the vestibular nucleus to the cerebellum, reticular formation, and spinal cord. These motor centers utilize vestibular information together with information from visual receptors and neck proprioceptors in controlling body orientation during movement.

Vestibular input to the visual system is critical for an important visual reflex called nystagmus. *Nystagmus* is the reflex control of eye movement that enables the eyes to fixate on a stationary object when the head is moving. During rotation of the head, the eyes move slowly in the opposite direction from that of the rotation, thus maintaining visual fixation on a constant point. If the rotation continues so that the limit of eye movement is reached, the eyes will rapidly flick in the direction of head rotation to fixate on a new point. This quick flicking movement is nystagmus. As the head turns, the eyes go through an alternating sequence of slow following and quick flicking movements.

Gustatory System

Gustation, or the sensation of taste, is produced by chemical stimulation of taste buds in the mouth. About 10 000 taste buds are located on the tongue, roof of the mouth, and walls of the pharynx.

Taste Sensation

The surface of the tongue is covered with many small projections, or *papillae*. Most of the taste buds are located on the apical ends of these papillae. There are as many as 100 taste buds per papilla. Each taste bud is composed of about 40 elongated cells arranged like the segments of an orange around the taste pore (Fig. 6.23). There are two kinds of cells—supporting and taste receptor. The apical surface of each receptor cell contains several hair-like microvilli that extend into the taste pore. Terminals of the gustatory nerve innervate the basal surface of the receptor cell.

To stimulate a taste bud a chemical substance must be dissolved in the fluid of the mouth. A taste sensation is not experienced when dry crystals are applied to the dry surface of the tongue. Taste stimulation is thought to result from the binding of the stimulus molecule to the membrane of the microvilli of the receptor cells. Stimulus binding causes a change in the membrane permeability of the receptor cell, which depolarizes the cell. It is believed that the receptor

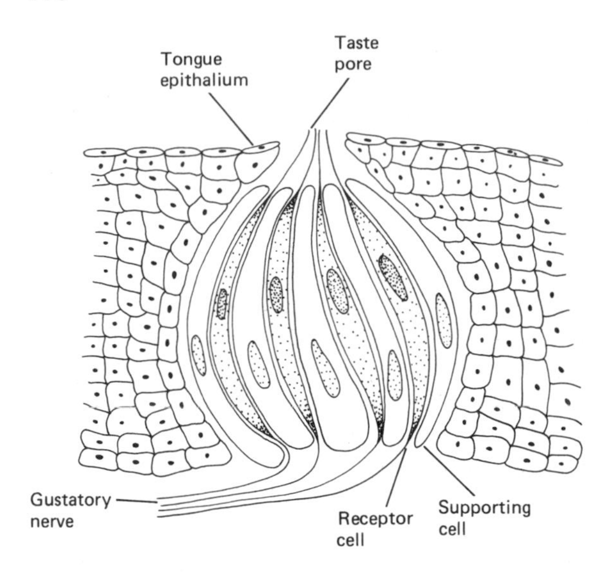

Fig. 6.23. Typical taste bud.

cell forms a chemical synapse with the terminals of the gustatory nerve. When the receptor cell is depolarized, it releases a chemical transmitter that excites the terminals of the gustatory nerve.

There are four taste sensations, namely, sweet, sour, salty, and bitter. Each is associated with stimulation of a different region of the tongue. The tip of the tongue responds best to stimulation with *sweet stimuli. Salty and sour stimuli* are most effective when applied along the sides of the tongue, whereas *bitter substances* produce their strongest effect when they make contact with taste buds on the back of the tongue.

The correlation between the chemical nature of the stimulus and the taste sensation is not understood. For example, a broad range of chemical stimuli evoke a *sweet response* when applied to the tongue. They include sugars, glycols, alcohols, aldehydes, ketones, amides, esters, amino acids, sulfonic acids, halogenated acids, lead salts, and beryllium salts. Saccharin, 600 times sweeter than sucrose (sugar), has a chemical structure markedly different from that of sugars. The other sensations are more readily related to chemical structure. Salty stimuli are always ionized inorganic salts, sour stimuli are acidic. The lower the pH of a solution, the more intense the sour sensation. Bitter stimuli are either alkaloids or long-chain organic molecules. The unpleasant nature of the bitter taste sensation is protective since many poisonous plants contain an alkaloid as their toxin.

Central Gustatory Pathways

Gustatory nerve fibers run from the tongue to the solitary nucleus in the medulla in three cranial nerves: the facial, the glossopharyngeal, and the vagus (Fig. 6.24). Fibers innervating the anterior two-thirds of the tongue run in the facial nerve, whereas the glossopharyngeal nerve contains fibers

Fig. 6.24. Central gustatory pathways.

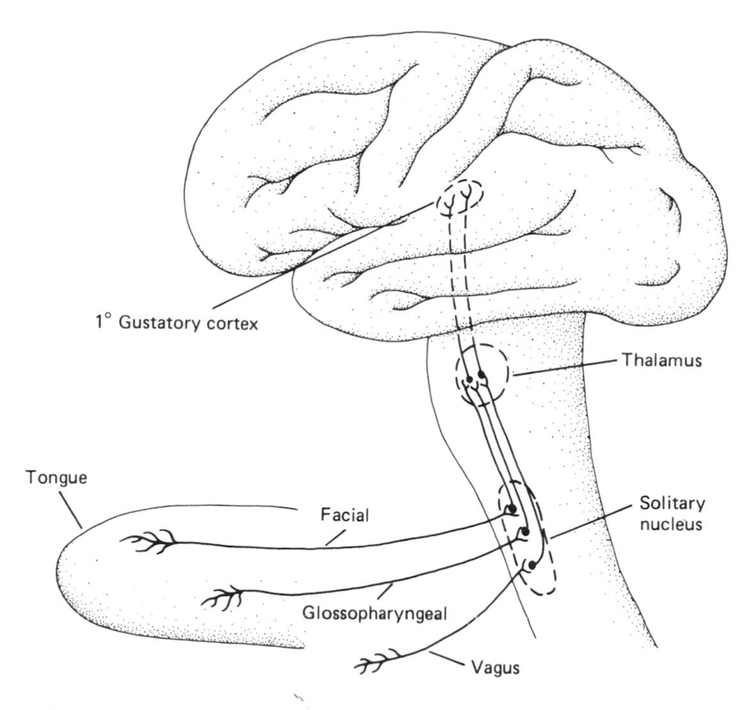

arising from receptors in the posterior third of the tongue. Fibers innervating taste buds in the pharynx run in the vagus nerve. Second-order gustatory fibers arise from the *solitary nucleus in the medulla* and run to an area of the thalamus immediately adjacent to the ventrobasal complex. Third-order fibers run from the thalamus to the postcentral gyrus of the cortex. They terminate in the *gustatory cortex,* immediately adjacent to the region of the somatosensory cortex, which latter is excited by somatesthetic stimulation of the tongue.

Olfactory System

Olfaction is the sense of smell. Olfactory receptors are located in the olfactory epithelium that lines the roof of the nasal cavity. The olfactory epithelium contains about 100 million *olfactory receptor cells* interspersed among sustentacular support cells (Fig. 6.25). The apical surface of each receptor cell contains several cilia, or olfactory hairs, projecting into a layer of mucus that covers the olfactory epithelium. The olfactory receptors are sensory neurons, and their axons form the olfactory nerve.

To be an effective olfactory stimulus, a chemical must be both volatile and water-soluble. Volatility permits the stimulant to be drawn up into the nasal cavity by sniffing. Water-solubility is required so that the stimulant can penetrate the mucous layer covering the receptor cells.

Olfactory stimulation is thought to involve an interaction between the stimulant molecule and specific receptor sites

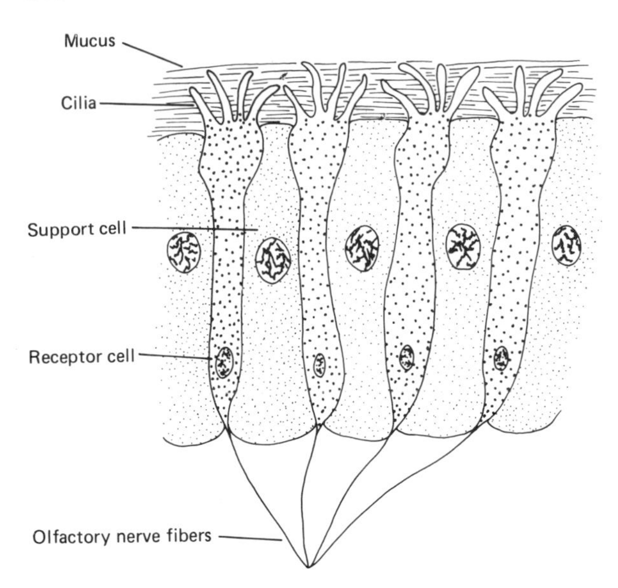

Fig. 6.25. Olfactory receptor.

on the ciliated receptor surface. Although the mechanism of this interaction is unknown, the receptor cell is depolarized to form a generator potential. The generator potential provides a source of depolarization that evokes the firing of action potentials in the olfactory nerve.

Evidence suggests that there are a number of primary olfactory sensations, perhaps 50 to 100. Many persons are odor-blind, or *anosmic*, to particular odors. More than 70 kinds of anosmia are described, suggesting that there are at least as many different olfactory receptors.

Fibers in the olfactory nerve terminate in the *olfactory bulb* (Fig. 6.26). Neurons from the olfactory bulb divide into lateral and medial tracts which project to various poorly de-

Fig. 6.26. Central olfactory pathways.

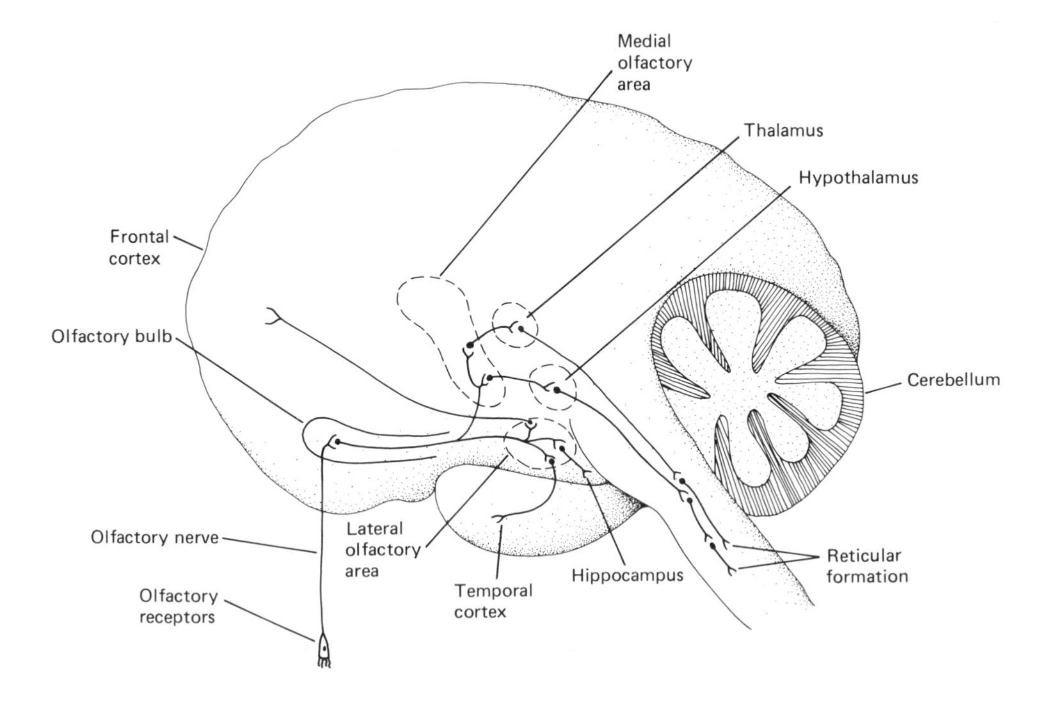

fined areas in the forebrain. They include the hippocampus and the frontal and temporal cortices. Olfactory information is also channeled to the thalamus, hypothalamus, and reticular formation. Most of the central structures receiving olfactory inputs are concerned with the integration of visceral and somatic behavior. These structures, which form the limbic system, regulate feeding behavior, sexual responses, and complex emotional patterns, such as fear and pleasure (see Chap. 10).

Selected Readings

Brindley GS (1970) Central pathways of vision. Ann Rev Physiol 32: 259

Daw NW (1973) Neurophysiology of color vision. Physiol Rev 53: 571

Eldredge DH, Miller JD (1971) Physiology of hearing. Ann Rev Physiol 33: 281

Eyzaguirre C, Fidone SJ (1975) Physiology of the nervous system, 2nd edn. Year Book Medical, Chicago

Goldberg JM, Fernandez C (1975) Vestibular mechanisms. Ann Rev Physiol 37: 129

Hodgson ES (1961) Taste receptors. Sci Amer 204: 135

Hubel DH (1963) The visual cortex of the brain. Sci Amer 209: 54

McIlwain JT (1972) Central vision: Visual cortex and superior colliculus. Ann Rev Physiol 34: 291

Moulton DG, Beidler LM (1967) Structure and function in the peripheral olfactory system. Physiol Rev 47: 1

Mountcastle VB (1974) Medical physiology, 13th edn. Mosby, St. Louis

Oakley B, Benjamin RM (1966) Neural mechanisms of taste. Physiol Rev 46: 173

von Bekesy G (1957) The ear. Sci Amer 197: 66

Wald G (1968) Molecular basis of visual excitation. Science 163: 230

Werblin FS (1973) The control of sensitivity in the retina. Sci Amer 228: 69

Witkovsky P (1971) Peripheral mechanisms of vision. Ann Rev Physiol 33: 257

Review Questions

1. Draw a cross section of the eyeball and label its component parts.
2. How is the size of the pupil regulated? Why is the control of pupil size essential to good vision?
3. Describe how the accommodation reflex controls the refractive index of the eye. Why is visual accommodation important for clear vision?
4. What is the fovea? Explain how it is associated with visual acuity.
5. Briefly compare and contrast the two kinds of photoreceptors. Consider their similarities and differences in morphology, photochemistry, and physiology.

6. Outline the sequence of steps in the synthesis and photodegradation of rhodopsin.
7. How is visual information encoded in the visual cortex? Consider the responses of simple, complex, and hypercomplex cells.
8. What are the two fundamental qualities of a sound? How are they measured?
9. Draw a diagram showing the arrangement of the bones of the middle ear. Describe their role in the transmission of the sound stimulus to the inner ear.
10. Make a cross section of the cochlear tube.
11. Briefly describe how the organ of Corti acts as an auditory transducer. Consider each of the steps in the transformation of sound pressure waves into cochlear nerve impulses.
12. Describe the mechanisms used by the inner ear to encode information about the pitch of the sound stimulus.
13. Draw a diagram of the vestibular apparatus.
14. Describe the morphologic and physiologic specializations that permit the saccule and utricle to detect both position and linear acceleration.
15. Diagram the structure of an ampulla. How is it specialized to respond to rotational acceleration of the head?
16. Describe the basis of the directional sensitivity of the vestibular hair cells.
17. Draw a diagram of a taste bud.
18. List the four taste sensations. Describe how each is correlated with chemical structure.
19. Describe the neural pathways that conduct sensory information from the taste buds to the cerebral cortex.
20. Draw a diagram of an olfactory receptor. How does it differ from a taste bud?
21. What chemical characteristic determines whether a particular substance will be an effective olfactory stimulus?

Chapter 7 Spinal Reflexes

A reflex is an invariant motor response, generated by the nervous system, to a specific sensory stimulus. Reflexes are the simplest responses of the nervous system and result from the sequential excitation of sensory, neural, and motor elements. These elements form the *reflex arc* (Fig. 7.1A). This is the neural pathway connecting the sensory receptor with the muscles that participate in the reflex response.

In the simplest reflex arcs, sensory neurons form their synaptic contacts directly on motor neurons (Fig. 7.1). These reflex arcs are *monosynaptic,* i.e., they contain only one central synaptic junction. Most reflex arcs are *polysynaptic,* including sensory neurons, interneurons, and motor neurons (Fig. 7.1B). The sensory neurons in polysynaptic reflex arcs do not end directly on motor neurons. Sensory excitation must be conducted through one or more layers of interneurons before affecting motor neuronal activity. Polysynaptic reflex arcs contain two or more central synaptic junctions.

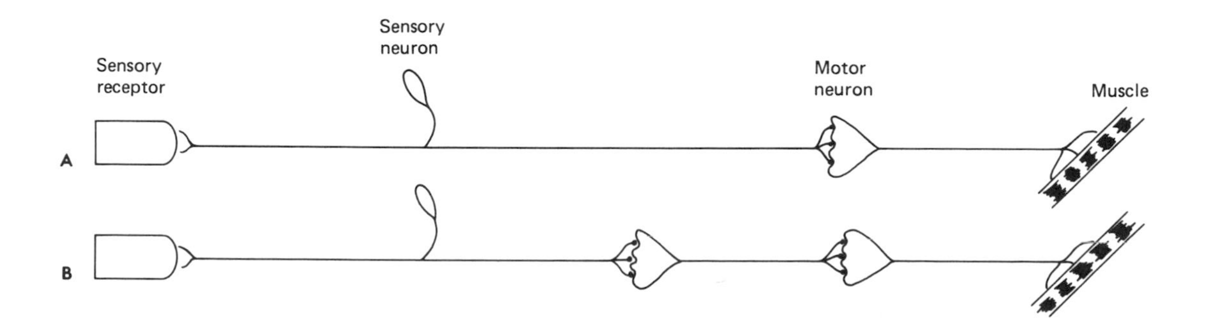

Fig. 7.1A and B. Reflex arc. **A** Monosynaptic reflex arc; sensory neuron has synaptic endings on motor neuron. **B** Polysynaptic reflex arc; one or more interneurons is interposed between sensory and motor neurons.

Types of Reflexes

Reflexes are found at many different levels of the nervous system. In the spinal cord, reflexes are important in the control of limb and trunk movement. They include reflexes regulating muscle length (stretch reflex), withdrawal from noxious stimulation (flexion reflex), and locomotion (crossed extension). Other reflexes, such as those enabling one to maintain upright posture and those regulating optical characteristics have their reflex arcs in the brain stem. Numerous visceral reflexes (sexual excitation, bladder emptying, and control of arterial pressure among others) are generated by reflex arcs in the autonomic nervous system.

Spinal Cord

Study of the reflex arcs of the spinal cord is the first step to understanding how the nervous system regulates motor activity. The motor capacity of the spinal cord can be illustrated by examining the responses in an animal whose spinal cord is surgically separated from its brain. After recovery from spinal "shock," *a spinal animal* can produce motor responses of surprising complexity. It will assume a standing posture when its feet come in contact with a firm surface, withdraw a limb from a noxious stimulus, scratch a point which is "tickled" on the surface of its limbs or trunk, and generate alternate walking movements of its limbs.

Spinal Organization

The motor capacity of the spinal cord is a result of the neural interconnections in the spinal gray matter. The spinal gray consists of sensory nerve terminals, interneurons, and motor neurons (Fig. 7.2). *Motor neurons* are located in the ventral horn, whereas interneurons are in both the dorsal horn and intermediate gray. *Sensory neurons* enter the spinal cord through the dorsal root and terminate on inter-

Fig. 7.2. Cross section of spinal cord. Contacts between sensory neurons, interneurons, and motorneurons in spinal gray matter.

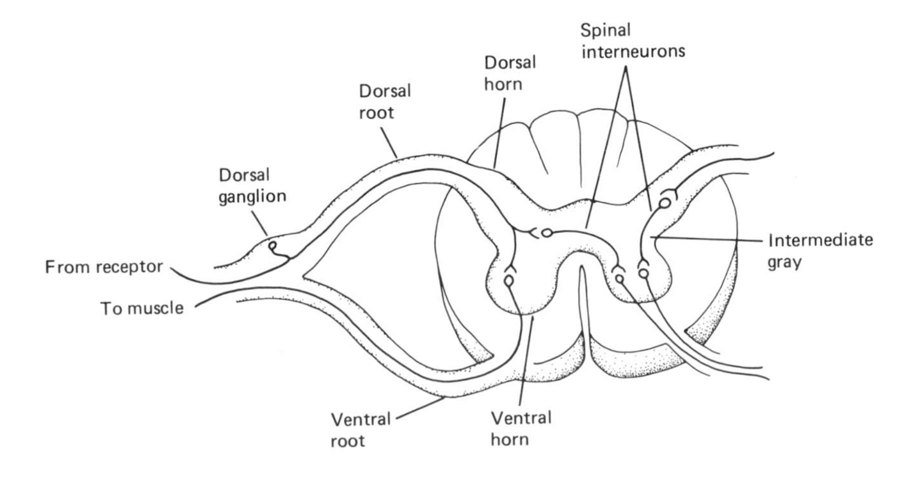

neurons in the dorsal horn and intermediate gray. The axons of some sensory neurons enter the spinal tracts and ascend or descend to other levels of the nervous system (see Chap. 5). Only one group of sensory fibers (Class IA), which arise from the muscle spindle receptors, have terminals that end directly on the motor neurons. *Spinal interneurons* receive synaptic contacts from sensory fibers, spinal interneurons, and ascending and descending fibers in the spinal tracts. Motor neuronal inputs come from spinal interneurons, descending and ascending fibers in the spinal tracts, and the Class IA sensory neurons.

A count of the neurons in one spinal segment of the dog (fifth lumbar) reveals that there are 2000 sensory neurons, 6000 motor neurons, and 360,000 interneurons. The true measure of spinal cord complexity, however, is not the neurons per segment but rather the number of synaptic contacts in that segment. The number is astronomical since the "typical" spinal interneuron receives 650 synaptic contacts and the "typical" spinal motor neuron has 5500 synaptic terminals ending on its surface.

Monosynaptic Stretch Reflex

The simplest spinal reflex is the monosynaptic stretch reflex. It can be evoked by applying a sudden stretch to a muscle. This is done by tapping a muscle tendon to provide stretch stimulation to the muscle. Muscle stretch stimulates stretch-sensitive receptors—the muscle spindles—located within the muscle. The sensory fibers innervating the muscle spindles enter the spinal cord through the dorsal root and terminate in the ventral horn. Here they form synaptic contacts on those motor neurons innervating the stretched muscle. Since there is only one central synaptic junction in its reflex arc (between the sensory neuron and the motor neuron), the stretch reflex is a monosynaptic reflex.

The central delay of the stretch reflex can be measured by placing recording electrodes on the ventral root to record

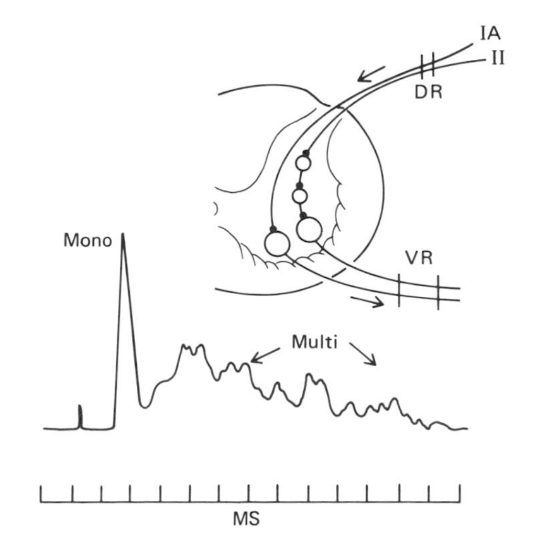

Fig. 7.3. Analysis of stretch reflex arc. Stimulation of sensory fibers in dorsal root (*DR*) gives rise to early and late responses recorded from ventral root (*VR*). Early response is monosynaptic; later response is multisynaptic. Stimulation of IA annulospiral fibers elicits monosynaptic response, whereas the II, III, and IV sensory fibers evoke multisynaptic response.

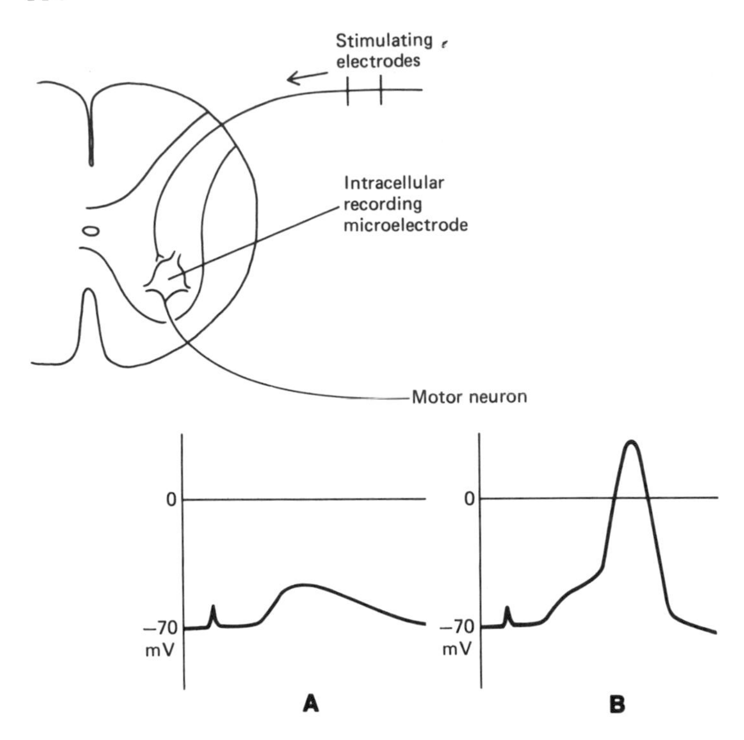

Fig. 7.4A and B. Microelectrode analysis of stretch reflex arc. **A** Weak IA stimulation produces an EPSP in motor neuron. **B** Increased stimulation increases EPSP size so that it exceeds motor neuron threshold and evokes an action potential.

the motor neuron discharge evoked by electrical stimulation of sensory neurons in the dorsal root (Fig. 7.3). Two components are observed in the responses recorded from the ventral root, namely, an early response with a short delay (about 1.5 ms) and a later response. If one adjusts the stimulus intensity so that only the IA fibers are stimulated (this can be done easily since the IA fibers are the largest sensory fibers), only the early response is seen in the ventral root recording. The 1.5 ms-delay for the early response is the sum of the *central synaptic delay* (0.5 ms) and the time required for conduction of an action potential in the sensory nerve and motor neuron (about 1.0 ms). The central synaptic delay is brief, just sufficient for an impulse to cross one central synaptic junction. Therefore some IA sensory neurons must end directly on the motor neurons which they excite. This conclusion can be verified by inserting a microelectrode into a ventral horn motor neuron (Fig. 7.4). Weak stimulation of IA fibers produces excitatory postsynaptic potentials (EPSPs) in the motor neuron. With increased stimulation (exciting more IA fibers) these EPSPs summate, becoming large enough to initiate an action potential in the motor neuron.

Sensory Receptors and Muscle Spindles

Muscle spindles are specialized striated-muscle fibers that are innervated by stretch-sensitive sensory endings (Fig. 7.5). These small, thin muscle fibers are the intrafusal fibers. Located within the main body of the muscle, they are in

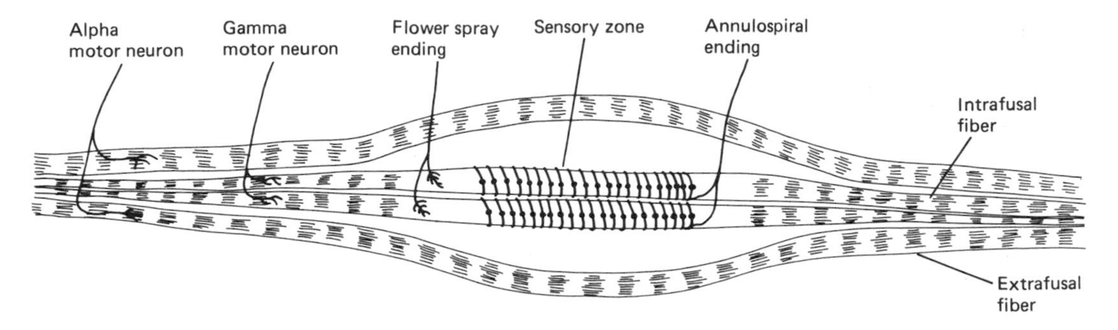

Fig. 7.5. Muscle spindle and innervation. Note that intrafusal muscle fibers are in parallel with larger extrafusal fibers.

contrast to the extrafusal fibers, which form the bulk of the muscle mass. Contractions of the intrafusal fibers are too weak to affect the entire muscle. The force generated by a contracting muscle is the result of contraction of the large extrafusal fibers. These two kinds of muscle fibers are innervated by different classes of motor neurons. Large alpha motor neurons innervate the extrafusal fibers, whereas, the intrafusal fibers receive their efferent innervation from small gamma motor neurons.

Each intrafusal fiber contains a noncontractile central sensory region in which are concentrated the nuclei of the fiber (Fig. 7.5). The sensory zone is innervated by two different sensory nerves: the annulospiral endings of the Class IA fibers (primary receptor endings) and the flower spray terminals of the Class II fibers (secondary receptor endings). The Class IA annulospiral endings are coiled around the nuclear region; the Class II fibers, which innervate the muscle spindle, form a series of flower-like terminal branches in the spindle.

Muscle-Spindle Stimulation

The primary and secondary sensory endings are excited by stretch of the central sensory zone of the intrafusal fiber. This zone can be stretched by either external forces (e.g., tapping a tendon) or contraction of the intrafusal fiber.

Contraction is evoked by the activity of the gamma motor neurons, which innervate the peripheral contractile zones of the fiber. The length of the intrafusal fiber is fixed because it is arranged in parallel with the much larger extrafusal fibers; it cannot shorten unless the extrafusal fibers contract. Therefore contraction of the contractile regions of the intrafusal fibers stretches the central sensory zone.

Excitation of the primary and secondary endings generates an increase in the discharge of impulses in the IA and II sensory fibers (Fig. 7.6). The Class IA fibers respond both to the rate of spindle stretch (dynamic response) and to the net change in spindle length (static response). Class II fibers are affected only by change in spindle length (static response).

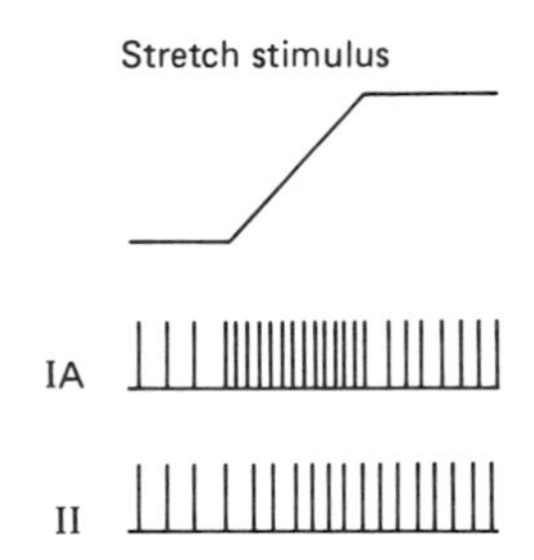

Fig. 7.6. Responses of muscle spindle afferents to stretch. Class IA fibers respond to rate of spindle stretch (dynamic response) and change in spindle length (static response); Class II fibers respond only to change in spindle length (static response).

Control of Muscle Length

The muscle spindle and its stretch reflex provide a mechanism for the regulation of muscle length. The primary function of the stretch reflex is to resist changes in muscle length imposed by forces external to the muscle. They include stretch of a muscle when (1) a load is being lifted, (2) the tendon is struck, or (3) an antagonistic muscle contracts (e.g., stretch of an extensor when a flexor at the same joint contracts). Because of their parallel mechanical arrangement, stretch of the muscle lengthens both extra- and intrafusal fibers. The resulting discharge of the primary and secondary sensory endings is conducted into the spinal cord where it evokes a reflex discharge in the alpha motor neurons innervating the extrafusal fibers of the stretched muscle. Contraction of the extrafusal fibers shortens the muscle, thereby compensating for the lengthening of the muscle by the externally imposed force.

Gamma Motor Control

The sensitivity of a muscle spindle to stretch is determined by the level of discharge in the gamma motor neurons innervating its intrafusal fibers. Increased gamma motor neuronal activity increases intrafusal contraction, thereby raising the level of tension in the intrafusal fiber and the stretch sensitivity of the muscle spindle. *Increased intrafusal tension* raises the level of sensory discharge, which elicits reflex discharge in the alpha motor neurons, thus increasing the contraction of the extrafusal fibers and shortening the length of the muscle. In similar fashion a decrease in gamma motor neuron discharge decreases intrafusal tension and muscle-spindle discharge, reducing alpha motor neuron activity, relaxing the extrafusal fibers, and lengthening the muscle.

Motor Neuron Coactivation

The nervous system is thought to control muscular contraction by modulating the discharge of both alpha and gamma motor neurons. Usually they are coactivated, i.e., increased alpha motor neuron discharge is accompanied by increased gamma motor neuron activity. *Coactivation* enables the muscle to contract without affecting the stretch sensitivity of the muscle spindle. In the absence of coactivation (only alpha motor neurons increase their discharge) the extrafusal fibers shorten, thus reducing intrafusal tension and decreasing the stretch sensitivity of the muscle spindle.

Tendon Reflex: Inhibition

When a muscle tendon is subjected to severe stretch resulting from, e.g., strong muscle contraction or external force,

Fig. 7.7. Reflex arc for tendon reflex.

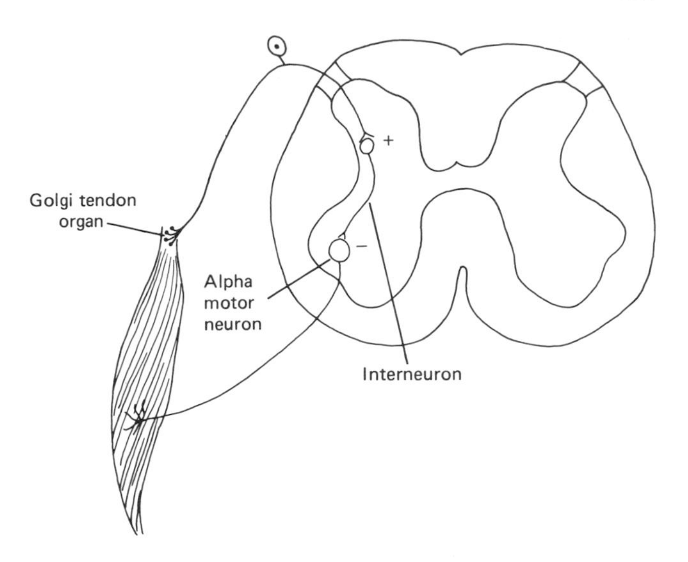

the muscle relaxes. The sensory receptors excited by severe muscle tendon stretch are the *Golgi tendon organs* (Fig. 7.7), strain-sensitive endings of Class IB sensory fibers present in the muscle tendon. Arranged in series with the extrafusal fibers, they are stimulated by an increase in muscle tension.

The Golgi tendon fibers enter the spinal cord and terminate on interneurons, which form inhibitory connections with the alpha motor neurons innervating the stretched muscle (Fig. 7.7). These *inhibitory interneurons* elicit inhibitory postsynaptic potentials (IPSPs) in the motor neurons. The IPSPs *inhibit the motor neuron* by blocking the initiation of action potentials at the axon hillock (Chap. 4).

The tendon reflex serves two purposes. By providing a pathway for rapid relaxation of a muscle, it *prevents muscle damage* during severe contractions. It also serves to *regulate muscle tension levels* much as the muscle spindle regulates muscle length. Whenever muscle tension increases so that the Golgi tendon organs are excited, the alpha motor neurons innervating the contracting muscle are inhibited, permitting both the muscle to relax and the level of tension to drop.

Flexion: Polysynaptic Reflex

The *flexion* or *withdrawal reflex* is a protective reflex. When a nociceptive or painful stimulus (e.g., pinching or contact with a very hot or cold object) is applied to the surface of a limb, the limb is rapidly flexed and withdrawn from the stimulus to prevent further tissue damage. Flexion results from the contraction of the flexor muscles and the relaxation of the extensor muscles, which span the flexed joint.

The strength of the response depends on the intensity of the stimulus. Increased intensity increases both the number of excited receptors and the level of excitation. A weak flexion can be elicited by applying tactile or light pressure stimuli to the skin. Intense nociceptive stimulation can have widespread effects. They include strong flexion of the stimulated, or ipsilateral, limb; extension of the opposite, or contralateral, limb (sometimes referred to as the *crossed extension reflex*) and spread up (or down) the spinal cord to the other pair of limbs—extending the ipsilateral and flexing the contralateral limb.

Whereas Class II touch and pressure-sensitive fibers can evoke a weak flexion response, strong flexion results only from stimulation of the smallest sensory fibers (Classes III and IV), which are sensitive to nociceptive stimuli (see Chap. 5). These sensory fibers terminate on a network of spinal interneurons that make synaptic contacts with the appropriate motor neurons (Fig. 7.8). No direct synaptic contacts exist between the sensory fibers and the motor neurons. The electrical stimulation of Classes II, III, and IV fibers in the dorsal root generates the late multisynaptic response observable in ventral root recordings (Fig. 7.3); there is no contribution to the early monosynaptic response, which results from IA stimulation.

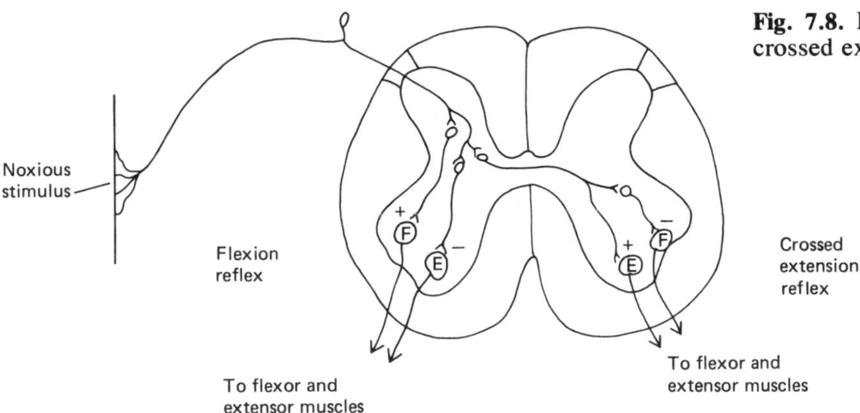

Fig. 7.8. Reflex arcs for flexion and crossed extension reflex.

Reciprocal Innervation

The network of interneurons excited by nociceptive sensory fibers is reciprocally wired so as to excite the ipsilateral flexor and to inhibit the ipsilateral extensor motor neurons (Fig. 7.8). If the intensity of the stimulus is sufficient, interneurons crossing the spinal cord are excited. They terminate on motor neurons in the contralateral ventral horn—exciting the extensors and inhibiting the flexors.

Reciprocal excitation and inhibition of antagonistic motor neurons are fundamental characteristics of the spinal wiring diagram. In general whenever motor neurons to a muscle are excited, the motor neurons innervating its antagonist are inhibited. The effect of reciprocal innervation is to prevent opposing sets of muscles from contracting simultaneously. Contraction of antagonistic sets of muscles would prevent any movement of a limb or joint.

Other Polysynaptic Reflexes

Stimulation of pressure receptors in the foot elicits the simultaneous contraction of extensor and flexor muscles thereby extending the limb. This is the *extensor thrust reflex,* important in regulating the rigidity of a limb so that it supports the weight of the standing body.

Light tactile stimulation of the skin elicits a *scratching reflex*. The response has two components, including (1) localization of the point of the body which was stimulated and (2) generation of a back and forth scratching movement.

If a spinal animal is supported in a sling, tactile stimulation of its foot can initiate *rhythmic alternate walking movements* in all four limbs. These movements result from alternate contraction and relaxation of the extensor and flexor muscles. The alternate excitation of one set of motor neurons and inhibition of its antagonists reflect the reciprocal innervation of motor neurons by the network of spinal interneurons. Since walking can be observed under conditions in which all sensory input is blocked, walking rhythm is generated by oscillations of spinal interneuronal activity.

Spinal Motor Neuron: Final Common Pathway

The spinal motor neuron forms the efferent path of all reflex arcs. All neural activity from the spinal cord and the brain must be channeled through the motor neurons to contract muscles and produce a behavior pattern. The level of impulse discharge in a motor neuron reflects the net level of synaptic excitation and inhibition impinging on it at the moment. It receives inputs from IA sensory fibers, excitatory and inhibitory spinal interneurons, and interneuronal fibers descending from the motor regions of the brain. The activity of a motor neuron is a result of the influences—excitatory and inhibitory—of all of its inputs. One particular input rarely dominates. For example, at the same moment that a motor neron in the hindleg is excited by input from muscle spindles, it may be inhibited by both cutaneous stimulation of the hindleg, and interneurons descending from the motor cortex and excited by noxious stimulation of the foreleg.

Selected Readings

Burke RE, Rudomin P (1977) Spinal neurons and synapses. In: Kandel ER (ed) Handbook of physiology: cellular biology of neurons. American Physiological Society, Bethesda, Maryland

Eyzaguirre C, Fidone SJ (1975) Physiology of the nervous system, 2nd edn. Year Book Medical, Chicago, Chaps. 13–17

Hunt CC, Perl ER (1960) Spinal reflex mechanisms concerned with skeletal muscle. Physiol Rev 40: 538

Merton PA (1972) How we control the contraction of our muscles. Sci Amer 226: 30
Mountcastle VB (1974) Medical physiology, 13th edn. Mosby, St. Louis, Chaps. 22–24
Pearson K (1976) The control of walking. Sci Amer 235: 72
Stein RB (1974) Peripheral control of movement. Physiol Rev 54: 215
Wilson VJ (1966) Inhibition in the central nervous system. Sci Amer 214: 102

Review Questions

1. What is a reflex arc? Draw diagrams of reflex arcs for the stretch reflex, flexion, and crossed extension reflexes.
2. What are the morphologic and functional differences between extra- and intrafusal muscle fibers?
3. Why is it important that alpha and gamma motor neurons be coactivated? What would result from activation of only alpha motor neurons? of only gamma motor neurons?
4. Briefly explain how one can prove that the stretch reflex is a monosynaptic reflex.
5. What is reciprocal innervation and why is it important?
6. Why is the alpha motor neuron called the final common pathway?

Chapter 8 Central Motor Control

Man is capable of performing a nearly infinite variety of movements, each of them generated by a specific pattern of motor neuron discharge. Only the simplest (e.g., withdrawal of a limb or scratching) can be generated by the isolated spinal cord. The motor repertoire of the spinal motor neurons and associated interneurons is limited to reflex responses (see Chap. 5). All other movements, ranging in complexity from coordinated walking to playing a Beethoven sonata, are generated by the central motor areas of the brain. The descending spinal tracts provide pathways for the central motor control of patterns of spinal motor neuron discharge. The central motor control systems include the motor cortex, which controls both pyramidal and extrapyramidal systems; the basal ganglia; and the cerebellum. Although at present many of the functions of these motor systems are not understood, each is thought to contribute in a specific manner to the generation of movement. The *motor cortex–pyramidal system* is associated with the voluntary performance of fine-skilled movements. Gross involuntary movements are generated by the *motor cortex–extrapyramidal system.* Both *basal ganglia* and *cerebellum* are concerned with coordination of movement. Coordination of slow, or ramp, movements is associated with the basal ganglia. Fast or ballistic movements are coordinated by the cerebellum.

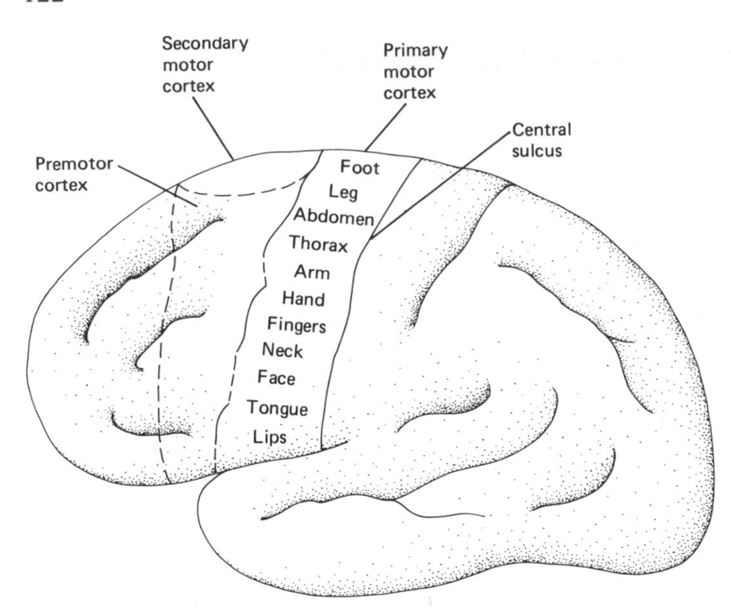

Fig. 8.1. Motor regions of cerebral cortex. Most of the secondary motor cortex is on medial surface of cortex out of plane of view in this figure.

Cortical Motor Organization

The motor regions of the cerebral cortex include primary motor, secondary motor, and premotor cortexes (Fig. 8.1). The *primary motor cortex* is immediately anterior to the central sulcus in the precentral gyrus. The premotor and secondary motor cortexes make up a broad band of tissue in front of the precentral gyrus. The *premotor cortex* occupies the entire lateral surface of this strip. The *secondary motor cortex* is confined to the medial surface, hidden from view in the intercerebral cleft.

Localized electrical stimulation of any point within these motor areas evokes a discrete reproducible motor effect. Each cortical region is characterized by the motor responses that are produced when electrically stimulated. The primary motor cortex controls the contraction of individual muscles. Stimulation of a point in the primary motor cortex always evokes a discrete muscle movement. The responses of the secondary motor cortex to stimulation are less discrete and localized; they include complex movements of head, neck, trunk, and limbs. The premotor cortex controls locomotory patterns, including those of mouth and tongue associated with word formation, coordinated movement of eyes and head, and fine movements of hands and fingers.

Primary Motor Cortex

The primary motor cortex is composed of an array of *vertical columns* of neurons. Each column controls either the excitation or inhibition of a single set of motor neurons which innervates a specific muscle. The columns are so arranged that excitatory and inhibitory columns affecting a particular set of motor neurons are adjacent to each other.

Many motor cortex neurons respond to the stimulation of somatesthetic receptors. Two major sources of sensory input to the motor cortex include: connections with the somatesthetic cortex and those from the reticular formation. Each column in the motor cortex receives inputs from those muscle proprioceptors and joint receptors affected by the motor activity that it generates. If, e.g., a column controls the contraction of a muscle flexing a joint, it receives inputs from both joint receptors and proprioceptors in the muscles spanning that joint. Those columns controlling movements of the hand also receive inputs from touch and pressure receptors in the hand.

Somatotopic Organization

The columns of the motor cortex are arranged in a somatotopic organization (Fig 8.1), an arrangement similar to that already described for the somatosensory cortex (see Chap. 5). A direct relationship exists between the position that a column occupies in the motor cortex and the location of the muscle which it controls. Columns controlling adjacent muscles are close to each other. The more anterior a muscle, the more ventrolateral the set of cortical columns that controls it. Stimulation of columns on the dorsomedial edge of the motor cortex affects muscles in the foot. Movements of the mouth and tongue can be elicited by stimulating columns at the ventrolateral border of the motor cortex.

The *size* of the cortical area controlling a set of muscles is proportional to the degree to which the muscles are involved in fine-motor control. Relatively large areas of motor cortex are involved in control of fingers, thumbs, lips, and tongue—all of which structures are capable of extremely fine movements. Only a small part of the motor cortex controls the postural muscles in the legs and back.

Fig. 8.2. Pyramidal or corticospinal tract.

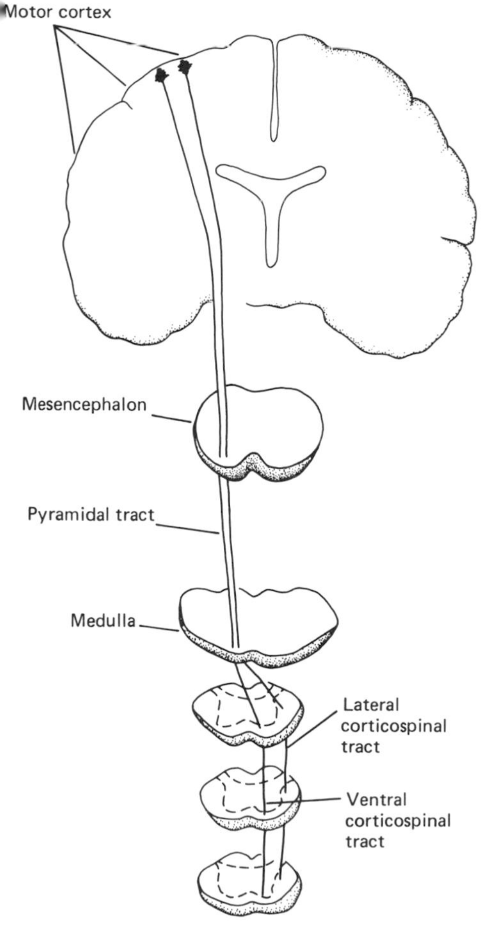

Pyramidal System

The pyramidal, or corticospinal, tract provides a *direct descending pathway* for the cortical control of spinal motor activity (Fig. 8.2). The tract runs from the motor cortex through the brain stem to the medulla where most of its fibers decussate, crossing to the contralateral side. The pyramidal tract then divides into lateral and ventral divisions before entering the spinal cord. The pyramidal fibers travel down the spinal cord in the *lateral and ventral corticospinal tracts* (Fig. 8.3). Although the vast majority of the fibers terminate on spinal interneurons, about 20% of the largest ones form direct synapses with spinal motor neurons.

All of the fibers in the pyramidal tract arise from either primary motor cortex (60%) or somatesthetic cortex (40%). Pyramidal tract neurons originating in the somatesthetic cor-

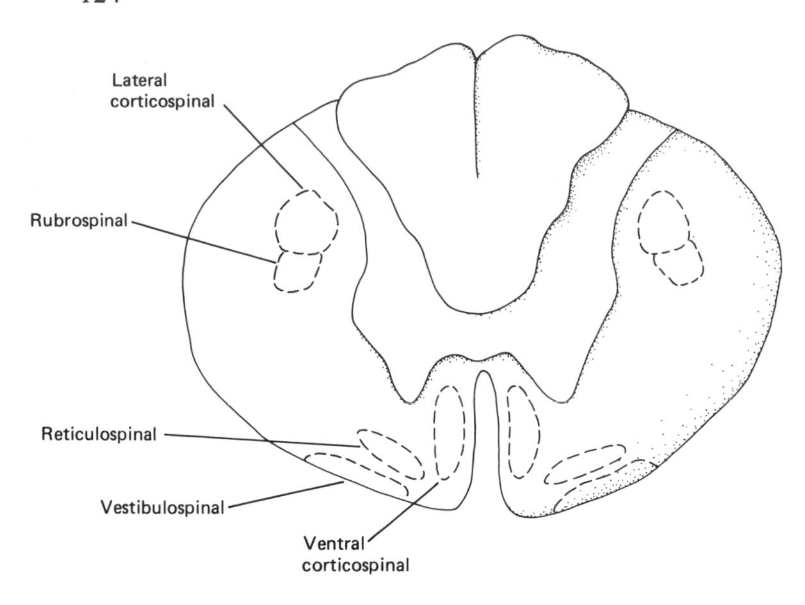

Fig. 8.3. Cross section of spinal cord showing principal descending spinal tracts of pyramidal (lateral and ventral corticospinal) and extrapyramidal (rubrospinal, reticulospinal, and vestibulospinal) systems.

tex provide a pathway for the cortical regulation of sensory activity in the spinal cord. They inhibit incoming sensory activity in the dorsal horn of the spinal cord.

Pyramidal Function

The principal function of the pyramidal system is to produce *fine-skilled movements,* e.g., playing a piano, threading a needle, jumping over a hurdle, or talking. These movements are believed to be initiated by activity in the neighboring premotor and secondary cortical motor regions. Once the concept for a movement is formed, the motor cortex generates the finely structured motor pattern required to perform the skilled movement. In the neural networks of the motor cortex, somatesthetic (touch, pressure, temperature, and pain) inputs from the reticular formation and from the somatesthetic cortex are correlated with information from basal ganglia and cerebellum to construct the appropriate motor pattern for producing a skilled movement.

Note that the pyramidal system is not required to produce most basic motor patterns, such as standing, walking, running, jumping, and eating. These activities are observed in monkeys whose pyramidal tracts have been surgically interrupted.

The pyramidal system is important in maintaining a tonic level of motor excitation, for section of the pyramidal tract produces a marked decrease in muscle tone. Most of the pyramidal activity descending to the spinal cord is excitatory. If this activity is interrupted, it depresses spinal motor activity.

Extrapyramidal System

The extrapyramidal system includes all motor nuclei and tracts—other than the pyramidal tract—over which corti-

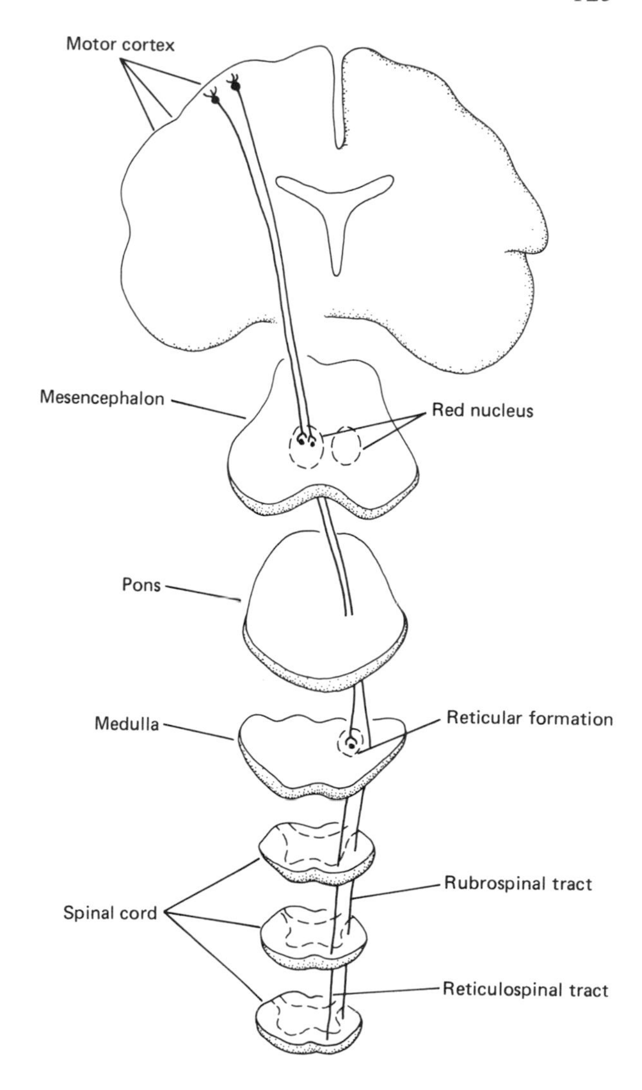

Fig. 8.4. Extrapyramidal system. Rubrospinal and reticulospinal tracts form part of multineuronal extrapyramidal pathway extending from cerebral cortex to spinal cord.

cal motor activity is conducted to the spinal cord. The extrapyramidal system is not a single discrete motor pathway like the pyramidal system. It is rather a complex network of motor nuclei and interconnecting tracts over which motor signals pass from the cerebral cortex to the motor neurons in the spinal cord (Fig. 8.4). It includes tracts connecting the cortex to the brain-stem motor nuclei—the three most important being the *red nucleus,* the *vestibular nucleus,* and the *reticular formation* (Figs. 8.4 and 8.5).

These intermediate nuclei give rise to a series of *descending spinal tracts*—rubrospinal, reticulospinal, and vestibulospinal—terminating on spinal interneurons and motor neurons (Fig. 8.3). Some neuroanatomists include the basal ganglia among the extrapyramidal nuclei. Present evidence suggests that the basal ganglia, like the cerebellum, are concerned with the coordination of both pyramidal and extrapyramidal movements.

Fibers running in the extrapyramidal tracts originate from almost every region of the cerebral cortex. Whereas most of them are axons of neurons located in either the motor cor-

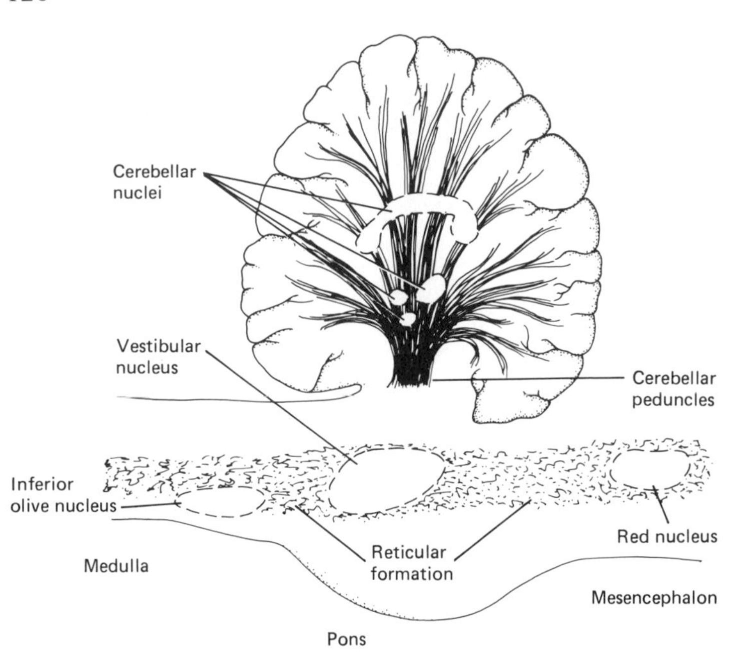

Fig. 8.5. Sagittal section of brain stem and cerebellum, showing organization of cerebellum and locations of major motor centers in brain stem.

texes (primary, secondary, and premotor) or the somatesthetic cortexes, many originate either in the other sensory areas of the cortex (auditory, visual, and gustatory) or in the higher associative areas of the frontal, parietal, and temporal cortexes.

Brain-Stem Motor Nuclei

The intermediate motor nuclei are concerned with postural control and the maintenance of an upright position. They are located in the core of the brain stem—extending from the medulla to the central thalamus (Fig. 8.5). Although some of their neurons are condensed into specific nuclei, most of them form part of the reticular formation—a diffuse network of interneurons extending throughout the core of the brain stem. The extrapyramidal neurons of cortex, basal ganglia, and cerebellum provide inputs to these nuclei (Fig. 8.6). These inputs from the higher motor centers are integrated within the reticular formation and associated nuclei with somatesthetic information from the spinothalamic tracts and data from the vestibular system to produce the motor responses required to maintain upright posture.

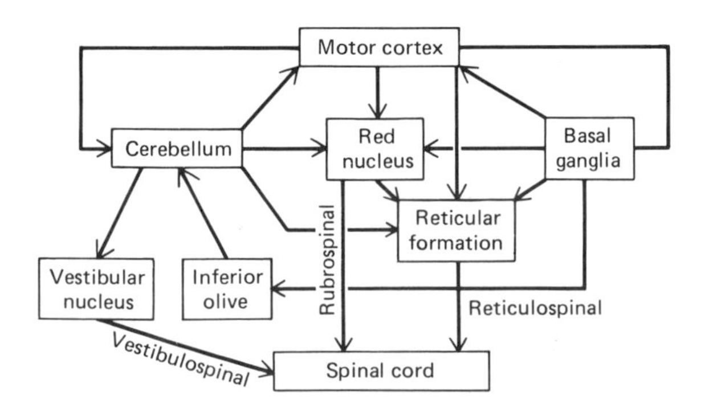

Fig. 8.6. Major interconnections between basal ganglia, cerebellum, brain-stem motor nuclei, and motor cortex.

To maintain an upright stance, gravity must be opposed by the contraction of the extensor muscles. The anterior two-thirds of the motor core of the brain stem is a source of intense facilitory activity to the extensor motor system. This *extensor motor drive* is normally modulated by inhibitory influences descending from the higher motor centers in cortex and basal ganglia. The intensity of this drive is revealed in an animal whose upper brain stem is sectioned to isolate the cerebral cortex and basal ganglia from the motor nuclei of the brain stem. Decerebration produces a specimen whose extensor motor sytem is hyperexcited. These animals characteristically show extensor rigidity, resulting from the intense excitation of the extensor musculature.

Extrapyramidal Function

The extrapyramidal system is concerned with the generation of *posture and locomotion,* such as walking, standing, jumping, running, and swimming. In the intact animal the formation of these motor patterns is initiated by the activity of cortical neurons terminating in the brain-stem motor nuclei. The intermediate motor nuclei integrate information about timing and coordination of movement (from cerebellum and basal ganglia) with inputs reflecting the sensory state of the body (from the reticular formation) to generate appropriate motor activity.
Cortical activity descending in the extrapyramidal pathways to the brain-stem motor nuclei is primarily inhibitory. This *descending inhibition* is important to the modulation of the brain-stem motor nuclei, since the output generated by these extrapyramidal nuclei strongly excites the spinal motor centers. In the absence of this cortical inhibition—as in an animal whose upper brain stem is sectioned—movement degenerates into a series of spastic muscle contractions.

Cerebellum

The cerebellum, formed by an extension of the dorsal surface of the brain stem, is connected to the latter by large tracts of nerve fibers, namely, the cerebellar peduncles (Fig. 8.5). The cerebellum consists of two parts, including (1) a large complexly folded cerebellar cortex overhanging most of the dorsal surface of the brain stem and (2) the small cerebellar nuclei, located deep within the peduncles.
Concerned with the coordination and timing of movement, the cerebellum is important as a *motor comparator*. Whenever a movement is initiated by the motor cortex, signals are sent to the cerebellum about the nature and desired result of the impending movement. The cerebellum stores and compares this information with the sensory inputs that it re-

ceives from the proprioceptors and other sensory systems affected by the movement. If during the movement the cerebellum receives sensory inputs indicating that the movement is awry, it furnishes the brain stem and cerebral cortex motor centers with information allowing them to make the necessary motor corrections.

The cerebellum is particularly important in the formation and execution of *ballistic movements*. These are movements whose rate of progression is too rapid to permit corrections during the action, such as rapidly playing a piano, throwing a ball at a target, leaping a hurdle and speaking. Correction during the movement is not possible because the time needed to (1) transmit sensory information to the cerebellum, (2) analyze the information, and (3) generate remedial action is much longer than the duration of the motor activity that produces the movement. Hence, ballistic movements must be *preprogrammed*. The cerebellum is critical to the programming because it stores information, both sensory and motor, which enables the pyramidal and extrapyramidal systems to choose the motor pattern that successfully generates a desired ballistic movement.

Input

Most fibers transmitting activity to the cerebellum terminate in both the cerebellar nuclei and cortex (Fig. 8.7). These include important proprioceptive inputs from spinal cord, vestibular system, and inferior olive nucleus. Other sensory inputs originate in the superior (visual) and inferior (auditory) colliculi. The cerebellum also receives major inputs from both motor and somatesthetic regions of the cerebral cortex.

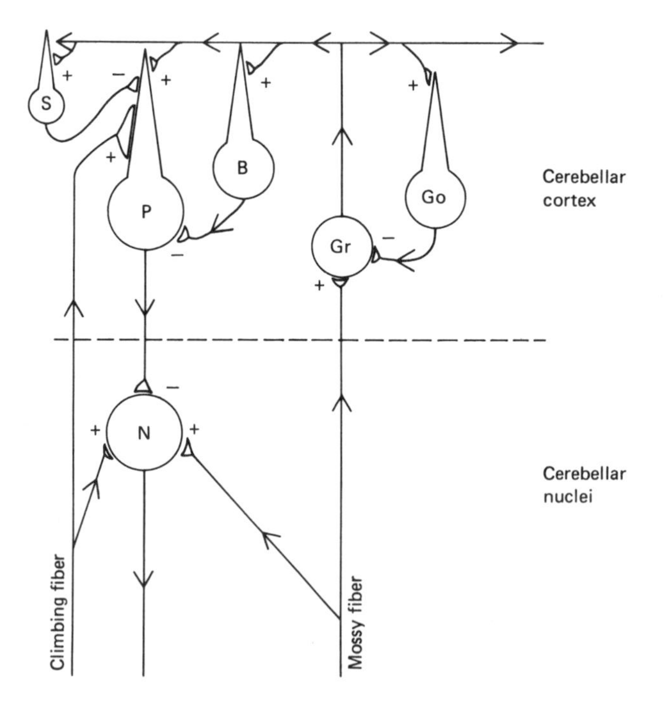

Fig. 8.7. Cerebellar circuit. Cerebellar cortex and nuclei receive excitatory inputs from climbing and mossy fibers. The only output from cerebellar cortex is Purkinje neuron (*P*), which inhibits discharge of neurons in cerebellar nuclei (*N*). Neurons intrinsic to cerebellar cortex include excitatory granule (*Gr*) and inhibitory basket (*B*), Golgi (*Go*), and stellate (*S*). Arrows indicate direction of action potential conduction. Synapses are either excitatory (+) or inhibitory (−).

These inputs reach the cerebellum over two different classes of fibers, namely, climbing and mossy (Fig. 8.7). The *climbing fibers,* affected by diverse proprioceptive stimuli, arise in the inferior olive nucleus. Their mode of activation is at present ill understood, although they are thought to be excited by activity generated in the cortical motor centers. *Mossy fibers* transmit most of the sensory input that reaches the cerebellum—proprioceptive, tactile, vestibular, and audio, visual—as well as inputs from the cerebral cortex. Mossy fibers supply the cerebellum with information about the body's general sensorimotor status.

Output

The four cerebellar nuclei (dentate, globose, emboliform, and fastigial) transmit cerebellar output. Neuronal discharge in the cerebellar nuclei is transmitted both to motor centers in the brain stem (red nucleus, reticular formation, and vestibular nucleus) and through the ventrolateral thalamus to the motor and somatesthetic areas in the cerebral cortex (Fig. 8.6).

The total output of the cerebellar cortex is transmitted to the cerebellar nuclei by the *Purkinje neurons*. These large inhibitory interneurons form the middle layer of the cerebellar cortex. Thus the final result of all activity in the cerebellar cortex is the inhibitory modulation of neurons in the cerebellar nuclei.

Function

The cerebellar cortex has a stereotyped structure. It is formed by five kinds of neurons, including Purkinje, granule, basket, Golgi, and stellate (Fig. 8.7). The *Purkinje neurons,* the principal integrating neurons, receive excitatory inputs from both climbing and mossy fibers. Each neuron is innervated by a single climbing fiber with which it forms large excitatory synapses. A single impulse in a climbing fiber excites the neuron to discharge. In contrast, mossy fiber excitation of Purkinje neurons is both indirect and weak. Mossy fibers do not terminate directly on the neuron. Rather they excite an intermediate neuron, the *granule neuron,* which forms small excitatory synapses on the dendrites of the Purkinje neuron. The inhibitory *basket, Golgi, and stellate neurons* modulate granule neuron excitation of the Purkinje neuron.

The cerebellar cortex is thought to function as a memory core, storing information about the proprioceptive and general sensorimotor states in a form available to both pyramidal and extrapyramidal systems so as to coordinate movement. The storage unit is the Purkinje neuron. It stores information, as small EPSPs, about the activity of its mossy fiber input. The climbing fiber is the readout device.

As already mentioned, a single impulse in a climbing fiber is always sufficient to elicit a response from the Purkinje neuron which it innervates. However, the intensity of the response—in both number of action potentials and rate of discharge—depends on the current level of Purkinje neuron excitability. If, on the one hand, the mossy fiber inputs to a Purkinje neuron are active, the neuron responds to a climbing fiber impulse with an intense discharge. On the other hand if mossy fiber activity is minimal, the neuron responds to climbing fiber activity with a single impulse.
There are 15 million Purkinje neurons in the human cerebellar cortex, each of which receives inputs from a somewhat different population of mossy fibers. Therefore each neuron serves as a unit for storing a particular bit of information about the sensorimotor conditions which are important in the coordination of a movement, e.g., the position of a joint, degree of contraction of a muscle, position of an image in the visual field, and the motor pattern that was just formed in the motor cortex. Through the inhibitory modulation of activity in the cerebellar nuclei, this information is relayed both to motor regions of cerebral cortex and to motor nuclei in the brain stem.

Cerebellar Lesions

Lesions of the cerebellum produce several characteristic motor effects, including intention tremor, dysmetria, and ataxia. The study of the motor effects of these lesions is a major source of insight into cerebellar function.

Intention Tremor

One function of the cerebellum is the coordination of the motor activities that brake or damp movements. Every successful movement involves the sequential activation of two sets of muscles—those moving the limb to the desired point in space and those stopping the movement when that point is reached. Injury to the cerebellum impairs or destroys smooth braking. In a subject with cerebellar lesions, movements overshoot and undershoot their goals. In trying to correct an overshoot, the subject, usually overcompensates and undershoots. The result is an intention tremor. To halt the movement, the subject puts the limb through a series of successive overshoots and undershoots of decreasing amplitude before the limb finally attains a constant position.

Ataxia

A second important function of the cerebellum is the coordination of movements requiring the sequential contraction of many muscles. Ataxia is the inability to perform a coordinated sequence of movements. In a subject with cerebellar lesions the timing of the muscular contractions is distorted; the result is that the individual is incapable of performing movements requiring precise sequences of muscular contractions. An ataxic person has extreme difficulty in performing many of the common motor activities, such as running, writing, and speaking.

Dysmetria

To perform rapid or ballistic movements successfully, the central motor system must be able to predict both course and duration of the movement produced by a particular motor pattern. This predictive capacity resides in the cerebellum. Lesions therein cause dysmetria, loss of the ability to predict the consequences of movement. The person overshoots when reaching for an object or when stepping over a barrier. Like the other cerebellar disorders just described, dysmetria has far-reaching effects on the subject's ability to generate coordinated movements. The effects of cerebellar lesions on ballistic movements are especially severe.

Basal Ganglia

The basal ganglia include the *caudate nucleus,* the *putamen,* and the *globus pallidus,* all of which are located beneath the cerebral cortex deep in the telencephalon (Fig. 8.8). Functionally they are closely related to several motor nuclei in the brain stem—the subthalamus, the substantia nigra, and the red nucleus. All of these structures are interconnected by a complex neuronal network to form the basal ganglia system.

The basal ganglia receive major inputs from both the motor regions of the cortex and cerebellum. The output of the basal ganglia is transmitted to motor cortex, cerebellum,

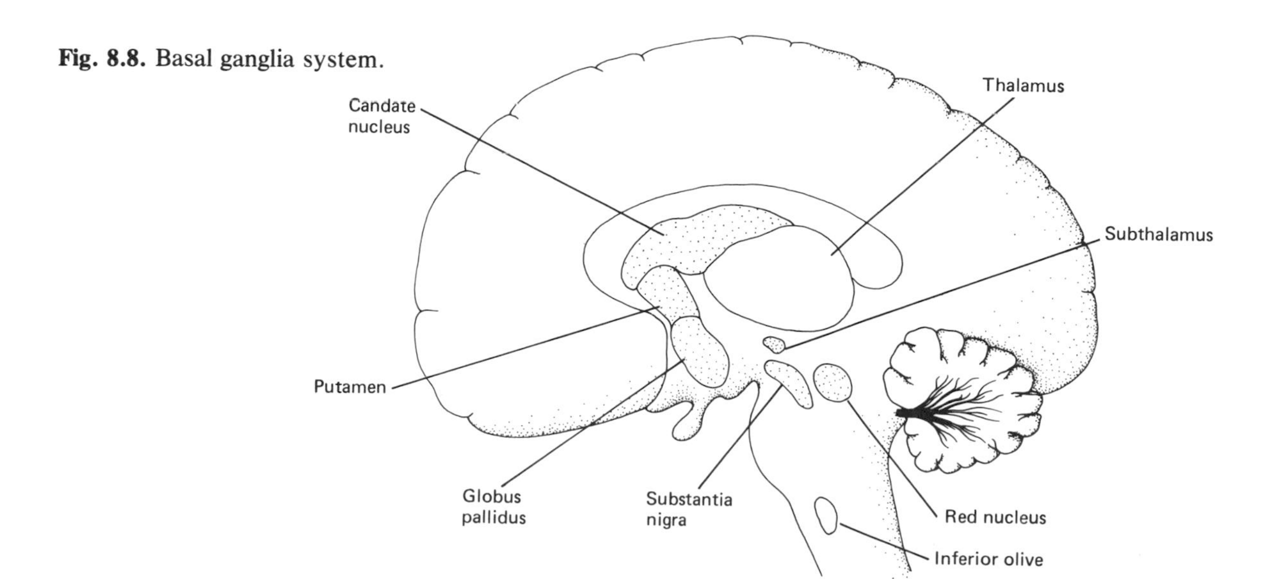

Fig. 8.8. Basal ganglia system.

and reticular formation (Fig. 8.6). Note that these interconnections form two important neural loops: one connecting with the motor cortex, the other with the cerebellum.

Role in Motor Control

The motor functions of the basal ganglia are at present not understood. They are thought to be involved in the generation of complex stereotyped movements like walking, eating, and maintaining posture. A current hypothesis suggests that the basal ganglia are concerned with the *generation of ramp* or *slow movements*. These are movements generated at a sufficiently slow rate to be subject to modification by sensory feedback from receptors that are excited during the movement. Examples of ramp movements are slow walking, stepping over a barrier, and threading a needle. When sensory feedback indicates that a particular movement is either overshooting or undershooting its goal, the basal ganglia transmit corrective signals to the motor regions of cortex and brain stem. Because of the relatively slow progress characteristic of ramp movements, the pyramidal and extrapyramidal systems can correct the ongoing motor pattern during the movement.

Lesions

Ironically the major source of definitive information on the role played by the basal ganglia in motor control comes from the disturbances in motor function produced by basal ganglia lesions. These disorders include athetosis, chorea, ballismus, and Parkinson's disease.

Athetosis—produced by lesions in the caudate nucleus and putamen and characterized by slow, writhing, involuntary movements of the limbs—involves continuous alternate jerky contractions of a limb's flexors and extensors. Like all other disorders of the basal ganglia, the subject cannot hold his or her limbs in a constant, maintained position.

Chorea is characterized by continuous involuntary jerky movements of the limbs and head and is associated with lesions in the caudate nucleus.

Lesions in the subthalamus produce violent uncontrolled movements of the limbs, the continuous generation of which is characteristic of *ballismus*.

In *Parkinson's disease* the dominant feature is a continuous low-frequency tremor accompanied by muscular rigidity. Because of the rigidity the subject has difficulty in initiating movement. The continuous involuntary tremor prevents the successful execution of ramp movements. Parkinson's disease is associated with lesions of the *substantia nigra*. Present evidence suggests that in the normal subject the substantia nigra inhibits the caudate nucleus, putamen, and globus pallidus. Destruction of the substantia nigra removes this inhibitory control loop. The result is an increase in excitatory output of the basal ganglia to the motor cortex and reticular formation and the characteristic rigidity and tremor of Parkinson's disease.

Selected Readings

Asanuma H (1975) Recent developments in the study of the columnar organization of neurons within the cerebral cortex. Physiol Rev 55: 143

Brooks VB, Stoney SD (1971) Motor mechanisms: the role of the pyramidal system in motor control. Ann Rev Physiol 33: 337

Eccles JC (1977) The understanding of the brain. McGraw-Hill, New York, Chap. 4

Evarts EV (1973) Brain mechanisms in movements. Sci Amer 230: 96

Eyzaguirre C, Fidone SJ (1975) Physiology of the nervous system, 2nd edn. Year Book Medical, Chicago, Chaps. 18–21

Kornhuber HH (1974) Cerebral cortex, cerebellum and basal ganglia: An introduction to their motor function. In: Schmitt FO and Worden FG (eds) The neurosciences III. MIT Press, Cambridge

Llinas RR (1975) The cortex of the cerebellum. Sci Amer 232: 56

Mountcastle VB (1974) Medical physiology. 13th edn. Mosby, St. Louis, Chaps. 26, 28, and 29

Review Questions

1. Briefly describe the organization of the columns in the primary motor cortex.
2. What are the major differences in the anatomic organization of the pyramidal and extrapyramidal systems?
3. Compare the functional roles of the pyramidal and extrapyramidal systems in motor control.
4. What is a ballistic movement and how does it differ from a ramp movement?
5. Construct a diagram showing the neural circuit in the cerebellum; label each of the neural elements.
6. Using the diagram from question 5, describe how the cerebellum is hypothesized to function as a motor comparator. Indicate the mechanisms proposed for both storage and readout of information.
7. Compare the motor disorders related to cerebellar and basal-ganglia lesions. Does this information support the hypothesis which associates the basal ganglia with ramp movements and the cerebellum with ballistic movements?

Higher Functions of Cerebral Cortex

Chapter 9

The highest functions of the nervous system—the power to reason and construct ideas, to be aware of one's sensory environment, to create abstract thoughts, and to store experiences—are to a large extent centered in the cerebral cortex. The total awareness of the relationship between oneself and the surrounding environment depends on excitation of the neural networks in the cerebral cortex. It is the site of both consciousness and intellect.

Organization

The cerebral cortex contains about 10 billion neurons of which most (about 90%) are arranged into six layers to form the neocortex. The *neocortex* is the highest integrative center in the somatic nervous system. It is concerned with both the processing and interpretation of sensory information (auditory, gustatory, somatesthetic, and visual) (and the control of complex movements generated by the skeletal muscles. It contains centers involved in the abstraction of ideas, speech, and the storage of experiences. Most of its functions are under voluntary control.

The *allocortex* forms the second major subdivision of the cerebral cortex. In contrast to the neocortex it has a simpler three-layered neural structure and is under involuntary control. It includes the cortical structures associated with the limbic system (see Chap. 10). It contains the highest centers of visceral control.

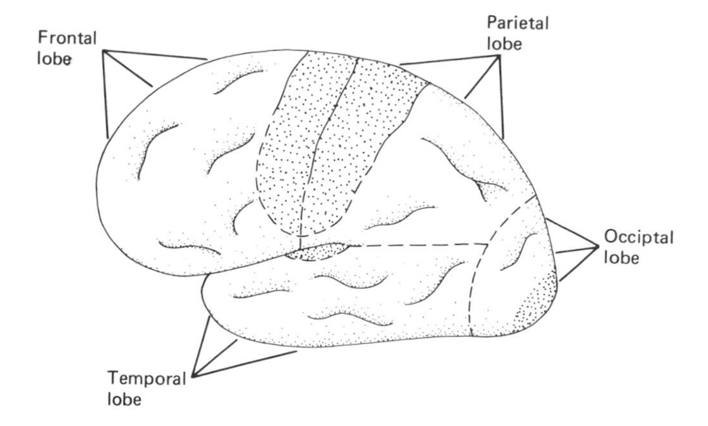

Fig. 9.1. Cerebral cortex is divided into specific (stippled) and association (nonstippled) regions. Specific regions include the motor area in frontal lobe, somatesthetic area in parietal lobe, visual area in occipital lobe, and auditory area in temporal lobe.

The neocortex is divided into four lobes: frontal, parietal, occipital, and temporal (Fig. 9.1). Each lobe is subdivided into specific and associative regions. The *specific regions of the cortex* receive their principal inputs from either the specific sensory nuclei of the thalamus—lateral geniculate (visual), medial geniculate (auditory), and ventrobasal complex (somatesthetic and gustatory)—or the specific motor nuclei—the anterior nuclei of the thalamus. The principal source of inputs to the associative regions of the cortex is from other cortical regions—specific and associative. Both specific and associative cortical areas also receive inputs from the nonspecific thalamic nuclei. These nonspecific inputs determine the level of cortical arousal (see this chapter, "Cortical Arousal").

Neocortex

The functions of the specific regions of the neocortex are stereotyped. In neurosurgical procedures electrical stimulation of a point in the sensory cortex produces a *specific sensory sensation,* such as (1) touch or temperature stimulation of the skin (somatosensory cortex), (2) specific taste (gustatory cortex), (3) flashing bars or spots of light (visual cortex), or (4) simple sounds and noises (auditory cortex). The specific sensory cortex determines the general characteristics of the stimulus (e.g., color, orientation and position of a visual image, or tonal pitch and duration). The motor cortex controls the generation of patterns of muscle contraction. The functional organization of these regions of the cerebral cortex is examined in Chaps. 5 (somatosensory), 6 (auditory, gustatory, and visual), and 8 (motor).

Association Cortex

The association cortex is divided into three regions: frontal, tempiral, and parietooccipital. Information about its functions is from two major sources. The first is obtained during some neurosurgical procedures, in which areas in the association cortex are stimulated electrically. Careful observations of the effects of this stimulation are a major source for unraveling the complex cortical functions. The second source is from the analysis of the behavioral deficits produced in patients with lesions inflicted by strokes or wounds to the association cortex.

The parietooccipital region is believed to be concerned with the interpretation of ongoing somatesthetic, gustatory, and visual activity. The physical characteristics (e.g., texture, weight, size, shape, and color) of the stimulus are correlated to develop a detailed image of the stimulus. Lesions in the parietooccipital cortex are characterized by the patient's inability to identify objects from somatesthetic and visual clues. He may be aware, e.g., that an object is orange and round and has an acidic sweet taste and still be unable to identify it as an orange.

Lesions in the region where parietal, occipital, and temporal lobes meet produce *word blindness*. Although the person can identify the letters forming a word, he or she is unable to explain its meaning. Other word-blind persons can identify each of the words in a sentence and yet cannot understand the meaning of the whole.

Patients with lesions in the posterior region of the temporal lobe are *word-deaf*. Although they can readily decipher the meaning of a written sentence, if the same sentence is read to them, they are unable to explain its meaning.

Large areas of *the temporal lobe* are thought to be concerned with the *storage of long-term memories* (see this chapter, "Memory"). Stimulation, especially of the posterior region of the lobe often elicits recall of complex memory patterns. They are usually associated with a specific moment in ones life, e.g., playing a game as a child, attending a concert, or talking with a specific person. The patterns are vivid and usually involve somatesthetic, audiovisual, and gustatory sensation surrounding the remembered event. A memory of a concert, e.g., may encompass the sounds of the music being played, the visual image of the orchestra on the stage, and the temperature of the hall.

The frontal cortex contains major centers for the control of *speaking and writing*. Lesions to the lateral posterior region of the frontal association cortex are correlated with the loss of linguistic ability, i.e., the patient cannot express ideas and thoughts in either written or oral form.

The most anterior part of the frontal lobe forms the prefrontal cortex, it is concerned with the regulation of *personality, creativity, and drive*. Lesions produce drastic changes in personality, attentiveness, and interest. Many of the normal social inhibitions are lost; the person loses interest in his appearance and his job and becomes uncommunicative.

Cerebral Dominance

Many of the higher functions of the association cortex are confined to a single hemisphere. The two hemispheres differ in their capacities to support the highest neural functions. The *categorical hemisphere* (usually the left) pertains to *interpretation and production of language*—written and spoken. Language centers are confined to the frontal and temporal lobes of the categorical hemisphere. The other hemisphere is concerned with *spatial construction and temporal relationships*. Called the *representational hemisphere,* it also contains centers for musical and graphical expression.

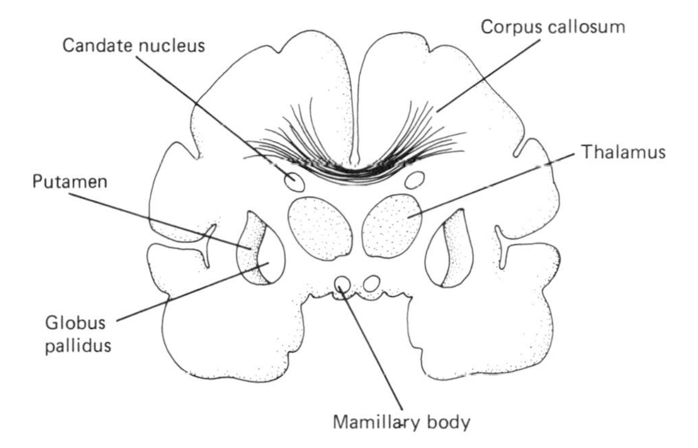

Fig. 9.2. Cross section through cerebral cortex. The two cerebral hemispheres are connected by large bands of white matter forming the corpus callosum.

Coordination of the activities of these two hemispheres is the primary function of the large bands of white matter—the *corpus callosum*—which interconnect the two hemispheres (Fig. 9.2). In a few instances the corpus callosum is surgically interrupted to prevent the spread of extreme epileptic seizures. The overall behavior of these "split-brain" patients appears to be normal on casual observations. However, careful testing shows that the two hemispheres function independently. Speech and writing are confined to the categorical hemisphere. The ability to recognize objects on the basis of visual and somatesthetic information is restricted to the representational hemisphere.

Activation of Cerebral Cortex: Consciousness

Consciousness is a state of attention to one's physical and mental environment. A reflection of the activity of millions of neurons in the cerebral cortex, consciousness depends on the level of excitation of these cortical neurons. Excitation results from a continuous barrage of impulses originating in the reticular formation of the brain stem; they are conducted to the cortex over a diffuse neural network—the reticular activating system. The level of consciousness is closely correlated with the level of excitation of this system.

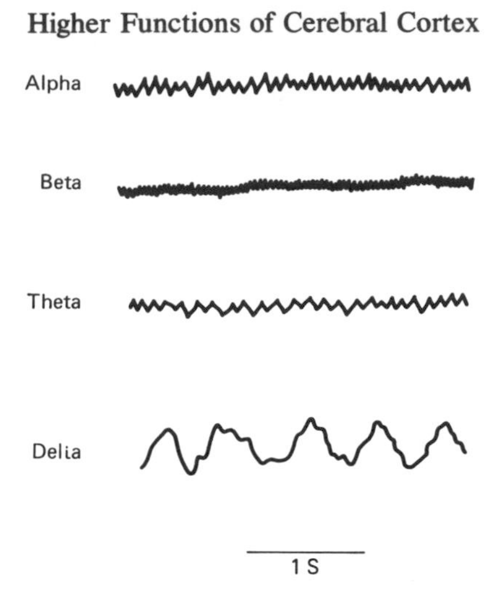

Fig. 9.3. Typical recordings of the four EEG rhythms.

Electroencephalogram

The level of cortical arousal, or consciousness, is usually measured by recording an electroencephalogram (EEG). An EEG reflects fluctuations in brain potential taken through the skull by attaching electrodes to the scalp. An EEG can also be obtained from the surface of the cerebral cortex during neurosurgical procedures or in experiments with animals.

The degree of cortical arousal is correlated with the frequency of the EEG recording. Traditionally EEG recordings obtained during the transition from active consciousness to unconsciousness or sleep are grouped into alpha, beta, theta, and gamma rhythms (Fig. 9.3).

The alpha rhythm is characteristic of the normal, awake individual who is resting. It is particularly prominent in EEG recordings made when the subject's eyes are closed. The frequency of the alpha rhythm varies from 8 to 14 waves/s. If a person at rest is suddenly exposed to a vivid sensory stimulus, e.g., turning on the lights in a dark room, a sudden sound, or tactile stimulation of the skin, the EEG recording shifts from the resting alpha rhythm to a higher frequency rhythm (15–50 waves/s), namely, the beta rhythm (Fig. 9.4). The *beta rhythm* is characteristic of the *awake, alert individual.* It is associated with the projection of one's conscious attention to the sensory stimuli in the environment. The beta rhythm is also observed in periods of mental concentration when one is solving a problem or forming a thought pattern.

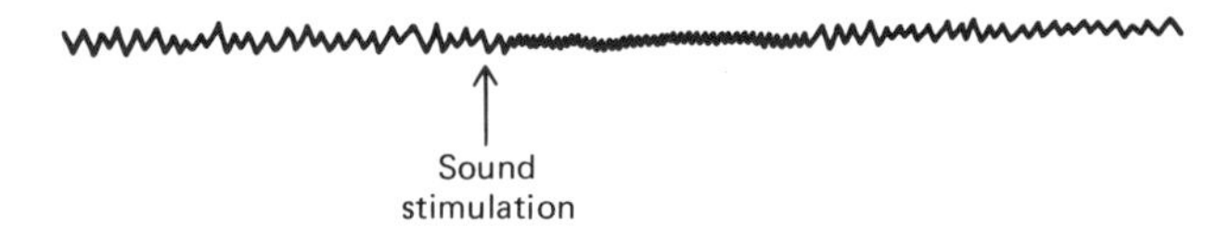

Fig. 9.4. Cortical arousal. Desynchronization of EEG rhythms when a resting, relaxed subject is exposed to a sound stimulus. The stimulus elicits a shift in EEG from an alpha to a beta rhythm.

In adults the shift from resting consciousness to sleep is correlated with a shift in the EEG from the alpha to the delta rhythm. The *delta rhythm* is characteristic of *deep sleep.* The lower frequency EEGs are also characteristic of the immature, awake nervous system. In EEGs from infants the delta rhythm (less than 4 waves/s) is dominant. The *theta rhythm* (4–8/s) is characteristic of *normal children.*

Neurophysiologic Basis

The pyramidal and stellate neurons are the principal cortical neurons (Fig. 9.5). Although the cell bodies of *pyramidal neurons* are located in layer 5 deep within the cortex, their

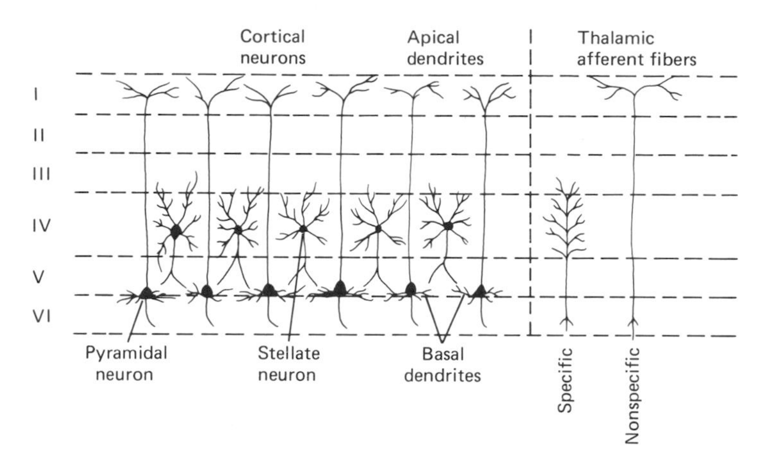

Fig. 9.5. Neuronal organization of cerebral cortex. Afferent fibers from specific thalamic nuclei terminate in layer 4 on stellate neurons. Pyramidal neurons have two sets of dendrites: basal dendrites in layer 5 are innervated by stellate neurons, whereas apical dendrites in layer 1 receive input from nonspecific thalamic fibers. Pyramidal neurons are cortical efferents; their axons transmit cortical output.

dendritic processes extend to the most superficial cortical layers (1 and 2). *Stellate neurons*—both their cell bodies and dendritic trees—are confined to layer 4. Inputs from thalamic neurons terminate at two different levels in the cortex. Inputs from the specific thalamic nuclei terminate in layer 4 on the dendrites of stellate neurons. Axons arising from the stellate neurons terminate on the basal dendrites of the pyramidal neurons. The apical dendrites of pyramidal neurons are innervated by inputs from the nonspecific thalamic nuclei, terminating in layers 1 and 2.

The EEG provides a measure of the synaptic activity in the outer layers of the cerebral cortex. The rhythm of the EEG reflects the intensity of the *synaptic activity,* EPSPs and IPSPs, in the outermost cortical layers. Most of the synapses in the superficial cortex are formed by incoming fibers from the nonspecific thalamic nuclei, ending on the apical dendrites of pyramidal neurons.

To generate rhythmic fluctuations of potential large enough to be recorded in the EEG, the activity of the neuronal elements in the superficial layers must be synchronized. *Synchronization* of the synaptic activities in the outer cortical layers is the result of two effects. (1) The neural networks in the outermost cortical layers have an intrinsic rhythmicity. (2) The neuronal discharge from the nonspecific thalamic nuclei, which primes the neural elements in the superficial cortex, is itself rhythmic.

The pyramidal neurons form the major pathway for the transmission of the cortical output—both to other regions of the cerebral cortex and to lower neural centers. As a result pyramidal neuron activity is essential to the maintenance of a conscious, alert state. Apparently the pyramidal neurons require a continuous input from the nonspecific thalamic nuclei to maintain excitability. In the absence of this input—following surgical interruption of the fibers from the nonspecific thalamus—the excitability of the pyramidal neurons is depressed; the experimental animal is rendered permanently unconscious.

Cortical Arousal: Reticular Activating System

If a person at rest is exposed to a strong sensory stimulus, he immediately alerts himself and directs his attention to the stimulus. This dramatic shift in mental activity from a relaxed to an active state is correlated with a shift in the EEG from an alpha to a beta rhythm (Fig. 9.4). This change, or *desynchronization of the EEG rhythm,* results from sensory excitation transmitted to the cerebral cortex from the nonspecific thalamic nuclei.

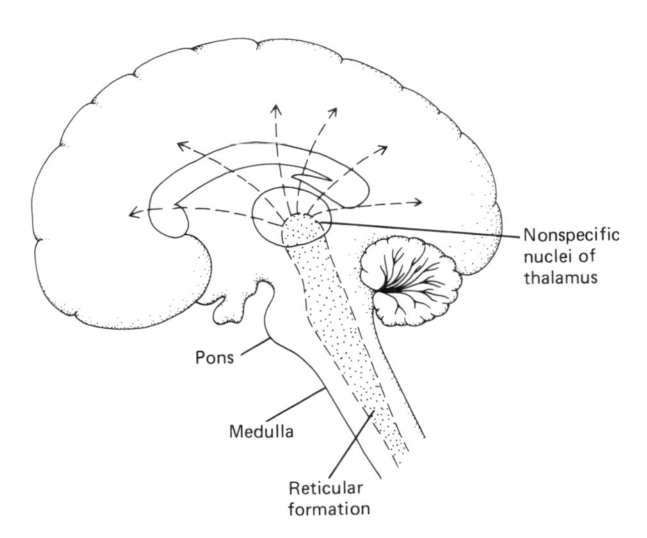

Fig. 9.6. Reticular activating system. This is the neural network that transmits sensory excitation through reticular formation of brain stem to nonspecific thalamic nuclei. Nonspecific thalamic fibers control level of cortical arousal.

The *nonspecific thalamic nuclei* form a diffuse neural network in the medial thalamus. They make up the most anterior extension of the reticular activating system (RAS), a diffuse neural network that extends from the reticular formation to the thalamus and which is concerned with the control of cortical excitability (Fig. 9.6). The RAS is excited by a large number of sensory inputs (auditory, olfactory, somatesthetic, vestibular, and visual) to the reticular formation. The RAS forms the pathway over which these sensory inputs are transmitted (via the nonspecific thalamic nuclei) to the superficial cortical layers.

Excitation of the RAS is required for *maintenance of consciousness*. Invariably lesions in the RAS of an experimental animal produce permanent unconsciousness. If the RAS of a sleeping animal is stimulated electrically, it awakens immediately.

Specific and Nonspecific Systems

Note the critical difference in roles of the *nonspecific* sensory system, i.e., the RAS, which forms a *diffuse pathway* to the superficial layers of the entire cerebral cortex, and that of the *specific* sensory systems, which form *specific projections* terminating in the middle layers of the appropriate sensory cortex. In contrast to the permanent loss of consciousness resulting from lesions in the pathways of the nonspecific system, lesions of a specific sensory pathway do not affect the level of cortical arousal. They produce a deficit in the subject's ability to interpret specific sensory information. Electrical stimulation of a specific sensory pathway does not affect the overall level of cortical excitation, although it will evoke activity in a specific sensory region of the cortex (e.g., audio or visual), which generates a specific sensation (e.g., flashing lights or noises).

Sleep

Sleep is a state in which the subject's awareness and attention to himself and to his environment is depressed. It is a state of decreased consciousness. Sleep is believed to have a restorative function. In sleep most body functions including muscle tone, heart and metabolic rate, respiration, sympathetic activity, and body temperature are depressed.
Sleep is essential in maintaining a normal mental state. Prolonged sleep deprivation is accompanied by general mental deterioration. A person who is sleep-deprived experiences decreased mental capacity for problem solving and task performance. Frequently this general depression is accompanied by increased irritability, which sometimes leads to abnormal behavior.

Sleep-Wake Cycle: EEG Patterns

The normal adult maintains a sleep-wake cycle of about eight hours of sleep to every 16 hours of awakeness. The cycle produces characteristic changes in the EEG potentials.
The approach of sleep is indicated by a shift in the EEG from a beta rhythm, reflecting an active, alert person, to an alpha rhythm (Fig. 9.7). The alpha rhythm indicates that the subject is relaxed and his eyes are closed. Although conscious, the overall responsiveness of the subject to sensory inputs is depressed.
Initiation of sleep with accompanying loss of consciousness is indicated by further slowing of the EEG rhythm from an alpha to a theta or even low-amplitude delta rhythm (Fig. 9.7). Short bursts, or spindles, of alpha waves are superimposed on the prevailing theta and delta waves. These spindles are characteristic of EEG recordings obtained during light sleep.
With time the depth of sleep increases. The increased depth is reflected by EEG recordings of *large-amplitude delta waves* (Fig. 9.7). These slow waves are characteristic of about 80% of the sleep cycle. *Slow-wave sleep* is characterized by general depression of body function and an absence of both dreams and rapid eye movements.

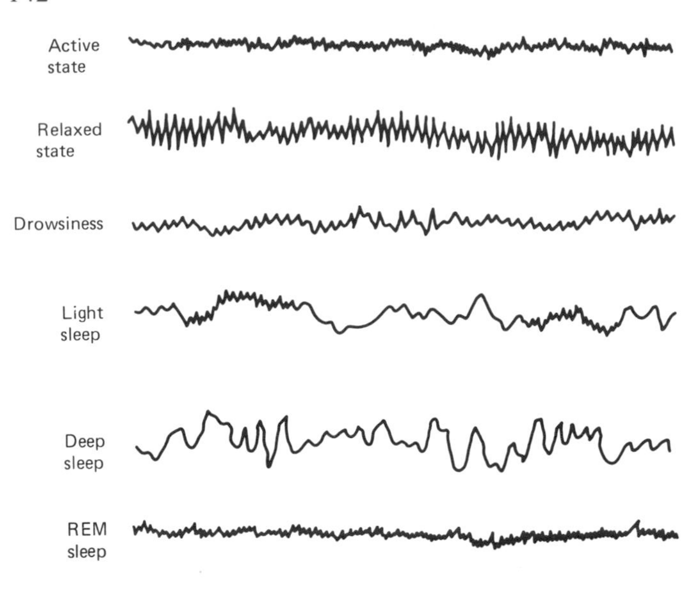

Fig. 9.7. Typical recordings of EEG during sleep-wake cycle.

Although consciousness is depressed in slow-wave sleep, a person can be readily aroused by stimuli that are particularly meaningful to him or her. One can be readily awakened, e.g., by the sound of footsteps, a door opening, or a child crying. In contrast insignificant stimuli of similar or greater intensity are ignored; persons living near a railroad track sleep undisturbed by the noise of passing trains.

Slow-wave sleep is sometimes called *nonrapid eye movement* sleep (NREM) to distinguish it from the second major kind of sleep, *rapid eye movement* (REM) sleep. Typically, five to six times a night the EEG patterns undergo a dramatic shift from the large-amplitude delta waves associated with slow-wave sleep to small-amplitude beta-like waves characteristic of REM or *paradoxical* sleep (Fig. 9.7). These periods of high-frequency EEG activity—5 to 20 minutes long—are associated with *active dreaming* and *rapid back-and-forth movements of the eyes.*

The high-frequency EEG patterns accompanying REM sleep are paradoxical because they resemble those of the alert, awake person. However, REM sleep is characterized by the strong suppression of consciousness and sensory awareness. It is much more difficult to arouse one from REM sleep than from deep slow-wave sleep. The threshold for awakening in response to sensory stimuli, e.g., footsteps or a child's crying, is much higher during REM sleep than during deep slow-wave sleep. The intense EEG activity (and rapid eye movements) that characterizes REM sleep presumably reflects the mental activity associated with dreaming.

Initiation of Sleep

The mechanisms controlling initiation of sleep are ill understood. At present there are two major theories for the initiation of sleep including, *neuronal fatigue* and *excitation of neural sleep centers*. Possibly both theories are correct. The complex neurologic phenomena that we call sleep may result from both the fatigue in the neural centers concerned with arousal and the excitation of those neural centers producing sleep.

Neuronal Fatigue

The onset of sleep is characterized by a shift in EEG pattern from a high-frequency beta, to a low frequency delta rhythm. This change reflects decreased activity of the RAS. The large delta waves, characteristic of slow-wave sleep, are the result of a *depression* of the RAS. Similar large slow-waves are observed in experimental animals when the neural fibers connecting the cortex with the RAS are destroyed. As was discussed earlier, stimulation of the RAS produces an immediate arousal in an intact sleeping animal.

The correlation between decreased excitation of the RAS and slow-wave sleep is thought to result from the neuronal fatigue in the RAS. In the spinal cord and elsewhere, high levels of neuronal discharge can produce fatigue—a depression of both synaptic transmission and neuronal excitability. Since the reticular formation is composed of diffuse multineuronal networks containing many synaptic junctions, continued high levels of reticular activity can be expected to lead to neuronal fatigue and consequent decrease in reticular system activity.

Sleep Centers

Several sleep-inducing centers have been identified in the brain. Stimulation of these centers (in the basal forebrain, thalamus, and caudal reticular formation) induces sleep. When the thalamic sleep center is stimulated at a rate of 8 to 10 pulses/s, a wide-awake animal falls into a deep slow-wave sleep. Lesions in the sleep centers of the basal forebrain or the caudal reticular formation produce insomnia.

The initiation of the two kinds of sleep—slow wave and REM—is associated with two different neurochemical systems located in the brain stem (Fig. 9.8). Slow-wave sleep is believed to depend on a system of *serotonin* (5-hydroxytryptamine), containing neurons located in the *median raphe nuclei*. Surgical destruction of these nuclei produces permanent insomnia. This state can be reversed by administering serotonin to the experimental animal.

A second neurochemical system, which is formed by norepinephrine-containing neurons in the dorsal pons is concerned with the control of REM sleep (Fig. 9.8). Surgical lesions of the *locus ceruleus* in the dorsal pons result in the loss of a capacity for REM sleep; REM sleep is thought to

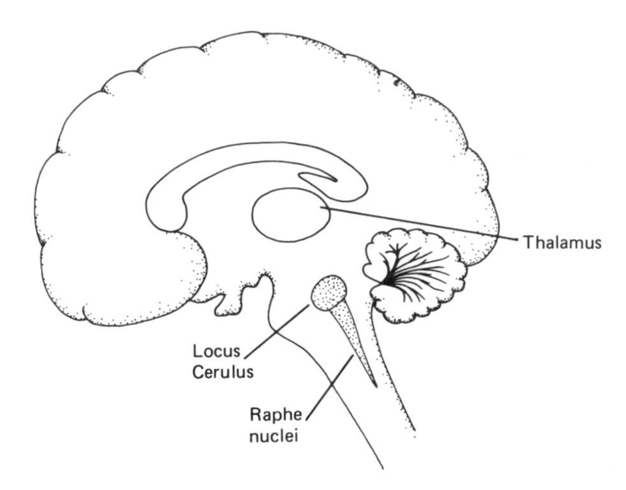

Fig. 9.8. Neurochemical systems in brain stem associated with sleep. The locus cerulus the neurons of which contain norepinephrine controls REM sleep. Serotonin containing neurons in the raphe nuclei are thought to regulate slow wave sleep.

be initiated by the periodic excitation of the norepinephrine-containing neurons in the locus ceruleus.

Memory

One of the most important functions of the nervous system is the capacity for storing experiences. A memory is a thought or experience stored by the nervous system in a form that is available for later recall. There are at least three kinds of memory, including short-term or sensory, intermediate, and long-term. At present the physiologic basis of memory is unknown. Although memory studies are an area of intensive neurophysiologic investigation, we possess little definitive information on the mechanisms that establish memory traces and the sites within the nervous system where they are stored. As a result the following discussion of memory and memory mechanisms is both tentative and conjectural.

Short-term memory lasts less than 1 s. It provides a capacity for the short-term storage of sensory experiences. The brief duration provides enough time for the cortical and subcortical centers to analyze the sensory experience and form a judgement of its significance. If the experience is important, it is entered into the intermediate memory; otherwise it is rapidly forgotten.

Intermediate memory is the capacity for detailed storage of information from several minutes to several hours. Examples of intermediate memory include conversation, word sequences and sentences, and number sequences. It also includes all significant events that occurred within the last several hours.

Long-term memories can last a lifetime and be surprisingly detailed, e.g., the arrangement of rooms and furnishings of the house in which one lived as a child, or words of a song memorized years before. Some long-term memories of significant events are vivid. For the author the memory of the moment when he first learned that John F. Kennedy had been assassinated—the time of day, the weather, details of the physical surroundings, and the people the author was with—is ingrained for life.

Mechanisms of Formation

At present there are two theories for the formation of long-term memories including, changes in neuronal connectivity and the formation of specific macromolecules (nucleic acids and proteins).

The theories are not mutually exclusive; alterations of synaptic connections may require the synthesis of specific macromolecules, or the production of specific macromolecules may induce changes in synaptic structure.

Repeated neuronal activation may result in the modification of the synapses within that network. These modifications can include the enlargement (regression) of synaptic contacts or the formation (loss) of new (old) synaptic contacts. Considerable experimental evidence suggests that the continued passage of neuronal activity across a synapse produces increased postsynaptic potentials. This potentiation of the synaptic potential is the result of increased release of neurotransmitter, from the presynaptic terminal.

Numerous experiments demonstrate that the establishment of long-term memories can be blocked by inhibiting either RNA or protein synthesis. Learning is associated with an increase in the *synthesis of both protein and RNA*. A small group of neurophysiologists supports the hypothesis that a memory trace is laid down in a specific sequence of amino acids; this configuration then forms a specific protein memory engram (or a specific sequence of nucleotides which forms an RNA engram). Most investigators, however, believe that the dependence of memory storage on protein and RNA synthesis reflects the importance of protein and RNA synthesis in the growth and formation of synaptic contacts.

Localization

The formation and recall of a memory trace requires the activity of thousands of neurons in the cerebral cortex, limbic system, thalamus, and other neural centers. Despite the known diffuseness of the memory trace, two areas of the cerebral cortex—the hippocampus and the temporal lobe of the neocortex—are associated with memory storage and recall.

It takes several hours to form a long-term memory. While the memory trace is being established, it is very sensitive to disruption as, e.g., from electroconvulsive shock, cooling of the brain, or surgical anesthesia. These disruptions are most effective if they are directed at the *hippocampus*. Persons

with bilateral hippocampal lesions are incapable of storing experiences as long-term memories, although those stored before the hippocampi were injured are unaffected.
The Neurons of the temporal lobe are closely associated with the *storage and recall* of long-term memories. Electrical stimulation of the temporal lobes can evoke vivid memories, such as childbirth, a classroom experience or a motion picture. Memory experiences that are evoked by temporal-lobe stimulation are vivid, e.g., a woman who remembers giving birth to her child reexperiences all aspects of the event including the conversation of the doctors and nurses, the appearance of the delivery room, the excitement, and the pain.
Although the temporal lobe is believed to control the recall of memory engrams, it is not thought to be the site of actual storage of the engram. Most neurophysiologists believe that such an engram results from the activity of large numbers of neurons that are dispersed throughout the cerebral cortex as well as extending into subcortical structures in the limbic system, thalamus, and hypothalamus.

Selected Readings

Agranoff BW (1967) Memory and protein synthesis. Sci Amer 216: 115
Atkinson RC, Shiffrin RM (1971) The control of short-term memory. Sci Amer 225: 82
Buser P (1976) Higher functions of thc nervous system. Ann Rev Physiol 38: 217
Eccles JC (1977) The understanding of the brain. McGraw-Hill, New York, Chapt 6
Eyzaguirre C, Fidone SJ (1975) Physiology of the nervous system, 2nd edn. Year Book Medical, Chicago, Chapt. 24
Gazzaniga MS (1967) The split brain in man. Sci Amer 217: 24
Harris AJ (1974) Inductive functions of the nervous system. Ann Rev Physiol 36: 251
Kandel ER (1977) Neuronal plasticity and the modification of behavior. In: Handbook of physiology: cellular biology of neurons. American Physiological Society, Bethesda, Maryland p. 1137
Luria AR (1970) The functional organization of the brain. Sci Amer 222: 66
Mountcastle VB (1974) Medical physiology. Mosby, St. Louis, Chaps. 8 and 19
Penfield W (1958) The excitable cortex in conscious man. CC Thomas, Springfield, Illinois
Sperry RW (1974) Lateral specialization in the surgically separated hemispheres. In: Schmitt FO and Worden IG (eds) The neurosciences. MIT Press, Cambridge, p. 1

Review Questions

1. Describe the functional differences between the specific and associative areas of the cortex.

2. What is the function of the corpus callosum?
3. Draw a diagram showing the interconnections of the principal neurons in the cerebral cortex? Compare the effects of specific and nonspecific thalamic inputs on the activity of these cortical neurons.
4. Describe the changes in EEG potentials that are recorded during sleep initiation.
5. What is REM sleep, and how does it differ from NREM sleep?
6. Identify the neurochemical systems associated with NREM and REM sleep.
7. List the three kinds of memory.
8. Discuss the current theories of the formation and storage of memory.

Neural Control of Visceral Activity

Chapter 10

Visceral Control Centers

The autonomic nervous system (ANS)—together with neural centers in the medulla, hypothalamus, and limbic system—regulates the visceral functions. These visceral control centers are concerned with the maintenance of a constant internal environment. All of the visceral systems—cardiovascular, endocrine, gastrointestinal, metabolic, renal, reproductive, and respiratory—are affected by the activity of these neural control centers.

The *visceral nervous system* regulates both the contraction of the visceral muscles (cardiac and smooth) and the secretion of gland cells. Its activity is under involuntary control. The "involuntary" visceral nervous system is often contrasted with the "voluntary" somatic nervous system, which controls the contraction of the skeletal muscles. The activities of the somatic nervous system are under voluntary control. The somatic nervous system includes the major sensorimotor pathways, their nuclei, the cerebral cortex, the basal ganglia, and the cerebellum.

Each subdivision of the visceral nervous sytem, such as the limbic system, hypothalamus, medulla, and ANS is concerned with a different level of visceral control. The *limbic system,* formed by the interconnection of several neural centers in the forebrain, is concerned with the control of complex behavioral patterns, e.g., feeding, maternal, sexual, and territorial. The *hypothalamus,* the major homeo-

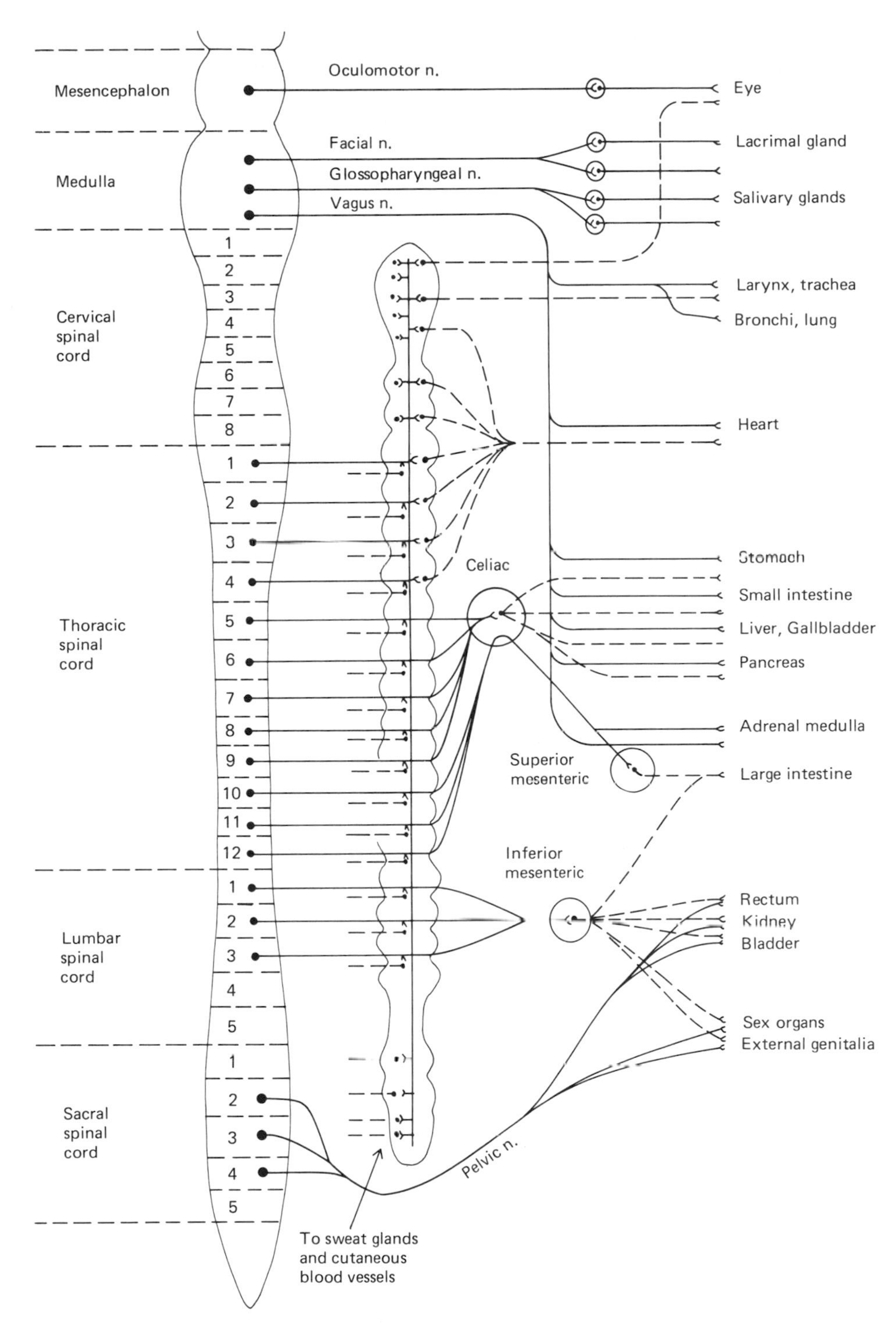

Fig. 10.1. Organization of autonomic nervous system.

static regulator, contains centers for the regulation of body temperature, water balance, feeding, and sexual and emotional behavior. It is also an important endocrine regulator because it controls the activity of the anterior and posterior pituitary glands. The *medulla* contains centers for the regulation of the cardiovascular and respiratory systems.

The visceral effector neurons originate in the ANS the primary function of which is to provide the neural pathways connecting the central nervous system (CNS) with the viscera. It is subdivided anatomically and functionally into two divisions, namely, the parasympathetic and the sympathetic (Fig. 10.1). The *parasympathetic division* is a *craniosacral system*. Its neurons arise from both the visceral nuclei of cranial nerves III, VII, IX, and X and the sacral segments of the spinal cord. Neurons belonging to the *sympathetic division* originate from the thoracic and first three lumbar spinal segments. Most visceral organs are innervated by both parasympathetic and sympathetic divisions. The effects of parasympathetic and sympathetic activity are usually antagonistic; a visceral muscle or gland cell that is excited by parasympathetic excitation is usually inhibited by sympathetic activity.

Autonomic Nervous System

The efferent pathway over which signals are transmitted from the CNS to the viscera is formed by two neurons, the preganglionic and the postganglionic autonomic neurons (Fig. 10.2). Together they form the efferent limb of the *autonomic reflex arc*. The preganglionic neurons originate either in the lateral horn of the spinal cord or in the visceral nucleus of a cranial nerve. They furnish a pathway for the transmission of impulses from the CNS to the postganglionic neurons located outside the nervous system in the *autonomic ganglia*. The postganglionic neurons terminate on their visceral target organs and are the final elements in the autonomic reflex arc.

Sympathetic Division

The sympathetic nervous system extends through 16 segments of the spinal cord, from first thoracic to fourth lumbar (Fig. 10.1). It includes both the preganglionic neurons originating in the lateral horn of the spinal cord and the postganglionic neurons located in the sympathetic ganglia. Most of the sympathetic ganglia are interconnected to form a lateral chain adjacent to the spinal cord. Postganglionic neurons arising from the lateral chain provide sympathetic innervation to the head (eyes, salivary glands, and blood vessels), to the thorax (bronchi and heart), and to the skin (blood vessels and sweat glands). Most of the abdominal visceral organs (stomach, small intestine, liver, large intestine, urinary bladder, and sexual organs) are innervated by postganglionic neurons arising from one of three *collateral sympathetic ganglia,* including the celiac, the superior mesenteric, and the inferior mesenteric, all of which are in the abdomen.

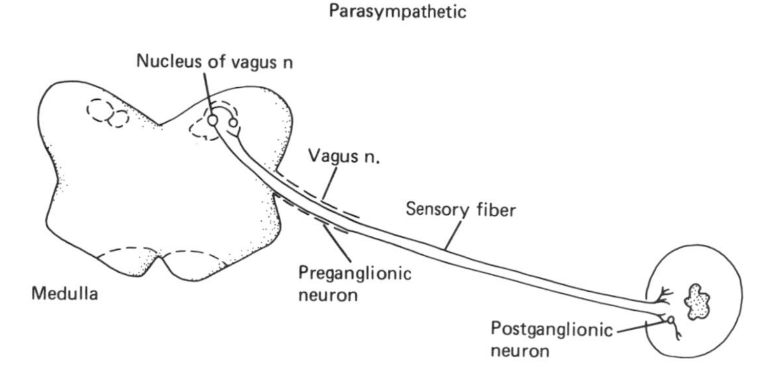

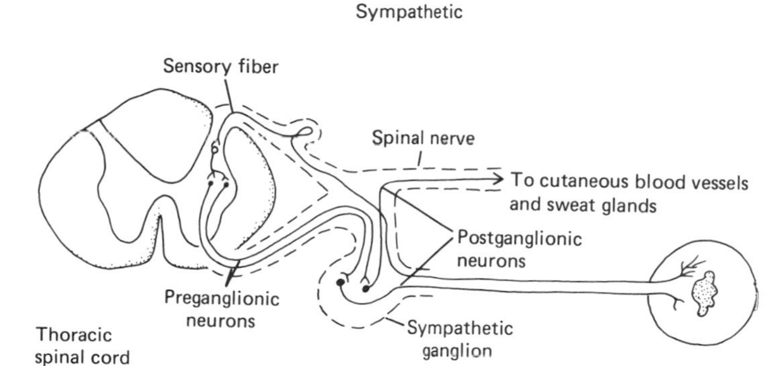

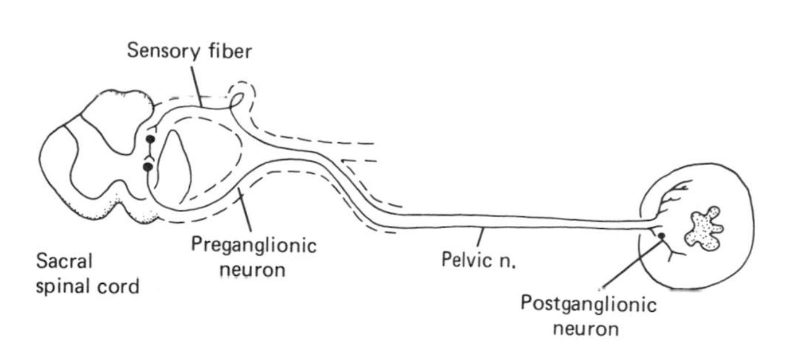

Fig. 10.2. Autonomic reflex arcs. Postganglionic neurons are either located in an autonomic ganglion (sympathetic and some parasympathetic fibers) or embedded in the wall of the visceral organ (most parasympathetic fibers).

Parasympathetic Division

The parasympathetic nervous system is divided into two divisions: *cranial and sacral* (Fig. 10.1). Parasympathetic preganglionic neurons are located both in the *visceral nuclei of the cranial nerves (III, VII, IX, and X)* and in the *lateral horn of the sacral spinal cord.* Postganglionic neurons are close to the visceral structures that they innervate. The postganglionic neurons are very short; many of them are embedded in the walls of the organ that they supply.

The fibers of the *vagus nerve* provide the major parasympathetic pathway. They connect with most of the thoraco abdominal visceral organs, including heart, trachea, bronchi, small intestine, and liver. Visceral structures in the head (salivary glands, tear glands, and blood vessels, among others) receive their parasympathetic innervation from fibers in the other cranical nerves (III, VII, and IX). The sacral division provides parasympathetic innervation to the rectum, kidney, bladder, and sexual organs.

Autonomic Neurotransmitters

Two major neurotransmitters—*acetylcholine* and *norepinephrine*—are released by neurons in the ANS. Those neurons secreting acetylcholine are *cholinergic;* neurons releasing norepinephrine are *adrenergic* (noradrenaline is the classic name for norepinephrine). Cholinergic autonomic neurons include all preganglionic fibers, both sympathetic and parasympathetic; all parasympathetic postganglionic fibers; and those sympathetic postganglionic fibers innervating blood vessels in the skin, skeletal muscles, and sweat glands. The vast majority of sympathetic postganglionic fibers are adrenergic, releasing norepinephrine as their neurotransmitter.

In addition to its release by most sympathetic postganglionic neurons, norepinephrine and its derivative, epinephrine, are also secreted by secretory cells in the *adrenal medulla*. These cells are supplied by preganglionic sympathetic neurons. They are closely related to sympathetic postganglionic neurons, and both have similar embryologic origins.

The actual effects of these neurotransmitters on their target organs are functions of the characteristics of the postsynaptic membranes of the innervated cells. Depending on the target organ, both acetylcholine and norepinephrine may excite or inhibit. If a visceral organ is innervated by both parasympathetic and sympathetic neurons, the effects of the two neurotransmitters are usually antagonistic.

Overall excitation of the sympathetic nervous system enables the body to cope with stress. *Sympathetic activity* increases heart rate, blood pressure, bloodflow to muscles, and blood glucose levels, while inhibiting gastrointestinal activity. Each of these effects increases the body's capacity to respond to stress. *Parasympathetic activity* is concerned with the recovery and restoration of body resources. Major effects of parasympathetic activity—decreased heart rate and blood pressure and increased gastrointestinal activity—are associated with body rest and restoration.

The specific visceral effects of parasympathetic and sympathetic activity are complex and numerous (Table 10.1). In the following section these effects will be briefly considered in relation to the major organ systems. For more detailed discussion, see those chapters dealing with the specific organ system.

Table 10.1. Effects of autonomic nervous system on visceral organs.

Organ	Sympathetic Activity	Parasympathetic Activity
Eye		
Pupil	Dilate	Constrict
Ciliary muscle	Relax for far vision	Contract for near vision
Lacrimal glands	—	Secrete
Salivary glands	Stimulate viscous secretion	Stimulate watery secretion
Lungs		
Bronchi	Dilate	Constrict
Bronchial glands	Inhibit	Secretion
Heart		
Muscle	Increase rate Increase contractile force	Decrease rate Decrease contractile force
Blood vessels		
Coronary	Dilate	—
Cutaneous	Dilate	—
Skeletal muscle	Dilate	—
Abdominal	Constrict	—
Stomach		
Motility and tone	Decrease	Increase
Sphincters	Contract	Dilate
Secretion	Inhibit	Stimulate
Intestine		
Motility and tone	Decrease	Increase
Sphincters	Contract	Dilate
Secretion	Inhibit	Stimulate
Pancreas	—	Secrete
Liver	Release glucose	—
Gall-bladder ducts	Relax	Contract
Urinary bladder		
Detrusor	Relax	Contract
Trigone and sphincter	Contract	Relax
Ureter motility and tone	Increase	—
Sex organs	Causes ejaculation	Causes erection
Adrenal medulla	Secrete epinepherine	—
Basal metabolism	Increase	—

—, No effect on organ.

The Eye

Parasympathetic fibers innervate the pupillary sphincter muscle of the iris, the ciliary muscle of the lens, and the lacrimal glands. Excitation of these parasympathetic fibers (1) constricts the pupil, thereby reducing the amount of light that enters the eye (*pupillary light reflex*), (2) focuses the lens to view a nearby object (*accommodation reflex*), and (3) promotes *secretion of the lacrimal glands*. The pupillary dilator muscle is inervated by sympathetic fibers. During stress, *sympathetic* excitation dilates the pupil.

Digestive System

The overall effect of the parasympathetic system is to *promote digestion*. Parasympathetic excitation increases gastrointestinal motility, promotes relaxation of the gastrointestinal sphincters, and stimulates the secretion of gland cells in the gastrointestinal tract. The sympathetic system *inhibits* gastrointestinal activity. It decreases gastrointestinal motility, increases sphincter contraction, and inhibits gland-cell secretion. Sympathetic excitation also reduces bloodflow to the abdominal viscera.

Cardiovascular System

Sympathetic excitation *increases cardiac output* by increasing heart rate, contractility of cardiac muscle fibers, and bloodflow in the coronary arteries. Excitation of the sympathetic system produces a shift in bloodflow from the body core and surface to the skeletal muscles. Sympathetic activity constricts the blood vessels supplying the abdominal viscera and the skin while dilating the blood vessels to the skeletal muscles. Stimulation of the parasympathetic system *decreases heart rate, cardiac contractility, and coronary bloodflow*. Excitation of the parasympathetic system has only a minimal influence on bloodflow in the systemic vessels supplying the abdominal viscera, skin, and skeletal muscles.

Visceral Sensory Pathways

Most visceral sensory receptors respond to nociceptive or pressure stimulation. Their fibers run in the sympathetic and parasympathetic nerves to reach the CNS and terminate either in the dorsal horn of the spinal cord or in the visceral sensory nucleus of a cranial nerve.

Although the CNS can determine the spatial location of a cutaneous stimulus with great accuracy, it is often unable to localize the site of visceral stimulation. Noxious stimulation of the viscera is experienced as painful sensations in quite unrelated areas of the body. Decreased coronary bloodflow, e.g., produces extreme pain in the neck, shoulders, and arms rather than at the level of the heart; stimulation of nociceptive receptors in the appendix (in the abdomen) produces intense pain in the lower thorax; noxious stimulation of the stomach produces "heartburn," which is felt in the thorax.

Referral of the site of noxious stimulation to a different region is called *referred pain;* it results from the pathway that most visceral sensory fibers follow in traveling to the spinal cord.

Most visceral sensory fibers travel in the sympathetic nerves from the viscera to the spinal cord. They then enter the lateral sympathetic chain and run several segments anteriorly in the chain before entering the spinal cord. As a result the fiber enters the spinal cord at a level several segments anterior to the location of its receptor ending in the viscera. Because the CNS has a segmental organization, it projects the sensory stimulus to a site in the segment of entry to the spinal cord. Since the spinal cord entry is several segments anterior to the visceral organ innervated by the sensory fiber, the CNS projects the stimulus to a site several segments anterior to the true position of the visceral organ.

Regulatory and Control Centers

The major neural centers for the control of the cardiovascular and respiratory systems are in the *medulla* (Fig. 10.3). These vital centers control the major cardiovascular and respiratory reflexes. They receive inputs from both baroreceptors and chemoreceptors in the cardiovascular and respiratory systems. Some medullary cells are sensitive to direct stimulation by CO_2 and H^+ levels in the blood. The cardiovascular and respiratory reflexes which these centers control are critical to the individual's survival. They are described in the chapters on cardiovascular and respiratory regulation.

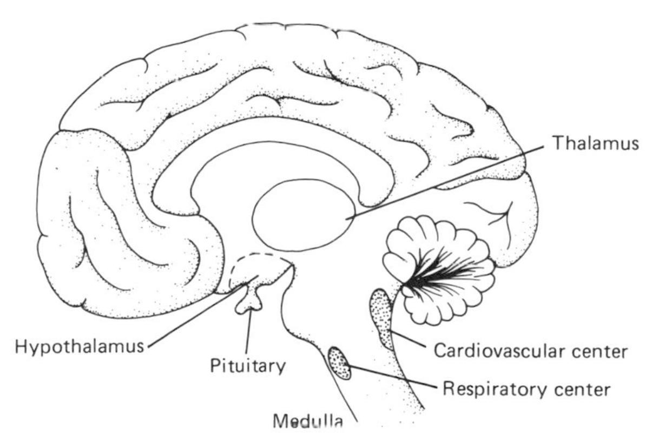

Fig. 10.3. Major visceral centers in brain stem.

In addition to the vital centers just mentioned, the medulla also has neural centers that control reflexes like coughing, sneezing, swallowing, and vomiting. These reflexes require coordinated activity of the pharynx, upper respiratory tract, and upper digestive tract.

Hypothalamus

The hypothalamus is the master visceral regulator. It directly controls both the neural activity in the ANS and the secretory activity of the anterior and posterior pituitary glands. It contains centers that regulate *body temperature, food intake, water balance, sexual activity,* and *emotional behavior.* Also, through its connections with the vital centers in the medulla, it modulates both cardiovascular and respiratory reflexes.

Anatomically the hypothalamus is formed by a cluster of nuclei located ventral to the thalamus in the diencephalon (Fig. 10.3). It may be divided into two zones (Fig. 10.4). The *anterior zone* contains the anterior, paraventricular, preoptic, and supraoptic nuclei. The dorsomedial, posterior, and ventromedial nuclei, together with the mamillary body, are in the *posterior zone*.

The hypothalamus receives its major inputs from the limbic system, the thalamus, and the reticular formation. Its neural outputs include pathways to the limbic system and through the reticular formation to the spinal cord. In addition to its neural output the hypothalamus secretes both the posterior pituitary hormones and the releasing hormones controlling anterior pituitary function.

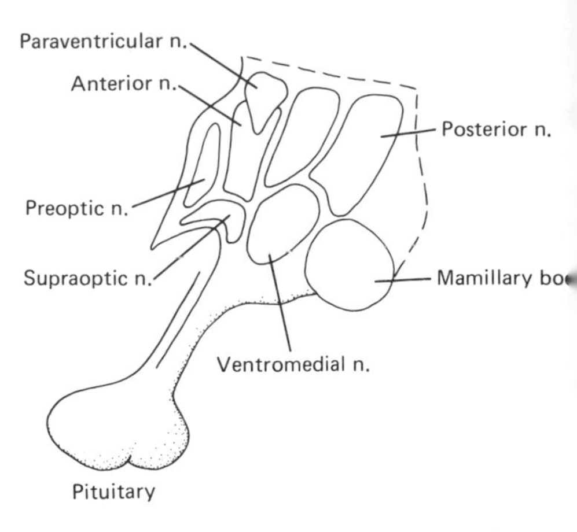

Fig. 10.4. Major hypothalamic nuclei.

Body Temperature

The hypothalamus contains centers for both regulation of heat loss and control of heat conservation and heat production (see Chap. 19). Electrical stimulation of the anterior hypothalamus increases the rate of heat loss; excitation of the posterior hypothalamus increases heat production.

These temperature-regulatory responses are initiated by the direct stimulation of *temperature-sensitive neurons* in the hypothalamus. Some of them are excited by an increase in blood temperature. Others are stimulated whenever blood temperature falls. Thermal stimulation of the warm receptors produces excitation of the heat loss center in the anterior hypothalamus. Cold stimulation of cold-sensitive hypothalamic neurons activates those posterior hypothalamic responses that increase the production and conservation of heat. Cutaneous thermoreceptors also supply inputs to these hypothalamic control centers. (See chap. 19 for a detailed discussion of temperature regulation and its control.)

Food Intake

The hypothalamus contains two centers that regulate feeding, a feeding center and a satiety center. Electrical stimulation of the lateral hypothalamus of an experimental animal induces feeding. If the feeding center is destroyed, the animal stops eating and will starve to death unless it is force fed. An antagonistic satiety center is located in the ventromedial nucleus destruction of which induces *hyperphagia* (continuous feeding). (See also Chap. 21.)

The feeding and satiety centers receive inputs from several kinds of sensory receptors. They include glucoreceptors, stomach stretch receptors (mechanoreceptors), and thermoreceptors. *Glucoreceptors* are hypothalamic neurons that are sensitive to changes in the blood glucose concentration. *Mechanoreceptors* in the walls of the stomach are sensitive to changes in volume of the stomach contents. *Thermoreceptors,* which respond to changes in body temperature, also provide an important input to the satiety and feeding centers. An animal that has not fed for some time has a low metabolic rate and a depressed temperature. The latter is usually lowest just before a meal. Following injection of a meal both metabolic rate and body temperature increase.

Water Balance

Stimulation of *the dorsal hypothalamus* in a dog or goat induces drinking. Destruction of this center completely inhibits drinking, causing dehydration leading to death. *Hypothalamic osmoreceptors* are thought to provide the major sensory input to the drinking center. These hypothalamic neurons are stimulated by changes in the osmotic pressure of the blood.

Emotional and Sexual Behavior

Stimulation of the *ventromedial nuclei* produces violent rage, often leading to an indiscriminate attack on any object in the environment. In contrast animals whose ventromedial nuclei are destroyed are markedly placid.

Electrical stimulation of several areas in the hypothalamus affects *sexual behavior*. Castrated animals exhibit normal patterns of sexual behavior following the implantation of testosterone in the hypothalamus. Lesions of the *anterior hypothalamus* inhibit sexual activity in both male and female animals. Implantation of estrogen into the hypothalamus of an ovariectomized animal elicits behavioral patterns characteristic of estrus, namely, active seeking and enticement of a mate.

Anterior and Posterior Pituitary Glands

The hypothalamus is the principal neural center for the control of pituitary function, regulating the secretion of hormones in both the anterior and posterior pituitary glands.

The *anterior pituitary* gland is connected to the hypothalamus by a vascular network, the hypothalamic-hypophyseal portal system (Fig. 10.5). This system provides a vascular pathway for the transport of releasing hormones from the hypothalamus to the pituitary. Through its regulation of anterior pituitary secretion the hypothalamus can control the release of the reproductive hormones (follicle-stimulating hormone, luteinizing hormone, and prolactin), as well as the levels of thyroid-stimulating hormone, growth hormone, and adrenocorticotrophic hormone (see chap. 25 for additional discussion of both the releasing hormones and the anterior pituitary hormones).

In contrast to the anterior pituitary, the *posterior pituitary* gland is innervated by hypothalamic neurons in the supraoptic and paraventricular nuclei (Fig. 10.5). These neurons provide pathways for the transport of hormones (oxytocin

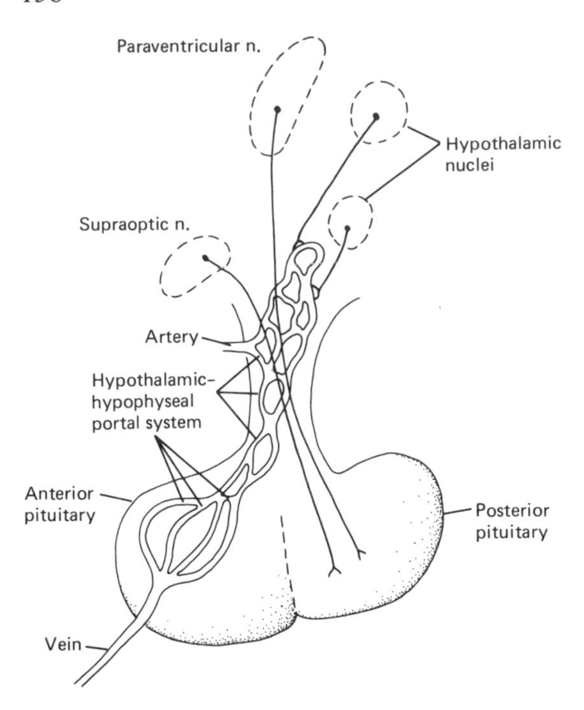

Fig. 10.5. Pathways connecting hypothalamus with pituitary. Hypothalamic-hypophyseal portal system transports releasing hormones from hypothalamus to anterior pituitary. Neurosecretory neurons originating in paraventricular and supraoptic nuclei of hypothalamus terminate in posterior pituitary.

and arginine vasopression) from the hypothalamus to the posterior pituitary (see chap. 25 for additional discussion of the function of the posterior pituitary).

Limbic System

The limbic system, containing the highest centers of visceral integration, is formed by the interconnection of several cortical and subcortical areas of the telencephalon. A close anatomic and functional relationship exists with the hypothalamus; indeed some authorities include the hypothalamus as part of the limbic system.

The limbic system is the *highest level of control of autonomic nervous activity and pituitary gland secretion*. It is the site of integration of three kinds of information. They are (1) visceral activity, (2) olfactory inputs, and (3) the activity of the association areas of the cerebral cortex—both motor and sensory. It is concerned with levels of motivation and the production of complex behavioral patterns that require coordination of both visceral and somatic motor activities for their successful execution. At present the precise roles of the cortical and subcortical areas, which make up the limbic system, are poorly understood. This is a result of the complexity of both their neural anatomic organization and the behavioral patterns that they control.

Anatomic Organization

The *limbic cortex* forms a ring at the base of the cerebral cortex which surrounds the upper end of the brain stem (Fig. 10.6); it includes the olfactory cortical areas and the

Fig. 10.6. Limbic system.

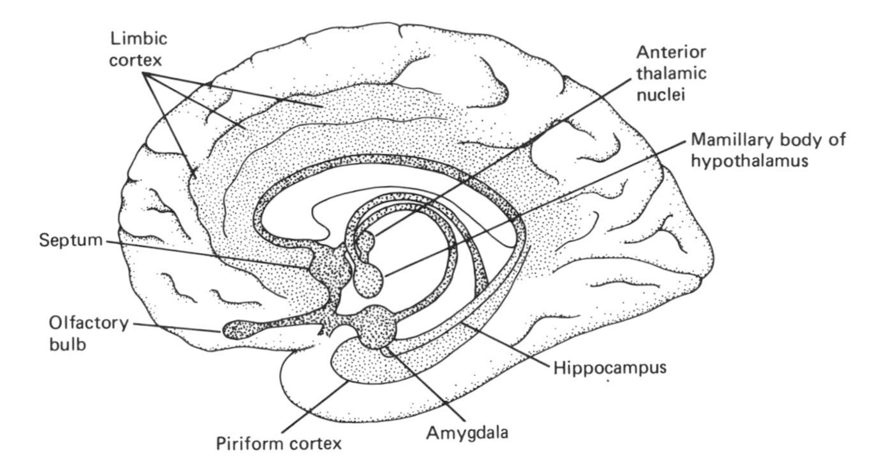

hippocampus. The limbic cortex is closely associated with the piriform cortex and orbitofrontal areas of the association cortex.

The limbic cortex is formed from a ''primitive'' kind of cortical tissue, the allocortex which contains only three layers of cortical neurons. In contrast, the sensory, motor, and association areas of the cerebral cortex are formed of neocortex, which contains six layers of cortical neurons.

The *major subcortical structures* associated with the limbic system include the amygdala and septum in the telencephalon and the anterior thalamic nucleus and hypothalamus in the diencephalon (Fig. 10.6). The limbic system also connects with the olfactory bulb.

Few direct connections exist between the limbic system and the remainder of the cerebral cortex; this may explain the difficulty of consciously controlling one's emotions. Inputs from the sensory, motor, and association areas of the cortex must pass through the *anterior thalamus* to reach the limbic system. The only exception is the connections between limbic system and frontal cortex. Other major inputs to the limbic system are olfactory (from the olfactory bulb) and viscerosensory (from the hypothalamus and visceral nuclei of the brain stem).

Outputs from the limbic system are directed primarily at the visceral centers in the hypothalamus, the pituitary gland (via the hypothalamus), and the ANS. The limbic system can affect somatic motor activities through its interconnections with the basal ganglia, anterior thalamus, and reticular formation.

Behavioral Patterns

A variety of complex behavioral patterns associated with feeding, reproduction, aggression, and flight is evoked when electrical stimuli are applied to different areas of the limbic system, including pleasure, rage, distaste, and fear—emotions that are associated with these complex behaviors.

The activity of the limbic system is closely correlated with that of the hypothalamus. There is evidence that various areas of the limbic system regulate the secretion of the releasing hormones in the hypothalamus. At present most of the functions—alleged or real—of the limbic system are poorly understood.

The behavioral patterns controlled by the limbic system are concerned with the preservation of either the individual or the species. The effects of stimulation and lesioning of regions in the limbic system indicate that the amygdala and piriform cortex are primarily concerned with the survival of the individual whereas the septum and hippocampus form centers regulating those behaviors essential to species survival.

Stimulation of amygdala produces behavior associated with feeding, e.g., chewing, swallowing, and licking. If other regions of the amygdala are stimulated, the animal initiates a violent attack either at other animals or at the experimenter. Lesions of the amygdala are associated with docility, increased feeding to the point of obesity and hypersexuality. During stimulation of the septum, an aggressive animal becomes docile. Lesions of the septum produce a furious and aggressive subject. The violent rage of rabies is associated with hippocampal lesions. Stimulation of many regions of the septum elicit sexual and maternal behavioral patterns.

Selected Readings

Axelsson J (1971) Catecholamine functions. Ann Rev Physiol 33:1

Barchas JD, Akil H, Elliott G, et al (1978) Behavioral neurochemistry: neuroregulators and behavioral states. Science 200: 964

Blackwell RE, Guillemin R (1973) Hypothalamic control of adenohypophysical secretions. Ann. Rev. Physiol. 35: 357

DiCara LV (1970) Learning in the autonomic nervous system. Sci Amer 222: 30

Eyzaguirre C, Fidone SJ (1975) Physiology of the nervous system, 2nd edn. Year Book Medical, Chicago, chaps. 22 and 23

Guillemin R, Burgus R (1972) The hormones of the hypothalamus. Sci Amer 227: 24

Schally AV, Arimura A, Kastin AJ (1973) Hypothalamic regulatory hormones. Science 179: 341

Snyder SH (1977) Opiate receptors and internal opiates Sci Amer 236: 44

Review Questions

1. What is an autonomic neuron? Describe the criteria used to determine whether it belongs to the parasympathetic or sympathetic nervous system.
2. Describe the effects of parasympathetic activity on the gastrointestinal tract. What are the effects of sympathetic excitation?
3. What is the morphologic basis of referred pain?
4. Which regulatory centers are located in the medulla? Which are located in the hypothalamus?
5. How does the hypothalamus control pituitary secretion?
6. Discuss the role of the limbic system in the control of visceral activity.

Chapter 11 Muscle

There are three types of muscles, including skeletal, cardiac, and smooth. Their general properties, distribution, and gross anatomic features are discussed in Chap. 1.

Skeletal Muscle

Skeletal muscle is made up of individual multinucleated fibers that are striated or banded (Figs. 1.8 and 11.1). The

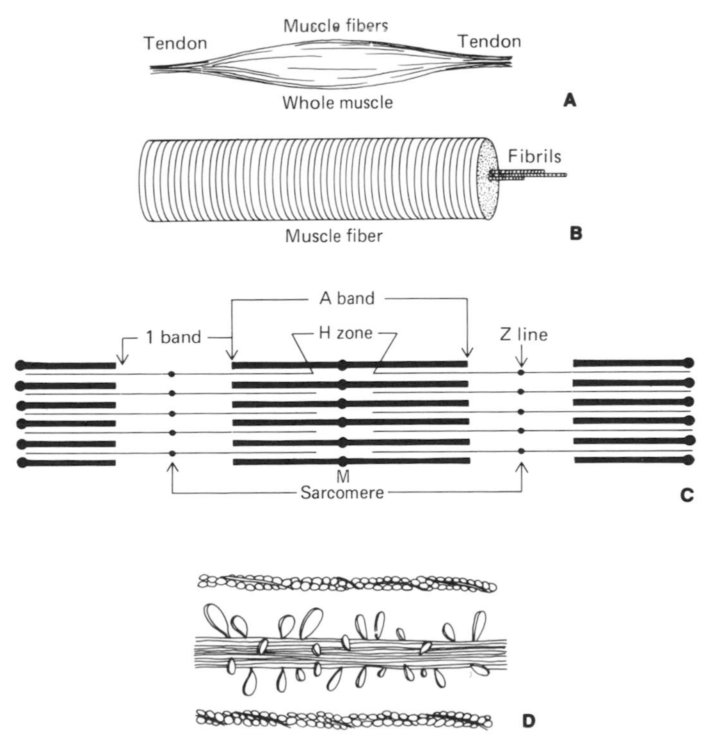

Fig. 11.1. **A** Skeletal muscle. Muscle fibers attached to tendon. **B** Individual fiber made up of fibrils. **C** Isolated myofibril exhibiting alternate light (I band-actin) and dark (A-myosin) bands. Within the A band is a lighter H-zone and in the middle of this one is a darker M band. Sarcomere extends from one Z line to another. **D** Arrangement of cross bridges between myosin (thick) and actin (thin) filaments.

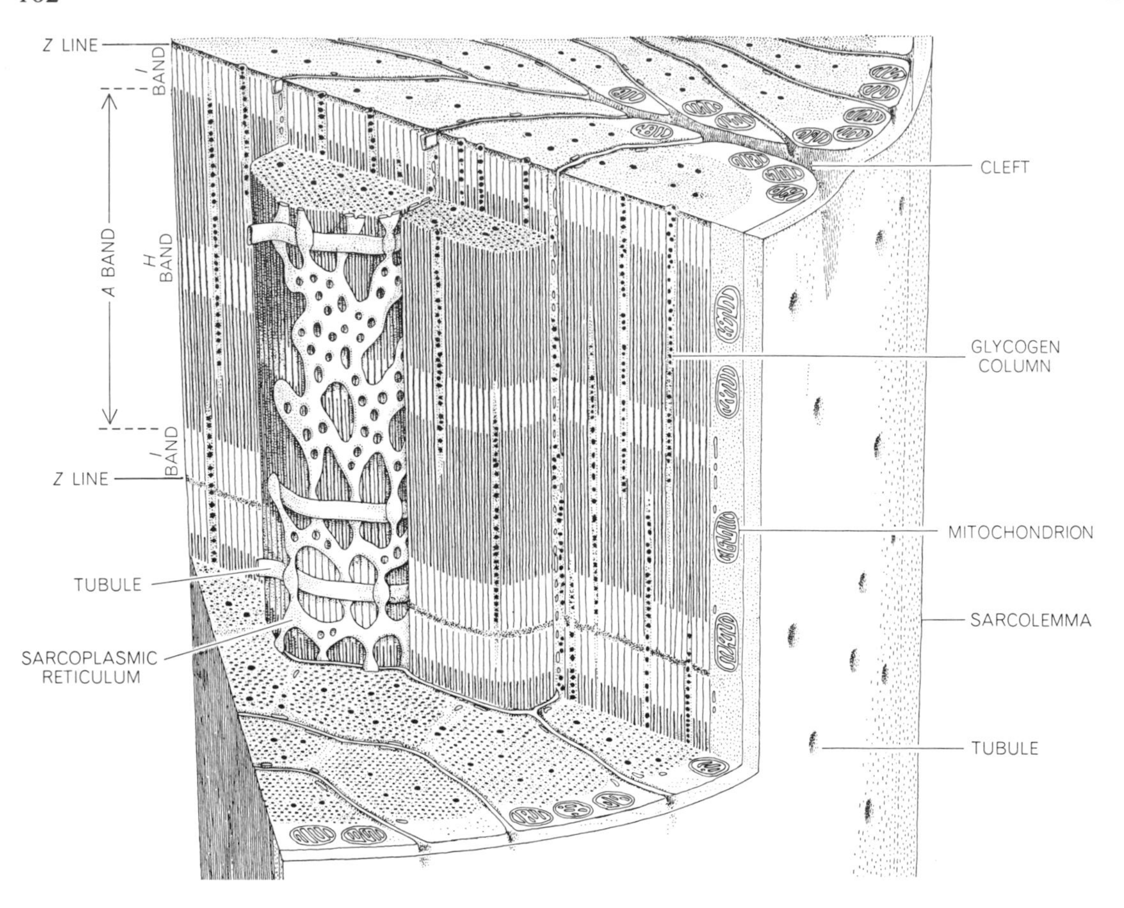

Fig. 11.2. Skeletal muscle in three-dimensional view. Fiber is made up of fibrils and surrounded by sarcoplasmic reticulum and T-tubules, which open up at muscle membrane. (Reproduced with permission from G. Hoyle: How is muscle turned on and off? Scientific American 222:84 (April, 1970). Copyright © by Scientific American, Inc.)

striations form alternating dark bands (anisotropic, or A bands) and light bands (isotropic, or I bands) (Figs. 11.1 and 11.2). The middle of one I band to the middle of another I band is the Z line, and this constitutes one sarcomere, the structural and functional unit of contraction. The fibers are made up of fibrils; when isolated and studied under the electron microscope, the A band is observed to have a lighter area called the H zone. A darker line bisecting the A line is called the M line.

Sarcoplasmic Reticulum and T-Tubules

Each skeletal muscle fiber is surrounded by a sarcolemma and is made up of fibrils surrounded by a sarcoplasmic reticulum and a septum of transverse T-tubules, forming a grid perforated by individual muscle fibrils (Fig. 11.2). The tubules are perpendicular to the fibrils; the sarcoplasmic reticulum is parallel to them. The area of contact between a T-tubule and the sarcoplasmic reticulum is a *triad,* so called because it is composed of a central small tubule and two large laterally situated cisternae of the sarcoplasmic reticulum. A triad occurs in skeletal muscle adjacent to each area where actin and myosin filaments (A and I bands) overlap. There are two triads per sarcomere.

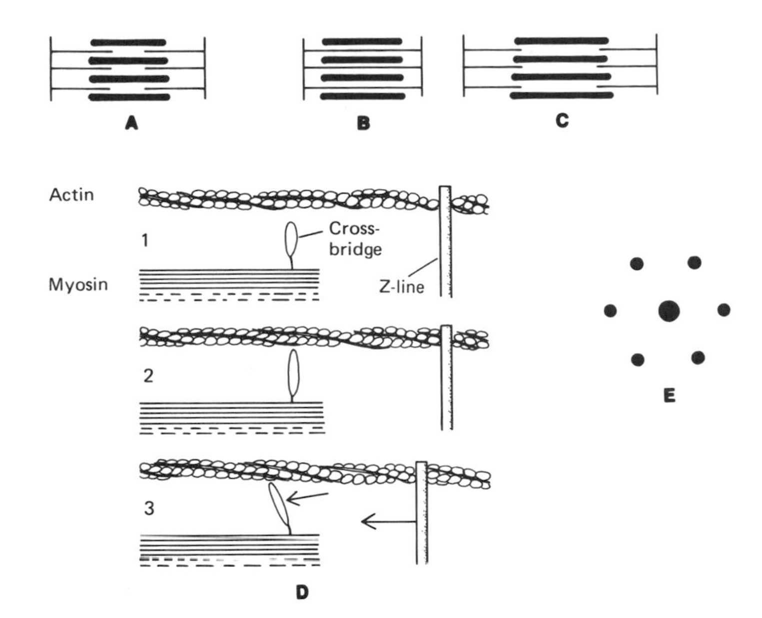

Fig. 11.3A–E. Sarcoplasmic reticulum and T-tubules. **A, B,** and **C** showing myosin (heavy lines) and actin (light) filaments during **(A)** rest, **(B)** contraction, and **(C)** stretched. Muscle shortens during contraction by the sliding inward of actin fibrils (see text). **D** Attachment of myosin cross bridges to actin fibrils which moves actin fibrils inward to shorten muscle (2 & 3) many of the cross bridges are involved in contraction. **E** cross section through A and I bands showing one large dot (myosin) filament surrounded by 6 small dots (actin) as revealed by electron microscope.

The main function of the T-system is the rapid transmission of the action potential from the cell membrane to muscle fibrils. The sarcoplasmic reticulum is also involved in the movement of calcium in the muscle.

Contractile Proteins

The alternating I and A bands comprise molecules of actin and myosin, respectively (Figs. 11.1C and 11.2). A cross section through the I and A bands (electron microscope) reveals a large dot (myosin filament) surrounded by six smaller dots (actin filaments, Fig. 11.3E).
Muscle also contains *tropomyosin* and *troponin,* and the latter is made up of three subunits of troponin, I, T and C. Strands of tropomyosin and troponin are attached to the much larger actin strands and are believed to be responsible for controlling the interaction between thick and thin filaments during muscle contraction. Troponin has a high affinity for calcium, and the interaction of this protein and calcium may be the trigger that initiates muscular contraction.

Mechanical Contraction

Mechanical contraction of muscle is preceded by electrical stimulation or discharge of motor neurons in the myoneural junction (motor end-plate) where nerve and muscle unit meet (see Chaps. 4 and 7). Acetylcholine is released at this junction, which crosses the post-junctional membrane to stimulate the electrical discharge of the muscle (action potential). The *action potential* initiates the release of calcium, which in turn triggers the contraction as follows (see also Chap. 4):

1. Release of Ca from the sacs or vesicles of sarcoplasmic reticulum and its spread to thick and thin filaments.

2. Binding of calcium to troponin.
3. Formation of cross linkages between myosin and actin and sliding of actin filaments on myosin resulting in shortening of muscle (Fig. 11.2 and 11.3).

Adenosine triphosphate (ATP) is the principal source of energy required in muscle contraction; thus, ATP → Energy + ATP + Phosphate.
The resting membrane potential is about 85 mV (see Chap. 4), and the *duration of action potential* is about 1–5 ms (longer than in nerves). *Velocity of conduction* in skeletal muscle is 3–5 m/s (much slower than in myelinated nerves).

Sliding Filament Theory

The sarcomere, as the contractile unit, changes the length of the muscle fibrils by the sliding action of actin filaments on the myosin filaments whose length do not change. Thus in a relaxed muscle the actin filaments are extended outward (longer sarcomere), more so when the muscle is stretched (Fig. 11.3A and C). In a contracted muscle the actin filaments slide inward and come together (Fig. 11.3B), or they may actually overlap, depending on the degree of contraction and shortening of the muscle. The change in the position of the myosin cross bridges and their role in moving the actin filaments are shown in Fig. 11.3D(2 & 3).
Isotonic (same force) contraction results in a shortening of the muscle. In *isometric* contraction the load may be too heavy to lift and the muscle cannot shorten, but it exerts a force (increased tension) and expends energy to maintain muscle fibers at the same length (isometric).
A muscle usually begins to contract isometrically and then isotonically as the load is moved and work is performed. Maximum efficiency of isotonic contraction is about 25%. A flexor muscle like the biceps (arm) shortens on contraction (isotonic), but during standing the quadriceps muscle (leg) tenses and remains rigid (isometric contraction).

Length, Force, and Velocity

The most important mechanical properties of muscle are: length, force, and velocity. In a contracting muscle there is an optimum length at which contraction is minimal. This can best be demonstrated in the laboratory by studying isometric contraction in which isolated muscles are fixed at different lengths (Fig. 11.4). If the muscle is short (fixed) it does not contract maximally, but if it is lengthened, as at *2* in Fig. 11.4A, it contracts maximally. If the muscle is stretched or overlengthened, (as in 3) it does not contract maximally. This length-force relationship is not very important in skeletal muscle, but it is important in cardiac muscle, which is the basis of the Frank-Starling law (see Chap. 15). The increasing

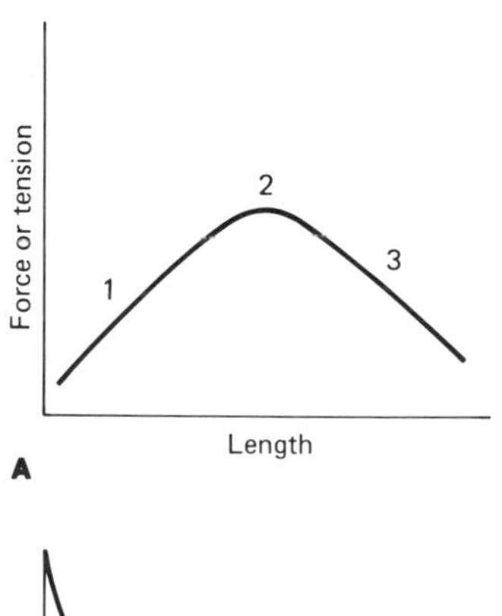

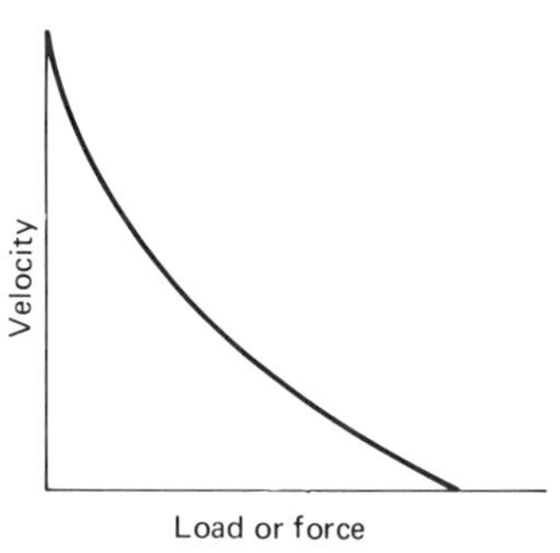

Fig. 11.4A and B. Force-tension-length relationship. **A** As muscle length is increased up to point *2*, there is increase in tension and force, which decreases as muscle is further lengthened (*3*). **B** Relationship of velocity of contraction to load or force. The heavier the load, the slower the contraction rate.

load on a muscle decreases the velocity of its contraction (Fig. 11.4B and Chap. 15).

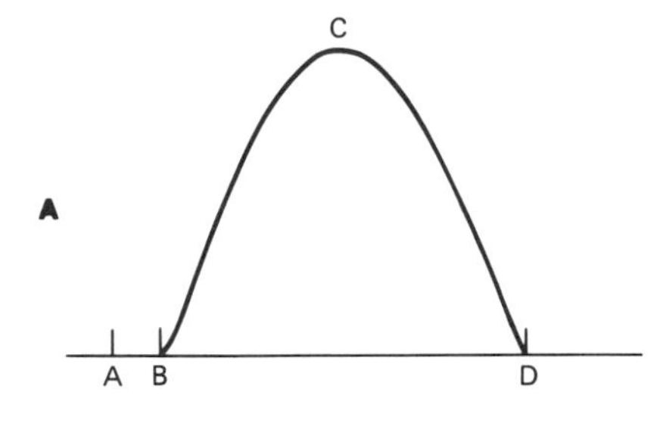

Fig. 11.5A and B. Stimulation and response. **A** Single muscle twitch (contraction). (*A-B*) latent period, stimulus applied to *A*; (*B-C*) contraction phase; (*C-D*) relaxation phase. **B** Treppe, or staircase effect, (increase in height of first 6 waves) produced by series of stimuli (just below tetanizing current) until constant and uniform wave appears (see text).

Stimulation and Response

When a muscle fiber receives a single stimulus, it responds by a single muscle twitch or contraction (Fig. 11.5A). The parameters of the stimulus to a muscle include (1) intensity (v or mV), (2) time and duration of stimulus (s or msec), and (3) frequency of stimulations (number/s). The duration of a single muscle twitch is about 0.1 s.

Treppe, or staircase phenomenon, occurs when a stimulus of constant strength and duration (but below tetanizing frequency) is repeated once or twice per second (Fig. 11.5B). It is characterized by increased contractions during the first stimulations, which finally reach a constant response. Treppe is believed to be caused by an increased availability of calcium for binding to troponin.

The electrical response of muscle to stimulation (action potential) exhibits a refractory period, but the contractile state of skeletal muscle does not have a refractory period (when it does not respond to stimuli). Therefore if repeated stimuli are applied to the muscle before relaxation of previous contractions has occurred, an additional activation or summation of muscle contractions is observed. The tension or force developed during summation is greater than that of a single twitch (Fig. 11.6). As the frequency of stimulations increases (but before tetanization occurs) the summation increases, but the wave forms are discernible. At higher tetanizing frequencies (60/s) the wave forms disappear and contractions fuse together. *Summation* is also caused by increasing the number of motor units contracting together (see this chapter, "Fast and Slow Muscles").

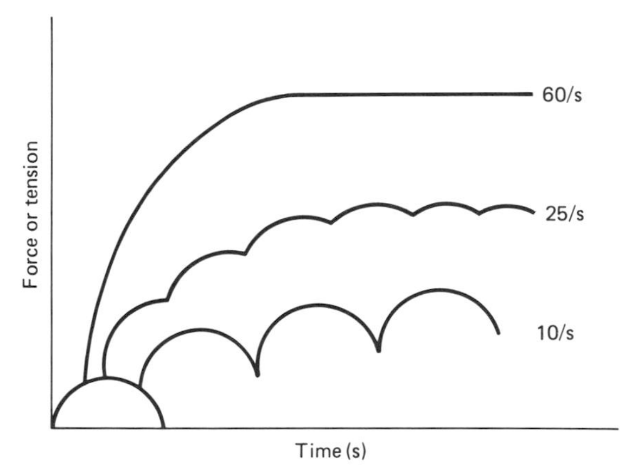

Fig. 11.6. Summation of waves of contraction (increased force or tension) at 10 and 25 stimuli/s. Tetanizing current (60/s) causes further increase in tension but fusion, or disappearance of, waves (see text).

Muscle Tone and Exercise

Tonus of a muscle refers to a state of partial contraction or to the tension in it. This tautness, tenses the muscle but produces no movement. When muscles lose tone, they become flaccid. Vigorous exercise increases muscle size by hypertrophy (increase in size of muscle fibers). Weak or slight muscular activity does not cause muscular hypertrophy. Strength and muscular hypertrophy can best be developed by short periods of *isometric* exercise.

Skeletal muscles are grouped according to the specific function they perform, including *flexion, extension, rotation* of a joint, *abduction,* and *adduction. Agonists* are muscles that move organs, such as the flexors of the arm and leg. *Antagonists* have the opposite effect, such as extension of arm and leg. When the tone of a flexor is increased, the tone in the extensor is decreased. *Synergists* are muscles that assist the agonists.

Fast and Slow Muscles

Muscles vary in their speed of contraction, depending on the required action. Ocular muscle (eyes) contract more rapidly than the gastrocnemius, which responds more rapidly than the soleus muscle, which in turn is concerned mainly with slow reactions. Fast muscles usually have a more extensive sarcoplasmic reticulum (involved in rapid release of calcium), are less vascular, and may be referred to as "white" muscle. Slow muscles have smaller fibers and are designed for prolonged performance. They are often called "red" muscles, because of the reddish tint imparted to them by their high content of myoglobin.

The muscle fibers innervated by a single motor nerve make up a *motor unit.* Small fast muscles tend to have fewer muscle fibers per motor unit than larger, slow muscles. Some of the former have as few as 3 fibers per unit, whereas large, slow muscles have as many as 1000 fibers per unit.

Muscular Fatigue and Disorders

Prolonged contraction of a muscle leads to fatigue and inability to do work. Nervous impulses to the muscle may be normal, but the mechanical response to neural stimulation is impaired because of the depletion of the principal energy source, ATP.

Nerve paralysis, as in poliomyeltis, leads to inactivity or *atrophy of muscles.* Many disorders of the CNS produce severe muscular symptoms, convulsions, muscle tremors, and tetanus, but the muscles are normal (see Chap. 8). *Muscular dystrophy* is a condition characterized by progressive muscular weakness and loss of contractility; its cause is unknown. In *myasthenia gravis,* a disease caused by a disorder of neuromuscular transmission, there is a deficiency of neurotransmitter (acetylcholine) at the motor end-plate, or receptors, such that transmission is poor or absent.

Cardiac and Smooth Muscle

Cardiac muscle is also straited (Fig. 1.8), behaving much like skeletal muscle except that there are points at the Z lines where individual fibers merge or interdigitate to form *intercalated discs;* these latter produce a muscular network (syncytium) that furnishes a more rapid spread of impulses from muscle fiber to fiber. Cardiac muscle is involuntary. The T-system of cardiac muscle is located at the Z lines, not at the junction of A and I bands as in skeletal muscle. Cardiac muscle responds according to the all-or-none law by contracting completely or not at all. Its contractile phase has a refractory period, i.e., it will not respond to stimuli, unlike skeletal muscle. Details on the structure and function of cardiac muscle is discussed in Chap. 15.

The structure of *smooth muscle* is discussed in Chap. 1. Although not striated, the muscle fibers are long and narrow. They are shorter than skeletal fibers and have one nucleus per fiber. Smooth muscle is found in the visceral organs and blood vessels.

There are two general classifications of smooth muscles: multiunit and visceral unit. *Multiunit* fibers operate independently of other fibers, and each fiber may be innervated by a single nerve ending. This type of muscle is found in the ciliary muscles of eyes, nictitating membrane, and smooth muscle of certain of the larger blood vessels. Erection of hairs is also produced by this type of muscle. Visceral muscle fibers are so crowded together that their cell membranes may be apposed or fused. Stimulation of a visceral muscle fiber causes a rapid spread of the impulse (action potential) to other fibers, because the electrical resistance between fibers is minimal. Visceral muscle is found in most of the organs of the body, including digestive tract, uterus, and ureters.

Selected Readings

Gauthier GF (1977) In: Weiss L and Greep RO (eds) Histology, 4th edn. McGraw-Hill, New York, Chap. 7

Guyton AC (1977) Basic human physiology, 2nd edn. Saunders, Philadelphia, Chaps. 9 and 10

Hoyle G (1970) How is muscle turned on and off. Sci Amer 222: 84

Julian FJ, Mos RL, Sollins MR (1978) Mechanism for vertebrate striated muscle contraction (a review). Circ Res 42: 1

Meiss RA (1976) In: Selkurt EE (ed) Physiology, 4th edn. Little, Brown, Boston, Chap. 3

Murray MM, Weber A (1974) The cooperative action of muscle and proteins. Sci Amer 230: 58

Review Questions

1. Name and characterize the general types of muscles based on anatomic differences.
2. Describe and discuss in detail the sliding filament theory of muscle contraction.
3. Name the principal contractile proteins and discuss their role in muscle contraction.
4. Define: (a) sarcomere, (b) sarcoplasmic reticulum, (c) T-tubules, (d) A band, (e) I band, (f) Z line, (g) flexor muscle, and (h) extensor muscle.
5. Which muscles are voluntary and which are involuntary? What do these terms mean?
6. What are the likenesses and differences between skeletal and cardiac muscle?
7. Where are skeletal and smooth muscles found? In what organs?
8. Define isotonic and isometric contraction.
9. What triggers the action potential in skeletal muscle?

Chapter 12

Body Fluids, Capillaries, and Lymphatics

Body fluids are found inside (intracellular), outside (extracellular), or between cells (interstitial) in various organs. Thus total body water is made up of intracellular and extracellular compartments, which may be subdivided as shown in Table 12.1. Intracellular water comprises 70% of total body water; however this varies depending on the subject's age, and the amount of lean body tissue (Table 12.2). Body water is higher in lean than in fatty tissues and is higher in infants and growing children and animals than in adults. More of the total water is found in the extracellular compartments in the very young subject. The size and distribution of these compartments in man are relatively constant and compare favorably with the same in other species. This ability to maintain constancy of the internal environment is known as *homeostasis,* a common but important concept in the study of body functions. Later we shall consider the mechanisms involved in the formation of these compartments (See also Chaps. 2 and 13).

Table 12.1 Intra- and extracellular fluid compartments.[a]

Fluid	Volume (liters)
Total body water:	40
Extracellular[b]	15
Plasma	3
Intracellular	25

[a] Volumes are for an adult person weighing 70 Kg.
[b] Extracellular compartment also includes lymph fluid of about 2 liters.

Table 12.2. Extra- and intracellular fluids in man.

Compartment	Percentage of Body Weight		Percentage of Total Body H_2O	
Extracellular:				
Plasma H_2O	4.3	17.5	7.3	30.0
Interstitial H_2O	13.2		22.7	
Intracellular		40.6		70.0
Total		58.1		100.0

Measurement of Body Fluids

Measurement is based on the dilution principle, or volume (V) distribution of an injected substance:

$$V = \frac{\text{Amount injected} - \text{amount excreted}}{\text{Concentration of substance in diluting fluid}}$$

For body fluids, the diluting fluid is water in one of three compartments, including total body water (intracellular and extracellular), extracellular fluid and blood plasma (see Chap. 13 for detailed discussion of blood plasma measurements). *Intracellular fluid* is calculated as total body water minus extracellular fluid. The substance injected depends on the compartment to be estimated. For total body water, a substance like antipyrine, when injected into the bloodstream, diffuses into both intra- and extracellular compartments. From the concentration of antipyrine remaining in the blood after this uptake of the substance, the volume of body fluid required to dilute the antipyrine injected can be calculated. The measurement assumes that none of the substance has been excreted in the urine, in which case a correction would have to be made.

Extracellular water determination involves the injection of a substance such as sodium thiocyanate, mannitol, or inulin, which is taken up or diluted by the extracellular compartment only.

Composition of Body Fluids

The principal constituents of intracellular, extracellular, and plasma fluids are shown in Table 12.3. Tissue or interstitial fluid has about the same composition as blood plasma except for the much higher level of proteins in the latter. The intracellular fluid differs mainly from the extracellular fluid in the concentration of Na^+ and K^+, which are high and low in extracellular fluid and low and high respectively in intracellular; Na^+ and K^+ figure prominently in cellular depolarization and repolarization.

Lymph fluid has the same composition as extracellular fluid except that it contains more protein (about 2 g %) than the latter.

Table 12.3. Principal constituents of body fluids (m Eq/liter).

Constituent	Blood Plasma	Extracellular (interstitial) Fluid	Intracellular Fluid
Na^+	152	143	12
K^+	5	4	157
Ca^{2+}	5	5	—
Cl^-	113	117	15
PO_4^{3-}	2	2	113
HCO_3^-	27	27	10
Mg^{2+}	13	3	26
Protein	16	2	74

—, Not applicable.

Formation

Formation of tissue fluids occurs by movement of fluid from capillaries to the tissues (filtration). Absorption takes place at the venule end.

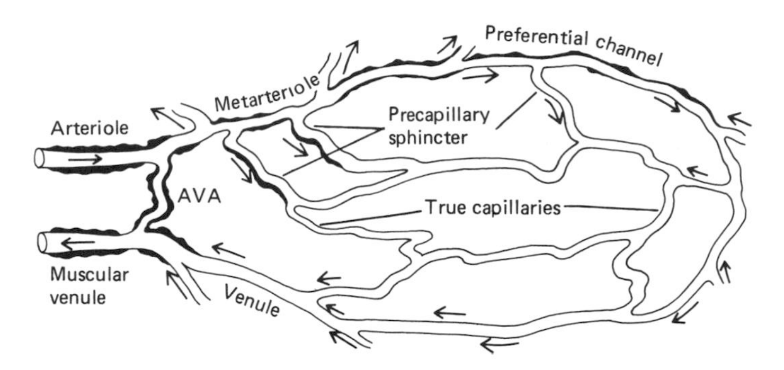

Fig. 12.1. Muscle capillary showing microcirculatory unit. (*AVA* = arteriole-venule anastomosis). (After Zweifach, B.W. Report from the Josiah Macy Jr. Foundation Conference, 1950.)

Anatomy of capillary. Diagrams of the capillary and its associated arterioles and venules are shown in Fig. 12.1. This is known as the microcirculatory unit (see also Chaps. 1 and 14). The true capillary is a thin-walled, endothelial-lined tube that is relatively permeable to fluids (Fig. 12.2) and small solutes but not to proteins and certain large molecules. This is so because the openings in the capillary wall (probably the junctions between the endothelial cells) are too small to allow their passage. Some capillaries, however, like those in the liver *are* permeable to proteins (see also Chap. 2). This fact suggests that there are pores large enough for macromolecules to pass through. *Electron microscopic studies,* however, do not reveal pores or clefts of sufficient size to allow such passage (Fig. 12.2). There are, however, numerous vesicles, and tagged proteins have been found in them, suggesting that they are transported out of capillaries by pinocytosis and exocytosis (see Chap. 2).

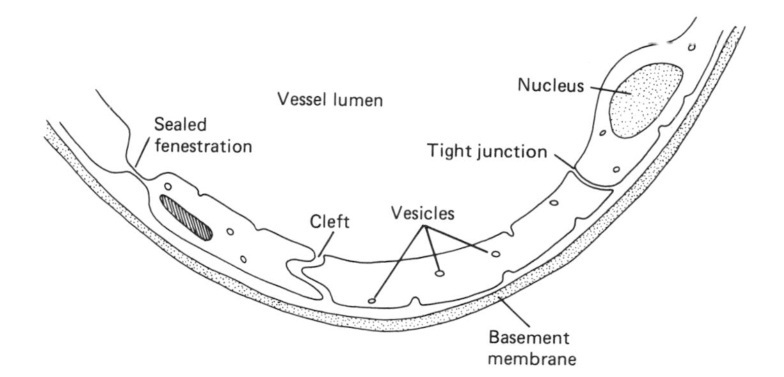

Fig. 12.2. Two epithelial cells making up part of capillary wall. The clefts provide for movement of fluid into and out of vessel lumen (See text).

Transcapillary Fluid Movement

Blood enters the capillary from the end of the arteriole and metarteriole and if the precapillary sphincters are open, it flows directly into the true capillary (Fig. 12.1). If they are closed, however, blood may flow in the preferential channel

and around to the venule without entering the true capillaries. Moreover, blood from the arteriole may be sent directly to the venule by an AVA shunt.
Transcapillary exchange occurs in the true capillary when fluid enters the tissue. Absorption of tissue fluid occurs in both venules and the venular ends of capillaries.

Starling's Hypothesis and Filtration

Formation of tissue fluid is regulated by the difference in capillary hydrostatic pressure (CHP) which tends to force fluid through the capillary wall (filtration) and plasma, osmotic pressure of the proteins (POP) which tends to hold fluid within the capillaries. These forces within the capillary (CHP and POP) are counteracted by tissue hydrostatic pressure (THP) and tissue osmotic pressure (TOP). Actually both of these forces are small and have little effect. Thus for tissue fluid formation or movement (TF):

$$TF = (CHP + TOP) - (THP + POP)$$

Starling's hypothesis is best illustrated by an example:

Capillary hydrostatic pressure (CHP) = 35 mm Hg
Tissue hydrostatic pressure (THP) = 1 mm Hg
Plasma osmotic pressure (POP) = 24 mm Hg
Tissue osmotic pressure (TOP) = 2 mm Hg

Movement of fluid from capillary to tissue (direction of force →)

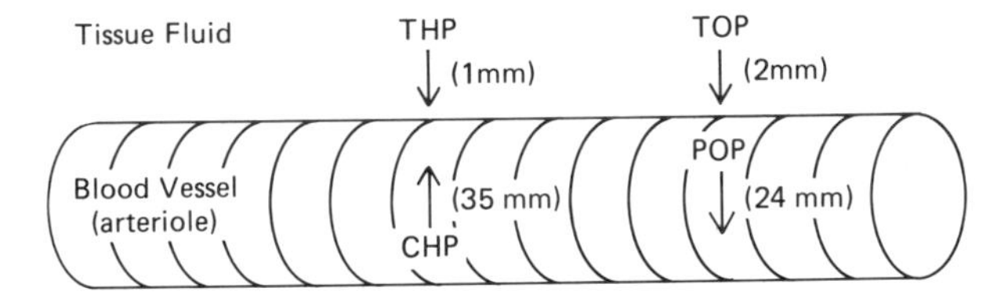

Difference: 34 mm Hg − 22 mm Hg = 12 mm Hg, which is the effective filtration pressure. Under these conditions fluid is forced out of the capillary into the tissues.
Movement of fluid from tissues to blood vessels, i.e., absorption:

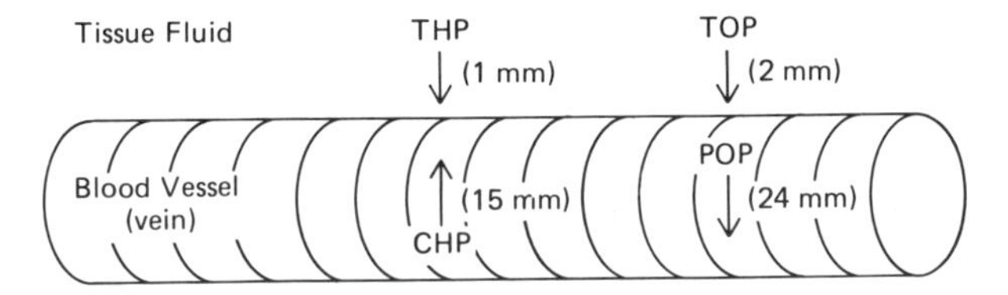

Difference: 14 mm Hg − 22 mm Hg = −8 mm Hg, which is the absorptive force; hence fluid is absorbed into venules and veins. Thus the main force in absorption is the osmotic pressure of the proteins, which is usually larger than CHP and therefore draws fluid into the vessels.

Permeability

Permeability depends on the effective filtration pressure, size of the pores, endothelial vesicles in the capillary wall, and the substance involved. The smaller the molecule, the more permeable. The permeability of the substances listed in Table 12.4 are arranged according to molecular size.

Table 12.4. Permeability coefficients.

Substance	Molecular Weight	Diffusion (cm^2/s)	Permeability (cm/s/100 g tissue)
Water	18	3.4×10^{-5}	28×10^{-5}
Sodium chloride	58	2.0	15
Glucose	180	9×10^{-1}	6
Inulin	5,500	2.4×10^{-1}	3×10^{-1}
Albumin	67,000	8.5×10^{-2}	1×10^{-3}

Filtration and Fluid Formation

Any factor that affects *capillary pressure* (Pc) affects filtration and tissue-fluid formation. Factors include:

Pa = Arterial pressure
Pv = Venous pressure
Pa = Resistance in precapillary (arterial end)
Rv = Resistance in postcapillary (venous end)
Pc = Capillary pressure

$$Pc = \frac{(Rv/Ra)\ Pa + Pv}{1 + (Rv/Ra)}$$

The formula indicates that:

1. An increase in Pa or Pv increases Pc
2. An increase in Pv is 5–10 times as effective as an increase of Pa on Pc
3. An increase in Rv increases Pc
4. An increase in Ra decreases Pc
5. Pc at the arterial end is 32–35 mm Hg in most capillaries and 15 at the venous end
6. An increase in central venous pressure of 10 cm water causes loss of 250 ml of fluid from plasma in 10 min (for adult person)

Fluid Exchange

Normally tissue fluid is formed and absorbed at rates such that there is no excessive accumulation in the tissues (edema). If, however, fluid is formed faster than it is absorbed, edema will result: (1) Decreased proteins in blood causes decreased POP and less absorptive force. (2) In-

creased venous pressure (CHP) above that of plasma osmotic pressure (POP) prevents absorption of tissue fluid and decreases lymphatic drainage and causes edema.
Factors favoring increased filtration and formation of fluid include (1) increased vasodilation at arterial end or decreased resistance (Ra), and (2) decreased proteins in blood or POP.
Factors favoring decreased filtration include (1) dehydration, and (2) vasoconstriction, or increased resistance (Ra) at arterial end. Tissues like the lungs are dry and form little or no tissue fluid because capillary pressure is lower than the plasma osmotic pressure.

Intake and Loss of Water

Because water intake and loss obviously influence the total body water, they tend to keep the water compartments in balance. The chief avenue for water loss is through the kidneys and water evaporation from skin and lungs. In excessive water intake, more water is excreted through kidneys; in thirst and dehydration little water is lost through the kidneys. Average daily water loss of an adult person amounts to about 1 liter through evaporation from lungs and skin and about 1.5 liters from the kidneys as urine, depending on ambient temperature and other factors.
Water intake is influenced by thirst, dehydration, and the need to replenish circulating plasma volume and ultimately other body fluids. Loss of water to the extent of about 0.8% of body weight causes thirst and an increase in water consumption. The daily average intake of water (drinking) is about 1.2 liters; about an equal amount of water is provided in the food that is eaten and metabolized.

Movement by Diffusion

Only 2% of plasma water is filtered out at the capillaries and normally a like amount is absorbed, but this relatively small amount is strategic and tends to keep the fluid compartment in balance (a homeostatic mechanism). Much more fluid is exchanged between the blood and tissues by diffusion (about 5000 times). Diffusion is a two-way process, and fluid continually moves back and forth from blood to tissue depending on the concentration of solute carried in the fluid. More details on the mechanism of diffusion are given in Chap. 2.

Osmolality of Intra- and Extracellular Compartments

If sodium chloride is ingested, there is an immediate increase in its osmolality (concentration) in the extracellular compartment (EC). Now fluid from the intracellular compartment (IC), which is less concentrated, will diffuse to the EC to dilute the higher concentrated compartment; water in the EC increases; water in the IC decreases.

Conversely, if hypotonic water is ingested there is an immediate decrease in sodium chloride-osmolality in the EC, and water moves out of it to the more concentrated IC (osmolality) to dilute it, with a resulting increase in water in the IC. Finally the concentration of sodium chloride of EC and IC will be the same at equilibrium (Fig. 2.3).

Lymph Formation

Lymph fluid is a type of tissue or interstitial fluid collected by special vessels or channels (Fig. 12.3) carrying the lymph to right and left lymphatic ducts that ultimately drain into the large central veins. Lymph differs from tissue fluid mainly

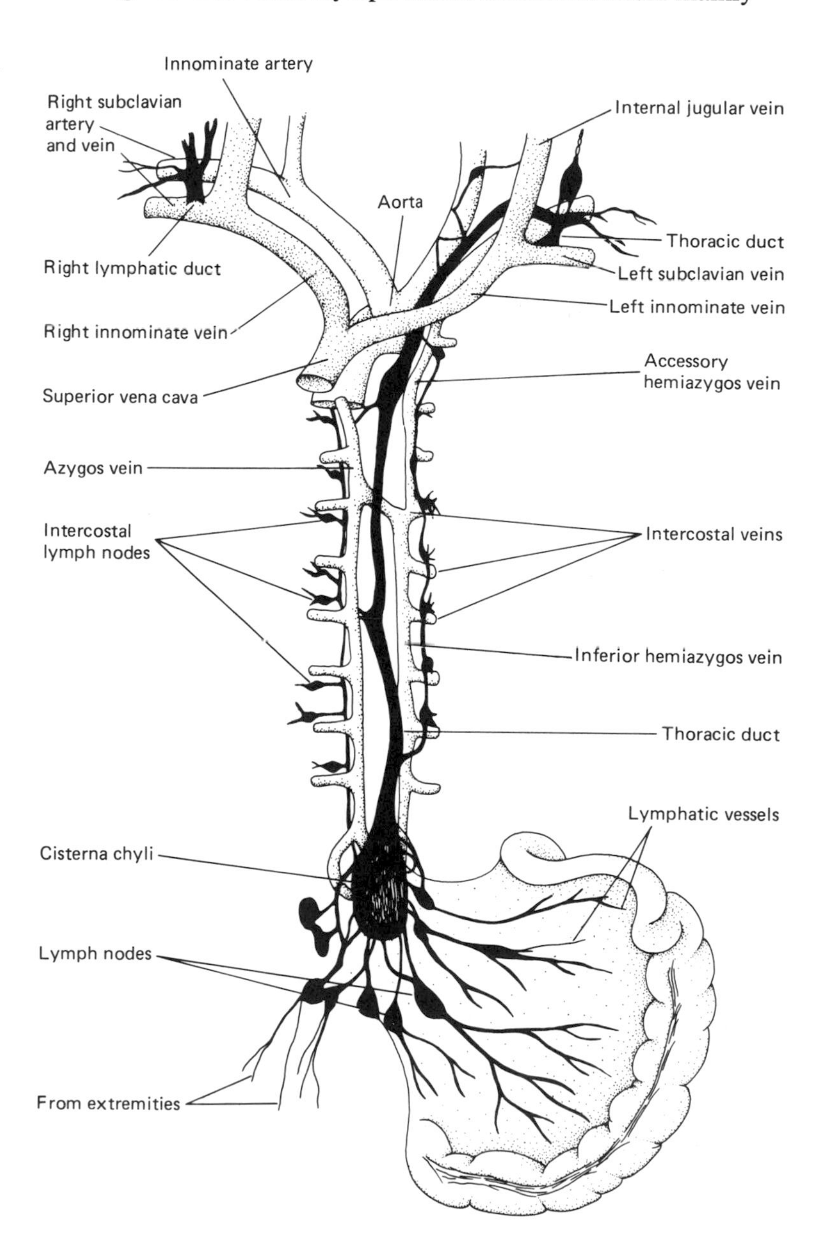

Fig. 12.3. Lymphatic drainage and vessels (see text).

in its content of protein, which is much higher (2g%) than the latter. The volume of lymph drainage is 2–4 liters/day.

The lymphatic vessels make up an essentially paravenous system, i.e., they run along with the veins and arteries. The vessels are deep and superficial and resemble the veins structurally except that they are thinner and more permeable than the latter. Like veins, they have valves to prevent the backflow of lymph.

Lymph vessels are located in nearly all organs except nails, cuticle and hair, cornea, and a few others. They are numerous in the liver and small intestine, are highly permeable, and contain high levels of protein (5–6 g%). Drainage is facilitated by lymphatic vessels (Fig. 12.3).

Lymph from the lower part of the body flows up the thoracic duct and empties into the venous system at the junction of the left internal jugular and subclavian veins. Lymph from the left side of the head and chest also enters the thoracic duct before it enters the veins.

Lymph from the right side of the head, neck, and arms enters the right lymph duct, which then empties into right subclavian vein. Average *lymph flow* in man is 1.4 ml/kg of body weight/h, or about 2 liters/24 h. One-fourth to one-half of the total plasma proteins formed are turned over in the lymph each 24 h.

Lymph flow is facilitated both by muscular contraction and by direct contraction or motility of the lymph vessels. Changes in tissue pressure and capillary pressure also affect flow.

Blockage of lymph vessels by a parasite causes a disorder known as elephantiasis characterized by a backup of lymph and tremendous swelling and enlargement of the leg muscles.

There is a pressure gradient in lymphatic vessels as in veins, with the highest pressure being at the periphery and the lowest at the central ends (thoracic and right lymphatic ducts).

Lymph nodes are small, oval bodies, varying in size from a pinpoint to an almond; they are located at areas around the lymph vessels (Fig. 12.3). The nodes are rather widely distributed, especially in the upper and lower extremities. They tend to serve as filters and as a defense against infection; they contain lymphocytes, a type of cell which combats foreign bodies and infections. Lymph nodes, such as those in the arms or legs (armpits and groin) may become enlarged and sore to the touch.

Selected Readings

Altman PL, Dittmer DS (1961) Blood and other body fluids. Fed Amer Soc Exp Biol, Washington

Altman PL, Dittmer DS (1974) Biology data book, 2nd edn, Vol III. Fed Amer Soc Exp Biol, Washington

Burke SR (1972) The composition and function of body fluids. Mosby, St. Louis
Guyton AC (1977) Basic human physiology, 2nd edn. Saunders, Philadelphia
Selkurt EE (1975) Basic physiology for the health sciences. Little, Brown, Boston
Williams WJ, Beutler E, Erslev AJ, et al (1972) Hematology; McGraw-Hill, New York

Review Questions

1. What are the compartments of body fluids? Name them and give the percentage distribution.
2. How would you measure blood volume, using the dilution principle?
3. How is tissue fluid formed? Show by example (Starling's hypothesis) how fluid (a) leaves blood and goes to the tissues and (b) how fluid is absorbed from tissues.
4. What determines the permeability of blood vessels?
5. What effect do the following have on formation of tissue fluid: (a) decreased protein content of blood, (b) decreased hydrostatic blood pressure, and (c) increased venous pressure?
6. What are the main lymphatic vessels that collect and drain the lymph?

Blood

Chapter 13

Whole blood is composed of a fluid portion (the plasma) which contains salts and other chemical constituents and certain formed elements, the corpuscles. These are the red (erythrocytes) and white (leukocytes) corpuscles which account for roughly 45% of total blood; the remainder (55%) is the plasma. The principal chemical constituents of blood are Na^+, K^+, P, Ca^{2+}, glucose, the plasma proteins (albumin, globulin and fibrinogen). These proteins amount to 6%-7% of the plasma. Fibrinogen is involved in blood clotting (coagulation), and albumin and globulin are larger molecular substances that do not readily pass through the semipermeable capillaries (blood vessels); thus they provide an osmotic pressure in the blood vessels (see Chaps. 2 and 12). This accommodation tends to prevent excessive passage of fluid into the interstitial tissues, and serves to control the fluid balance between blood and tissues (see Chap. 12).

Since there is almost twice as much albumin as globulin in the blood (A/G), the former has an important blood osmotic function and serves also to bind substances that are carried by the plasma (drugs, vitamins, hormones, and pigments). Globulin is a source of antibody production and a defense mechanism against diseases. Plasma proteins also aid in correcting or buffering changes in acidity (pH) of the blood.

The electrolytes Na, K, Ca, and P are important constituents of blood and most tissues and are involved in activation of the latter.

In addition to the special functions of the erythrocytes, leukocytes, and thrombocytes to be considered later, blood (1) absorbs and transports nutrients from the alimentary canal to the tissues; (2) transports blood gases to and from tissues; (3) removes metabolic waste products; (4) transports hormones; (5) regulates water content of tissues, pH, and body temperature; and (6) defends against infection by production of antibodies and other substances.

The *hematrocrit* represents the packed erythrocyte volume resulting from centrifugation of blood at 1500 g (1500 × 980

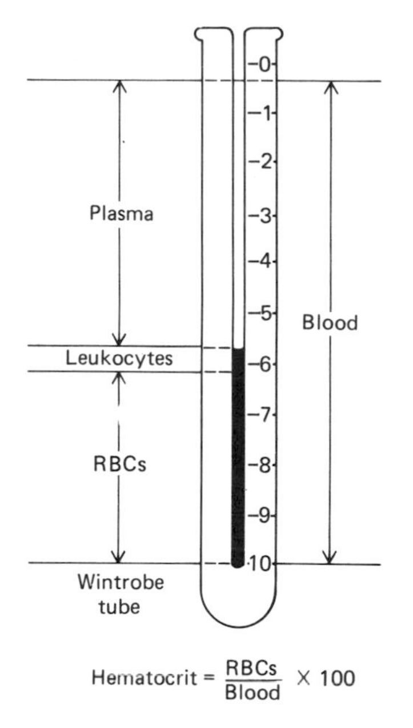

Fig. 13.1. Blood plasma and whole blood with red and white cells.

cm/s²) for 30 min or at higher speeds for less time. Blood is withdrawn from an artery or vein and prevented from clotting by treatment with an anticoagulant, such as sodium or postassium oxalate or citrate. The blood is placed in a hematocrit tube and centrifuged until the red blood cells (RBCs) settle to the bottom of the tube and plasma is above the cells. The proportion of each can be read off from the marked tube (Fig. 13.1). Since the red cell volume includes about 4% plasm (apparent hematocrit) the corrected hematocrit is obtained by multiplying the apparent hematocrit by 0.96. Factors affecting RBC size and number influence hematocrit.

In the hematocrit tube, just above the RBC layer is a thin buffy white or yellow layer containing the white blood cells (WBCs) or leukocytes.

Physical Characteristics

Blood is viscous, weighs more per given volume than water (specific gravity), and exerts osmotic pressure which influences the exchange of fluid between it and tissues.

Blood Volume

Blood represents the smallest of the body's water compartments except lymph, but it is one of the most important ones because it helps to regulate the other compartments and to influence blood pressure, the volume of blood returned to the heart, and the output of the heart.

The average volume of plasma in adult human beings is 4.3%-5% of the body weight, or about 3 liters; the volume of RBCs is about 2 liters. From 60%-80% total blood volume is distributed to the veins, and the remainder nurtures the arteries and capillaries.

Blood volume, like the other water compartments, remains fairly constant. Hemorrhage reduces the volume drastically but the effect is decreased by a shift of tissue fluids to the blood; other means of fluid conservation are also brought into play such as decreased evaporative losses and decreased excretion from kidneys. Losses in plasma or fluid volume can be corrected by ingestion of water, but loss of erythrocytes requires several days for replacement.

Measurement

Blood volume (plasma) is determined by injecting a known amount of a dye like Evans blue (T1824) into the bloodstream and determining its concentration after it mixes with the total circulating blood and before it disappears from the circulation.

$$\text{Thus, Plasma Volume} = \frac{\text{Amount of injected dye}}{\text{Concentration of dye in blood}}$$

The average mixing time of the dye is about 10 min for man.

$$\text{Total Blood Volume} = \frac{\text{Plasma volume} \times 100}{\%\ \text{plasma}}$$

A person with a plasma volume of 3000 ml and a cell volume (hematocrit) of 40% has:

$$\text{Total Blood Volume} = \frac{3000 \times 100}{60} = 5000 \text{ ml, or 5 liters.}$$

The *viscosity* of blood, about 5 times greater than that of water, is markedly influenced by the concentration of erythrocytes and plasma proteins. Increased viscosity increases both resistance to blood flow and the pumping action of the heart. The *specific gravity* of whole blood depends mainly on the number of erythrocytes and varies in human blood from 1.0520 to 1.0610; this is one of the factors influencing the rate at which RBCs settle in a test tube containing whole blood. The *sedimentation rate* is expressed in millimeters/hour and ranges from 2-10 mm/h. Certain diseases (rheumatic fever, tuberculosis, arthritis, and toxemias) increase sedimentation rate.

Formed Elements

The formed elements of the blood are erythrocytes, leukocytes, and platelets (thrombocytes) (Fig. 13.2). In whole blood there are roughly 5 million RBCs, 9000 to 10,000 leukocytes, and 300,000 platelets/mm^3 of blood (Table 13.1). The formed elements shown in Fig. 13.2 are derived from the bone marrow in several stages of cell division (cell lineages) (Fig. 13.3). Most investigators agree that the formed elements descend from one indifferent uncommitted stem cell (monophyletic view). This cell gives rise to four types of committed stem cells (CSC), which in turn produce the blast cell stages for the final formed elements. The blast cells produce the early stages of true cells (cytes) followed by the

Table 13.1. Normal values for formed elements of blood.

Element	Average (mm^3)	Range (mm^3)	% of total (average)
Total leukocytes	9000	5000–11,000	
Eosinophils	275	100–400	3.0
Basophils	25		0.27
Neutrophils	5000	3000–7000	55.50
Monocytes	400	100–600	4.44
Lymphocytes	2000	1000–3000	22.22
Erythrocytes	4.7×10^6	$4–6 \times 10^6$	
Platelets	300,000	200,000 to 400,000	

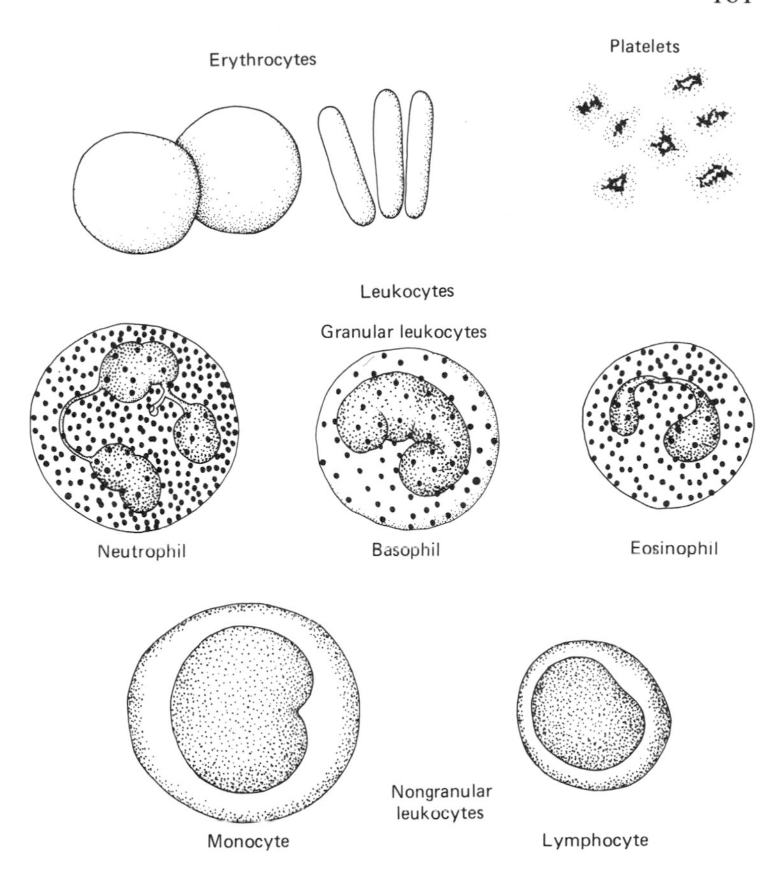

Fig. 13.2. Blood cells: erythrocytes, leukocytes, and platelets.

final stages. The mature human erythrocyte has no nucleus, but its progenitor, the normoblast, does, as has the erythrocyte of other species below mammals (birds and reptiles). Abnormal formation of erythrocytes may result in the disorder known as erythroblastosis fetalis, characterized by a

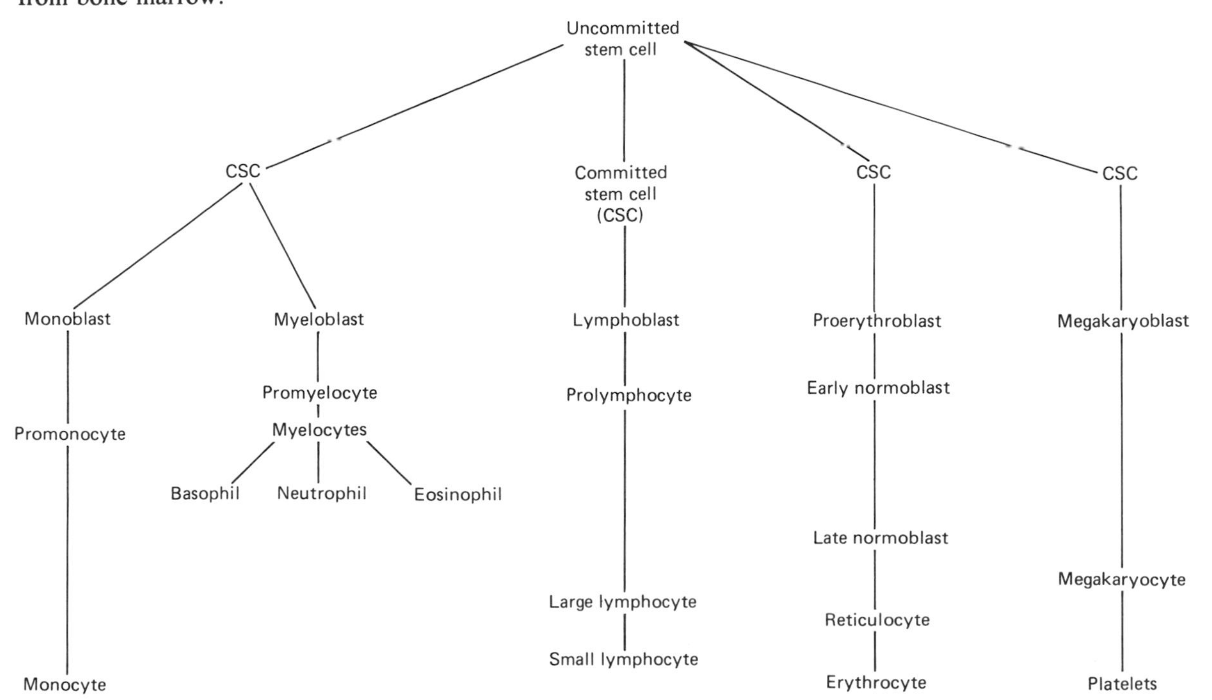

Fig. 13.3. Development of blood cells from bone marrow.

preponderance of erythroblast cells that fail to develop into mature erythrocytes in unborn or newborn babies.

Erythropoiesis is the process of red blood cell formation in the bone marrow. It is under humoral control; deficient oxygen supply (hypoxia) stimulates the production of a circulating substance in the blood termed erythropoietin (ERP), which is the erythropoiesis-stimulating factor. A glycoprotein formed in the kidney, ERP acts directly on bone marrow to increase both the production of erythroid cells (reticulocytes) and their conversion to mature erythrocytes.

Androgens and cortical hormones increase erythropoiesis; estrogens depress it. Formation of RBCs depends on an adequate supply of iron and protein for the synthesis of hemoglobin and other substances, such as B vitamins, particularly vitamin B_{12}. These deficiencies and other disorders cause various anemias.

Erythrocytes

The average life span of human erythrocytes is 120 days. The old dying RBCs become fragile, the cell membrane ruptures, and the cells fragment, but the iron that is released is used again in the bone marrow. The heme portion of the hemoglobin molecule is converted into the bile pigment, bilirubin, which is later secreted by the liver.

Resistance

Hemolysis is the act of rupturing of thc RBC and discharging of hemoglobin into the plasma. Freezing, thawing, and changes in osmotic pressure of blood are factors that produce hemolysis. Solutions with the same osmotic pressure as blood that do not cause hemolysis are isotonic. Solutions with lower osmotic pressure than that of blood are hypotonic, and those with higher pressures than that of blood are hypertonic. Hypotonic solutions cause the cells to burst by increasing water content; hypertonic solutions cause the cells to shrink because water is lost from them.

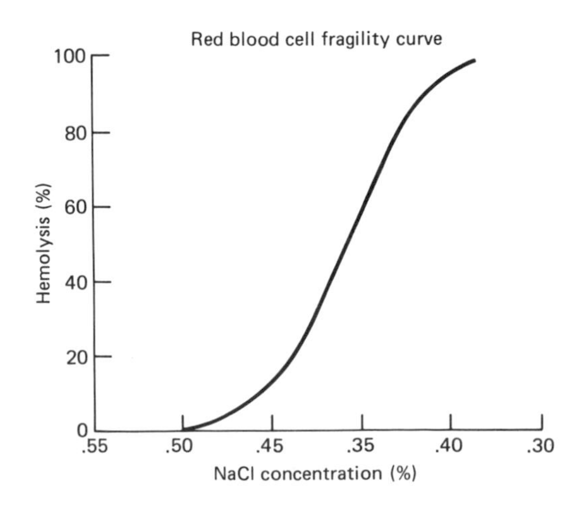

Fig. 13.4. Red blood cells fragility as influenced by different concentratons of NaCl. As concentration of NaCl decreases and solution becomes hypotonic, more blood cells burst or hemolyze.

The fragilty of RBCs may be determined by their resistance to solutions of known concentrations of NaCl where the solutions are hypotonic (isotonic NaCl is about 0.9%) (Fig. 13.4).
Erythrocytes are biconcave discs with a diameter of about 8.5 microns (Fig. 13.1). Shape and surface area of the cell favor maximum diffusion and transport of 0_2 and Co_2. The average number of RBCs in the normal male adult is 5 million/ mm^3 (range 4-6), and in adult females about 4.5 million (Table 13.1).
Several factors affect the number, such as adequate supplies of iron, B-vitamins, dietary proteins, altitude hypoxia, and exercise.

Anemias

Anemias may be classified according to number and size of cells and the hemoglobin content as shown in the following list.

1. Normocytic and normochromic:
   Decrease in number of cells, but normal size and hemoglobin content; they develop following acute hemorrhage because plasma volume is replenished long before RBCs are restored.
2. Microcytic and hypochromic (decreased pigment):
   Decrease in size and number of cells and amount of hemoglobin; also called iron-deficiency anemia.
3. Macrocytic and hyperchromic:
   Decrease in number but increase in size of cells with high concentration of hemoglobin; characteristic of pernicious anemia caused by lack of extrinsic factors (vitamin B_{12}) or intrinsic factors produced in gastric mucosa.
4. Aplastic anemia:
   Caused by malfunctioning bone marrow; may result from excessive X-radiation or from unknown cause.
5. Sickle cell anemia:
   Cells are abnormal in shape (sickle) and do not transport oxygen normally; hereditary and particularly prevalent in Blacks.

Anemias cause decreased efficiency of the transport of oxygen to the tissues because of decreased hemoglobin. Anemic persons fatigue easily and their endurance is weakened. *Polycythemia* is a disorder characterized by increased erythrocytes, and it may be caused by dehydration or by an overactive bone marrow. Hypoxia, or high altitude, markedly increases the number; however, after a normal subject adapts to high altitude, the numbers decrease and may approach those observed at sea levels.

Platelets and Coagulation

Platelets (thrombocytes) of mammals are cytoplasmic fragments without nuclei (Fig. 13.1) that have developed from megakaryocytes; in lower forms (birds and reptiles) the cells are nucleated. Mammalian platelets are 2-4 μm in diameter. The number of platelets varies from 200,000-400,000/mm^3 of human blood (Table 13.1). Platelets are believed to play a role in hemostasis or clotting of blood. They aggregate at the site of a ruptured blood vessel and, with thromboplastin, help to form a clot.

When whole blood coagulates, it forms a clot or solid plug made up of fibrin. This insoluble protein is derived from fibrinogen, which is activated by the enzyme thrombin, which in turn is derived from prothrombin, a material activated by several accelerating or converting factors. Details of the steps in blood clotting are shown in Fig. 13.5 and the factors are summarized in Table 13.2.

Blood serum that is clear and free of cells like plasma does not contain fibrinogen, which has been used up in the clotting process.

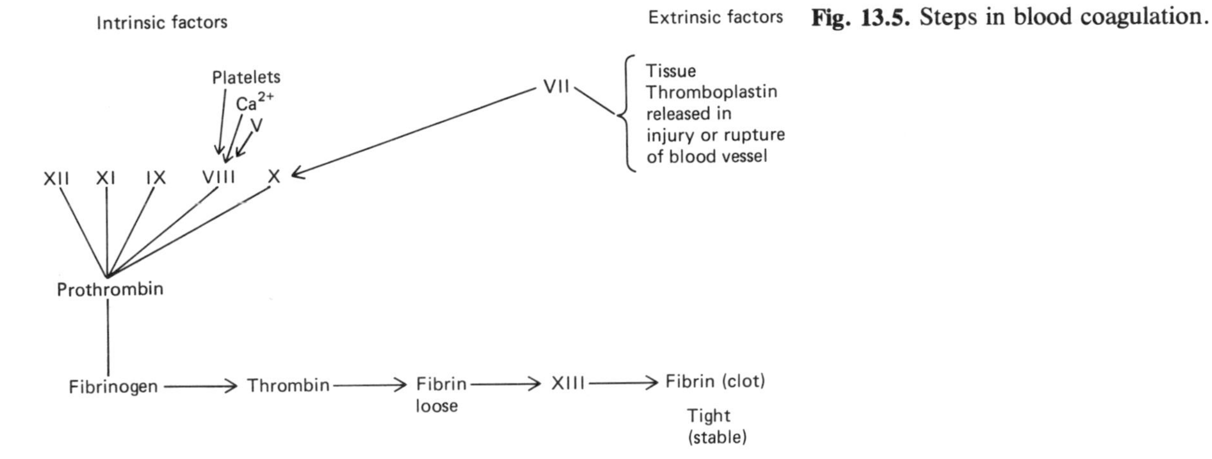

Fig. 13.5. Steps in blood coagulation.

Table 13.2. Factors in Coagulation.

I	Fibrinogen
II	Prothrombin
III	Tissue thromboplastin
IV	Calcium^{2+}
V	Labile factor, proaccelerin, accelerator globulin
VI	Stable factor, proconvertin; serum prothrombin conversion accelerator (SPCA)
VII	Antihemophilic factor A, antihemophilic globulin (AHG), platelet cofactor I, thromboplastinogen A
VIII	Antihemophilic factor B, plasma thromboplastin component (PTC), platelet cofactor II, Christmas factor (CF)
IX	Stuart (Prower) factor
X	Antihemophilic factor C, plasma thromboplastic antecedent (PTA)
XI	Hageman factor
XII	Fibrin stabilizing factor

Coagulation Time

Coagulation time represents the interval required for blood held in a test tube at 37°C to clot; normal time is 3-8 min. Prothrombin time is a better measure of the clotting process and refers to the time required for blood to clot in the presence of excessive thromboplastin and calcium added in known amounts to blood. Ca^{2+} had been removed previously from blood sample. Normal prothrombin time is 12-17 s. Increased time occurs in persons deficient in certain factors (V, VII, and X).

Anticoagulants prevent blood from clotting. In the test tube certain substances (oxalates or citrates) remove or precipitate the calcium in the blood and prevent clotting. *Heparin,* produced in the body, normally prevents coagulation by inhibiting the action of thrombin, prothrombin, and thromboplastin. Vitamin K is necessary for normal clotting. *Dicumarol,* a substance originally found in sweet clover, impairs production of prothrombin by inhibiting the utilization of vitamin K. Heparin and dicumarol may be injected without ill effects, but oxalates and citrates precipitate calcium and should not be injected because they are lethal.

Blood Groups

Blood from certain persons when transfused to other persons may cause a severe reaction leading to death. This is because human RBCs contain antigens or agglutinogens that produce reactions or antibodies, (agglutinins) and coagulation when transfused. The severity of the reaction depends on the blood group types. The common blood groups are O, A, B, and AB.

Table 13.3. Determination of blood cell groups.

Serum[a]	Reaction[b]	Cells Belong to Group
A and B	Ag	AB
A	Ag	B
B	Ag	A
A and B	none	O

[a] Known groups to unknown cells.
[b] Ag = agglutination.

Table 13.4. Blood groups and their reactions when cross transfused[a].

Blood Group	Cross transfusion	Reaction
A	Cells to serum B and D	Ag
	Serum to cells B and AB	Ag
B	Cells to serum A and O	Ag
	Serum to cell A and AB	Ag
AB	Cells to serum A, B, O	Ag
	Serum to cells A, B, AB, O	NE
O	Cell to serum A, B, AB, O	NE
	Serum to cells A, B, AB	Ag

[a] Cells of one group cross-transfused to serum of another.
[b] Ag = agglutination' NE = no effect.

Persons with type AB blood are universal recipients because they have no circulating antibodies (agglutinins) and can receive blood of any type without causing coagulation (Tables 13.3 and 13.4). Persons with type O blood are uni-

versal donors because their blood contains no antiagglutinins and can be given to others without reactions from ABO incompatability.
The percentage distribution of these groups in the United States is A, 41; B, 10; AB, 4; and O, 45.
An example of ABO incompatability is shown as follows:

B cells in B plasma—normal

A cells in B plasma—
clumping (agglutination)

It is possible to determine the blood group to which persons belong by testing their blood cells in the serum of known blood groups (Table 13.4).

Rh Factors

Another group of agglutinogens (Rh) is found in the blood of most persons; when present the blood is said to be Rh-positive. If the blood of an Rh-positive person is transfused to one who is Rh-negative, no immediate effect is evident, but the Rh-negative recipient develops antibodies to the positive agglutinogens, a condition that causes difficulties if such a transfusion is repeated.
The Rh-positive factor is dominant to the Rh-negative one, and offspring from Rh-positive and Rh-negative parents will be Rh-positive. If the man is Rh-positive and the pregnant mother Rh-negative, the fetus will be Rh-positive. The fetal agglutinogens diffuse across the placenta and into the mother who then develops antibodies (agglutinins) to Rh-position blood; if these agglutinins then rediffuse from mother across the placenta to the fetus, agglutination occurs leading to erythroblastosis fetalis (see this chapter, "Formed Elements"). Usually death occurs at about the time of birth. The first-born child of an Rh-negative mother may be sound, but subsequent fetuses will likely die before birth. The Rh-incompatabilty can be minimized by administering anti-Rh gamma globulin to the pregnant Rh-negative woman, thus neutralizing the fetal Rh-negative antigens.

M and N Factors

These agglutinogens give rise to three blood groups: M, N, and MN, but they do produce serum antibodies and hence are not problems in blood transfusions.

Hemoglobin

Hemoglobin is the oxygen-carrying pigment in the erythrocytes. It loads up with oxygen (becomes saturated) in the lungs and unloads its oxygen to the tissues (see Chap. 17).

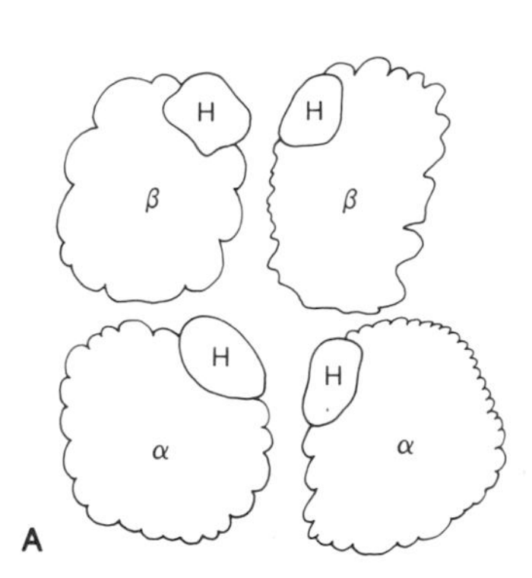

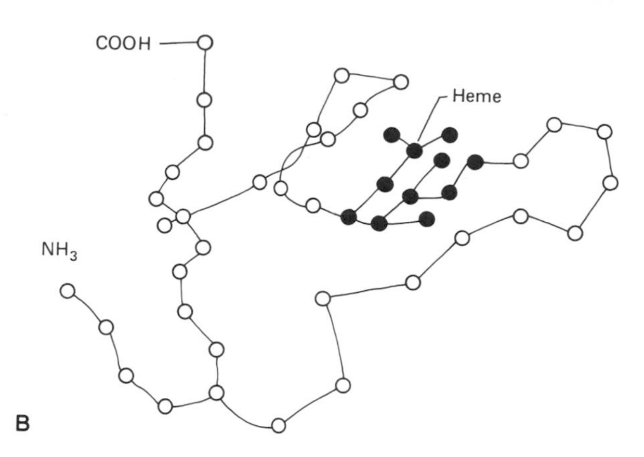

Fig. 13.6A and B. One molecule of hemoglobin (Hb). **A** Molecule comprises four subunits (alpha and beta), each consisting of one polypeptide chain and one heme (H). **B** One subunit of Hb with heme enfolded by a polypeptide which is a linear sequence of amino acids represented by open circles in chain. Chain begins with amino group (NH_3) and ends with carboxyl group (COOH). (Modified after Perutz, M.F. Scientific American 239, no. 6, p. 92, 1978.)

Hemoglobin is a protein with a molecular weight of over 60,000. The molecule is made up of four subunits; each subunit contains heme, the iron-containing porphyrin derivative. This derivative is conjugated, or attached to a more complex structure, a polypeptide, and the latter is often referred to as the globin part of the molecule. There are 2α and 2β-polypeptide chains (Fig. 13.6). Each of the chains contains more than 140 amino acid residues. Normal human blood contains mainly two types of hemoglobin; type A is found in adults and type F is found in the fetus. Other (abnormal) types are reported.

Hemoglobin (Hb) binds O_2 to iron (Fe^{2+}) in heme to form oxyhemoglobin (fully oxygenated). When hemoglobin gives up its oxygen, it is reduced or deoxygenated. Oxygen association and dissociation are discussed in greater detail in Chap. 17. Carbon monoxide reacts with hemoglobin to form carboxy-hemoglobin more readily than does O_2; it displaces O_2 on the Hb molecule and may lead to death. The hemoglobin content of normal human blood varies from 15–16 g%; the level is slightly higher in males. Each gram of Hb can combine with 1.3 ml of O_2.

Leukocytes

Leukocytes are derived from one uncommitted stem cell in the bone marrow which gives rise to committed stem cells (Fig. 13.3). The latter ultimately produce the monocytes (mononuclear) and basophils, neutrophils, and eosinophils, which are polymorphonuclear. Non-mammalian species like birds have heterophils instead of neutrophils. The polymorphonuclear cells have cytoplasmic granular particles that have a characteristic shape and stain a distinctive color. Basophils have large purple or violet granules (basic stain), and the eosinophil granules take on acid stain or eosin color (pink-yellow). Neutrophils are neutral in their staining reaction.

Monocytes have a large nucleus with a relatively small cytoplasm. Lymphocytes also have a single nucleus; some lymphocytes are found in the bone marrow, but most of them are formed in the lymph nodes, spleen, and thymus from the original bone-marrow stem cells.

The total number of leukocytes is variable, ranging from 5000–11000/mm³ normally and increasing greatly in certain diseases and disorders (leukocytosis) and from strenuous exercise. The distribution of cell types and percentages of the total leukocyte count are shown in Table 13.1.

The main function of the leukocytes is to combat foreign agents including toxemias and bacteria and to produce antibodies. They phagocytize (engulf) and destroy such materials. Leukocytes are attracted to areas of injury or infection. Neutrophils are the most numerous leukocytes and therefore the most important of the phagocytic cells. Eosinophils and basophils are few and their function is dubious.

Lymphocytes, making up more than 20% of the total leukocyte count are very important as a defense mechanism against disease. Their phagocytic activity is rare, but they are important in synthesizing and releasing antibodies. Monocytes also exhibit phagocytic activity.

In *leukocytosis* the total count may increase from fivefold to 20-fold. Leukemia is a form of leukocytosis in which there is a malignant proliferation and formation of granulocytes. *Leukopenia* represents a decrease in total leukocyte count and may be caused by a depression of bone-marrow activity, resulting from X-radiation or the ingestion of toxic drugs.

Selected Readings

Altman PL, Dittmer DS (1961) Blood and other body fluids. Fed Amer Soc Exp Biol, Washington

Altman PL, Dittmer DS (1974) Biology data book, 2nd edn, Vol III. Fed Amer Soc Exp Biol, Washington

Guyton AC (1977) Basic human physiology, 2nd edn. Saunders, Philadelphia,

Perutz, MF (1978) Hemoglobin structure and respiratory transport. Scientific American 239, no 6, p 92.

Selkurt EE (1975) Basic physiology for the health sciences. Little, Brown, Boston

Williams WJ, Beutler E, Erslev AJ, Rundles, RW (1972) Hematology. McGraw-Hill, New York

Review Questions

1. What are the main constituents of whole blood? What is plasma? What is serum?
2. How many erythrocytes are present in human male blood and in human female blood?

3. How much hemoglobin is in blood? What is the main function of hemoglobin?
4. What is the hematocrit?
5. What is erythropoiesis and erythropoietin?
6. What are: (a) microcytic and hypochromic anemia, (b) macrocytic anemia, (c) sickle cell anemia, and (d) pernicious anemia?
7. Define (a) polycythemia, (b) platelets.
8. What are the final and principal steps in coagulation of blood?
9. Name the principal blood groups.
10. Persons with blood group _____ are considered to be universal donors, and universal recipients have blood group _____.
11. More persons belong to groups A than O than other types (true or false).
12. If the blood cells of one group are incompatible with the serum of another blood group, what happens?
13. Define agglutinogens and agglutinins.
14. Define Rh factor. What are the dangers of being Rh-negative?
15. Name and define the leukocytes.
16. What is (a) leukocytosis, (b) leukopenia?

Circulation

Chapter 14

General Characteristics

The circulatory system consists of a central pump (heart) and the peripheral blood vessels (arteries, veins, and capillaries). The arteries carry blood pumped by the heart to the tissues by smaller arteries (arterioles) and capillaries. Blood is returned to the heart by smaller veins (venules) and larger veins. Figure 14.1 shows the circulation of blood to the principal organs and systems (see Chap. 1 for further details). Thus arterial (oxygenated) blood is pumped from the left ventricle through the aorta to the organs, and venous blood is returned to the right atrium, where it goes to the right ventricle, to the pulmonary arteries to the lungs where it is oxygenated. It then returns to the left atrium through the pulmonary veins. The pressure of the blood in the pulmonary arteries and veins is lower than systemic arterial pressure.

The arterial system is a high-pressure system, the venous system a low-pressure one, and structural changes in these vessels adapt them to these differences.

Anatomy

The arteries are made up of different tissues from inside (lumen) to outside. The innermost coat of the lumen is the endothelium; next to it is the elastic tissue coat (Fig. 14.2),

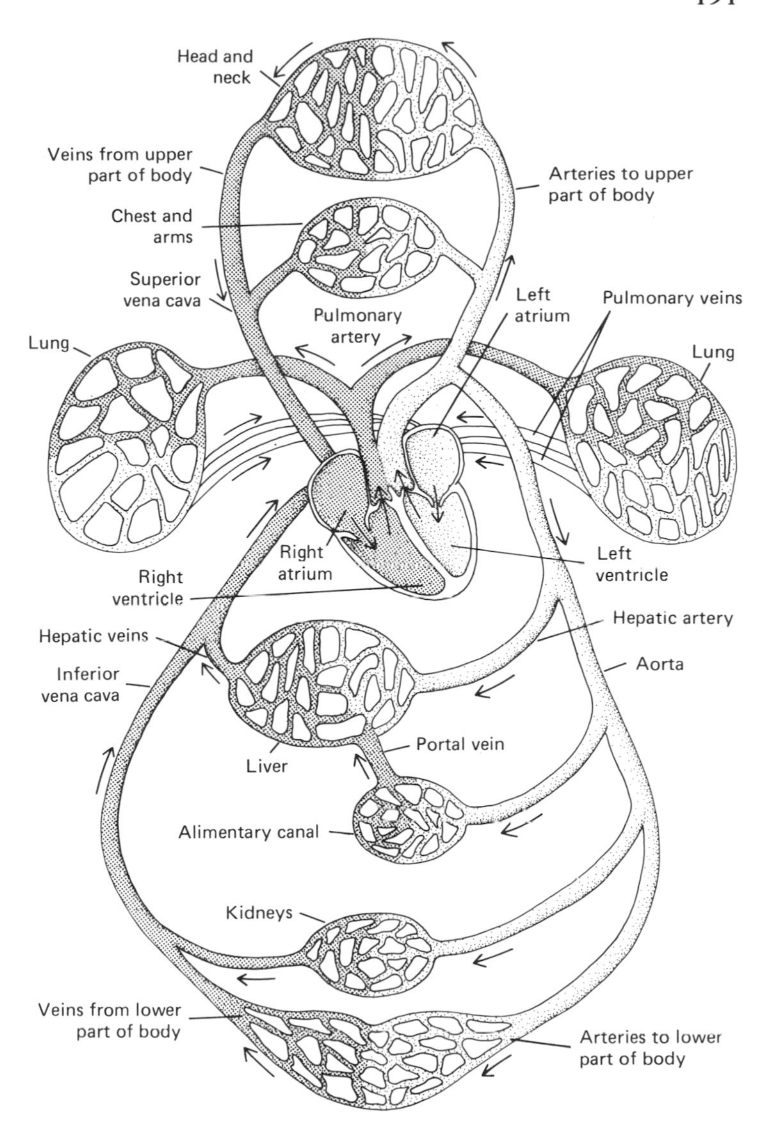

Fig. 14.1. Circulation of blood (arterial blood in white, venous blood in black).

which varies in amount and thickness in different blood vessels. The next layer is the smooth muscle tissue, which is critical in causing the vessel to constrict or dilate. There are two types of smooth muscles (circular and longitudinal). The circular muscles cause the vessels to constrict in shorter, limited regions, whereas the longitudinal muscles cause them to constrict over more extended areas. Blood vessels (vaso vasorum) and nerves are found under muscle layers. Stimulation of the sympathetic nerves causes contraction of the smooth muscles and vascular constriction. The outermost layer of tissue comprises collagen fibers; they tend to stretch but do not have the resilience of elastic tissues.

The size of the blood vessels and the distribution of the tissues in them vary in the different blood vessels (Fig. 14.2). In general, arteries have more elastic tissue and less collagen than veins, but the latter have more collagen than elastic fibers.

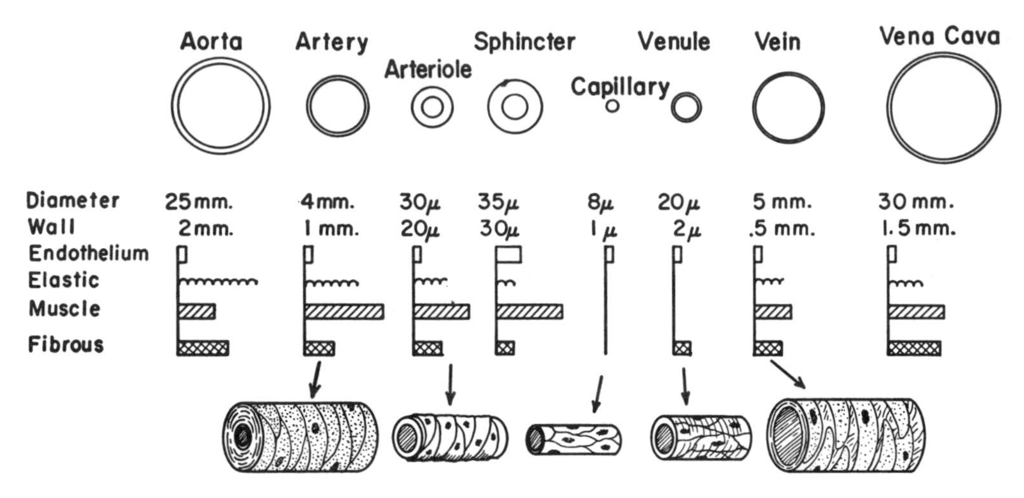

Fig. 14.2. Structure of blood vessels in various parts of vascular system. Entire vascular system is lined with sheath of endothelial cells, but amount of smooth muscle and connective tissue varies. (Reproduced with permission from Rushmer, R.F. Cardiovascular Dynamics, 3rd ed. W.B. Saunders Co., 1970.)

Capillaries have an endothelial layer but no muscle or connective tissue. They are relatively passive and their motility is governed largely by changes in the arteriole and venule ends (see Chap. 12).
Lymphatic vessels are specialized, similar to veins except that they are thinner and more permeable than the latter (see also Chap. 12).

Fluid Dynamics (Hemodynamics)

The heart pumps blood through blood vessels of varying size, distensibility, and resistance. Since blood contains formed elements, it is viscous and exerts a greater resistance to flow than does water. The volume of blood (flow) delivered to an organ in a given time (volume/time) is determined by the pressure difference (ΔP) exerted on the resistance (R) to flow as follows:

$$F \text{ (flow)} = P/R; \text{ therefore } P = FR \text{ and } R = P/F$$

Figure 14.3 illustrates these relationships where there is a pressure drop (difference) between points P_1 and P_2, which

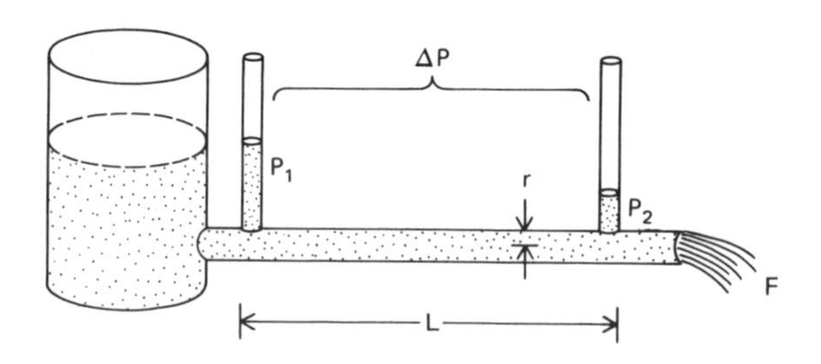

Fig. 14.3. Relationships between pressure, flow and resistance in tubes. Flow is proportional to the pressure gradient (ΔP), directly proportional to vessel radius (r^4), and inversely proportional to length (L) or resistance (see text).

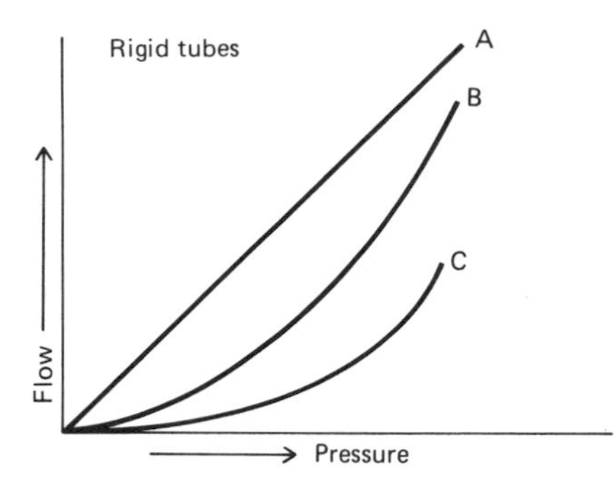

Fig. 14.4. Direct relationship of pressure and flow in rigid tube (A) and blood vessel which has elasticity and distensibility and which for a given increase in pressure (curve B) shows less than an equal (linear) change in flow. Curve C represents blood vessel whose smooth muscle is stimulated to contract or constrict, causing increase in pressure and a lesser increase in flow.

is determined by the distance between the points, or the length of the vessels (L). The radius (r) of the tube is fixed and exerts a constant resistance to flow. For rigid tubes, flow is directly proportional to pressure in a linear fashion, but not in a blood vessel which has elasticity and distensibility (Fig. 14.4).

Flow (F) is directly proportional to the radius of the vessel raised to the fourth power (r^4); this means that if the radius is increased 16%, the flow rate is increased 100%. Thus slight changes in diameter of the lumen of the blood vessel have pronounced effects on flow.

Resistance (R) is influenced inversely by the viscosity of the fluid. For water (a Newtonian fluid) this factor is constant. For blood the viscosity depends on the number of erythrocytes, plasma proteins, and other factors. The greater the viscosity, the less the flow.

The *Poiseuille equation* describes in greater detail the factors in the resistance to flow:

$R = \frac{P}{F} = 8 \times \eta \times \frac{L}{r^4}$, where η is the viscosity (in poise) and 8 is the factor of integration.

The interrelationships of pressure, flow, and resistance are summarized as follows:

1. Flow is directly related to pressure head and radius of vessel (r^4)
2. Flow is inversely related to the length of blood vessel (L) and the viscosity of fluid (η)
3. Resistance to flow is directly proportional to the length of vessel and the viscosity
4. Resistance to flow is inversely proportional to the radius of vessel (r^4).

Peripheral Resistance Units

The absolute units for calculation of resistance are dynes/s/cm^5, a value of about 1700 in man.

A more meaningful figure for comparing changes in relative resistance is the peripheral resistance unit (PRU) calculated as follows:

$$PRU = \frac{\text{Cardiac output (1 or ml/min)}}{\text{Blood pressure (mm Hg)}}$$

An increase in PRUs indicates an increase in resistance to blood flow and may also mean an increase in vasoconstrictor tone (although not always) in blood vessels.

Velocity of Blood Flow

Blood flow like the flow of water in a stream, may be laminar or turbulent. Flowing fluid may be regarded as thin layers (laminae) of fluid sliding past each other. The force tending to slow each of these laminae is the shear stress or force.

A blood vessel with laminar flow exhibits a flow profile which depends on the velocity of flow, which in turn is influenced by, among other factors, the size of vessel and resistance (Fig. 14.5). At a given pressure head and viscosity, the velocity of flow is inversely proportional to the vessel's radius or cross-sectional area (CSA). Thus, velocity tends to increase in a smaller vessel, and velocity of flow,

$$V = \frac{F \text{ (volume of flow)}}{CSA \text{ (cross-sectional area)}}$$

In the human aorta with a diameter of 2 cm, a CSA of 3 cm², and a flow rate (volume flow) of 84 ml/s, the velocity of flow is calculated as:

$$V = \frac{F}{CSA} = \frac{84 \text{ ml/s or } (cm^3/s)}{3 \text{ cm}^2} = 28 \text{ cm/s}$$

In smaller arteries the velocity is much higher, and in larger veins it is lower.

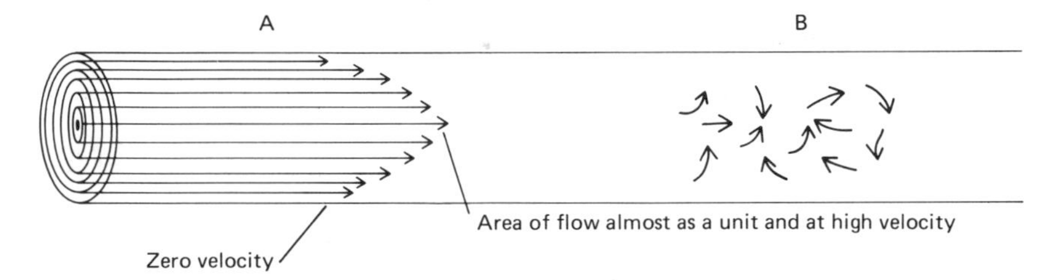

Fig. 14.5. **A** Velocity profile of laminar flowing stream. Velocity at wall of tube = 0, but velocity increases to a maximum at center of tube. **B** Turbulent flow characterized by flow in swirls and vortices.

When the velocity of flow of a laminar stream is increased to a certain point, the flow becomes *turbulent,* or flows in swirls, as in a swift stream of water (Fig. 14.5B), which becomes noisy. This point is referred to as Reynold's Number (Re), and $Re = \frac{VD\blacksquare}{V_i}$, where V = velocity of fluid, D = diameter of vessels, ■ = density of fluid, and Vi = viscosity of fluid.

This average number is about 2000, depending on the size of vessels; it is lower in a large vessel and higher in a small vessel. Flow is likelier to become turbulent at the branches of vessels than at the mainstream. Consequently vessels are likelier to develop structural abnormalities at these points (arteriosclerosis).

Law of LaPlace

It is well known that small arteries and even capillaries are less likely to rupture than are larger vessels, which may appear surprising. The main reason for this is their small diameter and the operation of the law of LaPlace. The law states

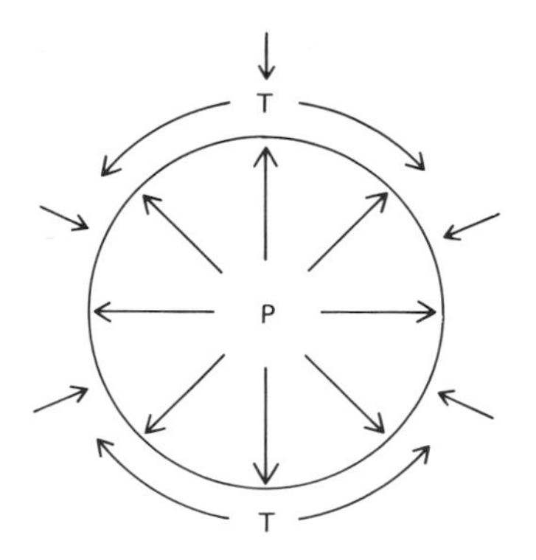

Fig. 14.6. Relationship between pressure (P) within blood vessel and tension in wall (T), or force required to prevent vessel from rupturing (law of LaPlace).

that the distending pressure (P) within a hollow vessel is equal to the tension in the wall (T) divided by the radius (R) (Fig. 14.6). For blood vessels, $P = \frac{T}{R}$; $T = PR$; $R = PT$

This means that:

1. Increase in pressure (P) causes increase in tension (T)
2. Since pressure (P) is indirectly related to radius, a vessel with smaller radius can withstand higher pressure because tension (T) is relatively greater
3. Radius (R) is directly related to tension (T) or ($T = PR$); an increase in R increases T; and a decrease in R decreases T.

Because of the operation of the law of LaPlace, smaller blood vessels and smaller hearts tend both to resist higher pressures and the tendency to rupture is less than larger structures.

The law of LaPlace is concerned with *inactive* tension, or tension created by inherent structural characteristics of the blood vessels, such as the amount of elastic tissue and collagen. *Active tension* is the tension caused by contraction of the blood vessel's smooth muscles, constriction of the vessel, and decreased blood flow. If nerves to these muscles are stimulated at increasing frequencies, the pressure in the vessels increases and the blood flow decreases (Fig. 14.4, curve C).

Transmural pressure represents the difference in pressure exerted on the blood vessels by forces outside the vessel, namely by the surrounding tissues and fluid and those within the vessel (blood pressure). A contracting muscle, e.g., may momentarily shut off blood flow because the constricting force outside the blood vessel is greater than the force or pressure within the vessel.

Cardiac Output and Blood Flow

Approximately 80%–85% of the total circulating blood volume is in the systemic circulation and the remainder in the pulmonary circulation (that going to and coming from the lungs). The actual distribution of this volume is shown in Figure 14.7; more than half of the blood of the systemic circulation is in the venous system.

Cardiac output represents the total output of blood from the heart in a given time and is usually referred to as the output in one minute (minute output). The output per heart beat is called stroke output or volume; minute output is determined by stroke volume × heart beat.

Cardiac output (CO) can be determined by several methods, including the direct Fick and the indirect Fick and which in-

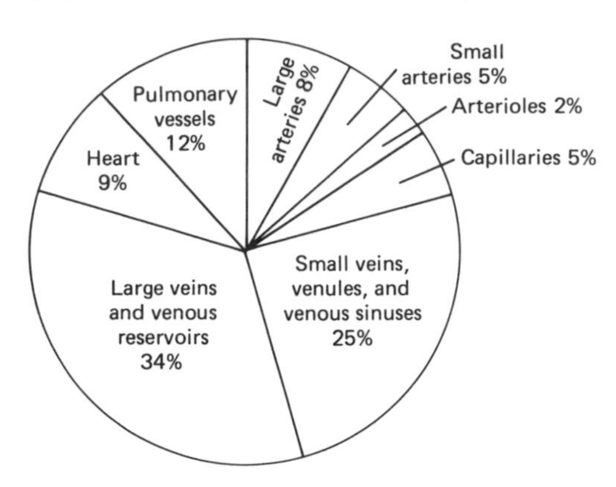

Fig. 14.7. Percentage of total blood volume in each portion of circulatory system.

volve dyes and the dilution principle, radioactive substances, and heat or thermodilution (see Chap. 2). Cardiac output may also be determined directly by electromagnetic probes which fit around an appropriate artery at right angles to the vessel. These probes detect and measure the velocity of blood flow from which volume can be calculated.

The *direct Fick method* involves measurement of O_2 consumed by the body (taken in by lungs) and dividing this by the difference in volume percent of O_2 content of arterial (AO_2) and mixed venous blood (VO_2). Thus

$$CO = \frac{O_2 \text{ consumption (ml/min)}}{AO_2 - VO_2}$$

The *indirect Fick method* (dye dilution) involves the injection of a known quantity of dye or other agent into a vein leading to the right heart, where it is mixed. An artery is cannulated and blood is collected as it passes the sampling point immediately after the injection of dye. The blood-containing dye may be run directly through a densitometer (photoelectric cell), which detects the dye concentration, and a dye curve is inscribed, or the samples may be collected directly and concentrations determined at given intervals. In either case a concentration curve is formed (Fig. 14.8). Before all of the injected original dye has passed the sampling point, the first part of dye leaving the heart has been recirculated; this causes the curve (concentration) to become elevated (see R on curve in Fig. 14.8). This part of

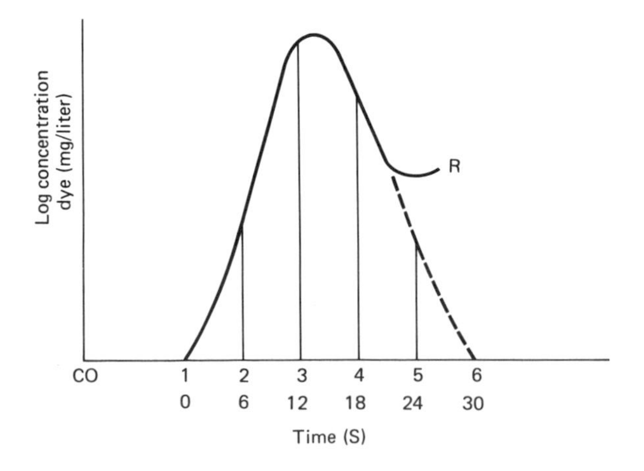

Fig. 14.8. Cardiac output curve (dye/dilution technique) plotted on semilog paper. R = recirculation (see text for further description).

the curve is ignored and the down sloping line is extrapolated to zero (dotted line on curve).

Cardiac output (CO) is determined by measuring directly (by planimetry) the area under the curve, or by determining the average or mean concentration of dye ($\overline{C}$) under the curve.

$$\text{Thus CO} = \frac{\text{A (amount of dye injected [mgs])}}{\overline{C}\ \text{time under curve (duration)}}$$

$$\text{CO} = \frac{\text{Amount of dye injected (12 mg)}}{\overline{C} \times \text{time (30s)}}$$

Concentration of samples (C_1, C_2, C_3, . . . C_n) are determined and plotted and coordinates erected manually (Fig. 14.8). The curve, however, may also be inscribed directly from a direct writer (densitometer). Thus,

$$\overline{C} = C_1 + C_2 + C_3 + C_4 + C_5 + C_6$$

$$= 0 + 4.2 + 8.5 + 7.5 + 3.8 + 0 = \frac{24}{6}\ 4\ \text{mg, and}$$

$$\frac{\text{12 mg dye injected}}{4} = \text{3 liters/30 s, or 6 liters/min.}$$

Factors Affecting Cardiac Output

The average CO for most adults is about 5 liters/min, which varies according to body size and build. A better index then of CO is the output related to body weight (in kg) or to surface area of body (in m^2), referred to as cardiac index. A human being of average size (70 kg) has a body surface area of about 1.7 m^2 and thus an average cardiac index of 3 liters/min/m^2.

As expected, *exercise* influences CO and heart rate and is outlined in Table 14.1.

Table 14.1. Exercise, cardiac output, and heart rate in man.

Condition	Heart Rate (beats/min)	Cardiac Output (liters/min)
Rest	60	5.5
Mild exercise	100	10.9
Strenuous exercise	138	15.0

Factors that increase heart rate, such as exercise or excitement, usually increase CO. However, adaptation to exercise as occurs in athletes may result in a relative decrease in heart rate but no decrease in cardiac output, or the stroke volume may increase with adaptation to exercise.

Ultimately the output of the right heart to the lungs and then back to the left heart (atrium) must equal the output from the left ventricle—and this is true for several heart beats—but the outputs of the two sides of the heart may vary on a beat-to-beat basis.

From infancy to about age 10 years, there is a rapid increase in cardiac index, which decreases progressively thereafter to old age (Table 14.2).

Diseases and disorders tending to decrease venous return affect output. Extreme decrease of blood volume (as in hemorrhage) decreases the blood returned to the heart (venous return) and subsequent output. Enlarged and weakened heart activity (as in congestive heart failure) also decreases output (decreased contractility).

Table 14.2 Cardiac Index.[a]

Age (years)	Liters
10	4.3
20	3.6
40	3.0
60	2.7
80	2.5

[a] Expressed in units of liters/min/m².

Venous Return

The contractile force of the heart muscle and the venous return indirectly determine the cardiac output. The contractile force and factors affecting it will be discussed in Chap. 15. The output of the normal contracting heart is mainly determined therefore by the volume of blood returned to it by the veins (venous return). Loss of blood volume (hemorrhage) results in a decreased venous return and cardiac output. Venous return is also affected by the pressure gradient for venous flow and the resistance to flow. The pressure gradient is represented by the difference in venous pressure at the periphery (higher) and central end (pressure in right atrium), which is lower. This atrial pressure might be 0 or slightly plus or minus. Increases in right atrial pressure above 0 reduce the pressure gradient and the venous return and cardiac output.

Pressure in the veins and venous return is also influenced by gravity, tone of the veins, and the skeletal muscle pump. Assuming the erect or standing position increases the hydrostatic pressure effect in the feet (periphery) and tends to decrease venous return; this however, is partially overcome by the skeletal muscle pump, which has the effect of decreasing hydrostatic pressure (Fig. 14.9). Contraction of the muscle tends to pump or force venous blood toward the heart, and the valves in the veins prevent backflow.

Respiration is also an important factor in maintaining venous return because the negative intrathoracic pressure,

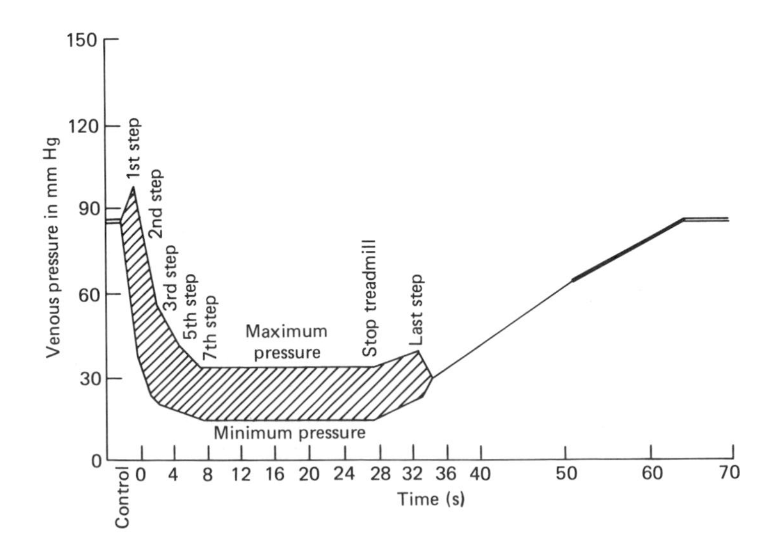

Fig. 14.9. Effects of walking on venous blood pressure in legs. When walking begins, there is slight increase followed by decrease in pressure to a level of about 30 mm Hg; it remains at this level until standing is resumed. (Modified from Pollack, A.A., Wood, E.H., J. Appl. Physiol 1: 649, 1949.)

which increases on inspiration, increases the venous pressure gradient by making the right atrial pressure less positive or more negative (0 or −2 or −3 mm Hg).

A persistent increase in central venous pressure may indicate congestive heart failure, characterized by a decreased venous return and cardiac output.

Distribution of Blood Flow

Distribution of blood flow to the organs and tissues of the body may be determined by measuring blood flow directly by flow meters or indirectly by the dilution principle and take-up of dyes or other agents by the organ concerned.

The distribution of the total flow to various organs in man, except the lungs, is shown in Table 14.3. All of the total CO passes through the lungs. Blood flow/100 g of tissue is highest in the kidneys, next highest in the liver, then heart, and then brain. Stated in another way, the percentages of total flow are greatest in the gastrointestinal tract (23), kidneys (22), muscle (18), brain (14), heart (5), and skin (4).

Blood flow is the means of transporting oxygen to the tissues, and the difference in the O_2 content of the arterial (A) and venous (V) blood to the organs (A − V O_2 difference) represents the actual amount of oxygen extracted from the arterial blood as it passes through the organs. This difference (Table 14.3) is greatest for the heart tissue (8 ml/100 g/min). The brain and liver are next in amount of O_2 extracted.

Exercise, which increases heart activity and blood flow to an organ, also increase oxygen extraction. Decreased resistance in blood vessels also tends to increase flow to the organs.

Table 14.3. Distribution of total cardiac output to various organs and tissues.[a]

		Blood Flow During Rest (Max. vasodil. = [1])			Oxygen Usage During Rest			
Organ	Weight (kg)	ml/min	ml/min/100 g	% total card. output	A-V O_2 difference, ml/100 ml blood	ml/min	ml/min/100 g	% total O_2 usage
Brain	1.4	750 [1500]	55	14	6	45	3	18
Heart	0.3	250 [1200]	80	5	10	25	8	10
Liver	1.5	1300 [5000]	85	23	6	75	2	30
G.I. tract	2.5	1000 [4000]	40					
Kidneys	0.3	1200 [1800]	400	22	1.3	15	5	6
Muscle	35	1000 [20,000]	3	18	5	50	0.15	20
Skin	2	200 [3000]	10	4	2.5	5	0.2	2
Remainder (skeleton, bone marrow, fat, and connective tissue)	27	800 [4000]	3	14	5	35	0.15	14
TOTAL	70	5500 ml/min or 5.5 liters		100		250		100

Adapted from Folkow B., Neil, E. (1971) from Circulation. Oxford University Press, London.

[a] The values in the table are "rounded" figures and roughly describe the situation in average man during rest. Figures within [1] give in very approximate terms organ blood flows at maximal vasodilatation of the respective circuits.

Certain substances when injected into the blood vessels of the brain do not pass through the blood vessel walls (*blood-brain barrier*). Only H_2O, O_2, and CO_2 cross the blood capillaries with ease; other substances cross slowly or not at all.

Blood Pressure

Flowing fluid exerts a pressure usually measured in mm Hg (torr) or more rarely in dynes/cm^2 across the wall of the vessel. Thus a pressure of 110 mm Hg means that if the vessel were attached to a mercury manometer, the pressure of the fluid (end pressure) would displace a single column of mercury to a height of 110 mm. If the vessel were attached to a water manometer, the displacement would be about 13 times as high as for mercury. A pressure of 1 mm Hg = 1330 dynes/cm^2, or about 0.002 pounds/in^2.

This internal pressure in the small thin-walled vessels may be influenced or partially counteracted by the pressure on the outside of the vessel, and this difference is called *transmural pressure* but is of little consequence in thicker walled arteries.

In the upper part of the lungs of a person in the upright position, the pressure in the air capillaries surrounding the blood capillaries may be greater than in the blood capillaries, and therefore the latter will collapse and cut off blood flow. Blood flow and pressure vary in the lung depending on position (see Chap. 17).

There is a gradient of pressure from arteries to arterioles and capillaries and from peripheral veins to central veins

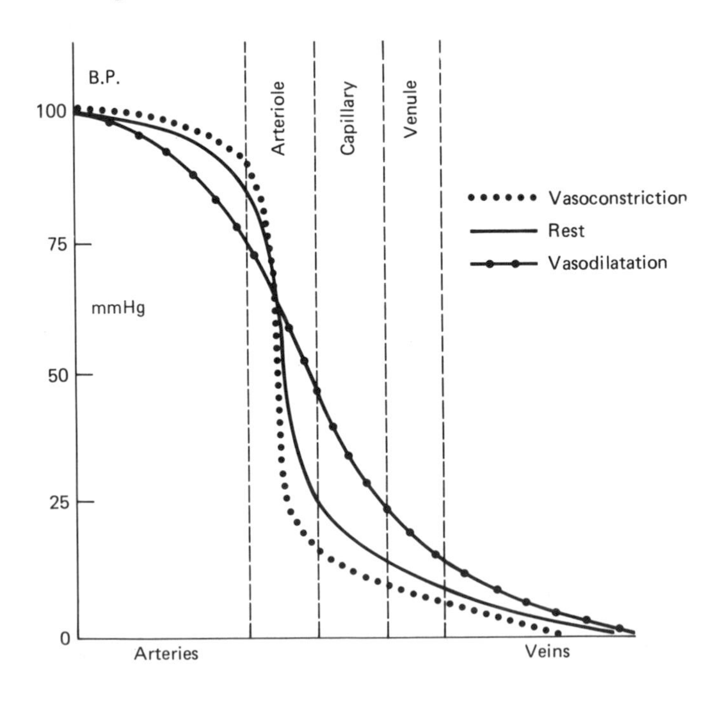

Fig. 14.10. Pressure changes across vascular circuit under conditions of rest, vasodilatation, and vasocontriction. Pressures are mean pressures. In large veins (vena cava) near heart, pressure may be slightly negative on inspiration. (Modified from Sampson Wright's Applied Physiology, Keele, C.A. and Neil, E. Oxford University Press, New York, 1971.)

(Fig. 14.10). Thus, this pressure in the aorta is greater than (>) arterioles > capillaries > venules > large veins > vena cava. It is this gradient of pressure that causes blood to flow out of the heart under pressure to the arterioles and capillaries and then to the venules, veins, and back to the heart (Figs. 14.1 and 14.10). The effects of vasoconstriction and vasodilation on this pressure gradient is also observed in Fig. 14.10).

Systolic and Diastolic Pressure

When blood is ejected from the heart into the aorta, the peak pressure reached is known as systolic pressure (SBP), and after systolic ejection when the aortic valves close and pressure decreases (diastole of the heart), it is referred to as diastolic pressure (DBP).

The differences between systolic and diastolic pressures is the pulse pressure as shown in Fig. 14.11. Mean or average pressure (MBP) may be approximated as $\frac{\text{systolic plus diastolic pressure}}{2}$, but it can vary considerably. In closer approximation, MBP $= \frac{(\text{SBP} + 2\ \text{DBP})}{3}$. A more accurate measurement is made by determining the area under the pressure pulse curve (Fig. 14.11) and dividing this by length of curve $\left(\text{Mean P} = \frac{\text{Area of curve}}{\text{Length of curve}}\right)$.

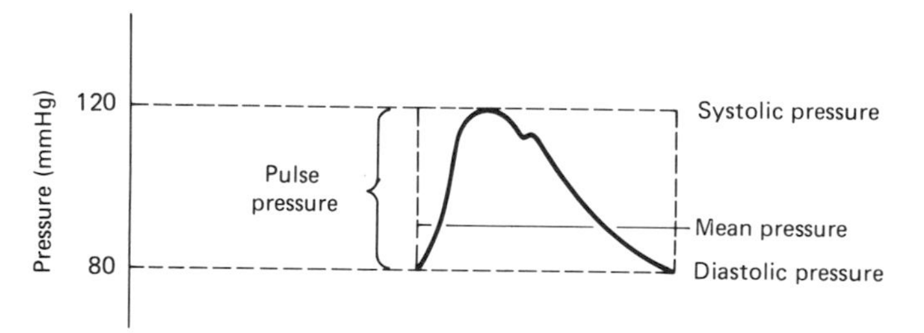

Fig. 14.11. Arterial pressure in aorta, showing systolic, diastolic, pulse pressure, and mean pressure.

The *pressure pulse* occurs because of the pulsating nature of blood flow in a highly elastic and distensible system (blood vessels). The shape and height of the pulse wave in man (but not in avian species) changes from the aorta to the peripheral arteries (Fig. 14.12); the amplitude is much higher at the periphery. The pulse flow curve decreases in amplitude from aorta to periphery. These changes are generally believed to be caused by reflection of waves, although some investigators believe that the changes are accounted for by the geometric elastic taper of the vessel without regard to reflections.

Velocity of the pulse wave depends on the size and stiffness of the vessel. In the aorta the velocity is 3 to 5 m/s, in the medium-sized arteries (subclavian and femoral) 7–9 m/s, and in the small peripheral arteries of the extremities 15 to 40 m/s.

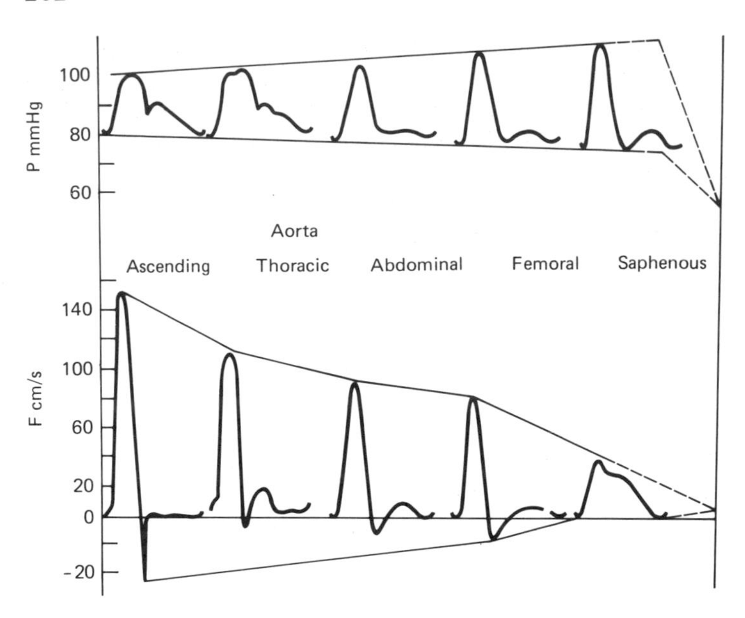

Fig. 14.12. Pressure (P) and flow curves (F) from central to peripheral arteries (in legs). Note that pressure curves increase in amplitude at periphery and flow curves decrease.

Measurement of Blood Pressure

Blood pressure may be determined directly by inserting a needle or catheter into the appropriate vessel and attaching it to the end of a suitable manometer which measures pressure by the displacement of a column of mercury (U-tube manometer), or the pressure is transmitted to a stiff membrane on an electronic strain gauge. More commonly blood

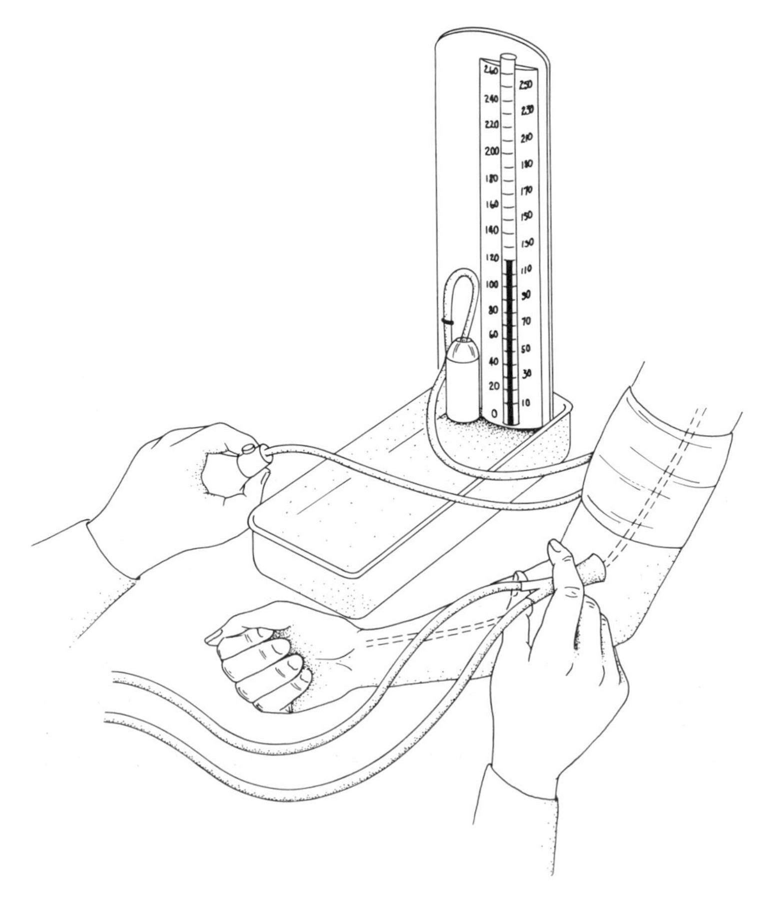

Fig. 14.13. Indirect measurement of blood pressure with sphygmomanometer showing pressure gauge, pressure cuff on upper arm and stethoscope to hear pulsating sounds of blood flow below cuff.

pressure is determined indirectly by using *a sphygmomanometer* (Fig. 14.13).

A rubber cuff of appropriate size is placed around the upper arm (Fig. 14.13) and inflated with a pressure bulb. When the pressure in the cuff becomes higher than the pressure in the brachial artery, the vessel collapses and no blood flows through it. The pressure is exerted on the arm and is also recorded on the manometer. The operator then places a stethoscope below the cuff and listens for sounds of flowing blood below the cuff. When the artery is collapsed, there are no sounds.

The pressure in the cuff is released (deflated) slowly and when the pressure on the outside of blood vessel is less than pressure inside the artery, blood begins to escape or flow back to the artery below the cuff. This first blood flow, or escape, can be heard with the stethoscope, and is called Korotkoff sounds. The pressure at this instant is observed in the manometer and represents systolic pressure. As the cuff is further deflated, more blood returns to the artery and the sounds vary and change; just before no sounds are heard (muffling sound) or when the sound completely disappears, the normal pressure has returned. This is the diastolic pressure.

The accuracy of the method depends on the operator and particularly the size of the cuff, which varies with the size of the person's arm. If the cuff is too narrow for the arm, pressure will be recorded too high, or too low if the cuff is too large. Obese persons with very large arms may reveal abnormally high pressures recorded by this method unless a correction is made for cuff size. The average cuff width for adults is 12 cm or 4.7 in.

Approximate corrections for the effect of upper arm size (circumference) when blood pressure is taken with a standard cuff are summarized in Table 14.4.

Table 14.4 Corrections for arm circumference in blood pressure determinations.

Arm Circumference (cm)	Systolic Blood Pressure Correction
15–18	Too low; add 15 mm Hg
27–30	No correction needed
35–38	Too high; subtract 10 mm Hg
42–45	Too high; subtract 20 mm Hg

Normal Blood Pressure Values

Blood pressure values of normal human subjects from age 15 to 60+ years are shown in Fig. 14.14. Both systolic and diastolic pressures increase linearly with *age* in males, but the changes in females are curvilinear, i.e., the pressure increases little with age from 20 to 40 years; after this age and the menopause the levels, which before this are lower than in males, rise sharply and actually exceed the levels in males.

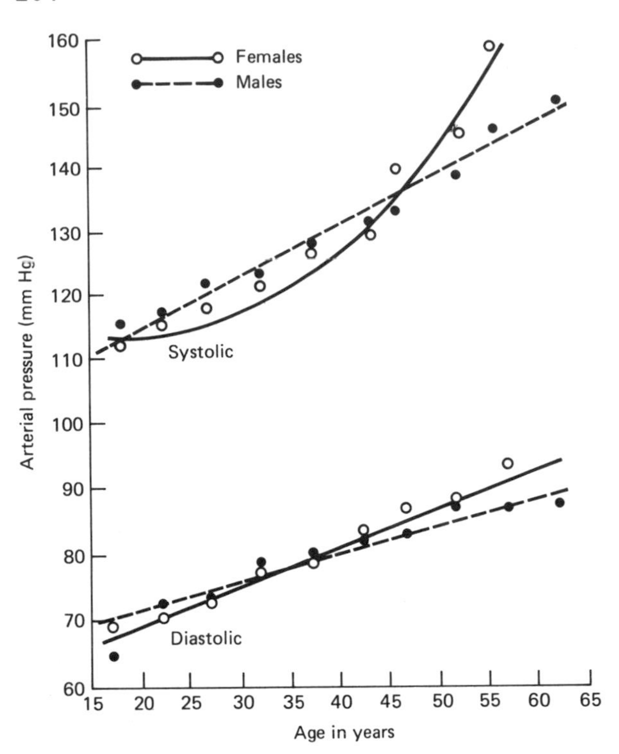

Fig. 14.14. Systolic and diastolic blood pressure as related to age and sex. (Modified from Morris, J.N. Modern Concepts of Cardiovascular Disease 30: 635, 1961.)

Although blood pressures are often overestimated in obese persons, their pressures are still higher than in persons of normal weight even after correction for the over estimation. When a person is lying down, the arterial pressure in the heart, feet, and head is practically the same, (95–100 mm Hg) because the hydrostatic, or gravity effect is 0, but the pressure in the veins is higher in feet (5 mm Hg) and head (5 mm Hg) than in the heart (2 mm Hg). This venous pressure gradient tends to force blood from feet and head back to the heart. In a standing, but not walking position the pressures that occur are shown in Table 14.5.

Table 14.5. Position and arteriovenous pressures.

Structure	Arterial (mm Hg)	Venous (mm Hg)
Head	51	−40[a]
Heart	100	2
Feet	188	90

[a] Note that −40 mm Hg pressure in veins of head actually may cause veins to collapse because pressure inside vessels is negative (less than outside pressure).

Hydrostatic pressure in the standing position prevents normal venous return to the heart, decreases CO, and may cause fainting in persons standing at attention (without moving).

Normally, walking or movement of the legs even a little causes the muscle pump in the legs to operate, which tends to minimize the hydrostatic pressure effect and the pressure differences in heart and feet, and this occurrence lowers venous pressure (Fig. 14.9).

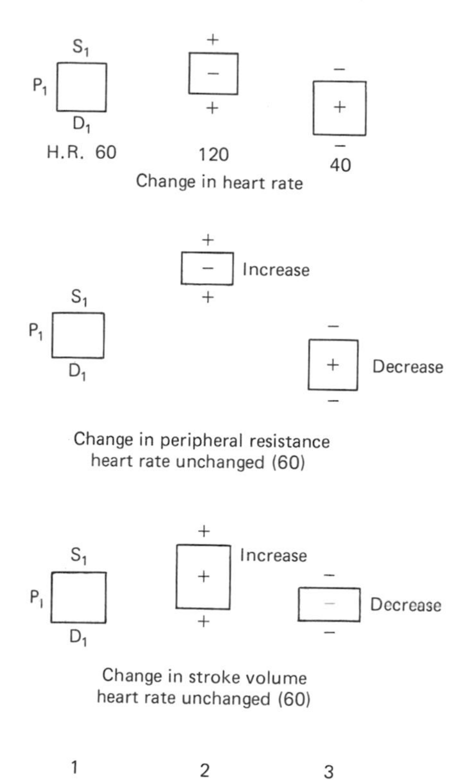

Fig. 14.15. Effects of changes in heart rate **(A)** (beats/min), peripheral resistance **(B)**, and stroke volume **(C)** on blood pressure values. Controls for systolic (S_1), diastolic (D_1), and pulse pressure (P_1) are indicated by squares in panel 1, and changes in panels 2 and 3 are also indicated. Increases or decreases in systolic and diastolic pressures are indicated by (+) and (−) at top and bottom of squares and in center of square for pulse pressure (also see text).

Exercise increases systolic and diastolic pressure, CO, and heart rate. Mild walking affects blood pressure (in mm Hg) as follows:

Pressure	Before Walking	After Walking
Systolic	117	132
Diastolic	74	67

Smoking cigarettes may increase systolic pressure by 10–20 mm Hg. Coughing increases blood pressure, as does sexual intercourse; blood pressure drops significantly during rest and sleep.

Respiration affects blood pressure in that pressure usually decreases on inspiration and increases on expiration. The changes are caused mainly by changes in CO. On inspiration the CO from the right heart to the lungs is increased, but the amount of blood returning to the left heart from the lungs is decreased and therefore the amount ejected from the left ventricle on inspiration is decreased; on expiration blood pressure increases and so does CO.

Behavioral patterns influence blood pressure. Anxiety and emotional stress without physical involvement increase blood pressure. Stressful factors involved in modern living tend to increase pressure. Sexual behavior increases blood pressure.

Blood pressure is consistently higher in blacks than in whites for all ages and sexes (race and social pressures). To what extent the higher level is attributable to stressful societal factors or inheritance is not known.

Inheritance is a definite factor in the level of blood pressure. Some families show an inherited tendency to high blood pressure (hypertension). This genetic tendency is observed not only in man but also in other species.

Norepinephrine (NE) produced by the sympathetic nerves and adrenal glands causes an increase in blood pressure. Tumors of the adrenal (pheochromocytoma) may produce excessive amounts of NE and cause hypertension. Abnormal production of *adrenal cortical hormones* also causes hypertension. In kidney disease, an abnormal amount of *angiotensin II* produces vasoconstriction and hypertension.

Physiologic Control

It was previously stated that blood pressure is influenced mainly by three factors, including (A) heart rate, (B) changes in peripheral resistance, and (C) changes in stroke output or CO (Fig. 14.15).

An increase in *heart rate* from 60 to 120 beats/min (panel 2A in Fig. 14.15) caused a slight increase in systolic and a greater increase in diastolic pressure with a resulting decrease in pulse pressure. A decrease in rate to 40 beats/min (panel 3) caused a decrease in both systolic and diastolic pressure from panels 1 and 2, but the pulse pressure was slightly higher than in panel 1. At slow heart rates, the

time elapsing between ejection of blood from the aorta and its flow or runoff to peripheral blood vessels is increased and increases pulse pressure. At high heart rates the runoff is less, because the time between heart beats is less as is pulse pressure.
An increase in *peripheral resistance* (B) at the same heart rate (60) causes a significant increase in systolic and diastolic pressure with a decrease in pulse pressure and runoff time. A decrease in peripheral resistance (3B) causes a significant decrease in systolic and diastolic pressure but with an increase in pulse pressure.
When *stroke volume* (SV) is increased (2C), there is a great increase in systolic with a slight increase in diastolic pressure and therefore a resulting increase in pulse pressure. Decreasing SV as in 3C results in a decrease in systolic and diastolic but mostly systolic with a significant drop in pulse pressure.

Neural Control

Changes in heart rate, peripheral resistance, and stroke volume, all of which influence blood pressure, are regulated by the peripheral and central nervous systems. To induce a change in peripheral resistance, the smooth muscles of the blood vessels must contract or constrict to increase resistance and dilate to decrease resistance (Fig. 14.16).
There are *sympathetic nerves* whose neurons originate in the lateral horn of the spinal cord and run in the thoracolumbar region and send preganglionic fibers to the chain or paravertebral ganglia (spinal cord) or visceral ganglia and synapse; postganglionic fibers then go to the smooth muscle of blood vessels (Fig. 14.16). Most of these nerves when stimulated either directly or by vasomotor centers in the medulla cause vasoconstriction by releasing norepinephrine at their ends and are called *adrenergic fibers*. There are also some sympathetic vasodilator fibers that innervate blood vessels in skeletal muscle and skin. Their fibers usually are mixed, and the constrictor fiber and the constrictor effect usually predominate when the sympathetic nerve is stimulated, but vasodilator fibers can be demonstrated with appropriate pharmocologic agents. These fibers (called cholinergic) release acetycholine when stimulated, which causes the vasodilation. These nerves are also known to innervate some veins.
Parasympathetic nerves, i.e., nerves of the cranial (head) division of the parasympathetic system (including many cranial nerves) supply fibers to the blood vessels of head and neck region; these nerves of the sacral division supply vessels to the genital organs, bladder, and large bowel. These long preganglionic fibers originate from nerve cells in craniosacral regions; those in the sacral region run in the spinal cord and emerge from appropriate vertebra and termi-

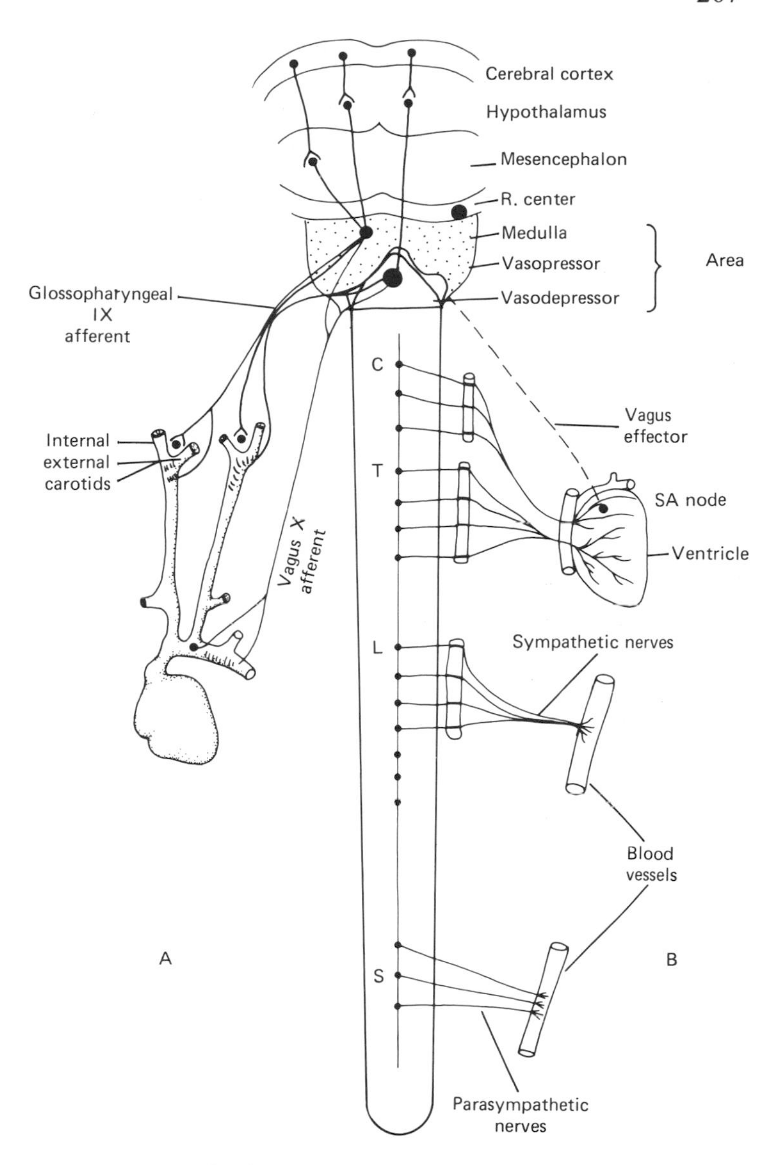

Fig. 14.16. Pressoreceptors in carotid sinus and aortic arch (′′′′) and afferent nerves IX and X, which receive pressure stimulus carry it to vasodepressor area (triangular and medial). Note also connections from cerebral cortex and hypothalamus which may also send stimuli to vasodepressor or vasopressor centers. Chemoreceptors (A) from carotid and aortic bodies (•) receive and respond to changes in $_pCO_2$ (increase) and $_pO_2$ (decrease) and transmit response through nerves IX and X to *vasopressor area.* Responses to stimuli from vasodepressor and pressor area may be traced from centers through spinal cord through appropriate sympathetic ganglia and nerves and vagus nerve (B) to heart, and blood vessels in lumbar (L) and sacral (S) areas. Two IX nerves shown, but only one nerve X in A, all afferent.

nate in ganglia in the organ innervated (Fig. 14.16). Thus the postganglionic fibers are short and release acetycholine and cause vasodilation. The cranial nerves emerge from the brain and run to the organ concerned. (See also Chaps. 3 and 10.)

Vasomotor Centers

The main vasomotor center is located in the medulla. There are two principal areas that are called vasodepressor and vasopressor located as indicated in Fig. 14.16. Stimulation of the larger more lateral area causes vasoconstriction in the blood vessels by sending motor nerve impulses from the brain down the spinal cord to the nerves. The triangular, *medially located vasodepressor* area, when stimulated, depresses blood pressure by decreasing or inhibiting vasoconstrictor activity or causing the release of acetycholine from vasodilator sympathetic fibers.

Although vasoconstrictor and vasodilator nerves change blood pressure by changing resistance, pressure is also changed by heart rate and stroke volume which are influenced also by sympathetic nerves that increase heart rate (*cardioaccelerators*) and by the parasympathetic nerve (the vagus) which decreases heart rate (cardioinhibitors). Sympathetic nerves to the ventricles of heart increase contractile force and stroke volume.

Vasopressors and Heart Rate

A reflex requires a receptor that receives or senses an impulse, and an effector that relays the impulse to the appropriate organ for stimulation (see Chap. 7). The reflex may involve only the receptor and the spinal cord (spinal reflex) or it may also involve the brain and the vasomotor center to influence blood pressure and heart rate. These receptors are (1) pressoreceptors (baroreceptors) located in the carotid sinus (outside curvature) and in the aortic arch (outside curvature); and (2) chemoreceptors located in the angle of the carotid sinus bifurcation (Fig. 14.16) and on the curve of the arch of the aorta (aortic bodies).
Stimulation evoking responses from the vasodepressor and vasoconstrictor centers are shown in Figs. 14.17 and 14.18, respectively.

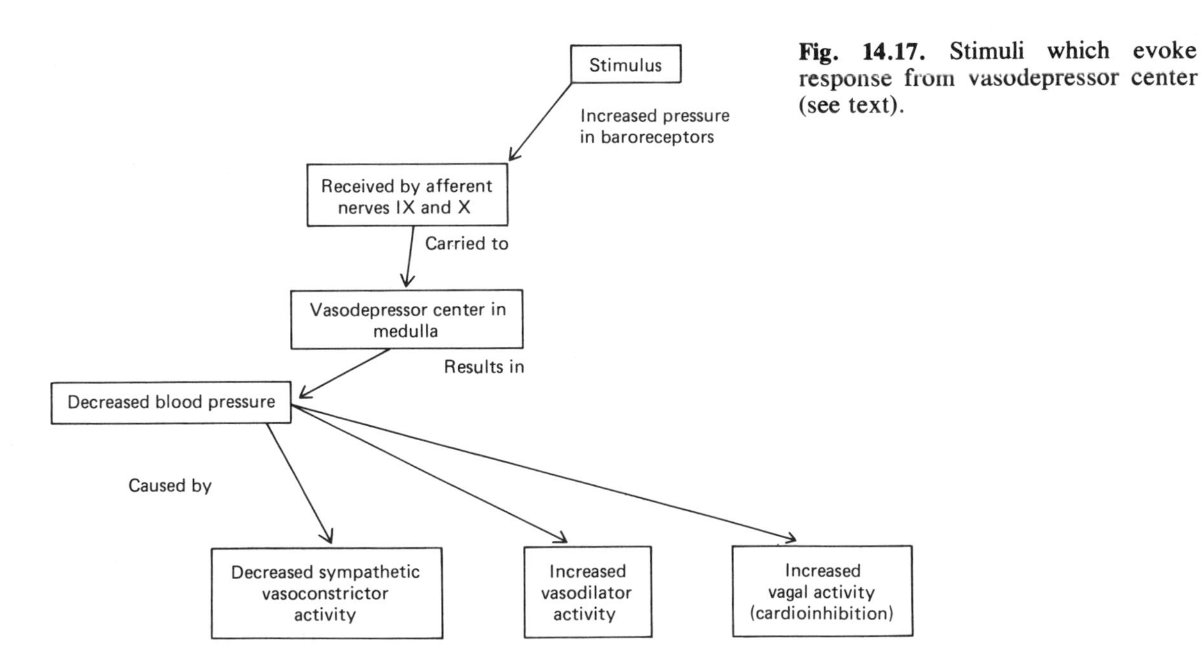

Fig. 14.17. Stimuli which evoke response from vasodepressor center (see text).

Alpha and Beta Receptors

The ultimate effects of sympathetic nerves on heart rate, contractility, and blood pressure are mediated by receptors in the organs and tissues involved.

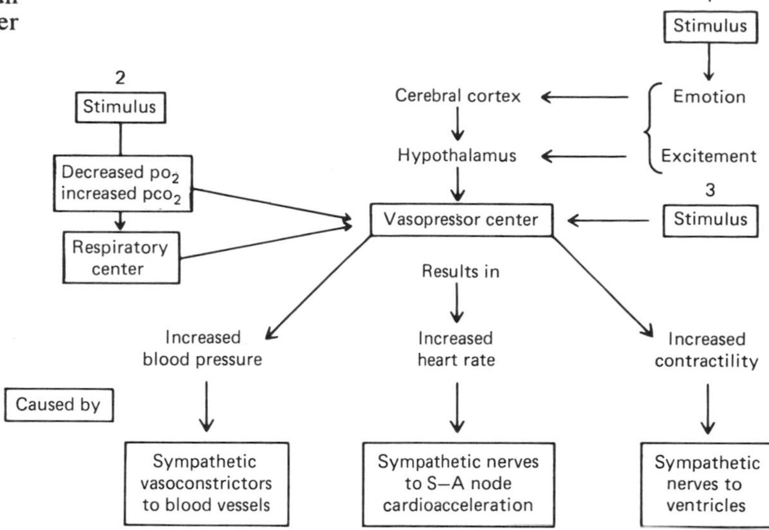

Fig. 14.18. Effects of certain stimuli on vasopressor or constrictor center (see text).

The effects of epinephrine (EP) and norepinephrine (NE) are an increase in blood pressure (vasoconstriction) in many of the blood vessels caused by stimulation of *alpha receptors* located in the blood-vessel walls. The effects of NE are always contrictor (alpha stimulation), but EP may stimulate some *beta receptors* in certain vascular beds (skeletal muscle) and cause vasodilatation. (See also Chap. 15.)
There are beta receptors in heart muscle which respond to EP, NE, and isoproterenol by increasing the heart rate and contractility.
Numerous pharmacologic agents block the effect of agents that normally stimulate alpha and beta receptors.

Local Control of Circulation

The local control of pressure and flow in peripheral vascular beds is determined by pressure relations inside and outside the vessels and the response of smooth muscles in the walls of arterioles and venules to vasoconstrictors and vasodilators. Some of them have been considered in relation to neuro-transmitters and alpha and beta receptors, and cholinergic ones.
Other agents, including hormones and chemicals produced

Table 14.6. Substances or conditions produced in body that constrict (C) or dilate (D) local blood vessels.

Substance	Substance
Serotonin (C,D)	Glucagon (D)
Histamine (C)	Cholecystokinin (D)
Acetylcholine (D)	Secretin (D)
Angiotensin (C)	Prostaglandins (C,D)
Kinins (D)	Hypoxemia (D)
Oxytocin (C,D)	Hypercapnea (D)
Vasopressin (C)	H^+ (D)
Adenosine (C,D)	K^+ (D)

in the body influence the tone of local blood vessels; some of them are listed in Table 14.6.

Normally in both physical systems and some blood vessels, flow increases with pressure, but in some organs and tissues of the body flow does not continue to change with a change in pressure. This phenomenon is known as *autoregulation* and figures prominently in the kidney. The responsible mechanisms are discussed in Chap. 24 and include (1) the effect of smooth muscle itself (myogenic activity) and (2) the influence of metabolites and chemicals mentioned previously (metabolic effect).

Selected Readings

Altman PL, Dittmer DS (1974) Biology data book, 2nd edn. Vol III, Fed Amer Soc Exper Biol

Berne RM, Levy MN (1977) Cardiovascular physiology, 3rd edn. Mosby, St. Louis

Folkow B, Neil E (1971) Circulation. Oxford University Press, London

Guyton AC (1977) Basic human physiology. Saunders, Philadelphia

Guyton AC, Young DB (1978) Cardiovascular physiology III. International Review of Physiology, Vol 18. University Park Press, Baltimore

Little RC (1977) Physiology of the heart and circulation. Yearbook Medical, Chicago

Selkurt EE (1975) Basic physiology for the health sciences. Little, Brown, Boston

Review Questions

1. How are the following related: pressure, flow, and resistance. Formulate.
2. Velocity of blood flow is determined by what factors?
3. What is the law of LaPlace which relates tension, pressure, and radius of blood vessels. Formulate.
4. Where is blood flow in what arteries likeliest to be turbulent?
5. What factors influence cardiac output? Name them.
6. What effect does exercise have on cardiac output?
7. Name two organs that receive large amounts of blood flow.
8. What is a sphygmomanometer and how does it operate?
9. Define a) systolic, b) diastolic, c) pulse, and d) mean blood pressure.
10. What happens to the blood pressure pulse wave from the aorta to an artery in the leg?
11. What is the systolic and diastolic blood pressure of a 20-year-old man and a 60-year-old man?
12. How does body weight affect blood pressure?
13. What are pressoreceptors and how do they operate to increase or decrease blood pressure?
14. What is meant by reflex control of blood pressure?

Chapter 15 The Heart As A Pump

Anatomy

The heart, a hollow, muscular organ about the size of an adult clenched fist, weighs 10 to 12 oz (160 grams) in adult males and 8 to 10 oz in females. It is located in the thoracic cavity in the space between the lungs. Most of the heart lies just to the left of the midline of body (Fig. 15.1). It consists of four chambers: two atria and two ventricles. The atria are thin walled, the ventricles much thicker, particularly the left ventricle, which is the high-pressure chamber.

Further details of the anatomy of the heart are shown in Figs. 15.2 and 15.3 where the entrance and exit of the princi-

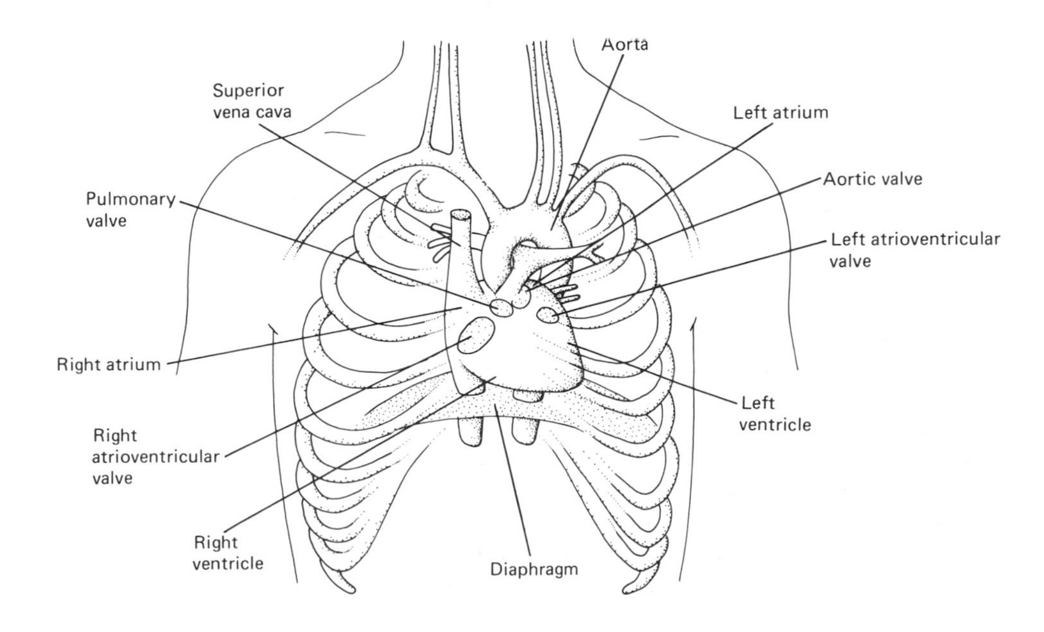

Fig. 15.1. Position of heart and valves in relation to thoracic cavity.

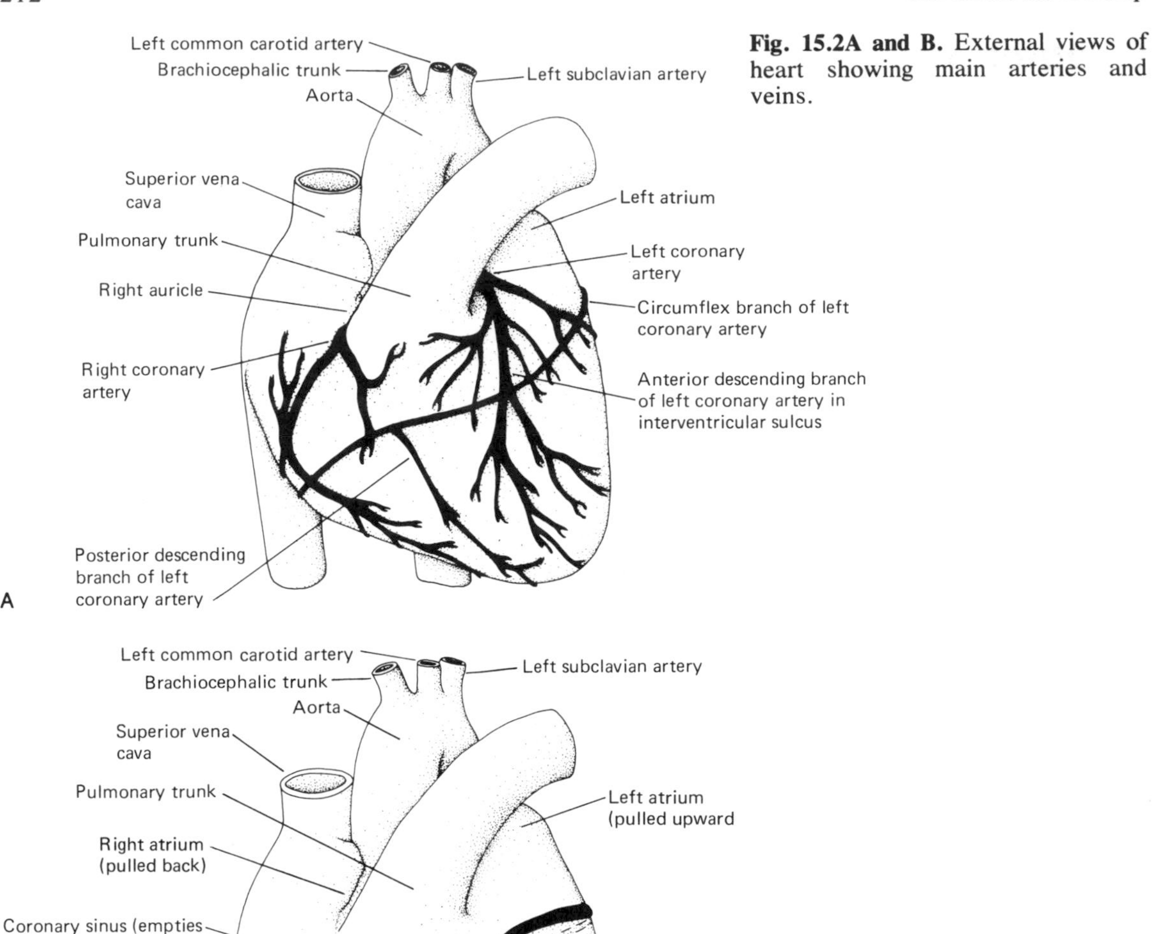

Fig. 15.2A and B. External views of heart showing main arteries and veins.

pal veins and arteries are revealed, as well as the coronary arteries and veins. The coronary arteries nourish the heart muscle, and the veins drain into the large coronary sinus, which empties into the right atrium. Occlusion or blockage of blood in the coronary arteries (coronary attack) results in decreased blood flow to the heart muscle (myocardial infarct) and malfunctioning of the heart.

The layers of muscle and tissue of the heart include the pericardial sac enclosing the heart (Fig. 15.4), the outer muscle layer (epicardium), the middle layer (myocardium) and the innermost layer (endocardium). For the detailed structure of cardiac muscle, see Chap. 11.

The heart valves include the right atrioventricular (AV) valve with three cusps or leaves (tricuspid) and the left AV

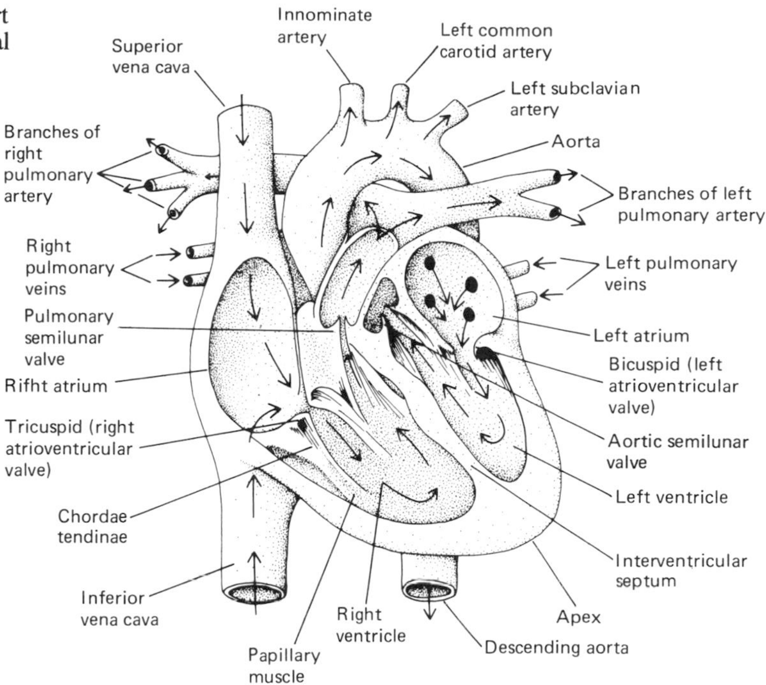

Fig. 15.3. Internal structure of heart showing entry and exit of principal veins and arteries and valves.

valve, which has two cusps (bicuspid) and is called the mitral valve. These valves separate atria from ventricles. The aortic valve is called the semilunar valve (Figs. 15.1 and 15.3; this type of valve is also in the pulmonary artery. These semilunar valves open in one direction only—when the pressure is high—and they close when the pressure is low. Closed and open tricuspid valves look something like this:

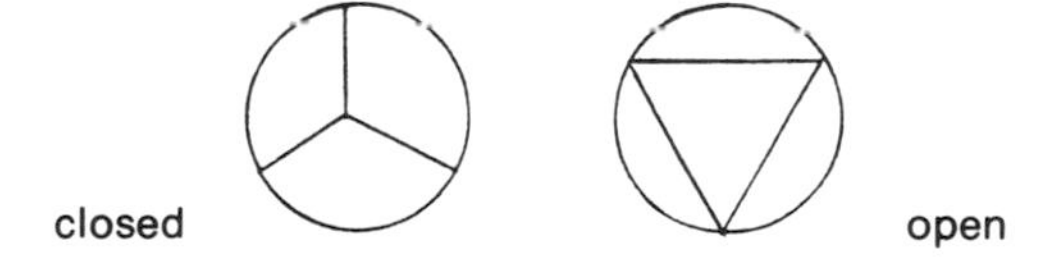

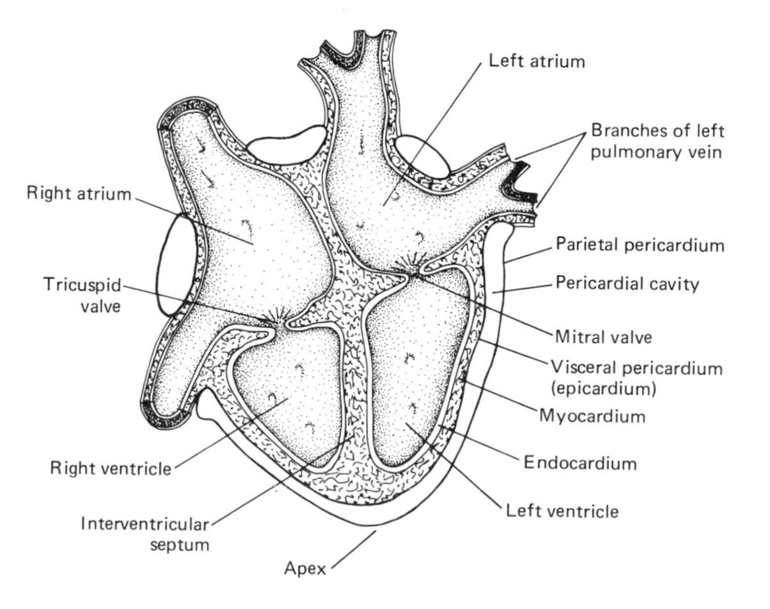

Fig. 15.4. Sectional view of heart showing pericardium and tissue layers, such as epicardium (outside, myocardium (middle), and endocardium (inner).

Cardiac Cycle

The events of the cardiac cycle (Fig. 15.5) include changes in (1) atrial systole, (2) isovolumetric contraction, (3) rapid ejection phase, (4) reduced ejection phase, (5) isovolumetric relaxation, (6) rapid ventricular filling, and (7) reduced ventricular filling or diastasis.

As indicated previously (Chap. 14), venous blood is returned to the right atrium from whence it flows through the right AV valve to the right ventricle, and from the latter by pulmonary artery and valve to the lungs. Here blood is oxygenated and is returned to the left atrium via the pulmonary veins, and from there through the left AV valve to the left ventricle. Finally, arterial blood is ejected through aortic valves into the arterial system.

Atrial systole (1), which represents the beginning of the cycle as indicated at top of Fig. 15.5. forces blood through the opened AV valves. Following this is the (2) isovolumetric contraction phase, characterized by closure of AV and aortic valves. The ventricle is full and contracts without changing its volume (ventricular volume curve), but its pressure rises rapidly to a point where the aortic valve opens (end of phase 2 and beginning of phase 3); this represents the rapid ejection phase and the further rise of pressure to a peak; then pressure begins to drop (phase 4-reduced ejection).

During rapid ejection (3), the volume of the left ventricle decreases rapidly, but the ventricle is never completely emptied. During reduced ejection (4), ventricular pressure falls and ventricular volume increases to its maximum, at which point the aortic valve closes and prevents both backflow of blood into the heart and a significant decrease in aortic pressure. This represents isovolumetric relaxation phase (5) or relaxation without a change in ventricular volume, but it also causes a drastic drop in ventricular pressure to a low point, at which time the left AV valve (mitral) opens and blood starts to flow from the atria to the ventricles (phase 6 or period of rapid filling). This phase is followed by reduced ventricular filling (7), or diastasis. With the beginning of phases 6 and 7, the left ventricular volume is rising (see curve in Fig. 15.5) and when it returns to the normal level, a new cycle begins.

Associated with these changes in ventricular pressure and volume are changes in heart sounds, venous pulse, and electrocardiogram.

There are usually four *heart sounds,* only two of which are audible with a stethoscope, but the sounds may be amplified and converted to waves to form a phonocardiogram (Fig. 15.5). The first sound occurs at the onset of ventricular systole and opening of valves (phase 2). It is the loudest, caused by many oscillations of blood against ventricular walls, and the latter vibrate. It begins at about the peak of

Fig. 15.5. Events of cardiac cycle showing changes in ventricular pressure, ventricular volume, heart sounds, venous pulse, and ECG during one cycle. Time relations (s) are indicated at bottom of curves.

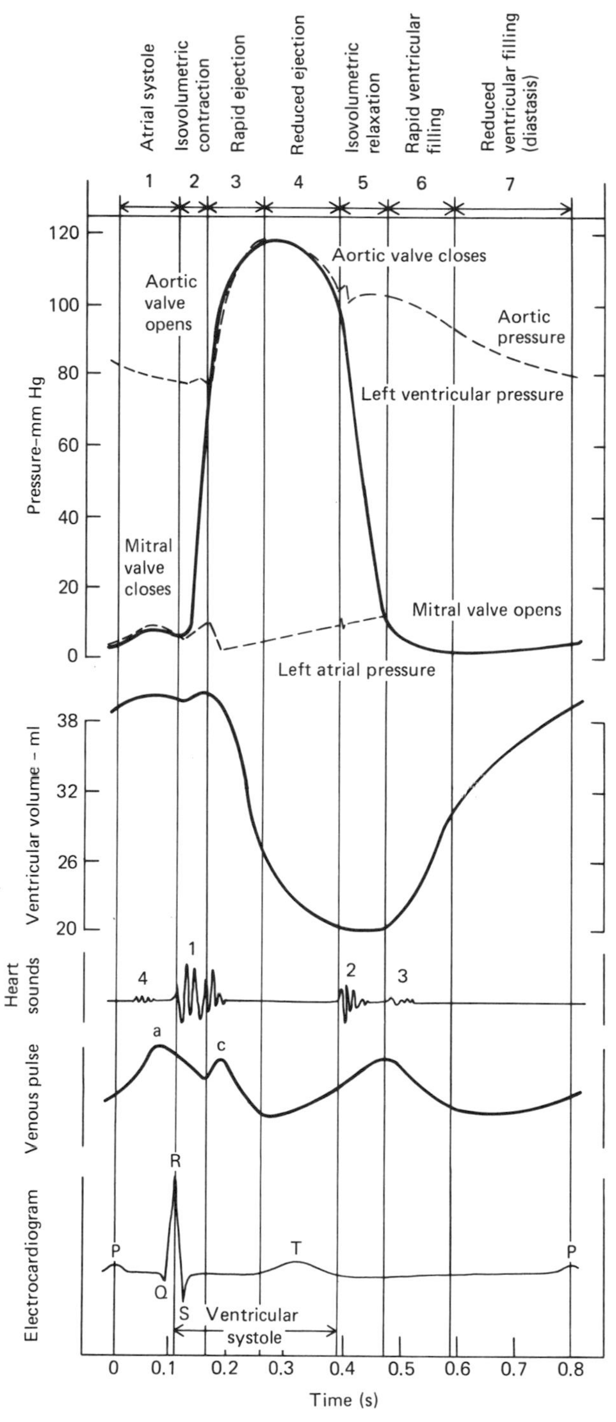

the R wave of the ECG. The second sound occurs on the closure of the semilunar valves, which initiates oscillations of blood caused by the recoil of the closed valve. This sound is of shorter duration and higher frequency than the first; it begins at the end of the T wave of the ECG.

The third sound occurs in early diastole: it is of low intensity and low frequency. The fourth sound may be heard in some subjects and is caused by oscillation of blood during atrial systole.

The QRS wave of the ECG precedes slightly the mechanical contraction phase of the heart. The T wave begins as the reduced ejection phase begins and is completed at about the point where aortic valve closes. The ECG, the electrical impulse, signals the onset of mechanical contraction.

Contractile Force

When heart muscle contracts, it tends to shorten (*isotonic* contraction). It does this because the muscle has a contractile element (CE) that shortens and a series elastic component (SE) that lengthens. When the muscle is not loaded (O load), it shortens rapidly to the maximum velocity of shortening (V max). This is illustrated in Fig. 15.6. As the load is increased from O to infinity, the velocity of shortening decreases and finally at a load that does not allow shortening of muscle (Po in the figure), the velocity is O. When this occurs, the muscle is said to be contracting *isometrically*. There is an increase in tension in the muscle. When the heart contracts with all valves closed, it contracts under a constant volume (load) (isometrically), but no blood is ejected (isovolumetric contraction phase).

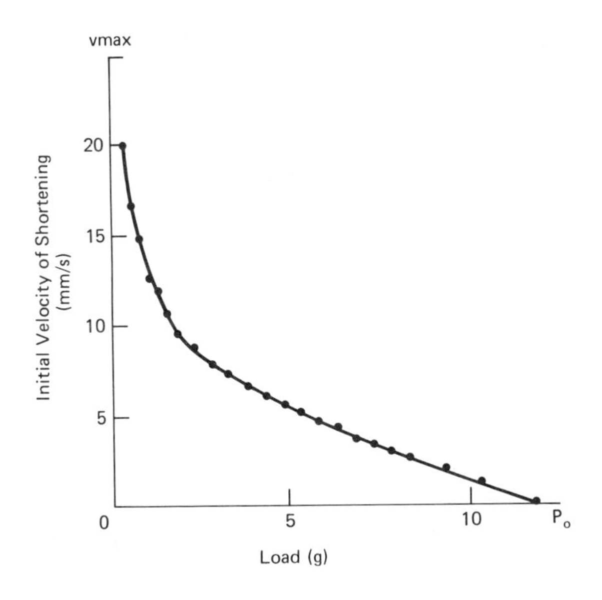

Fig. 15.6. Velocity of shortening of muscle and load. V_{max} represents isotonic contraction and Po represents maximal tension produced by isometric contraction.

Pumping Action

The pumping action of the heart depends on the force of contraction of the heart muscle. This force is greatest during the ejection of blood from the left ventricle to the aorta. The amount ejected, or stroke volume, depends on the amount in the heart at time of ejection and on the strength of contraction.

Frank-Starling Mechanism or Law

This law was first enunciated by a German, Otto Frank, and later by an Englishman, Ernest H. Starling. The law states that the amount of blood ejected from the heart increases as the amount of blood in the heart (volume) increases. As the latter increases, the heart muscle is stretched or distended, and the volume and pressure in the heart at diastole (end-diastolic pressure and volume) are also increased.

The law means that the heart can vary its output, depending on the amount of blood returned to it (venous return). It tends to regulate the outputs of right and left hearts, which may vary from beat to beat, but total output of each heart is the same over time. If, e.g., the output of the left ventricle on one beat is high because of high end-diastolic pressure or volume, then on the following beat the output will be less and may be equal to or even less than right-heart output.

Thus, if stroke volume (output) is plotted against (1) end-diastolic, pressure (2) left atrial pressure, or (3) length of muscle fibers (Fig. 15.7), it will show that as left atrial pressure or muscle-fiber length increases, stroke work or output increases. (Each curve is a Frank-Starling curve). This is known as heterometric (differing length or measurement) autoregulation. The heart regulates its own output.

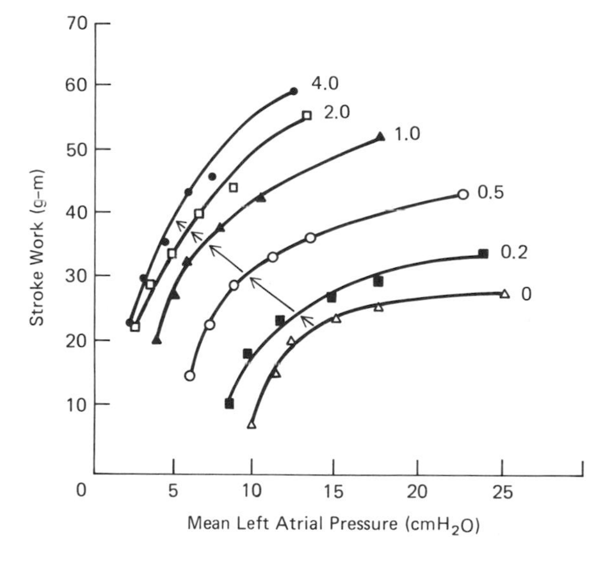

Fig. 15.7. Illustration of Frank-Starling law (heterometric autoregulation) and homeometric autoregulation (increased stroke work at constant atrial pressure or muscle fiber length). Each one of the curves is a Frank-Starling curve and shows that as atrial pressure and muscle length increase, there is an increase in stroke work or output. The heart nerves were stimulated with increasing strength (0–4 frequency) and the resulting curves, called Sarnoff curves, are shifted to the left (upward); or at the same atrial pressure there is greater stroke work (greater contractility) in curves from 0–4 frequency (homeometric autoregulation).

Homeometric Autoregulation of Contractility

Contractility refers to the ability of heart muscle to contract and do work at a definite stretch or muscle fiber length. This is homeometric (constant length, or measure) autoregulation and is shown in Fig. 15.7 by the different Sarnoff curves, which at the same atrial pressure or muscle-fiber length, are shifted to the left (greater contractility). Factors that affect contractility include increased sympathetic nerve stimulation or norepinephrine, increased calcium, and certain other agents like digitalis, used clinically to strengthen activity in weakened hearts. Increasing afterload on the heart, by increasing aortic blood pressure increases contractility as does increasing heart rate.

Work of Heart

The work of the heart consists of ejecting blood from the heart and propelling it (volume) under pressure to various parts of the body. A force or pressure (P) is required to push volume (V) a given distance. Thus W = Mean pressure × volume or W = PV; where P is in grams and V is in centimeters.

The work of the left ventricle is about 5 times that of the right ventricle because the pressure is about one-fifth that of the left ventricle, and the average blood flow (V) is the same for the two ventricles.

To obtain the figure for total work, it is necessary to multiply work for one stroke volume × heart rate.

The work of the left ventricle of a human heart pumping 5 liters (5000 cm)/min at a mean pressure of 100 mm Hg (135 g/cm) = 5000 × 135 = 675,000 g-cm or 6.75 km of work/min.

Mechanical Efficiency of Heart

The mechanical efficiency (ME) of the heart = $\frac{\text{mechanical work done}}{\text{total energy used}}$. The heart is only 14% to 25% efficient, which means that a great deal of the energy that is expended is wasted. Much of this is heat energy, which does not perform work. The heart uses oxygen as its fuel and a 300-g heart may consume about 27 ml O_2/min, but much of this is used in producing heat energy. With exercise and training the ME may increase.

Increased blood pressure increases the workload of the heart while decreasing its mechanical efficiency. It is desirable therefore to keep blood pressure relatively low and cardiac output relatively high to minimize the work of the heart.

Heart Rates

The resting heart rate is the rate of a person at rest or relatively free from anxiety. This is variable but averages about 70 beats/min. During sleep the rate drops to 50–60 beats/min, which is an average rate also for well-trained athletes.

Exercise increases heart rate as well as blood pressure depending on the strenuousness. It may range from 50–70 beats/min at rest to 100 at mild exercise, 120 at moderate exercise, and 138 at severe exercise. Sexual intercourse increases heart rate greatly.

Extrinsic Regulation

Heart rate is influenced by extrinsic and intrinsic factors. The extrinsic factors are the sympathetic and parasympathetic nerves to the heart, or cardioaccelerators and cardioinhibitors, respectively (Figs. 14.16 and 15.8). The cell bodies of the preganglionic sympathetic fibers arise from the intermediolateral columns and emerge from the spinal cord and vertebral ganglia at the level of cervical ganglia (superior, middle, and inferior) and from thoracic ganglia 1 to 5. Postganglionic fibers then run from these ganglia to a plexus on the heart. Actually the human superior, middle, and inferior cardiac nerves are derived mainly from the post ganglionic fibers from these ganglia, but branches from the thoracic ganglia may also contribute.

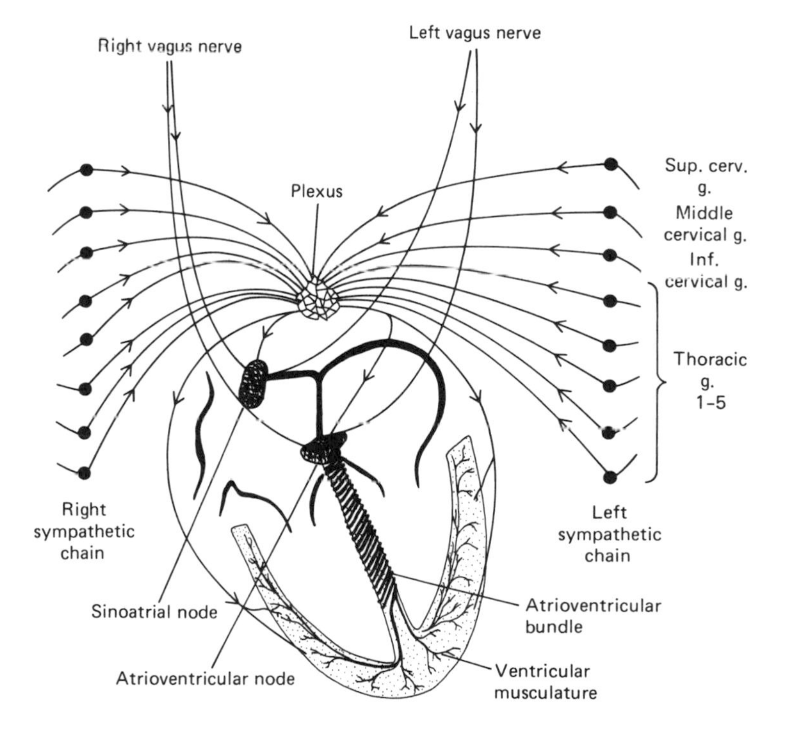

Fig. 15.8. Sympathetic nerves to heart showing origin of postganglionic fibers from cervical and thoracic ganglia. These fibers run to a plexus and on to the heart. Branches go to the SA and AV nodes and to the ventricles (see text). The parasympathetic nerves (vagus) run also to the SA and AV nodes.

Postganglionic fibers from the right side affect predominately the SA node and those from the left side the AV node; fibers from both left and right sides go to the ventricles. The fibers to the SA and AV nodes help to regulate rate and rhythm of the heart whereas those to the ventricles help to regulate contractility. These sympathetic fibers to the heart release NE (the neurotransmitter), which stimulates the *beta receptors of the heart* (see Chap. 14); the latter are responsive to beta agonists, such as isoproterenol, and are inhibited by beta antagonists, such as propranolol. *Beta receptors* are also stimulated by epinephrine and respond by increasing heart rate and contractility.

When heart rate and blood pressure are influenced reflexly (see Chap. 14), higher centers in the brain (cerebrum, hypothalamus, and medulla) participate and receive afferent stimuli and respond by sending motor impulses down the spinal cord to the appropriate sympathetic ganglia and nerves (Fig. 14.16).

Parasympathetic Nerves

These arise in the medulla in the dorsal motor nucleus as the vagus nerves, which emerge through the neck in proximity to the common carotid artery. The preganglionic fibers are long and their ganglia are located on the heart, from which postganglionic fibers emerge (Fig. 15.8). The right vagus nerve stimulates mainly the SA node to cause cardiac slowing or inhibition. The left vagus affects mainly the AV node and AV conduction, although it sends some fibers to the SA node. The vagal fibers are cholinergic (release acetylcholine) and exert an inhibitory tone or effect on heart rate because transection of the nerves or blocking the nerves with atropine causes an increase of heart rate, because vagal inhibition (inhibitory tone) has been abolished.

Sympathetic and Parasympathetic Tone

Since sympathetic nerves accelerate heart rate and vagal nerves inhibit or slow heart rate, the degree to which each influences rate is referred to as the tone or strength. When the parasympathetic nerves are transected or blocked with atropine, heart rate increases significantly but when the sympathetics are cut heart rate slows only slightly. This is evidence that the parasympathetics exert more effect in slowing heart rate than the sympathetics do in accelerating heart rate. When both sympathetics and parasympathetics are blocked or transected in adult humans, heart rate increases to about 105 beats/min, which is significantly higher than the normal resting rate. This is the *intrinsic heart rate*.

Reflex Control of Heart Rate

This was discussed previously (Chap. 14) where it was shown that changes in blood pressure stimulate pressoreceptors, which reflexly cause changes in heart rate by the vasomotor centers. Thus, if blood pressure is too high, the pressoreceptors signal a decrease in pressure and heart rate and if the pressure is too low, heart rate is increased. It was also shown that emotion and excitement which involve the higher brain centers also activate the vasomotor centers (Figs. 14.16 and 14.18).

Intrinsic Regulation

Intrinsic regulation is the ability of heart muscle to regulate its own fate independent of extrinsic effects, such as nerves and hormones. Experiments on completely denervated hearts and those isolated from the body demonstrate that such hearts are capable of regulating their rates depending on the demand. Greyhounds whose hearts were denervated were able to race almost as fast as normal dogs; their hearts were able to increase their rate although not as rapidly as the innervated heart. The Frank-Starling mechanism, whereby the heart responds to changes in fiber length by increasing output is another example of intrinsic regulation. Body temperature is another factor affecting intrinsic heart rate. Increasing body temperature and energy metabolism increase intrinsic rate and decreasing body temperature decreases heart rate.

Selected Readings

Altman PL, Dittmer DS (1974) Biology data book, 2nd edition, Vol. III. Fed Amer Soc Exper Biol

Berne RM, and Levy MN (1977) Cardiovascular physiology. Mosby, St. Louis

Guyton AC (1977) Basic human physiology. Saunders, Philadelphia

Guyton AC, Young DB (1978) Cardiovascular physiology III, International Review of Physiology, Vol. 18. University Park Press, Baltimore

Jensen D (1971) Intrinsic cardiac rate regulation. Appleton-Century Crofts, New York

Little RC (1977) Physiology of the heart and circulation. Year Book Medical, Chicago

Selkurt EE (1975) Basic physiology for the health sciences. Little, Brown, Boston

Sturkie PD (1976) Avian physiology, Chap. 5, Springer Verlag, New York

Review Questions

1. What type of valves are the right and left AV valves. Describe them.
2. At what stage in contraction of the heart are all valves closed?
3. At what stage are AV valves open?
4. What are isovolumetric contraction, isometric contraction, and isotonic contraction?
5. What is the Frank-Starling mechanism?
6. What is heterometric autoregulation?
7. What is homeometric autoregulation?
8. How efficient (in percent) is the work output of the heart?
9. How is oxygen consumption related to mechanical efficiency of the heart?

10. Name several factors that influence heart rate.
11. What is meant by extrinsic and intrinsic regulation?
12. What nerves are involved in heart rate regulation and how do they accomplish it?
13. Give an example of reflex control of heart rate.

Chapter 16 Electrocardiography

Conducting System

There is a specialized anatomic conducting system in man and other species (Fig. 16.1). It consists of the sinoatrial node (SA), the atrioventricular node (AV), and the bundle of His, with right and left branches and the Purkinjie fibers. This system is composed of specialized muscle cells that have the property of automaticity, and their rate of conduction of an electrical impulse is faster than ordinary atrial and ventricular muscle. The impulse is first picked up by pacemakers cells in the SA node (primary pacemaker), which normally controls heart rate. It then spreads (depolarizes) over the surface of the atria, and the impulse then spreads downward to the AV node (secondary pacemaker), which is then activated (depolarized). From the AV node the impulse passes down the bundle of His and spreads out (right and left) to activate the ventricular muscle. The specialized system makes up only a small portion of the total auricular and ventricular mass.

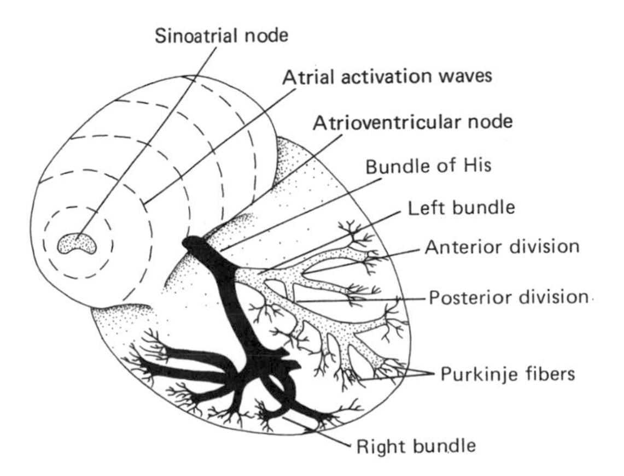

Fig. 16.1. Specialized conducting system of heart. (After Lipman, BS, Massie, E, 1959.)

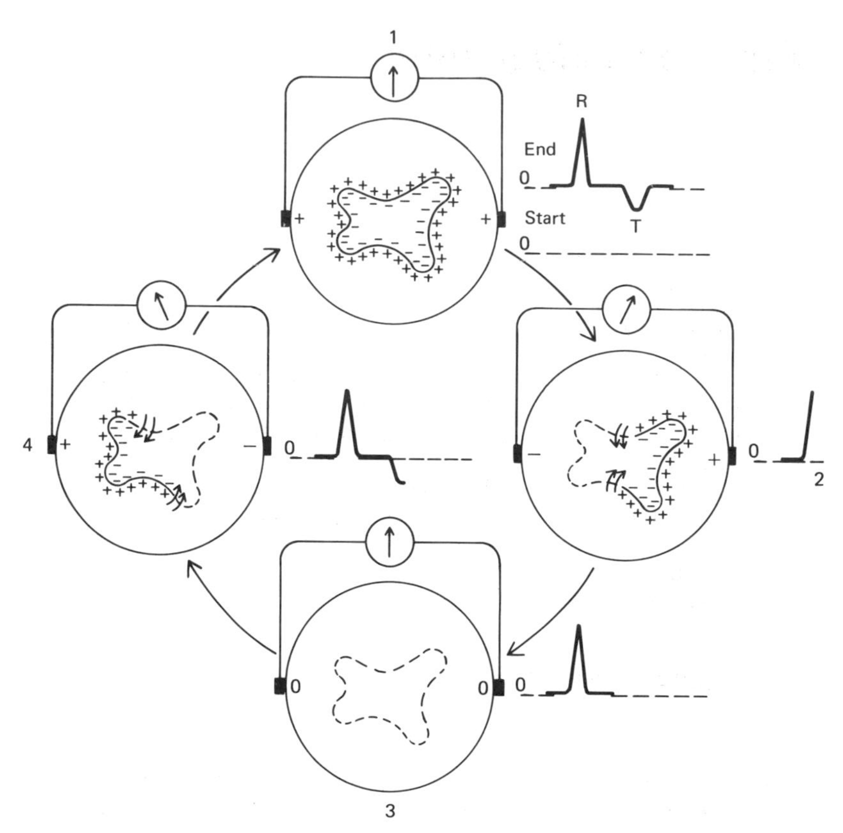

Fig. 16.2. Depolarization and repolarization of heart muscle (ventricle). Resting muscle at upper center (1) has (+) and (−) charges (balanced and inactive). The impulse now spreads to (2) (partially depolarized) and then to (3) where it is completely depolarized (QRS wave). The muscle begins to repolarize at (4) and is completely repolarized at (1). Note waves of ECG at various stages of depolarization/repolarization (see text). T wave in (1) is inverted as it should be if rates of depolarization and repolarization are the same; but in most human ECGs they are not the same and T wave is upright. (After Katz, L.N., Electrocardiography, Lea and Febiger, 1947.)

Depolarization and Repolarization

The spread of the electrical impulse (action potential) through the conducting system and atrial and ventricular muscle constitutes depolarization and repolarization. The resulting waves, or configurations, are referred to as waves of depolarization (QRS) of ventricles and repolarization (T wave of ventricles).

Resting or inactivated muscle is polarized; there are positive charges on the outside of the muscle membrane and negative charges in the inside (Fig. 16.2, position 1). The muscle begins depolarizing in position 2 (discharge or abolition of positive and negative charges), and in position 3 it is completely depolarized. A record at this point would show a completed depolarization wave. This is followed by the wave of repolarization (position 4) where it begins at the point where depolarization first began. Repolarization is completed at position 1, and this coincides with the negative wave (v) and the recharging of the muscle mass back to the resting stage.

Depolarization and repolarization precede the actual mechanical contraction of heart muscle (Fig. 15.5).

Electrocardiogram

The electrocardiogram (ECG) is a record of the electrical activity (depolarization and repolarization) of the heart picked up from electrodes attached to parts of the body other than the heart itself (leads), by an instrument called the electrocardiograph.

The ECG may be recorded directly by an ink-writing lever or pen, or a heated stylus on heat-sensitive paper. Because of the inertia of the mechanical stylus, its frequency response is low (about 80–100 Hz), but it is suitable for human subjects. The impulse may also be picked up by a high-frequency instrument (500 or more Hz) and recorded photographically or electronically; it records most faithfully the speed and shape of ECG waves of fast-beating hearts (over 300/min) as in rats, birds, and other small animals.

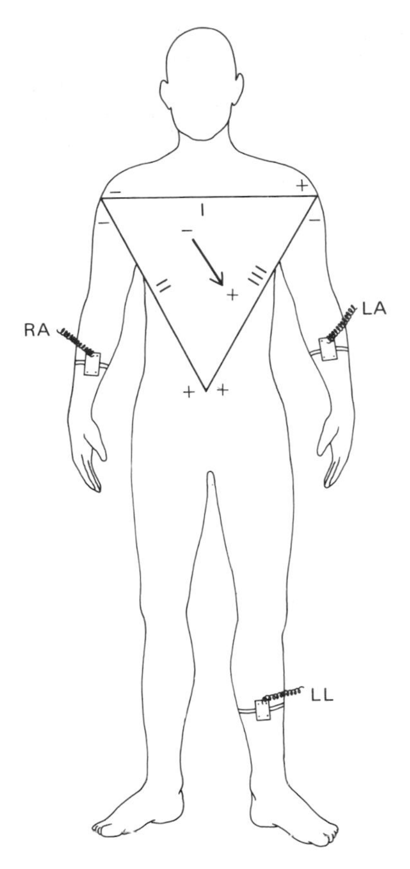

Fig. 16.3. Einthoven triangle and galvanometer connections for leads I, II, and III.

Leads

Leads taken directly from the heart produce records termed electrograms (EGs). Indirect leads are those taken at varying distances from the heart, including *limb* leads and *chest leads* (*V leads*). These leads may be *uni-* or *bipolar*. Bipolar electrodes are equally responsive to electrial impulses, but with unipolar leads one electrode is active and the other inactive or zero potential (indifferent electrode). The potential difference of unipolar electrodes is about one-half that of bipolar electrodes. Goldberger has augumented the standard unipolar lead so that the potential difference between electrodes is about 70% of bipolar electrodes. Such leads are designated *augumented unipolar* limb leads (aVR for chest and right arm, aVL for chest and left arm, aVF for chest and foot or for left leg).

Fig. 16.4. Precordial leads or chest leads (unipolar) designated as V leads placed at positions V-1, V-2, V-3, V-4, V-5, and V-6 as shown. The other electrode is the indifferent one.

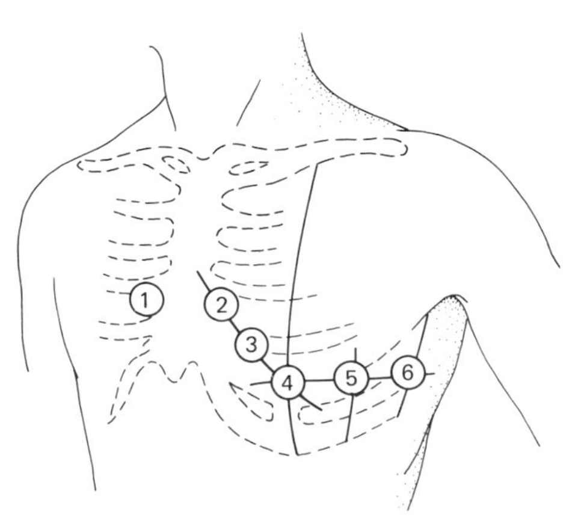

Standard Limb Leads

These (bipolar) leads are: lead one I, right arm and left arm (RA and LA); lead II, RA and left leg (LL); lead III, LA and LL. These three leads form a roughly equilateral triangle with the heart located near the center (Fig. 16.3).

Chest or Precordial Leads

These are unipolar, and the active or exploring electrode is placed at different numbered positions on the chest and designated as V1, 2, 3, 4, 5, and 6 as indicated in Fig. 16.4. Such leads tend to localize impulses and to detect certain heart disorders.

Chart Speed and Amplification

The standard speed for the paper of the electrocardiogram for man is 25 mm/s. The horizontal lines, 1 mm apart on the

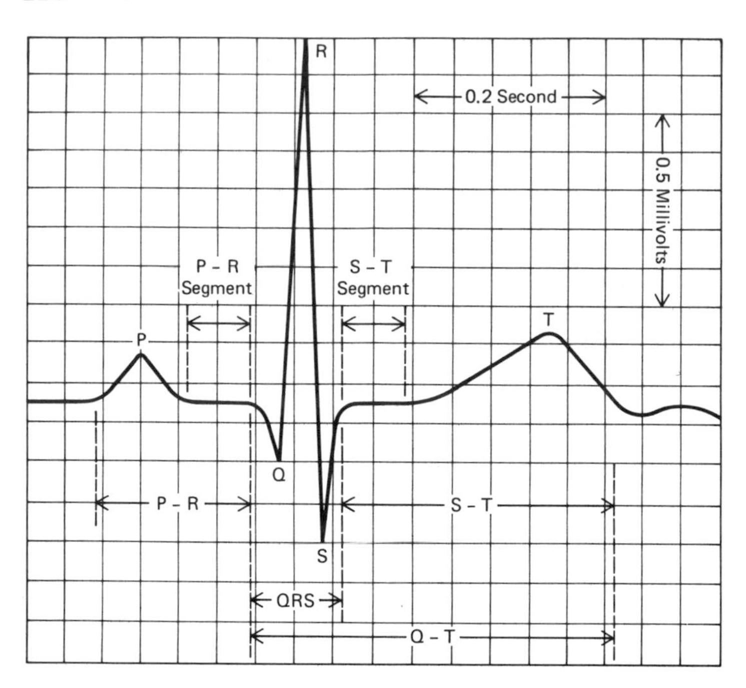

Fig. 16.5. Typical lead II tracing of human being in frontal plane showing the P, Q, R, S, and T waves and P-R and S-T segments above and intervals of P-R, S-T, and Q-T below (see text). These spaces on the paper are normally 1 mm, but here they are magnified.

paper, represent units of time (0.04 s). The vertical lines, also 1 mm apart, represent amplitude or voltage (Fig. 16.5). The electrocardiograph is standardized for man at 1 mV. When 1 mV is impressed on the instrument, it causes a vertical deflection of 10 mm. The electrodes (for the leads) are attached, or strapped to the desired positions, employing electrode paste which provides effective electrical contact, and the instrument has designations for the various lead positions.

Table 16.1. Features of typical ECG wave for man.

Wave	Description	Duration in Seconds, Range	Height (mm) Range for Leads I, II, and III
P	Atrial depolarization wave; usually positive	0.08–0.12	0.55–1.25
Q	First ventricular depolarization wave is negative and may not be present	—	0.36–0.61
R	Main wave of depolarization in limb leads is positive and upright	See QRS	5.5–11.5
S	Last ventricular wave of depolarization; negative and may not be present		1.5–1.7
QRS	Waves of depolarization	0.08–0.10	—
T	Wave of repolarization; positive in man; wave is upright	0.28 variable	1.2–3.0

—, Data not relevant.

Details of ECG

A normal (typical) human ECG in one of the limb leads is shown in Fig. 16.5, and the configuration, duration and amplitude of the waves are presented in Table 16.1. The P wave represents atrial depolarization, the QRS complex signals the onset of ventricular depolarization, and the T wave represents its repolarization. An atrial wave of repolarization (U wave) is usually absent.
Intervals of time between certain of these waves are important in analyzing the ECG and are presented in Table 16.2. Extreme changes in the duration of these intervals may reflect abnormal heart function.

Table 16.2. Features of wave intervals.

Wave[a]	Interval	Time (s)
P–R	From beginning of P to beginning of R or Q, whichever is first	0.18–0.20
QRS	From beginning of Q or R (whichever is first) to end of S, or R if S is absent	0.08
Q–T	From beginning of Q to end of T represents depolarization and repolarization of ventricle	0.38–0.31
S–T segment	From end of S to beginning of T	—

[a] Height or amplitude of waves varies, depending on lead and electrical axis (see also Fig. 16.6).
—, Data not relevant.

Electrical Axis

Einthoven in 1908 conceived of recording the action potentials of the heart from three remote limb leads arranged so that they form roughly an equilateral triangle with the heart located near the center (Fig. 16.3). In accordance with Kirchhoff's law, all electrical forces (EF) flowing in a circle to a given point should equal 0. In triangle A (Fig. 16.6), e.g., the EF in lead I flows (see arrows) from RA (negative) to La (positive). In lead III, EF flows from LA (negative) to LL (positive), and in lead II the force flows from RA (negative) to LL (positive). Thus, the force (EF) in I and III flowing to LL is equal to the EF in II (I + III = II) which is flowing to LL in the opposite direction, or I + III − II = 0. Thus, I = II − III and III = II − I. The magnitude of these electrical forces is proportional to or indicated by the height or amplitude of the waves in each lead. (See triangles A, B, C, & D of Fig. 16.6.)
In leads I, II & III of triangle C, all of the waves are upright (positive) because RA electrode is negative in I and II and picks up the impulse (EF) first. In lead (III, the LA electrode is also negative and picks up the impulse first, yielding an upright wave.
The electrical axis represents the average electrical or electromotive force acting in an average direction during the period of electrical activity of the heart. It is a vector and has direction and magnitude. An average axis means that the

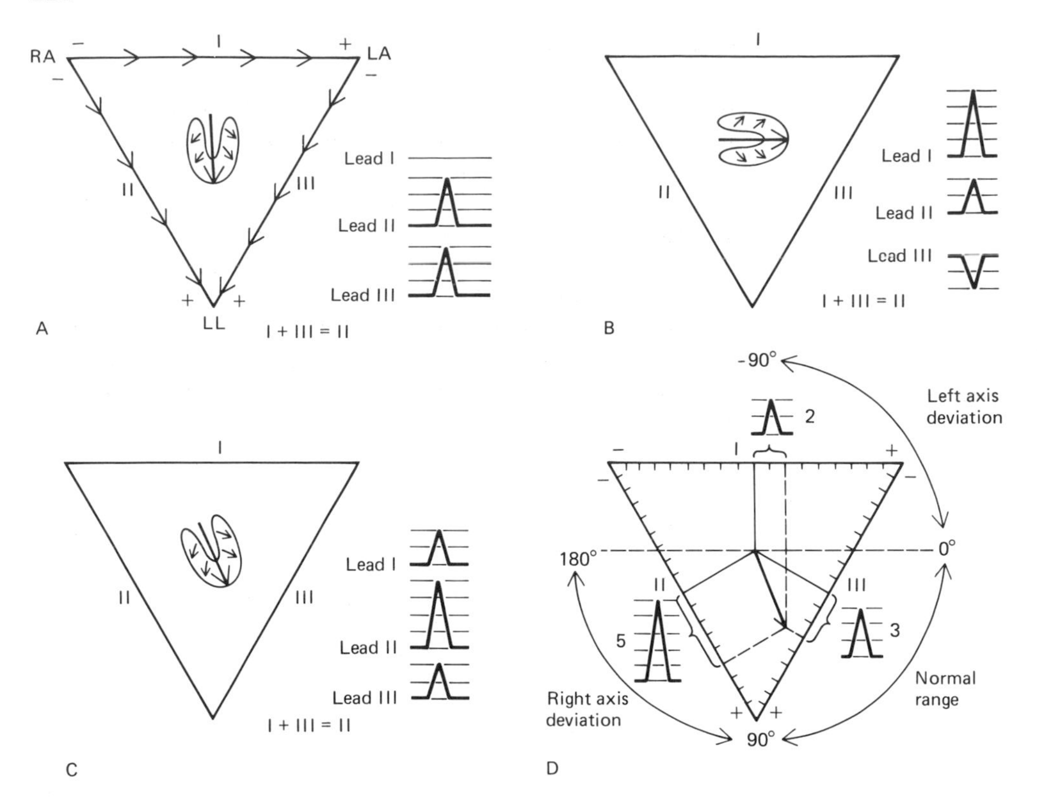

Fig. 16.6. Relationship of electrical axis of heart to recordings. To determine axis, see **D**. Here the QRS (mainly R) in lead I is 2 units in amplitude and upright; in lead III it is 3 units, and in lead II, 5 units. Thus II = I and III (or 5 = 2 + 3), and this is the case in any of the triangles shown in this figure (Einthoven's law). See text for details. (Modified from Greenspan, K, 1977.)

force spends more of its time and magnitude in a given direction. It is determined as shown in Fig. 16.6D. Any two leads (usually I and III) are used. In lead I (D) the QRS (mainly an R waves) is 2 units in amplitude and 3 units in III. These units are projected from 0 (solid lines from 0) on their lead lines, I and III.

Perpendicular lines (dotted) are dropped from the 2 and 3 unit lines on I and III. Where solid (zero) lines of I and III intersect, an arrow is projected to the point of intersection of dotted lines of I and III. This arrow (vector) points to between 0 and 90°, or roughly 65°. Note that the height of wave in lead II is 5 units and since II = I and III, 5 units in II = 2 and 3 units in I and III.

If the electrical axis is parallel or nearly so to the lead line in question (→), as in lead I of Fig. 16.6B, the amplitude of the wave will be greatest. If the axis is perpendicular to the lead line (I) as in Fig. 16.6A, there will be no wave in that lead. This is because the impulse from the heart strikes the lead line at about its center and it then spreads both toward the right and left arms; these impulses cancel out each other to give zero potential.

In Fig. 16.6C the amplitude is low in lead III, because the axis is nearly perpendicular to III.

Deviation of Axes

The normal range of electrical axes is from 0 to +90° (Fig. 16.6D). If the axis is shifted from 0 toward −90° it is called left axis deviation (Fig. 16.6D). This sort of shift often

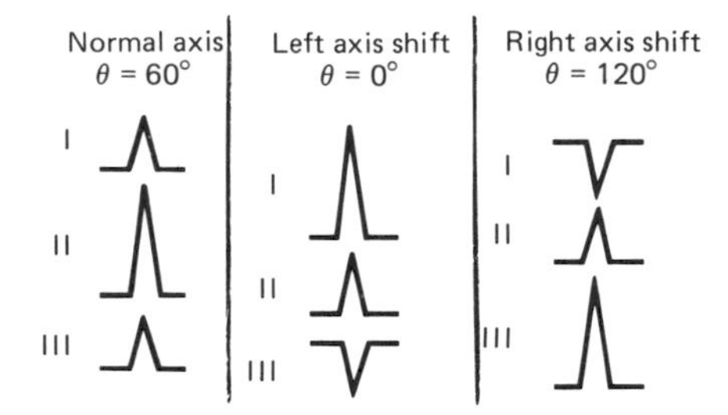

Fig. 16.7. Effects of right and left axis shift (deviation) on configuration of ventricular complexes in leads I, II, and III. Note the inverted (negative) wave in lead III in left axis shift and the inverted wave in lead I of right axis shift.

occurs in short, obese persons and is not abnormal but means rather that the anatomic axis of heart is located or shifted to the left.

If the shift is from +90° toward 180° (Figs. 16.6D and 16.7), it is a right axis deviation and is more likely to be abnormal. The effects of right and left axis shifts on the height of the ventricular waves in leads I, II and III are shown in Fig. 16.7.

Abnormalities of ECG

There are basically two types of disorders; one affects the rhythm and initiation of impulse and the other the conduction of impulse and the shape and configuration of the waves.

Cardiac arrhythmias (rhythm disorders) are characterized by irregular firing of the SA node to produce the arrhythmias. The rhythm and rate may be slow (bradycardia) and irregular or very fast (atrial tachycardia; Fig. 16.8D). Atrial premature beats (Fig. 16.8A) show a short P-P interval followed by a long P-P. There may be marked slowing of the heart, called sinus slowing or sinus arrest, characterized by long pauses, and an increased but irregular P-P interval. This is a disorder affecting the initiation of the impulse in the SA node. Premature ventricular beats exhibit a bizarre S wave as the premature beat, arising from an ectopic foci in the ventricule (Fig. 16.8C). Ventricular tachycardia is characterized by rapid and regular firing or discharge from an ectopic foci of ventricle (Fig. 16.8E).

Fig. 16.8. Cardiac arrhythmias (see text for details). (Modified from Greenspan, K, 1966.)

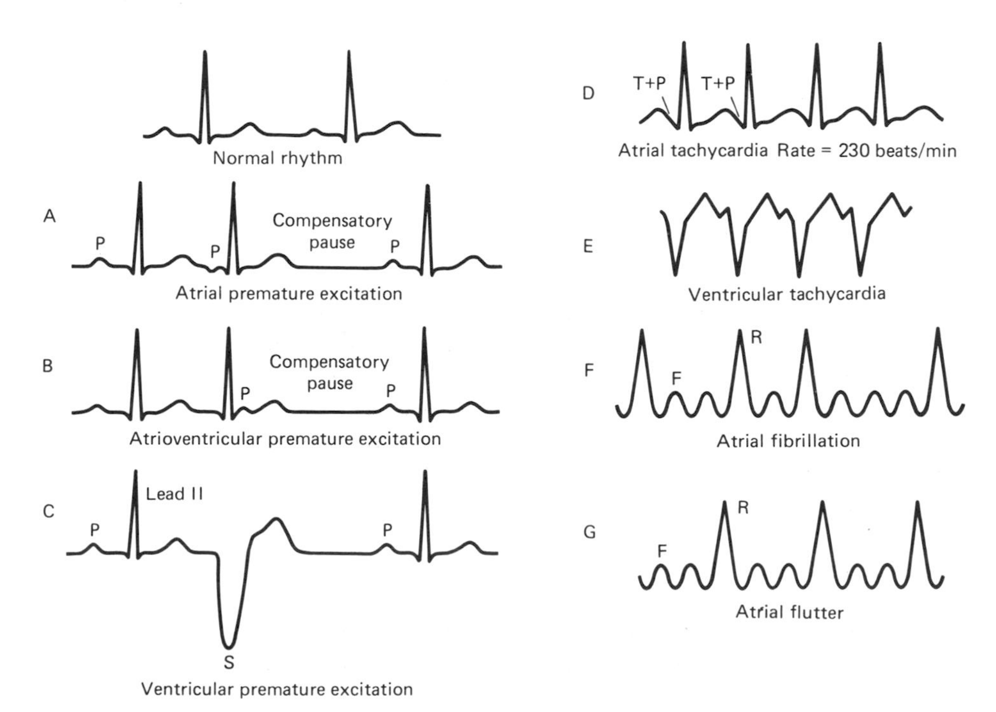

Fibrillation of atria or ventricles produces irregular and arrhythmic beats that are ineffectual in pumping blood. It is most serious when the ventricles are involved and may be fatal. Atrial fibrillation is less serious and is characterized by irregular and arrhythmic beats with atrial rates of 2 to 5 times the ventricular rate (Fig. 16.8F). Here are irregular P or F waves in a sequence of 2, 1, 3 F's to each R wave. *Atrial flutter* exhibits less rapid and more rhythmic atrial beats but are usually 2 to 3 times the ventricular rate (Fig. 16.8G).

Atrial fibrillation may result from the discharge of multiple ectopic foci in the atria, whereas discharge from a single ectopic foci probably causes atrial flutter. Atrial rate can be reverted to normal by appropriate drugs or by mild electric shock (cardioversion).

Conduction Disorders

Coronary artery disease, particularly acute myocardial infarction, is one in which there is a blockage of blood flow to heart muscle, which leads to the disorder. This is shown in Fig. 16.9, where changes in the QRS complex are revealed after an attack and after recovery has occurred some time later. In the acute phases, there are marked changes in configuration of Q, S-T, and T waves. Note particularly the elevated S-T segment and inverted T wave in certain leads.

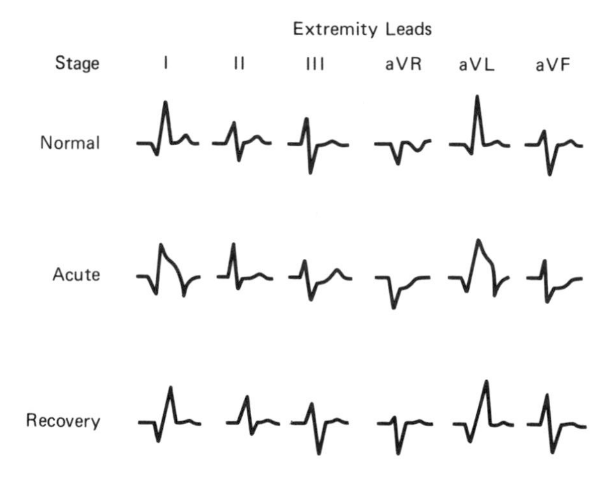

Fig. 16.9. Evolution of coronary disease (myocardial infarction) in leads I, II, III, aVL, and aVF. In acute stages, note abnormal Q and elevated S-T segment and inverted T in some leads. After several weeks, the record slowly returns to near-normal (recovery phase). (Modified from Lippman, BS, Massie, E, 1959.)

The first change in muscle is ischemia (lack of blood and anginal pain) and then injury followed by infarcted (dead) muscle. These changes in muscle circulation cause changes in the pathways of conduction to produce the modified configuration, or an injury current is produced. One other change also associated with coronary disorders is irregular rhythm of the beat.

Heart Blocks

Heart blocks are caused by disorders of the SA node (SA block), which should not be confused with sinus slowing or

Fig. 16.10. Atrioventricular heart block (see text). (After Greenspan, K, 1966.)

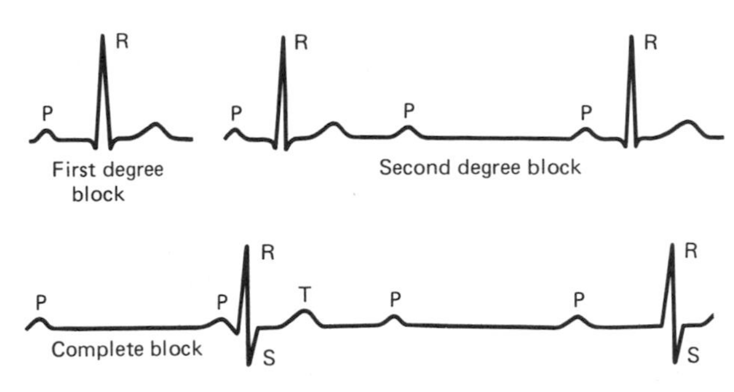

arrest. It is characterized by long pauses and an increase in the P-P interval, which is a multiple of the normal P-P interval, with dropping of atrial (P waves) and ventricular beats. The initiation of the P waves by SA node is normal; the beat is not conducted but blocked in the SA node.

Atrioventricular blocks are classified as first, second, and third degree (complete block) (Fig. 16.10). In first-degree block, there is a progressive increase in P-R interval until finally one ventricular beat may be dropped.

In second-degree block, the interval of the P waves is normal and an occasional ventricular beat is dropped. The rhythm or number of dropped beats may vary from every other beat to various other rhythms.

In complete block, there are more P waves than ventricular beats, indicating that the atrial and ventricular rhythms are independent of each other and that the ventricular regular beats from AV node or blocked and only ectopic ventricular beats are present. Persons with complete AV block often have electric pacemakers that stimulate the ventricles to beat with the atrial beats.

Selected Readings

Altman PS, and Dittmer DS (1974) Biology data book, 2nd edn, Vol. III. Fed Amer Soc Exper Biol

Berne RM, and Levy MV (1977) Cardiovascular physiology, 3rd edn. Mosby, St. Louis

Burch G, Windsor T (1972) A primer of electrocardiography, 6th edn. Lea and Febiger, Philadelphia

Greenspan K (1977) Physiology, 2nd edn (1966) and 4th edn (1977), edited by E. E. Selkurt, Little, Brown & Co., Boston

Lipman BS, Massie E Clinical scalar electrocardiography, 4th edn. (1959) and Lipman, Massie and Kleiger (1972) 6th edn. Year Book Medical Publishers, Inc., Chicago

Review Questions

1. Diagram the specialized conducting system in man.
2. What are the primary and secondary pacemakers and why are they so named?

3. Illustrate by diagram Einthoven's triangle and the relationship of the limb leads (I, II, and III) to the impulse from the heart.
4. What is Einthoven's law and its significance?
5. In taking leads I, II, and III of a person, it is found that the QRS in lead I shows practically nothing and leads II and III reveal waves of the same amplitude. What does this mean? What is the average electrical axis?
6. In the three limb leads, the R wave is tallest in lead II. What does this mean with respect to direction of impulse and lead line, or electrical axis?
7. In the ECG of a normal person the ECG paper has squares of 1 mm. What do these squares mean, vertically and horizontally?
8. How would you determine the average electrical axis of R wave which = 2 units in lead I and 3 in lead III?
9. Diagram a case of second-degree AV block.

Chapter 17 Respiration

Chemical reduction of molecular oxygen to form water is the primary source of energy for the mammal. The principal site for the reaction occurs within the mitochondrion. Without this reaction life cannot exist for more than seconds. Coincidental with the reduction of O_2 is the production of CO_2; this comes mainly from oxidative phosphorylation of glucose which also occurs principally in the mitochondrion.

The oxygen in CO_2 does not have its origin in molecular oxygen directly. The utilization of O_2 and the production of CO_2 are linked through intermediate metabolic reactions; theoretically each could continue for a short time in the absence of the other.

The exchange of O_2 and CO_2 between the organism and the environment is defined as respiration. In complex mammals a number of processes must occur in sequence for respiration to successfully meet the needs of the animal. Specifically (1) the gases must be moved between the environment and the lungs, usually referred to as "pulmonary ventilation." (2) The gases must be exchanged between the spaces in the lungs and the blood (pulmonary respiration). (3) Exchange of gases between the blood and the various tissues must occur (internal respiration). Finally transfer of gases must occur within the tissue to sites of utilization (for O_2) and from sites of production (for CO_2) (cellular respiration).

Table 17.1 Average gas tensions and contents in environment and human body assuming barometric pressure as 760 torr.

	Ambient		Alveolar		Systemic Arterial Blood		Tissue	Pulmonary Arterial Blood	
Gas	Pressure (torr)	Content (%)	Pressure (torr)	Content (%)	Pressure (torr)	Content (%)	Pressure (torr)	Pressure (torr)	Content (%)
Oxygen	159	20.94	100	13.16	95	20[b]	≤40	40	15[b]
Carbon dioxide	0.3	0.04	40	5.26	40	1.2[c]	≥46	46	1.4[c]
Water vapor	6	0.78	47[a]	6.18	47	—	—	47	—

[a] Vapor pressure at 37°C.
[b] O_2 content given as ml of O_2/100 ml blood.
[c] CO_2 content given in mM carbonic acid.
—, Data not relevant.

Failure in any of the four processes results in inadequate respiration and jeopardizes the existence of the animal.
The *movement of respiratory gases* is the result of pressure differences maintained within the respiratory system and between the environment and the tissues. Decreased pressure of O_2 in the tissue causes that gas to move toward the tissue. For CO_2, the pressure gradient is in the opposite direction and CO_2 moves into the environment (Table 17.1). The study of the physiology of respiration then is really the study of these gradients and how they are maintained. Note in Table 17.1 that water vapor pressure in the body is greater than the surrounding environment so water is lost from the body during respiration.

Anatomic Components

The respiratory system in mammals includes the tissues and organs associated with pulmonary ventilation and pulmonary respiration (air passages, lungs, and musculoskeletal system). *Air passages* consist of the nose, nasal cavity, nasopharynx, larynx, trachea, bronchi, and bronchioles. The *lungs* consist of pulmonary bronchioles and alveolar sacs as well as the arteries, capillaries, and veins of the pulmonary circulation. Bronchioles are of two types acting as (1) air conduits (simple pathway for gases to move through) and (2) both air conduits and sites of exchange between gases inside of the tube and the pulmonary capillaries, i.e., they contain alveoli, which are also involved in pulmonary respiration. The other set of structures associated with respiration, *skeletal and muscular,* consist of ribs, intercostal muscles, diaphragm, and accessory muscles.

Air Passages

The nose and nasal cavity serve as a passageway for air to be warmed, moistened, and filtered (Fig. 17.1A). The nasal cavity is also the site of the receptors for the sense of smell. The external nose is a triangular framework of bone and cartilage covered by skin with two oval openings on its undersurface called anterior nares, or nostrils. Each anterior naris

Fig. 17.1A. Anatomy of air passages. Nose, mouth, and pharynx shown in sagittal section.

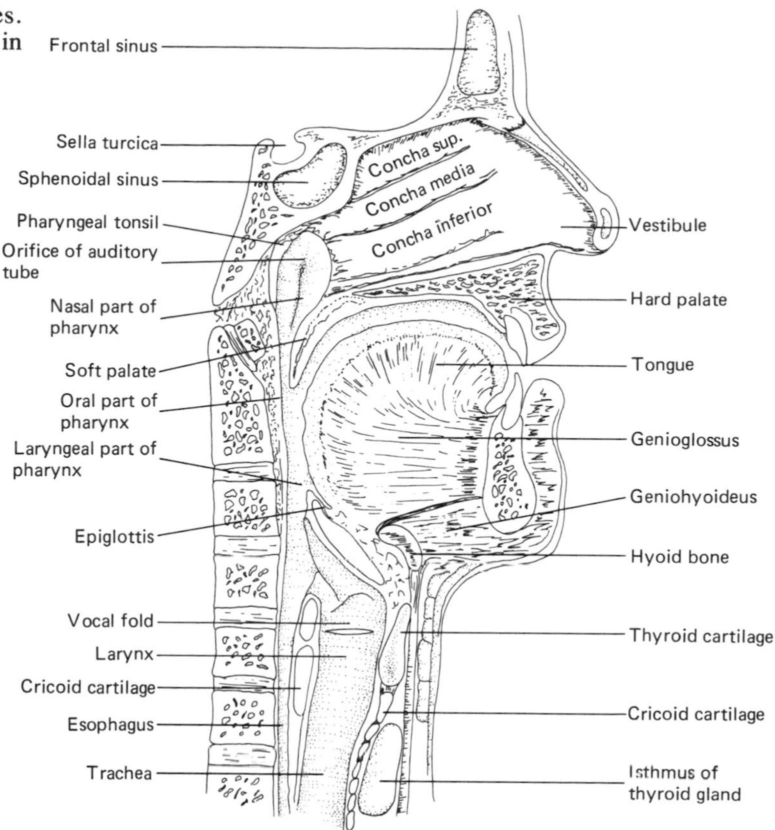

(sing.) opens into a wedge-shaped cavity, the nasal cavity. The cavities are separated from each other by a septum. Three light and spongy scroll-like mounds (conchae) project from the lateral walls of the cavities, partially dividing the cavities into four incomplete passages (meatuses). The cavities are lined with highly vascular mucous membrane at the entrance to the anterior nares. Sebaceous glands and numerous coarse hairs, together with the ciliated epithelial cells and goblet cells, serve to filter particles from the inspired air. In the upper portion of the cavity are olfactory cells, the receptors for smell. The cavities empty into the nasopharynx by the posterior nares.

The *larynx* is between the trachea and the base of the tongue. Two folds of mucous membrane, which do not quite meet at the mid-line, divide the cavity of the larynx. The midline tissue between the mucous membranes is referred to as the glottis and is protected by a flap of fibrocartilage, the epiglottis (17.1A). At the edges of the glottis in the mucous membrane are fibrous and elastic ligaments which are called the inferior or true vocal folds (cords). Above the vocal folds are two ventricular or false vocal folds. They serve to protect and keep the true vocal folds moist; they also serve in holding the breath and preventing the entrance of food into the larynx during swallowing. Specialized muscles tighten and slacken both true and false vocal folds. The action of these muscles are important in phonation as well as in preventing objects from entering the respiratory passageway.

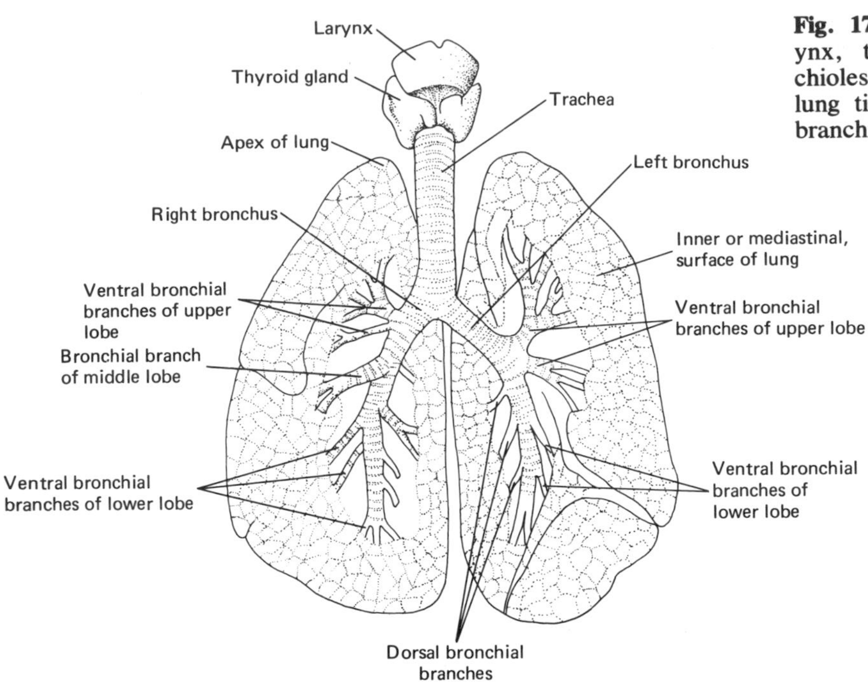

Fig. 17.1B. Anatomy of lungs, larynx, trachea, bronchi, and bronchioles. Portions of the overlying lung tissue were removed to show branching.

The *trachea* begins at the posterior end of the larynx (Fig. 17.1B) as a membranous and cartilaginous tube and continues into the thoracic cavity where it divides into a left and a right bronchus. These are air conduits to the lungs. In most mammals the rings of cartilage are not complete around the trachea. The sections next to the esophagus form essentially a fibrous ligament rather than a piece of cartilage. The right bronchus is usually shorter and wider than the left. After the major bronchi have entered each lung, they divide progressively into smaller and smaller tubes (bronchioles) until reaching the smallest tubes that do not exchange gas with the pulmonary blood, i.e., terminal bronchioles. From larynx to terminal bronchioles the tubes are lined with ciliated epithelium.

Lungs

The lungs appear grossly as spongy, porous, cone-shaped structures on either side of the thoracic cavity. The smallest unit structure of the lung, the lobule (Fig. 17.2) consists of a terminal bronchiole leading into the pulmonary bronchiole and a sac-like ending, the *alveolar sac*. Along the walls of the pulmonary bronchiole and the alveolar sac are pouch-like projections from the lumen. These are the alveoli. The wall structure of the alveolus is a single layer of type I epithelial cells. Pulmonary microcirculation crosses the outside wall of each alveolus. The inner surface of the alveolus is covered with a surface active material (surfactant) composed of phospholipoproteins or lipid-polysaccharide molecules, or both. *Surfactant* is believed to be secreted by granular pneumonocytes (type II cells). Because of their close contact with other structures, the individual alveolus is roughly polygonal shape with dimensions of 80 to 250μ. The total alveolar surface area for exchange is considered to be expo-

Fig. 17.2. Lung lobule showing bronchiole and alveolar structures and their relationship to pulmonary and systemic blood supply.

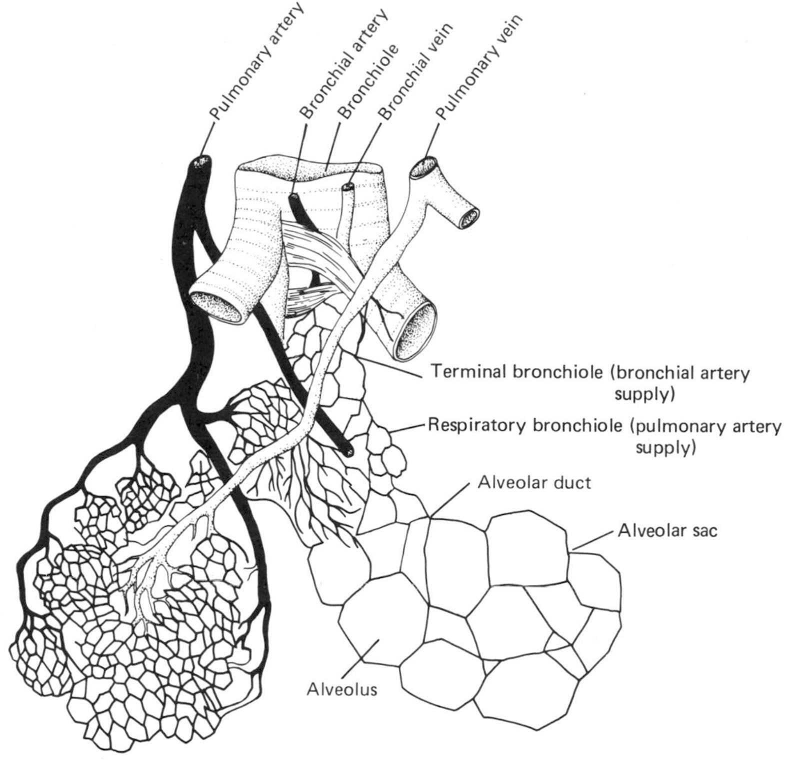

nentially related to body weight. Alveolar surface area tends to decrease with age in the adult.

Pleura

Surrounding each lung is a serous membrane sac, the *pleura* (Fig. 17.3). The outer (parietal) layer adheres to the inner thoracic wall and the diaphragm. The inner (visceral) layer covers the lung. The potential space between the two layers

Fig. 17.3. Intrapulmonary and intrapleural pressures during inspiratory-expiratory sequence. Air flow rate and lung volume changes are also shown.

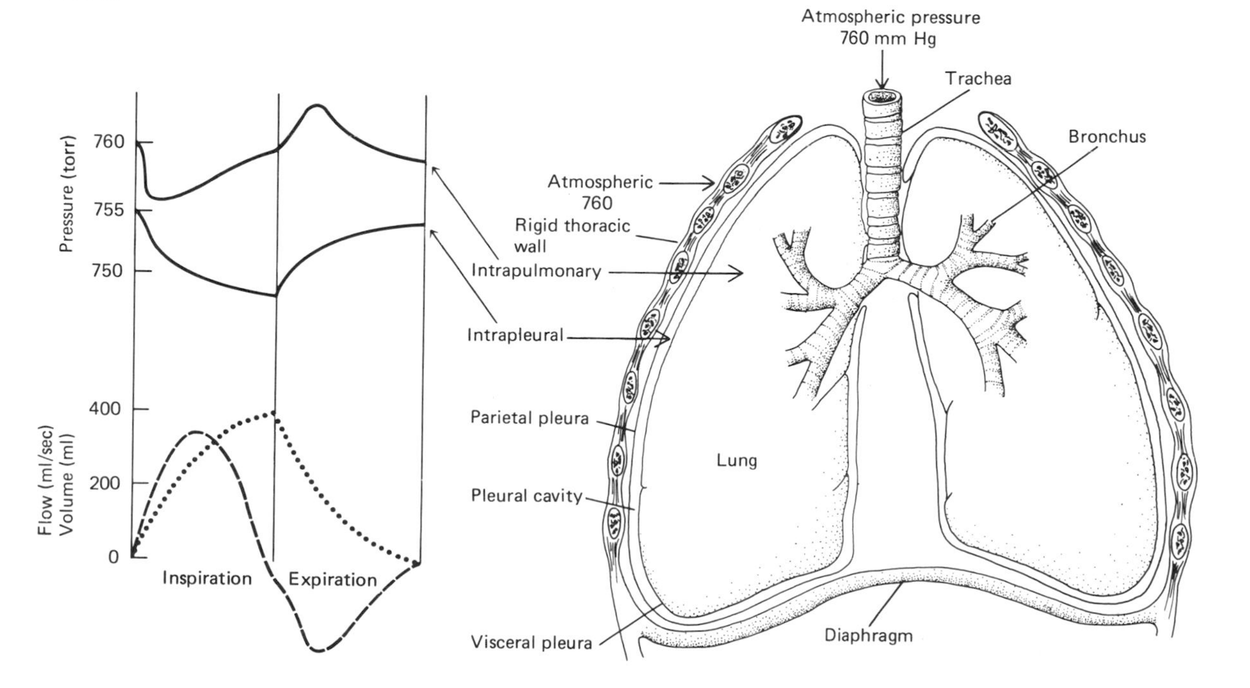

is called the *pleural cavity*. The layers normally move easily over each other with movement of the thorax. Pressure in the intrapleural space (intrapleural or intrathoracic pressure) is always negative. When the respiratory muscles are at rest, intrapleural pressure in human beings averages 4.5 torr (mmHg) less than atmospheric pressure (−4.5 torr). The interpleural space between the lungs is called the mediastinum, in which are located the trachea, the thymus, and the heart (together with all the major vessels entering and leaving it), the lymph vessels and nodes and the esophagus.

Pulmonary Blood Vessels

The pulmonary artery carries blood from the right ventricles of the heart. It divides into right and left branches which pass into the respective lungs. These arteries arborize (branch), following the pattern of the bronchi and supply the gross structures of the lung and alveoli (blood capillaries) (Fig. 17.2).

The air in the alveolus is separated from blood in the capillary by (1) the alveolar wall, (2) the capillary wall, and in some cases (3) an interstitial layer between the alveolar and capillary walls. Drainage is from capillaries into small veins that ultimately unite to form the pulmonary veins that empty into the left auricle.

The bronchial arteries of the systemic circulation also supply blood to the lungs. Specifically they supply the bronchi and bronchioles, lymph nodes, wall of the blood vessels, and the pleura. Most of this blood is drained by the bronchial veins, which in turn are drained on the left by the hemiazygos vein and on the right, by the azygos vein. A very small amount of bronchial arterial blood drains into the pulmonary veins.

Thoracic Cavity

The functioning thoracic cavity is limited on the back by the first 10 thoracic vertebrae (Figs. 1.8–1.11). Two additional thoracic vertebrae are normally indicated as part of the thorax. However, they are functionally part of the abdominal cavity and do not participate actively in respiration (see also Chap. 1). The front of the thorax is formed by the sternum—a flat, narrow bone situated on the median line of the thorax. The lowest portion of this structure is called the xiphoid process (Fig. 1.10). The rest of the cavity is formed from ribs and costal cartilages. Ribs are arches of bone articulated with the thoracic vertebrae in the back. The ribs are arranged as lateral pairs. For the first seven pairs, they increase progressively in length and attach by costal cartilage. The next three pairs of ribs, numbers 8, 9, and 10, decrease in length and join the costal cartilage by bars of cartilage. Rib pairs 11 and 12 are unattached in the front (floating ribs). The articulation for ribs 1, 10, 11, and 12 is with a single vertebra (Fig. 1.11). For the other ribs the articulation is formed by the bodies of two adjacent vertebrae. Each rib slopes downward from its vertebral attachment so that its sternal attachment is lower than its vertebra. The spaces between the ribs are the intercostal spaces.

Respiratory Muscles

Respiratory muscles are those whose contractions move the ribs. Those running from the head, neck, arms, and some upper thoracic and lower cervical vertebrae as well as the external intercostal muscles (rib to rib) raise the ribs and increase the dimensions of the thoracic cavity. The diaphragm which is the musculotendinous sheet attached to the vertebrae, ribs, and sternum serves to separate thoracic and abdominal cavities (Figs. 1.8 and 1.9). It is a principal muscle structure involved in normal inspiration. Additional muscle groups can be called on during a more forceful inspiration. During forced expiration, muscles attached between ribs (internal intercostal muscles) and to the ribs and lower thorax and upper lumbar vertebrae, as well as muscles of the abdominal cavity, act to depress the ribs and to force the abdominal contents against a relaxed diaphragm, thus decreasing the thoracic cavity dimensions.

Pulmonary Ventilation

As long as intraplueral pressure remains less than atmospheric pressure, the dimensions of the lungs follow closely the dimensions of the thoracic cavity. Lung movements result from contraction of respiratory muscles associated with the movement of structures in the thoracic cavity wall and of the diaphragm.

External Respiratory Movements

Relaxation of all muscles associated with respiration causes the thorax to assume the position of passive expiration. By appropriate muscular activity it is possible either to inspire or to expire further from this position.

Inspiration results from expansion of the thoracic cavity and is always an active process. Because of their articulation with the vertebrae, the ribs are moved upward and outward, increasing the distance from the vertebral column to sternum as well as the lateral dimensions of the thorax. Movement of the ribs is referred to as costal breathing. Contraction of the diaphragm changes its shape from a domelike structure to a more flattened plane that increases the length of thorax from the head to abdomen (diaphragmatic or abdominal breathing). Normally diaphragmatic breathing is the major contributor to inspiration. Since humans are bipedal, each time the ribs and sternum move there is a change in the center of gravity of the body and a need to adjust the muscles associated with posture.

During quiet breathing in human subjects the elastic properties and the weight of the displaced tissue are normally sufficient to return the tissues to their preinspiratory position. Thus expiration results passively from a graded decrease in inspiratory muscle activity. *Active expiration* can result from contraction of the internal intercostal muscles in addition to the other muscle groups that depress the ribs and decrease the lateral and sternum-vertebrae distance. Active expiration can also result from contraction of the abdominal muscles, which push the viscera against a relaxed diaphragm and decreases the head-to-abdomen dimension of the chest.

Increasing the lung size decreases (temporarily) the total intrapulmonary (alveolar) pressure. Intrapulmonary pressure is equal to atmospheric pressure when air is not moving and the glottis is open. It is less than atmospheric pressure until the lungs are filled during inspiration and greater than atmospheric pressure during expiration. Intrapleural pressure also varies during respiratory movement; however, it is always less than atmospheric pressure and is always negative. (Fig. 17.3).

Lung Volume Changes

Volume changes are not the same in all parts of the lung during inspiration. There are three major reasons for this. First, the thoracic cavity does not increase to the same degree in all directions. Second, all parts of the lung are not equally distensible. Third, there is believed to be a gravitational effect tending to pull the lung down (Fig. 17.4).

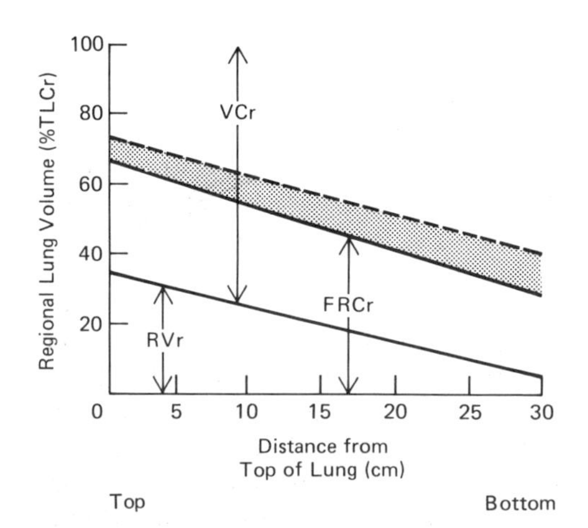

Fig. 17.4. Fraction of lung region volume (%TLC_r) available for expansion, i.e., vital capacity for region (VC_r). Note top of lung is to left of graph and VC_r is greater in lower regions than in top regions. (Adapted from Milic–Emili et al. J. Appl. Physiol. 21:749, 1966.)

The volume of air that is inspired or expired in one breath is called *tidal volume*. The volume of air that can be moved with a maximum respiratory effort is called the *vital capacity*. It does not represent the entire volume of air in the lung (total lung volume) because it is not possible to completely collapse the lungs. The volume of air that cannot be removed without collapsing the lungs is called the *residual volume*. There is an additional volume of air that can be inspired with maximum effort after a normal inspiration. This volume is called *inspiratory reserve volume*. The volume of air that can be expired with maximal effort after normal expiration is called *expiratory reserve volume*. The *functional residual capacity* consists of the expiratory reserve volume and residual volume. It is the volume of air in the lungs into which the normal tidal volume is diluted (Fig. 17.5). Thus the composition of the gas in the lungs is normally not changed drastically after one tidal volume.

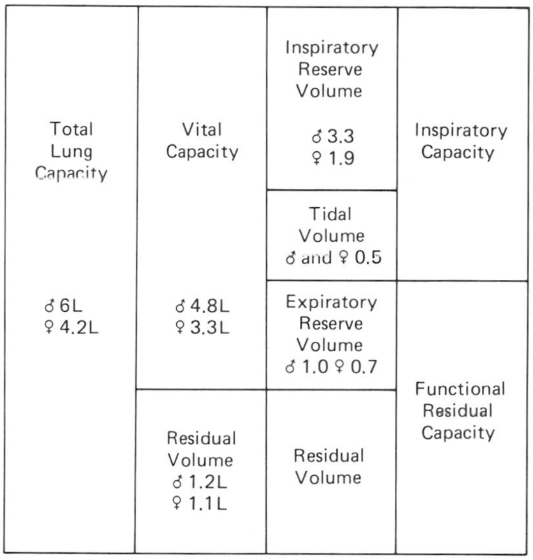

Fig. 17.5. Partition of lung volume and capacities in adult human subjects (average values given for male (♂) and female (♀) human subjects (liters)).

<table>
<tr><td rowspan="4">Total Lung Capacity
♂ 6L
♀ 4.2L</td><td rowspan="3">Vital Capacity
♂ 4.8L
♀ 3.3L</td><td>Inspiratory Reserve Volume
♂ 3.3
♀ 1.9</td><td rowspan="2">Inspiratory Capacity</td></tr>
<tr><td>Tidal Volume
♂ and ♀ 0.5</td></tr>
<tr><td>Expiratory Reserve Volume
♂ 1.0 ♀ 0.7</td><td rowspan="2">Functional Residual Capacity</td></tr>
<tr><td>Residual Volume
♂ 1.2L
♀ 1.1L</td><td>Residual Volume</td></tr>
</table>

Minute volume ($\dot{V}$) is the volume of air inspired in one minute. It can be calculated by multiplying the average tidal volume (V_T) by the number of breaths per minute (f) or $\dot{V} = fV_T$. A portion of V_T does not come into contact with active pulmonary circulation, e.g., air in trachea and bronchi down to terminal bronchioles and nonperfused alveoli. Gases in this portion of the inspired air do not exchange with the gases in pulmonary blood, and this part of V_T is called dead space (V_D). The portion of V_T that does exchange gases with pulmonary blood is called alveolar volume (V_A). From a physiologic point of view, alveolar ventilation ($\dot{V}_A$) is a most important consideration in external respiration ($\dot{V}_A = f(V_T - V_D)$) because it represents the volume of inspired air per minute that exchanges gases with pulmonary blood.

Respiratory Mechanics

Moving air into and out of the lungs requires work. Three types of forces must be overcome to bring air into the lungs, namely, (1) elastic resistance, (2) airflow resistance in the tracheobronchial tree, and (3) nonelastic tissue movement, e.g., raising the ribs and supporting tissue.

Compliance

The work required to overcome *elastic resistance* of the lungs and the chest walls (thorax) is considered to be independent of time. The maximum work performed occurs when tidal volume is also maximum. This form of resistance can be estimated by determining the pressure required to change the lung and chest volume. This measure is called compliance (C)

$$C = \frac{\Delta V}{\Delta P},$$

where ΔV = change in volume (L) and ΔP = change in pressure (cm H_2O).

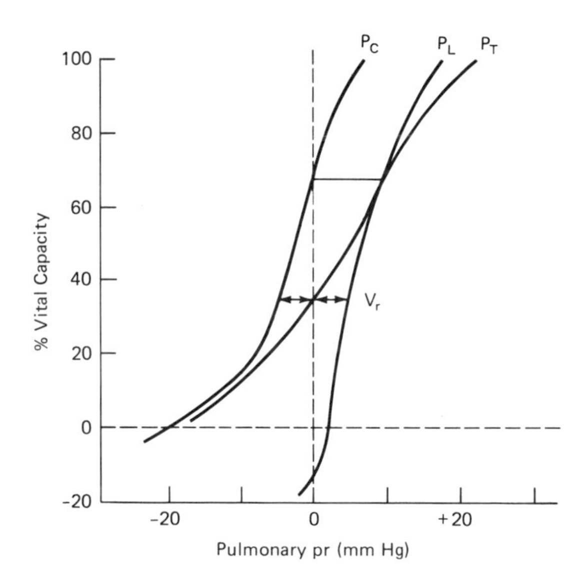

Fig. 17.6. Total pressure (P_T) contributed by sum of chest (P_c) and lung (P_L) elastic properties at different levels of chest expansion (%VC). The slopes of the lines represent compliance. Note that resting end tidal volume (V_r) occurs at point that negative P_c equals positive P_L. If elastic properties of either lung or chest change, V_r would be shifted. Any volume other than V_r requires muscle tension to create force needed (P_T).

The compliance of the lung and chest wall together can be estimated by preparing a graph relating the intrapulmonary pressure required to maintain known volumes of gas in the lung. Experimentally, this is done by filling the lungs with known volumes, allowing all respiratory muscles to relax, and measuring the pressures in the mouth (nostrils blocked). The compliance of the lung is equal to intrapleural pressure and thus can also be determined (Fig. 17.6).
It is estimated that three-fourths to seven-eighths of the total elastic resistance is attributable to surface tension, and

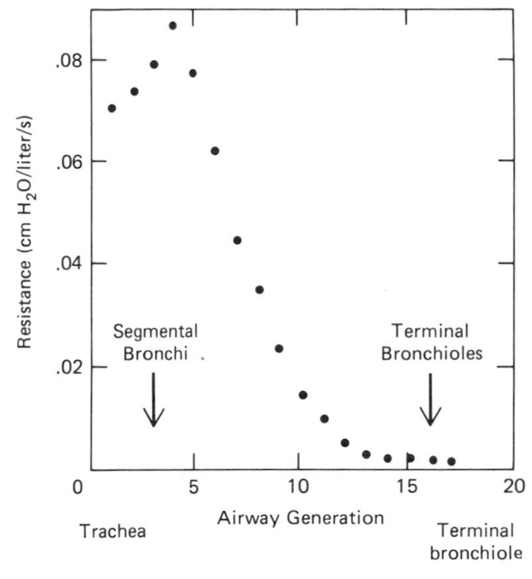

Fig. 17.7. Resistance in various parts of air passages. Note resistance is highest in major bronchi and lowest in small bronchioles. (Adapted from Pedley et al., Resp. Physiol. 9:387, 1970.)

the remainder is accounted for by elastic properties of the tissue. Surface tension is the force in any one direction holding the surface of an enclosed volume together. The higher the surface tension, the more energy required to make the surface larger (increase the volume). As the surface gets smaller in the lungs (decreasing volume), the surface tension decreases because of the surfactant (see section on "Lungs" in this chapter). Surfactants are also believed to be responsible for stabilizing the lung alveoli so that they do not collapse during expiration.

The principal site for *airflow resistance* is shown to be the medium-sized bronchi (Fig. 17.7). On the basis of Poiseuille's equation (see Chap. 14), one would expect the smallest bronchioles to be the site of highest resistance, but this is not true. Less than 20% of the measured airflow resistance is contributed by airways less than 2 mm in diameter. The plentiful supply of small airways represents a larger cross-sectional area available for airflow. At very low lung volume a phenomenon of "airway closure" is reported, i.e., reversible collapse of small bronchioles. Under such circumstances some energy is spent during inspiration opening the collapsed bronchioles. Airflow resistance is time related and is greatest at rapid breathing rates. During breathing, it reaches a maximum even though the inspiratory volume has not reached its maximum.

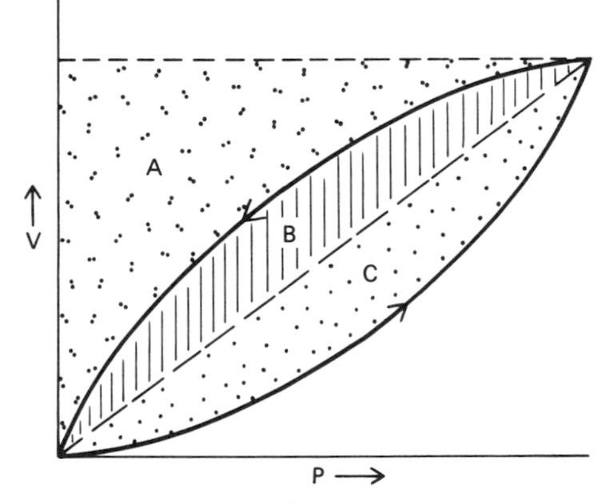

Fig. 17.8. Idealized pressure-volume relations during one tidal volume (solid line on increasing values indicates inspiration, solid line pointing to decreasing volume indicates expiration). Integrated areas A and B equal total *elastic* work done during inspiration. Area C represents all *nonelastic* work done during inspiration, i.e., tissue and air flow. Area B is nonelastic work done during expiration.

The *nonelastic tissue resistance* work of moving the chest wall and the lung is also time related. In younger adults it amounts to about 20% of the total cost of breathing.

Total work required to move air into and out of the lung, which includes moving the chest wall, is estimated from a pressure-volume graph (Fig. 17.8). $W = \int_0^V Vdp$. The work can be apportioned into elastic work (areas A and B on Fig. 17.8) and nonelastic work (area C, Fig. 17.8). For a given minute volume there is a rate of work at which the sum of elastic and the time-related non-elastic components is min-

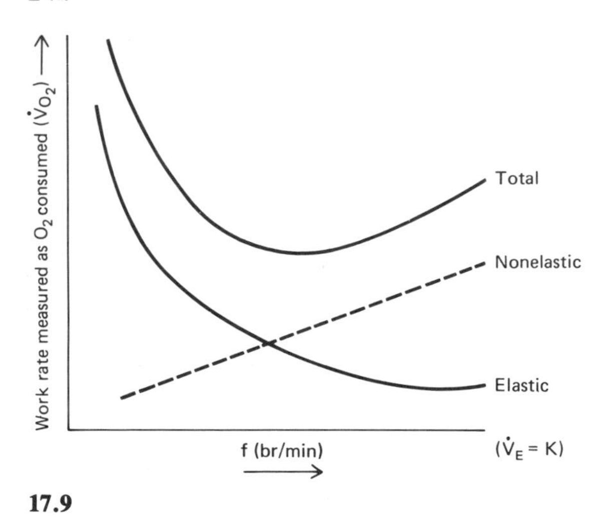

17.9

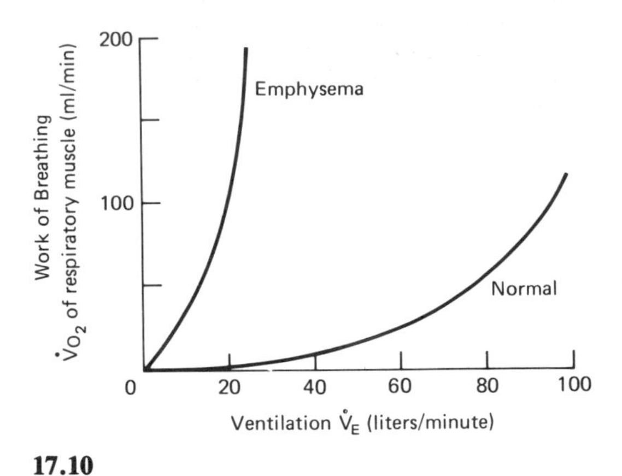

17.10

imum (Fig. 17.9). During normal respiration less than 5% of total O_2 consumption is required to move the air into and out of the lungs (Fig. 17.10).

Fig. 17.9. Hypothetical graph of elastic, nonelastic, and total work done at different respiratory rates (f) in human subject when total minute volume was kept constant. Note optimum f when sum of elastic and nonelastic work contributions results in minimum of total work required. The work rate is measured as O_2 consumed (V_{O_2}).

Fig. 17.10. Oxygen consumed by respiratory muscles at various rates of breathing in normal human subject and subject with decreased lung compliance (emphysema).

Pulmonary Circulation

During pulmonary circulation, blood leaves the right ventricle by pulmonary arteries, passes through capillary beds located between two alveoli, and returns to the left atrium by pulmonary veins. In addition to its role in gas exchange, the pulmonary circulation also serves as a blood reservoir for the left heart; it provides nutrition for alveolar duct and alveolar tissue; it removes excess fluid from the alveoli, and it filters particles from the systemic venous blood. It may also activate and inactivate a number of pharmacologic compounds.

Its role in gas exchange is the primary concern of this discussion. The pulmonary circulation is a low-pressure system when compared to the systemic circulation. Average pressure in the pulmonary arteries is 25/10 mm Hg with a mean pressure of about 15 mm Hg. Since the cardiac output of the right and left hearts must be equal, resistance in the pulmonary circulation $\left(R = \frac{Pr}{Flow}\right)$ is approximately one-fourth to one-fifth that of the systemic circulation (see also Chap. 14).

Both arteries and veins in the pulmonary circulation can increase their capacity to hold blood with very small changes in blood pressure (high compliance). Because of their high compliance, the capacity of the vessels is greatly affected by gravitational forces, i.e., the height of the column of blood. During normal ventilation in humans, the lower part of the lungs gets a relatively greater bloodflow than does the upper portion because of the gravitational effect (Fig. 17.11).

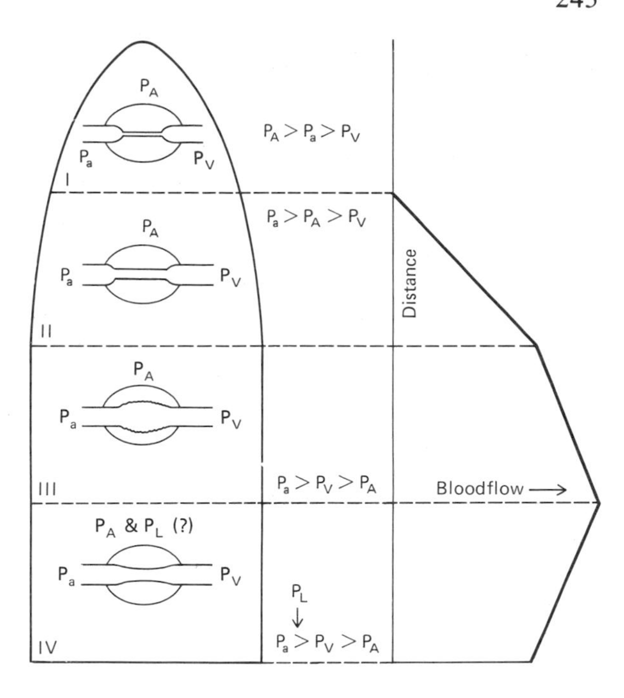

Fig. 17.11. Pattern of blood flow distribution in lung is influenced by distance from bottom of lungs. In zone I (top) intrapulmonary pressure (P_A) exceeds arterial blood pressure (P_a) during some portion of respiratory cycle, and blood flow is restricted. In zone II P_a is greater than P_A and P_A is $\geqq$ pulmonary vein (P_v). Blood flow is controlled by the $P_a - P_A$ difference (slightly increased blood flow compared to zone I). In zone III P_A is less than P_a and P_v. Blood flow in this zone is determined by $P_a - P_v$ gradient (greatly increased blood flow). At bottom of lung, zone IV, flow is restricted again and interstitial pressure (P_L) is suggested as one possible cause of restricted blood flow.

Capillaries in the lungs form a meshwork on the alveolar surface so that several capillaries can cross a single alveolus. There is some question as to whether the capillaries are arranged so that several individual tubes are in contact with an alveolar membrane or whether the exchange surface is in the form of a sheet with cartilagineous supports. Blood in the "capillaries" is in contact with alveolar gas approximately 0.75 s at rest and 0.34 s during mild exercise. Capillary blood pressure is low, in the range of 7–9 mm Hg. Using Starling's hypothesis of bulkflow, fluid movement between the capillary and the tissue space (see also Chap. 12), there is a net pressure to move fluid from the alveoli back into the blood.

Nerves and Drugs

Pulmonary vessels are well supplied with sympathetic vasoconstrictor nerve fibers. Stimulation of these fibers are believed to be involved in mobilization of blood in the pulmonary reservoir. Sympathetic and parasympathetic vasodilator nerve fibers are found in limited numbers. Their physiologic role is not fully understood at this time. *Epinephrine* and *angiotensin* II constrict pulmonary arterioles. *Serotonin* (5HT) and *histamine* constrict pulmonary veins. In contrast to the response of systemic vessels, local increase in CO_2 or decrease in O_2, or both causes vasoconstriction. Thus bloodflow to poorly ventilated alveoli is restricted, and better ventilated alveoli are preferentially perfused.

Pulmonary Respiration

Gas is a form of matter that distributes itself uniformly throughout a confined volume. There is little interaction between molecules in the gaseous form. Their movement can exert a force when they strike the sides of the confined volume. This force applied to a unit of area is called gas pressure, given in units of mm Hg or torr, and is proportional to the number of molecules and the mean velocity of molecules. At room temperature the pressure exerted by an individual molecular species, e.g., O_2 or N_2 is independent of the other molecular gases present (see also Chap. 2). Total gas pressure measured is the sum of the pressures of individual molecular species present (spoken of as partial pressure) or $P_B = P_{N_2} + P_{O_2} + P_{H_2O} + P_Z$ where P_B = barometric pressure. The fraction (F) of a given gas (x) in the dry gas mixture can be calculated from the following relationship:

$$F_x = \frac{P_x}{P_B - P_{H_2O}\,(\text{vapor})}$$

Conversely, the partial pressure of a given gas (x) can be calculated from its fraction: $P_x = F_x\,(P_B - P_{H_2O})$. Dry atmospheric air contains 20.94% O_2. $P_{O_2} = \frac{20.94}{100} \times$ 760 torr (sea level) = 159.1 torr (see Table 17.1).

Respiratory gases exchange between the alveoli and blood in the pulmonary capillary by diffusion. *Diffusion* results from the continuous movement of gas molecules and effects the net transfer of molecules from a region in which they are in higher concentration to one in which their concentration is lower (Fick's law; see Chap. 2).

Gas Laws

There are purely physical factors that influence the diffusion rate between alveoli and blood. (1) *Density of the gases.* Graham's law applies. It states that in the gas phase under the same conditions the relative rate of diffusion of two gases is inversely related to the square root of the density of the gases. (2) *Solubility of gases in a fluid media.* Henry's law applies. It states that the mass of a dissolved gas in a given volume of liquid, with temperature remaining constant, is proportional to the solubility of the gas in the fluid (Bunsen solubility coefficient) and the partial pressure of the gas in equilibrium with the fluid. (3) *Temperature.* Increased temperature increases the mean velocity of the molecules (increased pressure) and decreases the solubility of the gas in a fluid at a given pressure. (4) *Pressure gradient.* For the gases in the respiratory system, Fick's law applies.

Diffusion Coefficients

On the basis of solubility and molecular size, the diffusion coefficient for CO_2 is approximately 20.7 times the diffusion coefficient for O_2. Since this is constant and the temperature in the lungs normally remains constant, only the partial pressures of these gases are responsible for the direction of gas exchange between the lungs and the alveoli (Table 17.1). In considering the physiological aspects of gas exchange in the lungs, one must consider the (1) pulmonary circulation in the alveoli, (2) surface area available for diffusion, (3) characteristics of the alveolar and capillary tissue, and (4) distance the gases must diffuse.

An estimate of the diffusion capacity of the lungs, referred to as transfer coefficient (T_{Lx} or D_{Lx} of some investigators) can be determined by measuring the quantity of gas (x) transferred each minute for each torr difference in partial pressure between alveoli (P_{AX}) and mean capillary blood (P_{cap_x}) or $T_{Lx} = \frac{\dot{V}_x}{P_{A_x} - P_{cap_x}}$; T_{Lx} varies with the type of gas studied and its location in the lung. Oxygen T_{Lx} in the whole lung of a resting person ranges from 19–31 ml/min/torr. During mild exercise it is elevated to 43 ml/min/torr.

Ventilation-Perfusion Ratios

The effectiveness of pulmonary respiration varies in different parts of the lung. This variability is explained in large part by the concept of a ventilation-perfusion ratio ($\dot{V}_A/\dot{Q}$). The ratio represents the number of alveoli that are ventilated and in contact with well-perfused pulmonary capillaries. During quiet respiration (eupnea) in human sub-

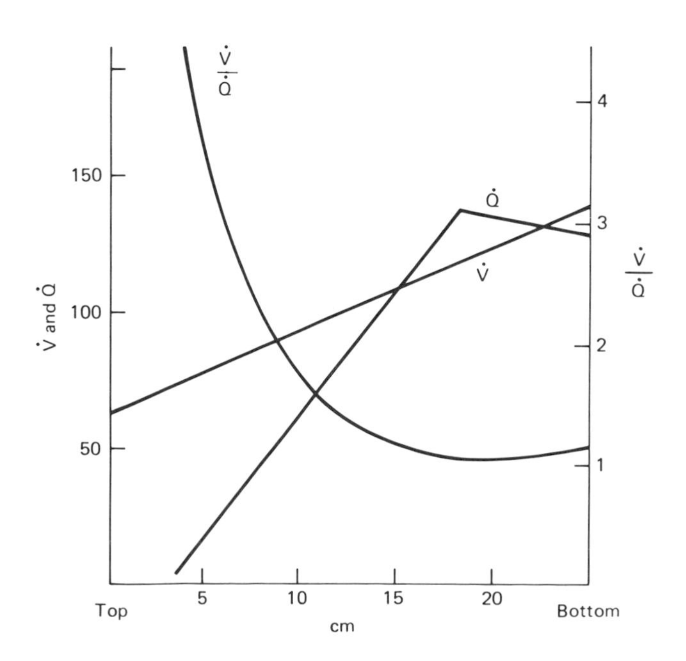

Fig. 17.12. Regional ventilation ($\dot{V}$) and blood flow ($\dot{Q}$) shown as percentage of total for lungs (from data on Figs. 17.4 and 17.11) and their ratio ($\dot{V}/\dot{Q}$) at different distances from top of lungs of erect human subject. Note change in $\dot{V}/\dot{Q}$ caused by decrease in zone IV blood flow.

jects, the upper portions of the lung are more fully expanded than the lower portion (Fig. 17.4), but the lower portions of the lungs are better perfused with blood than are the upper sections in the upright individual (Fig. 17.11). As tidal volumes increase, the lower portions of the lung are increasingly used and better perfused. The $\dot{V}/\dot{Q}$ tends to approach a minimum value of 1 in the lower portion of the lung (Fig. 17.12).

Transport of Respiratory Gases

Approximately 1.5% of the O_2 carried in the systemic arterial blood at normal P_{O_2} is dissolved in the plasma. The rest is held in loose chemical combination with the hemoglobin (Hb) found in the red blood cell. Hemoglobin is a conjugated protein consisting of a ferrous iron-porphyrin compound (heme) joined to the protein globulin (see Chap. 13 for further details). Each Fe^{2+} molecule of Hb combines loosely and reversibly with one molecule of O_2. Fully oxygenated hemoglobin carries 1.39 ml of O_2/g of Hb (some sources indicate 1.34 ml O_2/Hb). If the Fe^{2+} is oxidized to Fe^{3+}, the compound is no longer able to transport O_2.

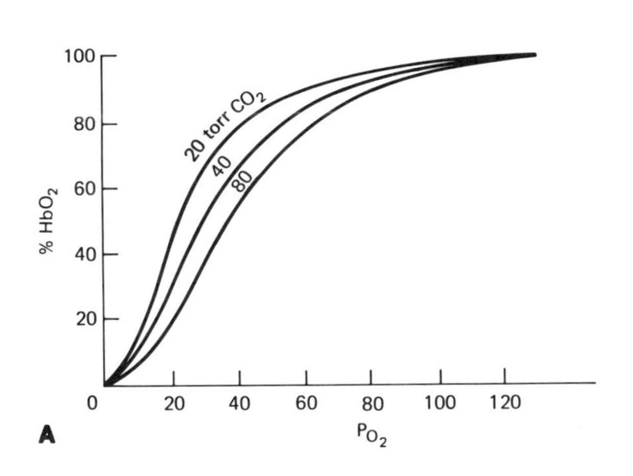

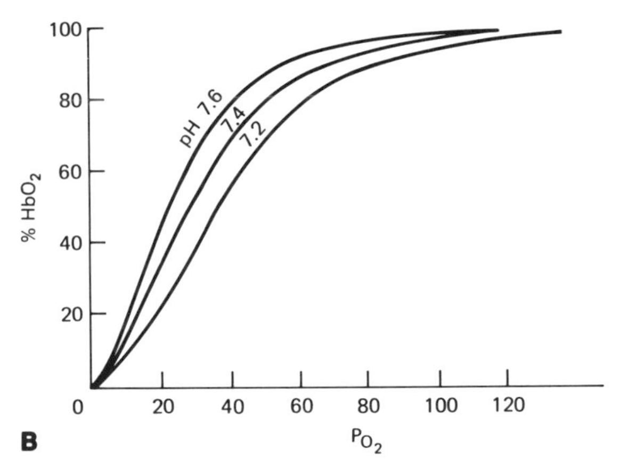

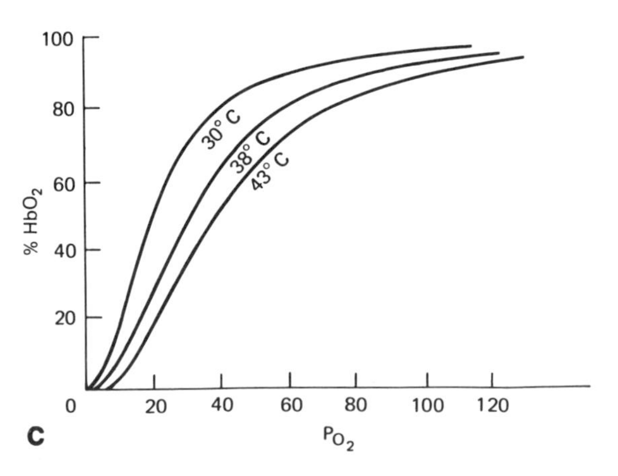

Fig. 17.13A–C. Oxygen dissociation curves for hemoglobin, relating pressure of oxygen (P_{O_2}) and percentage of hemoglobin that has bound oxygen (%Hb O_2). **A** effect of varying CO_2 pressure on blood at 38°C; **B** effect of varying pH on blood at 38°C; **C** effect of temperature at constant P_{CO_2} of 40 torr.

Fully oxygenated hemoglobin (HbO_2) releases more protons (is more acidic) than the deoxygenated hemoglobin (Hb). As a result, in a solution at pH 7.25, the release of 1 mM of O_2 from HbO_2 permits 0.7 mM of H^+ to be taken up without a change in pH; thus the delivery of O_2 aids in buffering acid produced locally in the tissue.
The relationship between the number of molecules of O_2 free to exert a force (P_{O_2}) and the number of molecules of O_2 attached to hemoglobin (HbO_2) is known as the *O_2 dissociation curve* (Fig. 17.13). HbO_2 can be represented in one of two forms: either in the fraction of Hb that is oxygenated (% HbO_2) or as the volume of O_2 per hundred ml of blood in the sample (volume percent). In either case the shape of the O_2 dissociation curve is the same.

Effects of pH, Carbon Dioxide and Temperature on Oxygen Transport

The ease with which O_2 combines with hemoglobin to form oxyhemoglobin ($Hb + O_2 \rightleftarrows HbO_2$) is affected by the P_{CO_2}, temperature, and concentration of H^+ (or pH) and organic phosphates, particularly 2, 3 diphosphoglycerate (DPG) in the blood (Fig. 17.13). Increased $[H^+]$ (decreased pH), P_{CO_2}, DPG, or temperature results in a shift toward the deoxyhemoglobin form in the dissociation reaction (shown previously, or to the right of the line represented by the lower value of $[H^+]$, P_{CO_2}, DPG, or temperature (and is referred to as a "shift to the right"). Conversely, a shift to the left represents a higher HbO_2 for a given P_{O_2}. Stated another way, a lower P_{O_2} is required to obtain the same level of HbO_2.
Since arterial blood has a lower P_{CO_2} and a higher pH, it will have a higher HbO_2 content than venous blood with the same P_{O_2} (Table 17.1 and Fig. 17.13). At the tissue level, the increased P_{CO_2} and decreased pH means that even without the change in P_{O_2}, some of the O_2 combined with hemoglobin in arterial blood will be released for use by the tissue. Thus the shift to the right with increased H^+ or P_{CO_2} (Bohr effect), or both, improves O_2 delivery at the tissue level.
The *coefficient of oxygen utilization* is defined as the difference in concentration of O_2 in the arterial and mixed venous blood divided by the arterial concentration of O_2 or $\frac{(Ca_{O_2} - Cv_{O_2})}{Ca_{O_2}}$. Normally this average value is about 0.25 and is based on a pooled sample of all systemic circulating blood. Coefficients of utilization of local areas of tissue may vary around this number. The overall coefficient of utilization would be lowered by shunting in the circulation.

Carbon Dioxide Transport

More than 90% of the CO_2 transported in the blood is in chemical combination. The rest (<5%) is transported in physical solution in the plasma. The two chemical combinations for transport of CO_2 are as bicarbonate ion (60%–70%) and on amino groups of blood proteins, e.g., globin of hemoglobin (10%–30%).

Bicarbonate is formed in the blood as a result of hydration of the CO_2 into carbonic acid and dissociation of carbonic acid into hydrogen and bicarbonate ion as follows:

$$H_2O + CO_2 \rightleftharpoons H_2CO_3 \rightleftharpoons H^+ + HCO_3^-$$

The rate of formation of carbonic acid is slow. It can increase several thousand times in the presence of the enzyme carbonic anhydrase. Since carbonic anhydrase is found in the red blood cell (RBC) and not in plasma, most of the association and disassociation of H_2O and CO_2 occurs in the RBC. At the tissue level, when CO_2 is added to the blood, the concentration of bicarbonate inside the RBC is increased. The bicarbonate freely diffuses through the RBC membrane into the surrounding plasma as a result of the concentration difference for this anion. Because the cell is selectively permeable, the diffusion of bicarbonate results in a charge difference inside the cell compared to the outside (more cation than anion present). Chloride ion diffuses in to rebalance the charge, in accordance with the Gibbs-Donnan equilibrium (see Chap. 2). In the lungs, when CO_2 is lost from the blood, and O_2 added, the reverse process occurs, i.e., bicarbonate diffuses in and chloride diffuses out. The movement of chloride ion to maintain the electrochemical equilibrium is referred to as the "chloride shift."

The formation of carbamino compounds, e.g., $\text{Protein} - NH_2 + CO_2 \rightleftharpoons \text{Protein} - NHCOO^- + H^+$ occurs rapidly without the apparent need for an enzyme. For hemoglobin the deoxygenated form can bind more CO_2 than the oxygenated form, which facilitates the transport of both O_2 and CO_2.

Respiration in Acid-Base Balance

Since the hydration of CO_2 results in carbonic acid exchange, transport of CO_2 has a great effect on the acid-base status of the circulating blood as well as on the body as a whole. The relation between the solution of CO_2 in blood and the pH of the blood is given by the *Henderson-Hasselbalch equation* (see Chap. 2):

$$pH = pK_a + \log \frac{\text{Salt}}{\text{Acid}}$$

where pK_a is the negative log of the dissociation constant for the entire reaction.

$$H_2O + CO_2 \rightleftharpoons H_2CO_3 \rightleftharpoons H^+ + HCO_3^-$$

At 37°C pK_a is approximately 6.1; $[HCO_3^-]$ and $[H_2CO_3]$ are given in mM/liter. Since the concentration of H_2CO_3 is proportional to the concentration of dissolved CO_2, we can calculate $[H_2CO_3]$ by determining the pressure of CO_2 in the solution (P_{CO_2}). At 37°C, 0.03 mM/Liter of carbonic acid are formed for each torr P_{CO_2}, and the Henderson-Hasselbalch equation can be written.

$$pH = 6.1 + \log \frac{[HCO_3^-]}{0.03\ P_{CO_2}}$$

An example of this relationship: if arterial blood pH is 7.4 and Pa_{CO_2} is 40 torr, $[HCO_3^-]$ is calculated to be 24 mM/liter. If P_{CO_2} is changed to 20 torr, there is a commensurate change in pH calculated as follows:

$$pH = 6.1 + \log \frac{24 - (0.03 \times 20)^1}{(0.03 \times 40) - (0.03 \times 20)}$$

$$= 6.1 + \log \frac{23.4}{0.6}$$

$$= 6.1 + \log 39$$
$$= 6.1 + 1.6$$
$$= 7.7$$

In actuality the buffers present in blood modify the relationship between P_{CO_2} and pH. Specific responses are best determined from available graphic data that describe the relationship between P_{CO_2}, HCO_3^-, and pH for the various body fluids.

[1] It is necessary to remove from the bicarbonate concentration the number of mM/liter contributed by carbonic acid.

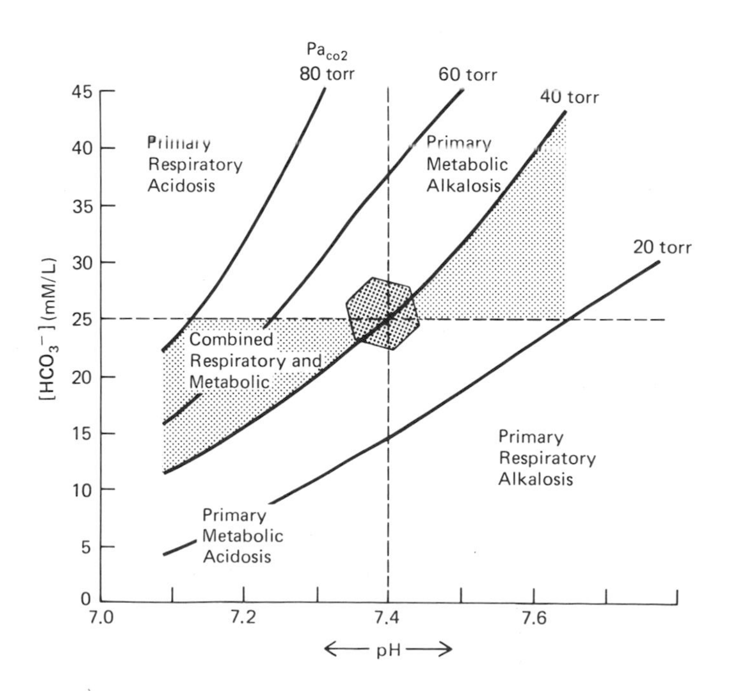

Fig. 17.14. Graph relating $[HCO_3^-]$ and pH for aqueous solution with P_{CO_2} indicated by isobars. Dark hatched hexagonal area indicates range of values found in blood of healthy persons at sea level. Displacement of values from normal range represents alteration in acid-base balance of individual. If pH is less than 7.4 and $[HCO_3^-]$ is greater than 25 mM/liter, condition is considered to represent primary respiratory acidosis. If pH is greater than 7.4 and $[HCO_3^-]$ is less than 25 mM/liter, condition represents primary respiratory alkalosis. Primary metabolic acidosis is represented by portion of graph below 40 torr Pa_{CO_2} isobar and pH less than 7.4. Primary metabolic alkalosis is represented by portion of graph above 40 torr Pa_{CO_2} isobar and pH greater than 7.4. All other portions of graph (lightly hatched areas) represent both respiratory and metabolic contributions as primary causes of imbalance. Primary cause of acid-base imbalance can be compensated by acid-base imbalance in opposite direction from other source, e.g., respiratory alkalosis compensated by metabolic acidosis. (Adapted from Woodbury, JW (1974) In Ruch T. Patton H (eds): Physiology and Biophysics Vol II, pp. 480–524, Saunders, Philadelphia.)

It should be kept in mind that the HCO_3^- present in blood is associated with other cations in addition to H^+. Changing the $[H^+]$ does not necessarily change the concentrations of other cations. That is accomplished by a balance between absorption and excretion of the specific cations. Cation excretion is controlled mainly by the kidney (see Chap. 24). If blood CO_2 is increased or decreased, the resulting change in pH (alkalosis for increased, acidosis for decreased, pH) is referred to as "respiratory." If the bicarbonate concentration is changed, the resulting pH change is referred to as "metabolic" (Fig. 17.14). Respiratory acidosis can be compensated for by metabolic alkalosis.

Tissue Oxygenation

Transport of O_2 from blood into the tissue sites of utilization results from simple diffusion (see Chap. 2). Since most of the utilization occurs in the cell mitochondria, the diffusion distances in tissue can be considered long by comparison to exchange in the lungs. In muscle tissue, the presence of myoglobin is believed to facilitate O_2 diffusion. Theoretical models incorporating factors affecting delivery and utilization of O_2, such as intracapillary distance, capillary blood flow, and tissue metabolism have been devised to estimate tissue P_{O_2}. The lowest P_{O_2} is found at the venous end and halfway between the capillaries, assuming equal blood flow in the capillaries and that capillaries are parallel.

Regulation of Respiration

Central Nervous Control

Arterial blood levels of CO_2, $[H^+]$, and O_2 are normally regulated within close limits by control of pulmonary ventilation. Specific areas within the central nervous system (CNS) are involved in initiating each pulmonary ventilatory effort as well as regulating the overall performance of the respiratory system. Participation by the CNS comprises two functionally separate elements: (1) automatic respiration associated primarily with structures in the brainstem and (2) voluntary respiration associated with suprabrainstem structures particularly the cortex.

Those areas of the CNS, particularly in the brainstem, concerned with respiration have been referred to as the "respiratory centers," a misleading term. Rather than a "center," it is an anatomical site through which pass a number of neurons concerned with respiration. There are several such sites in the brainstem. The response obtained from each center is affected by neuronal afferents from several locations in the body as well as by the local level of metabolites in blood and in the cerebrospinal fluid (CSF).

The spinal cord processes some intranuncial connections that can modify respiratory effort. However, automatic respiration does not occur if the brain is separated from the spinal cord.

The most caudal part of the brain that controls rhythmic respiration is the medulla. *The medullary centers* when in contact with the spinal cord are capable of maintaining a rhythmic respiratory effort without any connection to the CNS above it. The medulla has two aggregates or groups of respiratory neurones: (1) primarily, inspiratory neurons called the *doral respiratory group* (DRG) and (2) both inspiratory and expiratory neurons termed the *ventral respiratory group* (VRG). The origin of respiratory rhythmicity is far from clear. The rhythm may result from automaticity of specific medullary neurons, e.g., DRG or from a neural network connected in a way so that its output oscillates. Such networks have been demonstrated. The DRG is suggested as either being the source or lying close to the source of rhythmic generation. Thus DRG drives the VRG as well as many of the spinal respiratory motorneurons.

Cephalad to the medullary respiratory areas, in the caudal two-thirds of the pons, is an area termed the *apneustic*

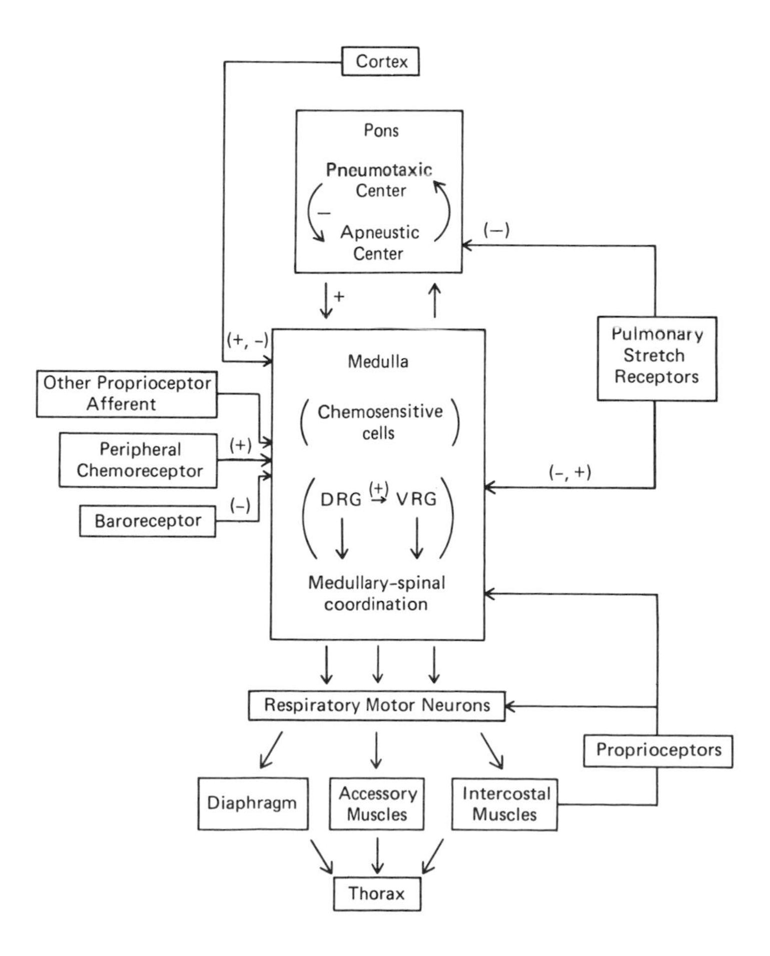

Fig. 17.15. Main functional components of respiratory control system. Some aspects of excitatory (+) and inhibitory (−) influences are indicated. Interrelationship between dorsal and ventral respiratory groups (DRG and VRG) in medulla is tentatively suggested on basis of current information.

center. Separation of the brain cephalad to the center combined with bilateral vagotomy results in an almost continuous inspiratory effort (apneusis). Thus the apneustic center is the site of a continuous inspiratory drive and normally is inhibited by neuronal activity from (1) an area in the extreme rostral pons called the *pneumotaxic center* or from (2) afferent vagus nerves, or both (Fig. 17.15). The pneumotaxic center is assumed to fine tune the automatic respiratory pattern, either by control over inspiratory time or by modulating CNS response to a number of afferent inputs that influence respiration, such as blood P_{CO_2} and degree of lung inflation.

In the high CNS the cerebral cortex influences respiratory rate and depth. Several specific areas in the cortex when stimulated either increase or inhibit respiratory effort. These sites are under voluntary control and are demonstrated whenever we speak or eat.

Sensory Receptors

There are numerous sensory receptors in the respiratory tract; receptors are also in contact with blood vessels, CSF, CNS tissue, and in joints and muscles of the limbs modifying the respiratory pattern.

Receptors in the nasal passages are innervated by the olfactory (I) and trigeminal (V) cranial nerves and are sensitive to various chemical agents as well as to mechanical stimulation. Response to stimulation can vary from apnea to sneezing. The pharyngeal area is innervated by a branch of the glossopharyngeal nerve (IX). Stimulation of this area results in sharp inspiratory efforts (e.g., sniffing). Receptors in the larynx and trachea are of several types which respond to chemical and mechanical stimulation. Innervation for these receptors come predominantly from branches of the vagus (X) nerve. Effects from stimulation are variable and include coughing, apnea, and slow deep breathing. Bronchoconstriction is also observed.

Lung Mechanoreceptors

In the lungs there are three types of receptors innervated by the vagus nerve, including: (1) slow-adapting pulmonary stretch receptors, (2) irritant receptors, and (3) type "J" stretch receptors. Inflation of the lungs stimulates *slow-adapting stretch receptors,* which act on the CNS to inhibit inspiratory efferent nerve discharge and cause passive expiration. This self-limiting pattern first reported in the 1860s is termed the *inspiratory-inhibitory Hering-Breuer reflex*. In humans the Hering-Breuer reflex does not normally operate until lung volume increases 1.5 to 2.0 times resting tidal volume. Under normal conditions pulmonary stretch receptor neurons in the vagus modify the respiratory rate by both their effect in the medulla and their inhibition of the apneustic center.

Irritant receptors are rapidly adapting receptors sensitive to chemical as well as to mechanical irritants, such as ammonia and particulate matter. Reflex responses to this type of receptor are increased inspiratory effort as well as bronchoconstriction.

Type J receptors do not normally respond to lung movement per se and are believed to be located in the wall of the pulmonary capillary. They can be stimulated by hyperinflation of the lungs. Their exact function is not resolved at this time.

There are a number of *other mechanoreceptors* that, when stimulated, affect the respiratory pattern. Among them are the pressure receptors located in the systemic arterial and venous systems. Stimulation of these receptors gives responses varying from temporary apnea to marked increases in respiratory rate. Movement of the joints and stretch of the muscle in the limbs increase both respiratory rate and tidal volume. Pain also acts as a respiratory stimulant.

Chemoreceptors

All three of the main products of tissue metabolism, viz., decreased P_{O_2}, increased P_{CO_2}, and increased $[H^+]$ in arterial blood stimulate respiration.

The most prominent respiratory receptors sensitive to the chemical composition of the fluid around them, *chemoreceptors,* are located in the aortic arch (aortic bodies), the region of the bifurcation of the common carotid into the internal and external carotid arteries (carotid bodies) (see Chap. 14) and the CNS near the surface of the fourth ventricle, or in ventrolateral areas of the medulla, or both. The aortic bodies are innervated by branches of the vagus nerve; carotid bodies are innervated by branches of the glossopharyngeal nerve. The aortic and carotid bodies are referred to as peripheral chemoreceptors. The CNS receptors are referred to as central chemoreceptors.

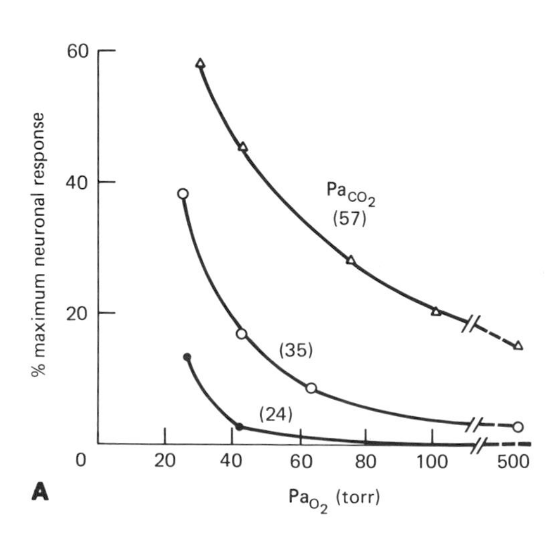

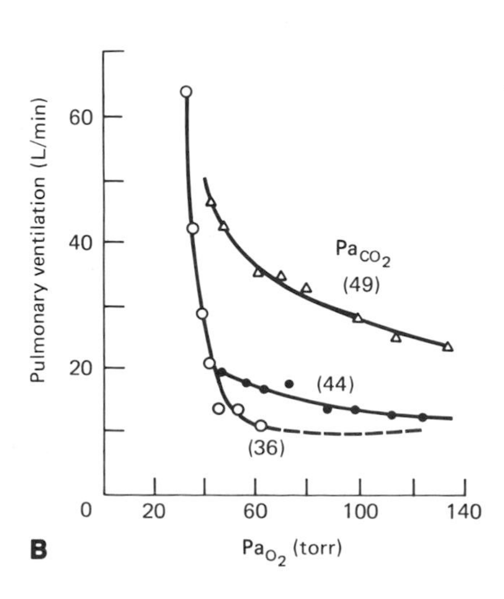

Fig. 17.16A and B. Response to changing Pa_{O_2} with change in Pa_{CO_2}. **A** average neuronal discharge from carotid body of cat; **B** ventilation response of human subjects (Pa_{CO_2} isobar indicated, in torr, by number in parentheses).

The rate of nerve impulses in the afferent nerve from the peripheral chemoreceptors is inversely proportional to the arterial P_{O_2} in the range 30 to about 500 torr (Fig. 17.16A). Marked increases in pulmonary ventilation are not normally seen unless arterial P_{O_2} decreases from its normal 90–100 torr to the range 60 torr or less (Fig. 17.16B). The degree to which P_{O_2} influences minute volume is influenced by the other metabolic influences, e.g., CO_2 and pH (Fig. 17.16B). The central chemoreceptors are not sensitive to low P_{O_2}. The effect of low P_{O_2} on respiration is to increase respiratory rate or tidal volume, or both.

Both peripheral and central chemoreceptors are affected by CO_2. About 80% to 85% of the total response observed results from central stimulation by CO_2. Increased arterial blood CO_2 tension (Pa_{CO_2}) above 30 torr increases the rate of nerve impulses from the receptor. The response is influenced by the P_{O_2} level (Fig. 17.16B). Increased Pa_{CO_2} always increases tidal volume; its effect on respiratory rate varies and is probably related to the Hering-Breuer reflex.

An increase in $[H^+]$ in arterial blood (decreased pH_a) with Pa_{CO_2} kept constant results in an increase in respiration; increased pH depresses ventilation. The $[H^+]$ of CSF as well as at specific receptors in the medulla also affects respiration. Since a change in P_{CO_2} changes pH (Fig. 17.14), the changes in respiratory pattern with Pa_{CO_2} are caused in part by the effect of $[H^+]$ on nerve activity. Movement of H^+ across the blood-brain barrier and into CSF is much slower than is diffusion of CO_2.

Hypoxia

Hypoxia results when there is inadequate supply of oxygen to sites of utilization in the tissue. A brief outline of the various causes of hypoxia may be used as a short review of all the processes of respiration. In the following outline each item represents failure in one or more of the processes listed at the beginning of the chapter. Systematic arrangement of the problems allows one to consider all the processes at one time.

I. Inadequate transport of O_2 in blood (anoxemic hypoxia) (systemic arterial blood does not contain enough O_2)
   A. Decreased P_{O_2}
      1. inadequate O_2 in inspired air
      2. decreased lung ventilation
      3. decreased exchange between alveoli and blood
      4. admixing of pulmonary and systemic blood
   B. Normal P_{O_2}
      1. decreased hemoglobin content (anemia)
      2. decreased ability of hemoglobin to combine with O_2

II. Inadequate transport of blood (hypokinetic hypoxia)
  A. Failure to deliver blood
    1. entire system (heart failure)
    2. local failure (specific arteries blocked)
  B. Failure to remove blood
    1. specific veins blocked
  C. Delivery of blood inadequate to meet an increased requirement

III. Inability of tissue to use O_2 delivered (histotoxic hypoxia)

Selected Readings

Bouhuys A (1977) The physiology of breathing.—Grune & Stratton, New York.

Slonim NB, Hamilton LH (1976) Respiration physiology, 3rd edn. Mosby, St. Louis.

West JB (1974) Respiratory physiology—the essentials. Williams & Wilkins, Baltimore.

Review Questions

1. Define: a) tidal volume, b) inspiratory reserve volume, c) functional residual capacity, d) minute volume, e) alveolar ventilation, f) compliance, g) oxygen dissociation curve, h) Hering-Breuer reflex, i) Bohr effect, and j) apneustic center.
2. What muscles are involved in inspiration?
3. What three major components have to be considered when considering the work of breathing?
4. Of what physiologic importance is the ventilation-perfusion ratio ($\dot{V}/\dot{Q}$) in the lungs?
5. Name the three forms by which CO_2 is transported in the blood.
6. How does the uptake of O_2 by the blood in the lungs facilitate the loss of CO_2 from the blood?
7. Which receptors are stimulated by decreases in oxygen pressure?
8. At what level in the brain is the basic rhythm of respiration established?

Energy Metabolism

Chapter 18

Living organisms differ from inanimate objects in the complexity, variety, and order of their components. The natural tendency of all systems is toward a less-ordered or more random state, i.e., their entropy increases. To counteract this natural tendency, organisms must continuously expend energy. According to the first law of thermodynamics, energy can neither be created nor destroyed. Therefore organisms must obtain energy in useful form from the environment and return equivalent amounts of energy in less useful form to the environment.

Over a century ago the French physiologist, Claude Bernard, recognized that the animal and the environment formed an inseparable pair, because matter and energy are continuously exchanged between organism and environment. Organisms exist in a *steady state* such that the rate of transfer of matter and energy from the environment into the organism is exactly matched by the rate of transfer of matter and energy out of the organism. Normal body function can be maintained through regulation of internal components, which requires energy. The total chemical energy utilization of an organism is called its *energy metabolism*, which provides an index to the overall condition and physiologic performance of the organism.

Definitions and Units

The traditional unit of measurement for energy used by most biological scientists is the calorie (cal). This unit is defined as the quantity of energy required to raise the temperature of 1 ml of H_2O by 1°C. However, this quantity of energy is so small that it is convenient to use larger units in studies of human energetics. One kilocalorie (Kcal or Cal) is the quantity of energy required to raise the temperature of one liter of water by 1°C.

Although cal and Kcal are used in biology with little ambiguity, the units become awkward, particularly in interdisciplinary studies. Physicists, chemists, and engineers use a different set of arbitrarily defined units in studies dealing with energy flux, and communication between physical and biological sciences has suffered partly due to a lack of common usage of different units. The scientific community has recently adopted a consistent set of units that are freely interconvertible throughout all scientific disciplines. The *International System of Units* (S.I.) defines standards for the seven physical quantities: *length, mass, time, electric current, luminous intensity, thermodynamic temperature,* and *amount of substance*. All other units can then be directly derived from these units. For example, force = mass × acceleration or, mass × length per unit time (velocity) ÷ by time. Work (energy) = force × length; power (energy flux) = work per unit time.

The S.I. Units appropriate to the study of energetics are the joule (1 joule = 4.187 cal) for energy units and the Watt (1 Watt = 1 J/sec) for measuring energy flux. Table 18.1 summarizes basic units for energetics measurements and provides conversion factors for units traditionally used by physiologists.

Table 18-1a-c. Conversion factors for some common units of measurement in energetics.

a Work, Energy, Heat

Into ↓ To convert→	Cal	kcal	Joule	cm^3 O_2[a]	liters O_2[a]
Calorie	—	1.0000(+3)	2.3885(−1)	4.8000(0)	4.8000(+3)
Kilocalorie	1.0000(−3)	—	2.3885(−4)	4.8000(−3)	4.8000(0)
Joule	4.1868(0)	4.1868(+3)	—	2.0097(+1)	2.0097(+4)
Cc O_2[a]	2.0833(−1)	2.0833(+2)	5.0073(−2)	—	1.000(+3)
Liter O_2[a]	2.0833(−4)	2.0833(−1)	5.0073(−5)	1.0000(−3)	—

b Power, Energy Consumption

Into ↓ To convert→	W	kW	kcal min^{-1}	kcal h^{-1}	kcal day^{-1}
Watt	—	1.000(+3)	6.9780(+1)	1.1630(0)	4.8458(−2)
Kilowatt	1.0000(−3)	—	6.9780(−2)	1.1630(−3)	4.8458(−5)
Kilocalorie/min	1.4331(−2)	1.4331(+1)	—	1.6667(−2)	6.9444(−4)
Kilocalorie/hour	8.5985(−1)	8.5985(+2)	6.0000(+1)	—	4.1667(−2)
Kilocalorie/day	2.0636(+1)	2.0636(+4)	1.4400(+3)	2.4000(+1)	—

c Metabolic Rate per Unit Mass Specific Power

Into ↓ To convert→	W kg^{-1}	kcal g^{-1} h^{-1}	kcal g^{-1} day^{-1}	kcal kg^{-1} h^{-1}	kcal kg^{-1} day^{-1}
Watt/kg	—	1.1630(+4)	4.8458(+1)	1.1630(0)	1.6440(+3)
Kilocalorie/gram-h	8.5985(−4)	—	4.1667(−2)	1.000(−3)	4.1667(−5)
Kilocalorie/g-day	2.0636(−2)	2.4000(+1)	—	2.4000(−2)	1.0000(−3)
Kilocalorie/kg-h	8.5985(−1)	1.000(+3)	4.1667(+1)	—	4.667(−2)
Kilocalorie/kg-day	2.0636(+1)	2.4000(+4)	1.0004(+3)	2.4000(+1)	—

[a] Conversion factors for O_2 consumed vary with RQ, value of RQ used here is ~0.79.

Energy Transformation and Utilization

Energy is obtained from the environment by ingestion of chemical potential energy stored in the chemical bonds of fat, carbohydrate, and protein molecules. Complex organic molecules are progressively oxidized with the liberation of energy resulting from the breaking of chemical bonds. Molecules are broken down to three carbon compounds for entry to the Krebs citric acid cycle (see Chap. 22) where they are further oxidized to CO_2 and H_2O. Protons and electrons liberated in these oxidation reactions enter the electron transport chain where molecular oxygen provides the terminal electron acceptor to yield water. All energy-generating processes requiring the participation of molecular oxygen are termed *aerobic* metabolism. Energy generated without oxygen, such as in glycolysis where glucose is broken down to lactic acid, is termed *anaerobic* metabolism.

All along those pathways, energy is liberated as the chemical bonds are broken. This energy is stored in high-energy phosphate bonds of adenosine triphosphate (ATP) and to a lesser extent in other phosphate compounds. Biologic oxidation of fuels are essentially low-temperature combustions. Since heat cannot be used as an energy source in the body, some of the energy released in fuel combustion is conserved by the action of ATP formation; ATP also serves as a method of energy transport since it can diffuse to those sites where energy is required. The formation and degradation of ATP is coupled to energy-demanding processes. When energy is needed, the bond of the terminal phosphate group is broken by hydrolysis and stored chemical energy is released. Energy in this form is available to the cells to do work.

Energy transfer is an inefficient process, and all reactions result in the liberation of a certain amount of heat. Energy is said to be "degraded" to heat because in this form it cannot be converted to useful energy in the body. About one-half of the chemical energy available in the metabolc pool is lost as heat in the formation of ATP molecules.

Muscle contraction is an even more inefficient process. About 80% of the energy utilized in muscle contraction is lost as heat due to low efficiency of energy conversion and only 20% of the energy is converted into work (muscle movement). Unless a person is doing work against the environment, virtually *all* energy generated is lost from the body as heat. When a man is swimming, e.g., he imparts a certain amount of energy to the water to propel himself through it. This results in a wave, which will itself eventually lose all the energy imparted to it as heat and the surface become calm. The energy thus expended in swimming represents work done in addition to the heat lost from the swimmer's body. If we consider a man lying in bed, essentially all work in the body is dissipated as heat as long as he does no external work. Consequently the rate of heat production provides an accurate indication of the metabolic rate of a resting person.

Energy Equivalent of Food

The energy available through combustion of a given compound is independent of the number of intermediate steps involved in its breakdown. Therefore the complete combustion of one mole of glucose yields 2867 KJ/mole (686 Kcal/mole) regardless of whether it is burned in a test tube or broken down in the body by catabolic processes. As a result, it is a simple matter to determine the energy content per unit mass of various types of foods.

Energy content of food is measured in a bomb calorimeter (Fig. 18.1), a closed chamber immersed in a water bath. A precisely weighed sample is placed in the chamber and the atmosphere inside is composed of pure O_2 at a pressure of 20 atmospheres. The sample is ignited electrically by a platinum fuse and explosive combustion of the sample occurs. Heat liberated by the reaction is precisely measured by sensitive calibrated thermometers as a change of temperature of a measured volume of water surrounding the chamber; energy of the sample is calculated as the product of the heat capacity of the water, its volume, and the change in temperature.

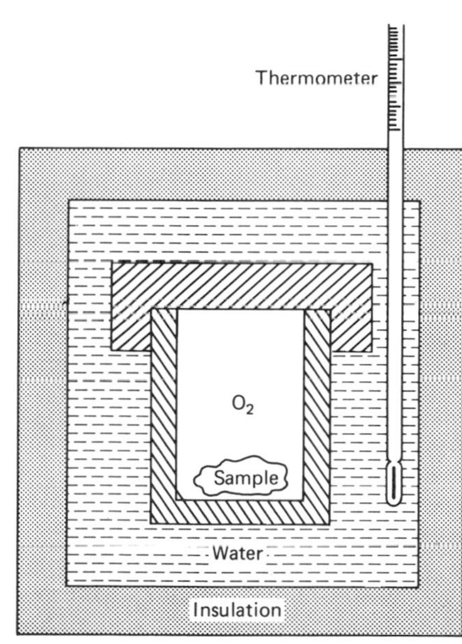

Fig. 18.1. Bomb calorimeter used to measure the energy content of a substance. Explosive combustion of the sample within the chamber liberates heat to the surrounding waterbath where it is measured with a sensitive thermometer.

Carbohydrates, on the average, yield 17.16 KJ/g (4.1 Kcal/g) of material burned (Table 18.2). Lipids are the most highly concentrated energy sources, yielding more than twice the energy of carbohydrate for each gram oxidized. Lipids represent the most economical method of long-term storage of energy in the body because less bulk is acquired per unit energy stored. Proteins are not completely broken down by cellular oxidation. In the body, amino groups are removed from protein and excreted as urea. As a result, the energy yield of protein burned in a bomb calorimeter is greater than the energy obtained by protein catabolism in the body. The amount of urea formed by the body from a given mass of protein may be burned in the bomb calorimeter to estimate the amount of energy unavailable to the body. Typical values obtained for the energy content of pro-

Table 18.2 Energy, respiratory equivalent, and volumes of O_2 produced and consumed versus type of food.

	Energy				Respiratory Equivalent					Volume	
	Bomb calorimeter		Human oxidation		O_2		CO_2		RQ	O_2	CO_2
Food	(Kcal/g)	(kJ/g)	(Kcal/g)	(kJ/g)	(K cal/ liter)	(kJ/ liter)	(Kcal/ liter)	(kJ/ liter)	$\left(\frac{VCO_2}{VO_2}\right)$	(Liter/g)	(Liter/g)
Carbohydrate	4.1	17.2	4.1	17.2	5.05	21.1	5.05	21.1	1.00	0.81	0.81
Protein	5.4	22.6	4.1	17.2	4.46	18.7	5.57	23.3	0.80	0.94	0.75
Lipid	9.3	38.9	9.3	38.9	4.74	19.8	6.67	27.9	0.71	1.96	1.39

tein by bomb calorimetry are 22.61 KJ/g (5.4 Kcal/g), whereas human oxidation yields about 17.17 KJ/g (4.1 Kcal/g; Table 2).

Measurement of Metabolic Rate

Since the energy expended by man at rest is rapidly transformed into heat either through inefficiency of energy conversion or by subsequent molecular interactions, the rate of heat produced is equivalent to the energy expenditure. Therefore, measurement of heat production represents a direct measurement of the metabolic rate and is theoretically more correct than estimates based on indirect methods.

Direct Calorimetry

Such direct measurements of energy metabolism were performed as early as 1788 by Lavoisier and LaPlace. They enclosed an animal in a chamber that was surrounded by a jacket of ice and an insulating layer of ice and water at a temperature of 0°C. Since no heat could penetrate the ice layer from the outside, any heat uptake in the ice jacket resulted from heat produced by the animal in the chamber. Knowing the quantity of heat necessary to melt a given quantity of ice, the investigators obtained a direct measurement of heat production from the amount of melted water collected from the ice jacket.
Modern direct calorimetry systems substitute a circulating fluid for the ice jacket (Fig. 18.2). Knowledge of the heat capacity of the fluid, total volume of fluid flowing through the insulated chamber per unit time, and the temperature difference between the fluid entering and leaving the chamber are used to calculate heat production. Direct calorimetry is

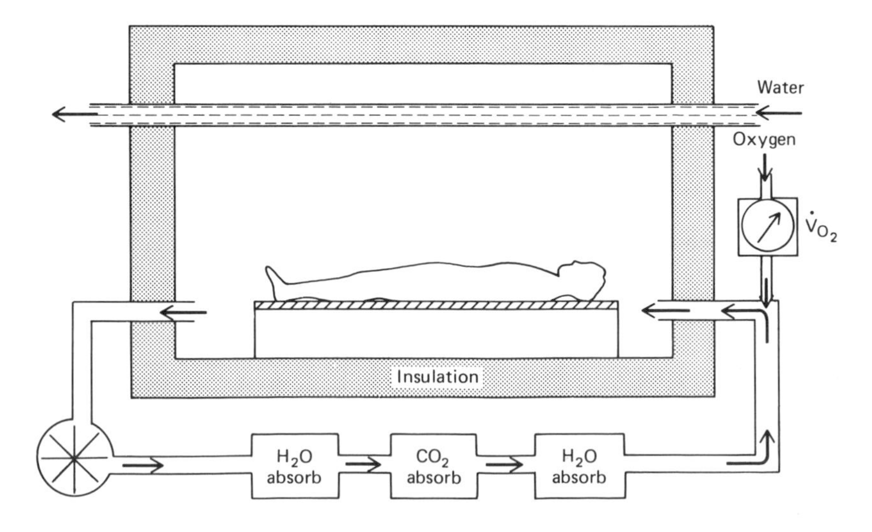

Fig. 18.2. Human calorimeter. Total energy output is sum of (1) heat evolved (measured from temperature rise of water flowing in coils through chamber), (2) latent heat of vaporization (measured from amount of water vapor extracted from circulating air by first H_2O absorber), and (3) work performed on objects outside chamber; CO_2 must be absorbed to prevent its accumulation within chamber. This process evolves water, so a second H_2O absorber is needed; O_2 consumption can be measured by noting rate at which O_2 must be added to keep chamber in a steady state.

rarely used except in special circumstances because accuracy is a problem except in extremely sophisticated apparatus, and indirect methods are accurate, inexpensive, and easy to operate.

Indirect Calorimetry

Energy metabolism can be estimated from the quantity of oxygen consumed or carbon dioxide produced under controlled conditions. *In closed systems* (Fig. 18.3) the subject rebreaths a quantity of air from an airtight system and the decrease in volume or partial pressure of oxygen is measured (gas exchange). *In an open system* the subject is placed in a continuously ventilated chamber and the oxygen consumption is determined from the difference in gas concentration of the inlet and outlet air multiplied by the flow rate of air throught the chamber. In both systems the water vapor added to the air expired by the subject must be absorbed because water vapor affects the pressure of other gases in relation to barometric pressure (see Chaps. 17 and 19).

The quantity of energy obtained per unit of O_2 consumed or CO_2 producced dcpcnds on thc typc of fucl. Carbohydratcs yield about 21 KJ for each liter of oxygen consumed (5 Kcal/liter O_2), proteins yield 18.7 Kj/liter O_2 (4.5 Kcal/liter O_2), and fats yield 19.8 Kj/liter O_2 (4.74 Kcal/liter O_2). Consequently conversion of measured values of oxygen consumption into their energy equivalent requires knowledge of the type of fuel. Simultaneous measurement of oxygen consumption and CO_2 production under steady-state conditions provides such information. The ratio of the rates of CO_2 produced to O_2 consumed, the *respiratory quotient* (RQ), is characteristic of the fuel being used (Table 18.2). Typically a resting man burns a variety of fuels and the RQ under these

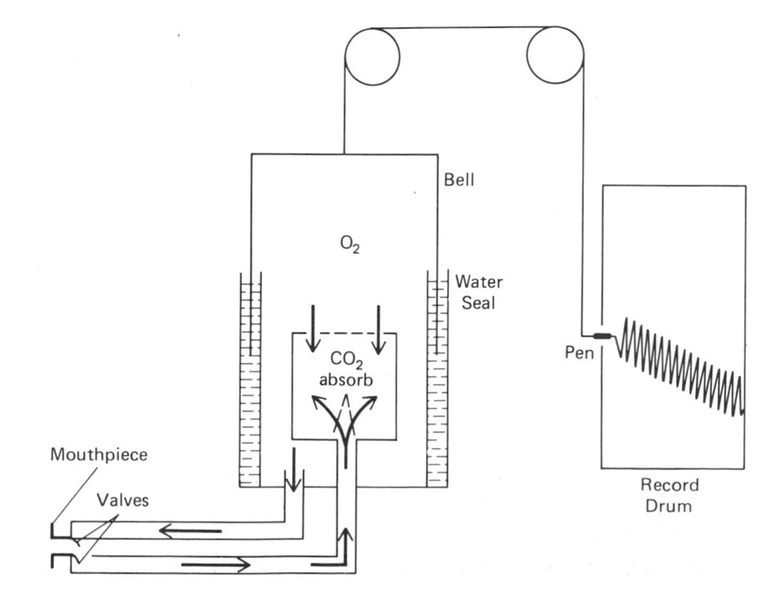

18.3. Spirometer arranged for O_2 uptake measurement. Mouthpiece is placed between subject's lips and teeth; extraneous gas exchange is prevented by noseclip (and ruptured tympanic membrane, ear plugs). To avoid greatly increasing effective dead space, separate tubes with valves to prevent mixing are used for inspired and expired gas; CO_2 absorber prevents accumulation of expired CO_2 in closed system. Volume of O_2 remaining in spirometer is recorded by a pen writing on paper attached to rotating drum.

circumstances is 0.83. At this RQ an energy equivalent of 20.22 kJ/liter O_2 (4.83 Kcal/liter O_2) is assumed to convert oxygen consumption into energy expenditure (Table 18.3).

Table 18.3. Energy equivalent of 1 liter O_2 at various respiratory quotients.

Respiratory Quotient	Kilocalories	Kilojoules
0.707	4.686	19.62
0.75	4.739	19.84
0.80	4.801	20.10
0.85	4.862	20.36
0.90	4.924	20.62
0.95	4.985	20.87
1.00	5.047	21.13

Other Indirect Methods

Several physiologic variables are correlated with rates of oxygen uptake and used with varying degrees of success to estimate the energy metabolism. Respiration or breathing frequency, ventilation volume (respiration frequency × tidal volume), heart rate, and cardiac output (heart rate × stroke volume) are all related to oxygen consumption and therefore are an index to energy expenditure. However, these methods lack precision in most cases except when the subject is at rest.

Radioisotopes, such as cesium and zinc, have been monitored to estimate metabolic rates. In general, the higher the metabolic rate, the more rapidly the label is removed. Doubly labeled water (D_2O^{18} or T_2O^{18}), containing labels of both hydrogen and oxygen, may provide accurate information on long-term metabolic rates.

The techniques of correlating *radiotelemetered physiologic* variables, or isotope washout rates, with energy metabolism are attractive because they provide long-term information from unrestrained subjects. Recently there has been a dramatic increase in the use of these techniques. However, their major drawbacks, including simplifying assumptions and lack of precision, has so far limited their usefulness.

Basal Metabolic Rate

Rates of energy metabolism vary considerably and depend on many factors. Basal metabolic rate (BMR) is measured under carefully controlled conditions.

The subject should be: (1) placed in a thermoneutral temperature to negate the effect of increased energy expenditure for thermoregulation. (2) resting comfortably, but awake, as sleep depresses energy metabolism by about 10%. (3) mentally relaxed because emotional stress may increase metabolism; and (4) fasted at least 12 hrs, and in a postabsorbtive state to reduce the calorigenic (heat-producing) effects of digestion of food.

The basal metabolism of a man of given height and weight can be predicted to within 5% to 10% of actual measurements. Larger deviations from predicted values are often indicative of metabolic disorder usually associated with hormonal imbalances.

Body Size

The basal metabolism of large persons is obviously greater than for small ones. Energy metabolism may be expressed as total metabolic rate (Kcal/hr, S.I. units = Watts), as mass-specific metabolic rate, i.e., the rate of energy expended per unit of body mass (Kcal/Kg/hr., S.I. units = W/Kg), or the rate of energy expended per unit of surface area (Kcal/m²/hr, S.I. units = W/m²).

In *mammals* the relation of BMR to body mass is remarkably uniform (Fig. 18.4). When plotted on log-log coordinates, the energy metabolism of mammals, including man, is described by the equation

$$\text{Log E} = 0.75 \text{ Log M} + 7.84.$$

where E is energy metabolism (Watts) and M is body mass in Kg.

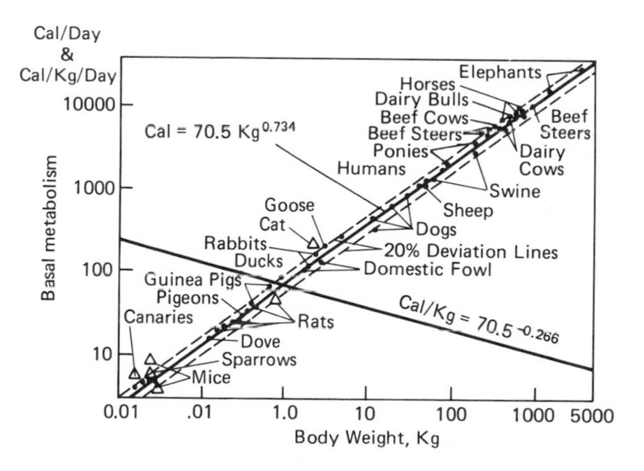

Fig. 18.4. Relationship of basal metabolism to body size in mammals. After Brody (1945).

Clinical physiologists often report metabolic rates per unit of surface area. This is traced to Rubner's classic study in 1868 on the metabolism of dogs of different size. Rubner found that metabolism varied approximately with the two-thirds power of body mass. He reasoned that heat loss is obviously related to the surface area and that heat production should also vary with body weight to the same power. However, the acquisition of more data from a variety of mammalian species did not support this idea. Therefore heat loss and surface area do not appear to be the major determinants of metabolism in animals of different size. Nevertheless the practice of reporting energy metabolism per unit surface

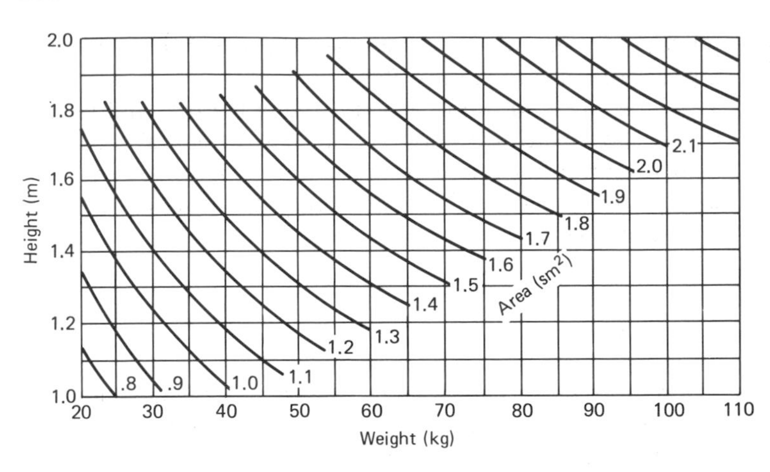

Fig. 18.5. Chart for determining surface area of man in square meters from weight in kilograms and height in centimeters according to formula area (m^2) = $W^{0.425} \times H^{0.725} \times 0.202$. (After DuBois and DuBois, Arch. Intern. Med., 1916.)

area is firmly entrenched in the physiologic literature. In man it is fortuitous that basal metabolism per unit surface area seems to be relatively constant. Tables for estimating surface area of human subjects from height and weight facilitate the conversion of data to surface specific metabolism (Fig. 18.5).

Exercise causes the metabolic rate to vary considerably. Rates of metabolism 20 times the resting level have been measured in trained athletes during short-term intense exercise and in excess of 10 times the resting level during prolonged work (Table 18.4). The O_2 uptake during exercise does not indicate the total energy expenditure because some energy is expended by glycolysis (anaerobically) and requires no oxygen.

Oxygen consumption for a submaximum exercise regimen is shown in Fig. 18.6. A given amount of energy is required to support the exercise, but the O_2 consumption does not immediately match the demand. With time, steady-state conditions occur during which the demand for O_2 is met by the supply. The difference between the O_2 demand and O_2 consumption represents the energy expended anaerobically and is termed the *oxygen debt*. Theoretically the energy expenditure should return to the resting level after exercise is terminated yet the O_2 consumption remains high and only slowly returns to the resting level. During this time the O_2 debt is said to be repaid (Fig. 18.6).

Table 18.4. Effects of activity on metabolic rate.[a]

Condition	Kcal/m²/h
Rest	
Sleeping	35
Lying awake	40
Sitting upright	50
Light activity	
Writing, clerical work	60
Standing	85
Moderate activity	
Washing, dressing	100
Walking (3 mph)	140
Housework	140
Heavy activity	
Bicycling	250
Swimming	350
Lumbering	350
Skiing	500
Running	600
Shivering	to 250

[a] From A. C. Brown. Energy Metabolism. In T. C. Ruch and H. D. Patton (Eds.), *Physiology and Biophysics* (vol. 3) (20th ed.). Philadelphia: Saunders, 1973.

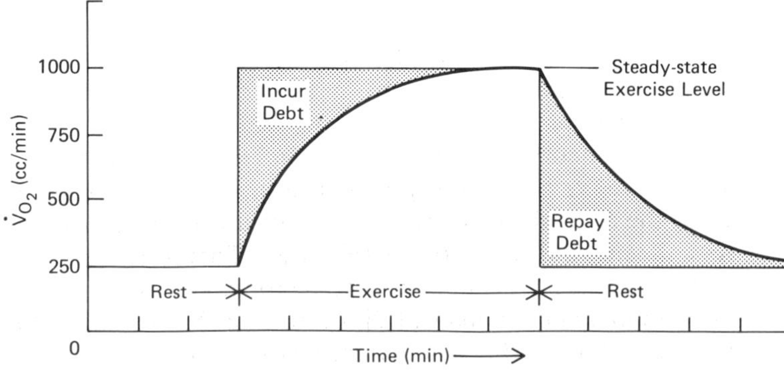

Fig. 18.6. Consumption of O_2 during modest exercise. Magnitude of O_2 debt is estimated by measuring total O_2 consumption above resting level after exercise has ceased. Ideally, area under the two sections is identical.

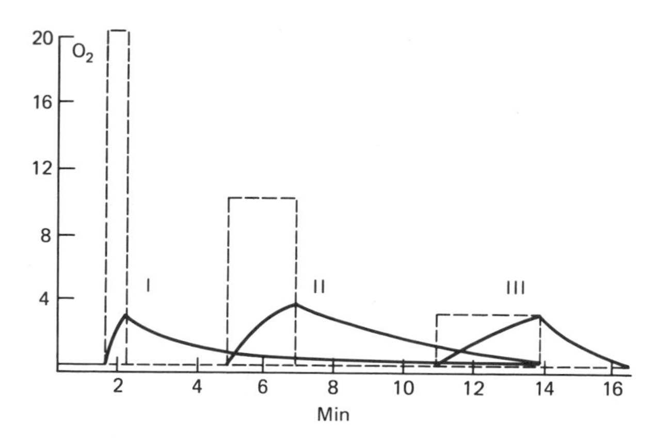

Fig. 18.7. Requirement for O_2 (broken lines) and O_2 consumption (solid lines) of man in three intensities of exercise: I, sprint lasting 30 seconds; II, 2 minutes ran to exhaustion in which O_2 consumption was maximal; and III, less intense work in which O_2 consumption was equal to O_2 requirement. Area of O_2 requirement rectangle is equal to sum of O_2 consumed before and after exercise.

The O_2 debt is divided into two components. The *lactacid debt* represents the added energy expended to convert the major byproduct of anaerobic metabolism, lactic acid, back to pyruvic acid, which can be used in the Krebs cycle for aerobic metabolism. Lactic acid in high doses is harmful to the body and is partially responsible for muscle fatigue immediately after running. The *alactacid debt* represents the added energy necessary to rephosphorylate high-energy storage compounds, such as creatine phosphate, and to replenish O_2 stores to the depleted muscle myoglobin.

Anaerobic metabolism is of greatest importance in short-term, maximum exercise. Running a 100-yard dash in about 10 seconds requries about 85% anaerobic power. As the duration of maximum effort increases, the ratio of aerobic/anaerobic power increases. If the maximum activity is sustained for two minutes, aerobic and anaerobic power are equal, each providing 50% of the energy generated. Maximum effort for 60 minutes requires 90% aerobic power. As duration of sustained activity increases the total metabolic rate (aerobic and anaerobic) decreases (Figs. 18.7 and 8).

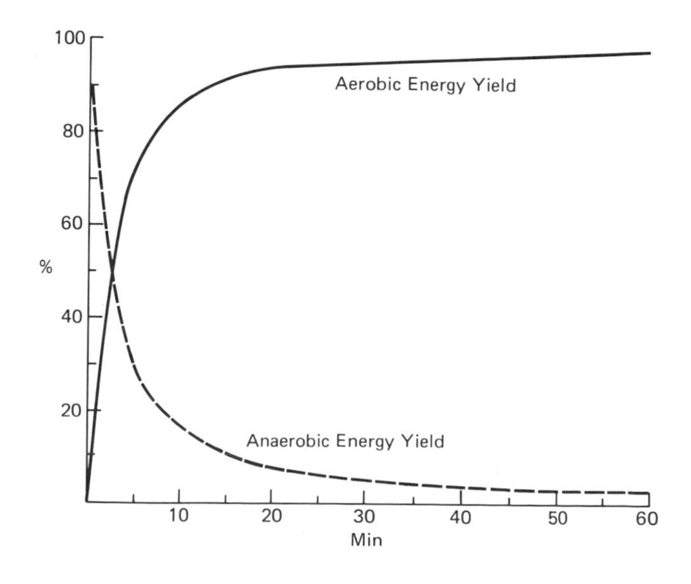

Fig. 18.8. Relative contribution of aerobic and anaerobic processes during maximum effort. At maximum activity of about 2 minutes duration, aerobic and anaerobic work each account for about one-half energy expended.

Other Factors and Metabolic Rate

Man maintains a constant core body temperature regardless of the temperature of his surrounding (*ambient temperature*). Therefore to control body temperature, heat production or heat loss, or both, must be regulated. At rest the rate of heat production (and therefore metabolic rate) is directly related to the difference between the core body temperature and the ambient temperature. Added energy expenditure (heat production) counteracts greater heat loss at low ambient temperature. (For further details see Chap. 19.)

As a result of ingestion of food, the energy metabolism increases. This process is the result of the calorigenic effect of the food and is termed *specific dynamic action* (SDA). The physiologic explanation for this phenomenon is uncertain. The magnitude of SDA depends on the food eaten. Protein diets yield a metabolic rate about 25% to 30% above the BMR. Carbohydrates and lipids, however, increase the metabolic rate only by 10% or less.

Table 18.5. The Mayo foundation normal standards of basal metabolic rate (kcal/m²/h).[a,b]

Males		Females	
Age	BMR	Age	BMR
6	53.0	6	50.6
7	52.5	6-½	50.2
8	51.8	7	49.1
8-½	51.2	7-½	47.8
9	50.5	8	47.0
9-½	49.4	8-½	46.5
10	48.5	9–10	45.9
10-½	47.7	11	45.3
11	47.2	11-½	44.8
12	46.7	12	44.3
13–15	46.3	12-½	43.6
16	45.7	13	42.9
16-½	45.3	13-½	42.1
17	44.8	14	41.5
17-½	44.0	14-½	40.7
18	43.3	15	40.1
18-½	42.7	15-½	39.4
19	42.3	16	38.9
19-½	42.0	16-½	38.3
20–21	41.4	17	37.8
22–23	40.8	17-½	37.4
24–27	40.2	18–19	36.7
28–29	39.8	20–24	36.2
30–34	39.3	25–44	35.7
35–39	38.7	45–49	34.9
40–44	38.0	50–54	34.0
45–49	37.4	55–59	33.2
50–54	36.7	60–64	32.6
55–59	36.1	65–69	32.3
60–64	35.5		
65–69	34.8		

[a] After Boothby WM, Berkson, J, Dunn HL, (1936) *Am J Physiol* 116:468.
[b] Normal limits are usually taken as ± 10% and divergence beyond these limits indicates an abnormal BMR for a subject of this age and sex.

Basal metabolism of people varies regularly as a function of *age*, and clinical standards have been derived (Table 18.5) to predict metabolism. Normal variation from such standard is ±15%. It is greater in the young, decreases at puberty and is lowest in old age.

During *sleep*, rates of metabolism are about 10% lower than the BMR. Presumably the difference between resting persons awake or asleep is due to the more complete state of muscle relaxation in the latter.

A *fever*, which elevates the mean body temperature, results in increased energy metabolism. It is well known that increases in temperature cause increases in the rate of chemical reaction. Therefore a higher mean body temperature tends to increase the chemical reactions associated with energy metabolism. Moreover, increased muscular activity, analogous to shivering, often occurs. Peripheral vasoconstriction and reduced capacity for sweating decrease heat loss, resulting in a greater quantity of heat storage in the body, causing further increases in body temperature and heat production (see Chap. 19).

Selected Readings

Bartholomew GA (1977) Energy metabolism. In: Gordon MS (ed) Animal physiology: principles and adaptations, 3rd edn. MacMillan, New York

Benedict FG (1938) Vital energetics. Carnegie Institution of Washington Publications, No. 503, Washington

Brody S (1945) Bioenergetics and growth. Reinhold, New York

Hemmingsen AM (1960) Energy metabolism as related to body size and respiratory surfaces and its evolution. Rep Steno Mem Hosp Nord Insulin Lab. 9 7-110

Kleiber M (1961) The fire of life. Wiley, New York

Lefebvre EA (1964) The use of D_2O^{18} for measuring energy metabolism in *Columbia livia* at rest and in flight. Auk 81: 403–416

Lusk G (1928) The elements of the science of nutrition, 4th edn. Saunders, Philadelphia

Mechtly EA (1973) The international system of units: physical constants and conversion factors. NASA-7012. National Aeronautics and Space Administration, Washington

Miller AT Jr (1954) Energy metabolism and metabolic reference standards. Meth Med Res 6: 76–84

Passmore R, Durnin J (1955) Human energy expenditure. Physiol Rev 35: 801–840

Stahl WR (1967) Scaling of respiratory variables in mammals. J Appl Physiol 22: 453–460

Review Questions

1. Under what circumstances does measurement of O_2 consumption yield total energy expenditure and what additional information must be known?

2. Why would measurement of CO_2 production alone give less accurate information on the energy metabolism of a subject than measurement of O_2 consumption alone? (Hint: see Table 18.2.)
3. What is the basis for a given value of O_2 consumption yielding different values of heat production when different fuels provide the energy?
4. Describe the pattern of aerobic and anaerobic energy generation in relation to energy demand for a long distance runner a) just off the starting line, b) in middle distance, and c) approaching the finish line.
5. How many different kinds of energy are there? What forms of energy represent "useful" energy in the body?

Chapter 19 Temperature Regulation

The temperature of an animal's immediate surroundings may profoundly affect its physiologic performance. On earth, the ambient temperatures may vary from about −50°C (−58°F) in the arctic winter to almost 60°C (140°F) during the summer in some deserts. The range of temperatures at which living cells can function, however, is only about 50°C.

Living cells freeze at a few degrees below 0°C. Freezing per se is not necessarily fatal to living tissues, and scientists have frozen tissues that maintained their viability for some time. However, as tissues freeze under natural conditions, the formation of ice crystals results in destruction of delicate cellular machinery. At the high end of the temperature scale, proteins denature at temperatures in excess of 45°C. Since proteins are responsible for virtually all regulatory functions in animals, their structural and functional integrity is crucial to normal body function. Temperature has an important metabolic effect on living tissue because biochemical reaction rates are sensitive to the temperature of the surrounding medium and normally increase two to three times for each 10°C increase in temperature.

There are several basic patterns of response to external temperatures in the animal kingdom. The *poikilotherms,* including most invertebrates and lower vertebrates, in the absence of a radiant heat source exhibit body temperatures that de-

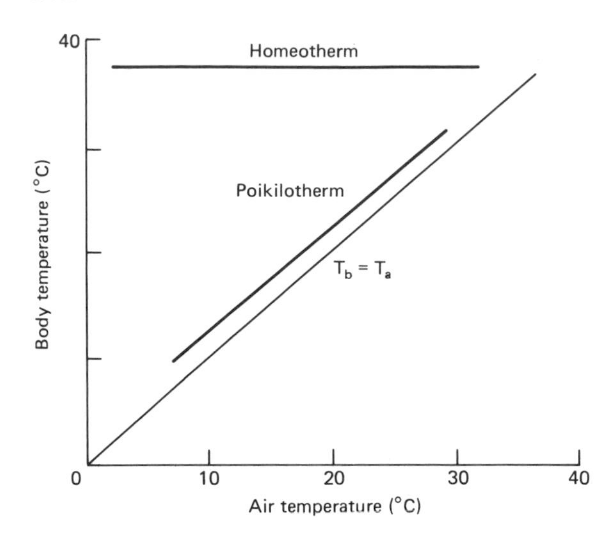

Fig. 19.1. Relationship of body temperature to air temperature in animals. Body temperature of poikilotherms conform rather closely to air temperature. Homeotherms regulate T_b at relatively constant level over wide range of air temperatures.

pend on the ambient temperature (Fig. 19.1). The alternative to thermal conformation is thermoregulation. Regardless of the environmental temperature, the core body temperature is controlled at a constant level. Birds and mammals employ this pattern and are termed *homeotherms*. Regulation of body temperature allows them to operate under thermal conditions that insure optimum enzyme function. *Heterothermy* is a special condition in which homeotherms temporarily abandon thermoregulation, allowing their body temperature to cool within about 1°C of the ambient temperature.

Many terrestrial animals, such as reptiles and some insects, maintain relatively constant body temperatures over a range of air temperatures during the daytime by exploiting heat sources and heat sinks in their environment. Consequently, during the day such animals fit the description of a homeotherm rather than a poikilotherm (Fig. 19.1). Animals can also be classified according to the source of heat used to regulate body temperature. *Ectotherms,* such as reptiles, are organisms utilizing external heat to control body temperature. *Endotherms,* such as man, utilize heat generated by their metabolism.

Body Temperature and Measurement

The temperatures throughout the body of an endotherm depend on both external environment and activity state. At rest, the body temperature of man is about 37.0°C, although it varies predictably during 24 hours with a slight drop during sleep. During exercise, the body temperature is often elevated several degrees as a result of increased heat production.

Temperature is not constant throughout the body and the internal temperature distribution is a complex function of heat

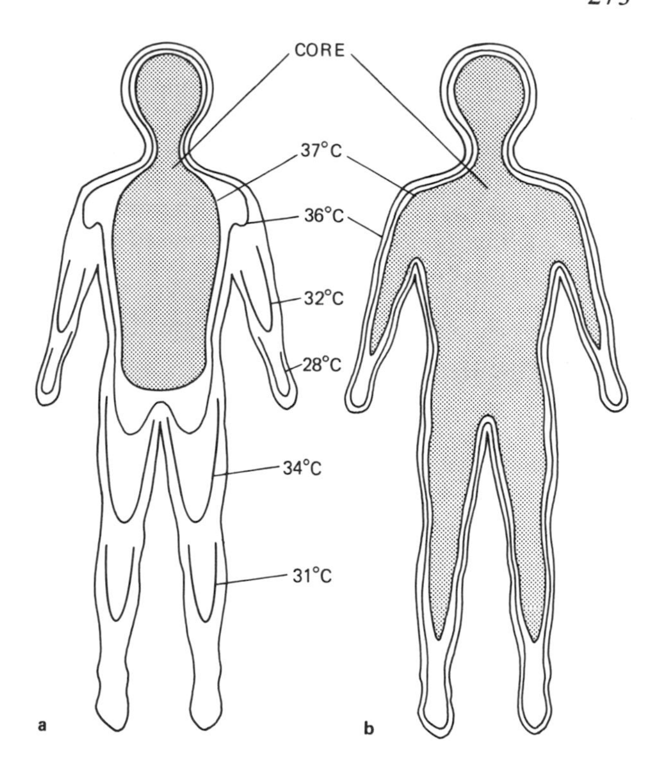

Fig. 19.2. Distribution of temperature throughout the human body at room temperatures of 20°C (A) and 35°C (B). Arrows indicate isotherms (areas of equal temperature). At 20°C large temperature gradients occur between the core (stippled area) and shell, and the core is restricted to the head and trunk. At 35°C, the core extends well into the extremities.

production by various tissues, heat transport by circulation, and local temperature gradients. Except during strenuous exercise, most of the metabolic heat production occurs in deep organs, such as the heart, viscera, and brain. Heat produced in the deep body region, or *core,* must be transferred to the body surface. Temperatures within the periphery are determined by the heat transferred from deep body regions and the air temperature. Thus the body can be visualized as having a core of relatively constant temperature and an insulating layer, the *shell,* which is highly variable, depending on the energy balance of the body. In a cold environment, blood flow to the periphery is reduced, causing lowered shell temperatures. At a high air temperature, heat is not lost as readily to the environment, and the core may include almost the entire body (Fig. 19.2).

The *temperature of a tissue* is a reflection of its heat content. Characterizing the body temperature by a single temperature measurement is therefore misleading but is more practical than attempts to characterize integrated measures of many temperatures throughout the body. The latter approach yields more information about the body's energy balance but is technically difficult to measure. Moreover, interpretation is also difficult due to the dynamic nature of temperature changes in the shell.

Body temperature is usually measured rectally or orally. Rectal temperature is the safest and most convenient method of measurement in animals and is generally considered to yield reasonable average core temperature. However, this temperature is not universally recognized. Other temperatures, including those of mouth, armpit, heart, or liver are often reported. From the standpoint of temperature regulation studies, rectally obtained core temperatures may change too slowly to provide accurate information. In such studies the temperature of the brain is desirable and is nicely approximated by the temperature of the tympanic membrane of man and other mammals. The temperature of the blood leaving the left ventricle is also used as a dynamic indicator of the body's energy balance.

Heat Balance

Heat gain by the body (through metabolic heat production or from the environment) must equal heat loss if body temperature is to remain constant (Fig. 19.3). If gain exceeds loss, the excess heat will be stored in the body, causing an increase in the body temperature (hyperthermia). This balance is conveniently described by the equation:

$$M \pm E_R \pm E_C \pm E_G - E_E \pm S = 0$$

where M = metabolic heat production, E_R = radiation, E_C = conduction, E_G = convection, E_E = evaporation, S = heat storage, (+) = gain, and (−) = loss of heat in the body. Heat can be either gained or lost through radiation, conduction, and convection, depending on environmental conditions. Heat is always produced as a by-product of chemical reactions in the body (see Chap. 22), and therefore metabolism is always positive; *evaporation* is always negative. The opposite reaction, *condensation*, is ordinarily of little consequence to the heat balance of man.

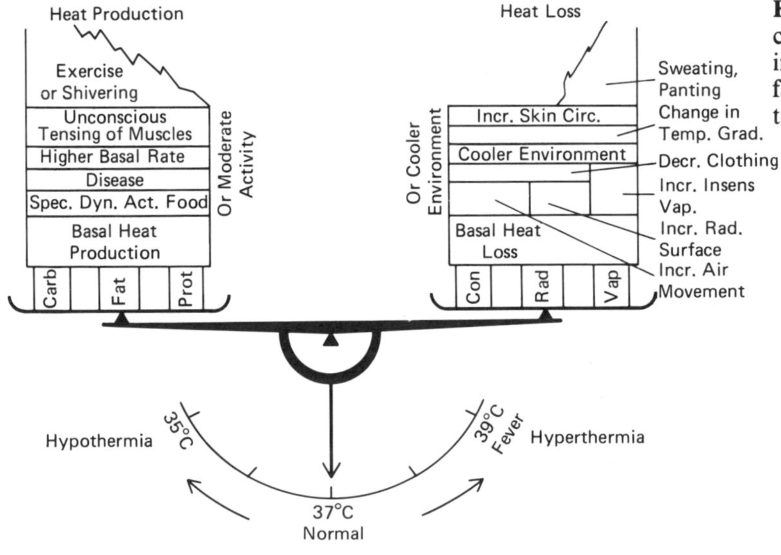

Fig. 19.3. Balance between factors increasing heat gain and heat loss resulting in constant body temperature. If factors do not balance, body temperature changes.

Radiation

All objects having temperatures greater than absolute zero (−273°K) lose energy by radiation. Radiation occurs in the form of electromagnetic waves and needs no medium to propagate it. Therefore radiation will travel through the near-vacuum of space or through the atmosphere on a thermal gradient from warmer to cooler objects.

Emissivity

The surface temperatures of objects and the type or quality of an object's surface is important in determining energy flux by radiation. *Emissivity* of a surface refers to its properties as a radiator. A surface absorbing all but reflecting no radiant energy has a maximum emissivity of 1. When a surface reflects all radiant energy, the emissivity of that surface is 0. Obvious examples of such a surface is a mirror or a highly polished metal. It is common for objects to absorb some wavelengths almost completely yet be highly reflective in other wavelengths. Human skin, e.g., whether white or pigmented, absorbs almost all infrared radiation. In the visible region, however, black skin absorbs significantly more visible radiation. Under most circumstances the surrounding objects radiating to the body—and to which the body radiates—are within 20°C of the body's surface temperature. Energy flux (Watts) can be approximated by the equation:

$$E_R = K_R (T_s - T_a)$$

where K_r is a coefficient of radiant heat transfer (units = W/°K) and the term $(T_s - T_a)$ represents the difference in temperature between the skin and the surroundings. The $T_s - T_a$ term is often characterized as the driving force since there must be a temperature difference for radiant energy flux to occur.

Conduction

Heat flow between two objects whose surfaces are in contact with each other is *conduction*. Heat moves down a thermal gradient from the warmer to the cooler object. In this process there is actually no transfer of material between the two surfaces; the energy is transferred through layers of molecules without any molecular transfer. The sensation of touching a "hot" or "cold" object relates to the direction of heat flow via conduction. If the temperature of the object touched is lower than the finger touching it, the direction of heat flow is from finger to object, whereas the converse is true if the object is warmer than the finger.
Heat flux via conduction (W) is a linear function of the difference in temperature between the two objects in contact and is described by the equation

$$E_C = K_c (T_1 - T_2).$$

The value K is the rate of conductive heat transfer per °K difference between the two objects (W/°K). Often, how-

ever, the heat retention of a substance rather than the rate of heat flow through it is the parameter of interest. Efficient insulators are poor conductors of heat and vice versa.
Metals, such as silver and copper, are among the best heat conductors, and various gases are among the poorest. Most biologic tissues insulate about as well as water, but fat is about twice as effective as muscle or bone. Consequently animals, such as seals, utilize a thick layer of fat to insulate the body core from the skin, which is in contact with the icy water of the polar oceans. As a result, skin temperature is maintained within a few degrees of water temperature. The driving force is only a few degrees in this case. Without the fatty layer, skin temperature would approach body temperature (37°C), in which case heat loss would be more than 10 times greater.
Under most circumstances conduction is not a major source of heat exchange in mammals because the amount of surface area in contact with solid objects is minimal. Ectotherms, such as reptiles and insects, often press their ventral surfaces on warm substrates when ambient temperatures are cool to increase their body temperatures by conduction.

Convection

Heat transfer by convection occurs as a result of a fluid or gaseous medium surrounding an object. Convective heat exchange, unlike conduction, results in an exchange of molecules as well as energy. This occurs because there is an "unstirred" boundary layer of the fluid (air or water) surrounding all objects. The thickness of the unstirred layer depends on environmental conditions. Direction of heat flow is from warmer to cooler regions. If air temperature exceeds body temperature, direction of heat flow will be into the body.
When the body is surrounded by still air, warm air rises from the skin and passes into the surrounding air, transferring both molecules and energy. This process is known as *free convection*. When air from the surrounding environment is in motion, the thickness of the boundary layer will be affected by the air velocity. A boundary layer of several millimeters in still air can be reduced to a thickness of only a few microns under windy conditions. Heat exchange of this type varies considerably with air movement and is called *forced convection*.
Rates of heat transfer by convection are described by the equation

$$E_C = h_c (T_s - T_a)$$

where E_C is the rate of convective heat transfer (W), T_s and T_a are skin and air temperatures, and h_c is the convective heat transfer coefficient. The coefficient h_c is sensitive to the size of the object and the wind velocity. For smooth cylinders the formula describing h_c is

$$h_c = KV^{1/3} D^{2/3}$$

where V is air velocity and D is the diameter of a cylinder (Fig. 19.4). The effect of wind velocity on the convective coefficient—and therefore on heat exchange—is demonstrated by the "wind-chill index." As a result of the wind's decreasing the boundary layer of air near the skin, heat loss is greater than in still air. Consequently the air "feels colder" than it actually is when the wind is high. Weather stations often report the actual air temperature and the apparent temperature that the wind-chill factor causes.

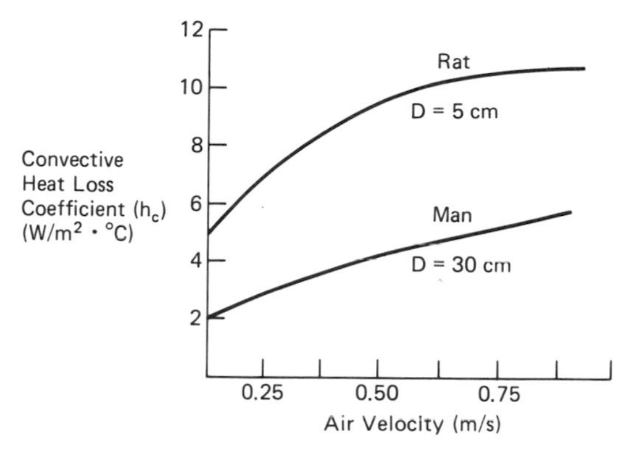

Fig. 19.4. Convective coefficient (h_c) of two cylinders as function of wind velocity. D = diameter; 5 cm, equivalent to rat; 30 cm, equivalent to man.

Convective heat transfer within the body is important for controlling the core body temperature. Most of the major heat-producing muscles are in the body core. As mentioned in the previous section, body tissue is a poor conductor of heat, and the blood carries a large amount of the heat produced in the core into the periphery, i.e., arms and legs where larger relative surface areas and cooler temperatures facilitate the loss of such heat to the environment.

Evaporation

When water evaporates from a surface, the surface will be cooled because energy is used to transform the liquid into the vapor or gaseous state. For each gram of water evaporated, about 0.59 watt-h of heat is lost. Water continually evaporates from the body under most environmental conditions and constitutes an important source of heat loss from the body. The *major pathways of evaporation* are from skin and respiratory passages as a result of ventilation. The volume of water loss depends on environmental conditions, particularly temperature and humidity.

If the air is saturated with water vapor (100% relative humidity), evaporation from the skin can not occur. Air that is breathed during inspiration is warmed to core body temperature in the lungs and saturated with water vapor. Even if the outside air is saturated, some water is lost during expiration because as the air temperature increases, a greater volume of water is held by a given volume of air.

Evaporation becomes extremely important at high air temperatures because when the air temperature equals the body temperature (or more precisely the skin temperature), there can be no heat loss from the body by radiation, conduction, or convection.

The major form of evaporation for temperature regulation in man is *sweating*. In other animals, such as many birds (with no sweat glands), some mammals, and a few reptiles, rates of evaporative water loss are increased at high temperature by panting. *Panting* increases the airflow across moist respiratory surfaces, thereby enhancing evaporation.

Heat Transfer Coefficient

Heat exchange by radiation, conduction, and convection is approximately linear to the difference between body temperature and environmental temperature. The radiative, conductive, and convective coefficients can be lumped together into an overall heat transfer coefficient (thermal conductance), which describe the rate of heat exchange for every degree kelvin difference between body temperature and air temperature according to the formula

$$dH/dt = C\,(T_b - T_a)$$

where dH/dt is the rate of heat loss (W), C is the heat transfer coefficient (W/degree kelvin °K), and T_b and T_a represent core body temperature and air temperature, respectively. This equation is a simplification and is useful in this form only under carefully controlled environmental conditions.

Metabolism and Air Temperature

The metabolic response of mammals to air temperature is shown in Fig. 19.5. The minimum level of metabolism occurs over a range of air temperatures variously known as the zone of *physical regulation, thermoneutral zone,* or *comfort zone*. Within this range there is no change in heat production, and the body temperature can be controlled by small changes in the animal's conductance by varying posture, blood flow from the core to the periphery, or increased sweating. At the lowest temperature of the thermoneutral zone, the *lower critical temperature,* the animal has minimized all avenues of heat exchange. If ambient temperature falls below the lower critical temperature, body temperature can no longer be maintained by reducing heat transfer, and therefore the heat production of the body must increase to prevent body temperature from falling. The range of temperatures below the lower critical temperature is known as the *zone of chemical regulation* or the *zone of metabolic regulation*. Man can artificially change this relationship by varying the weight of his clothing. Adding a winter coat in-

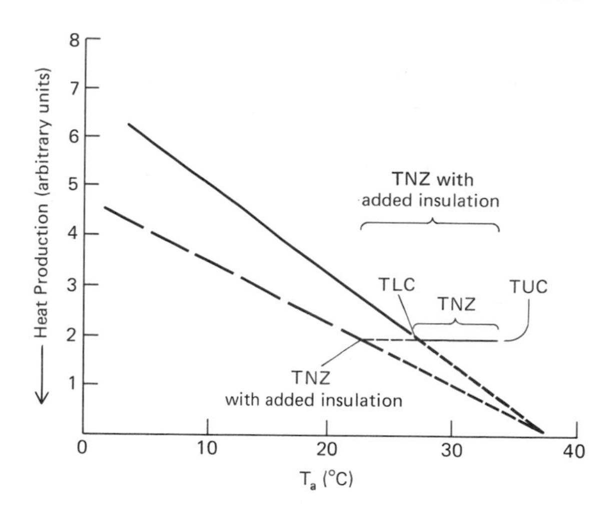

Fig. 19.5. Relationship of energy metabolism to air temperature in a mammal. Interrupted line indicates the same animal with greater insulation (in animals by molt of thicker fur, in man wearing a coat). Minimum heat transfer coefficient equals the slope of the relations of metabolism vs. temperature.

creases insulation and therefore decreases rate of heat loss (Fig. 19.5). A similar response is shown seasonally in animals that molt thicker, longer fur (better insulation) in winter.

Thermoregulatory System

The thermoregulatory control system is composed of a series of elements whose functions are interrelated. Thermal information is obtained by peripheral or deep-body temperature sensors. The output from these sensors is carried by afferent nerves to the thermoregulatory control center in the hypothalamus. The latter then activates various effectors that control either the rate of heat production or of heat loss. Feedback to the control system by the nervous system

Fig. 19.6. Thermoregulatory system of a mammal. Thin lines represent neural pathways, thick lines represent circulatory pathways. (After Bligh, J, 1973.)

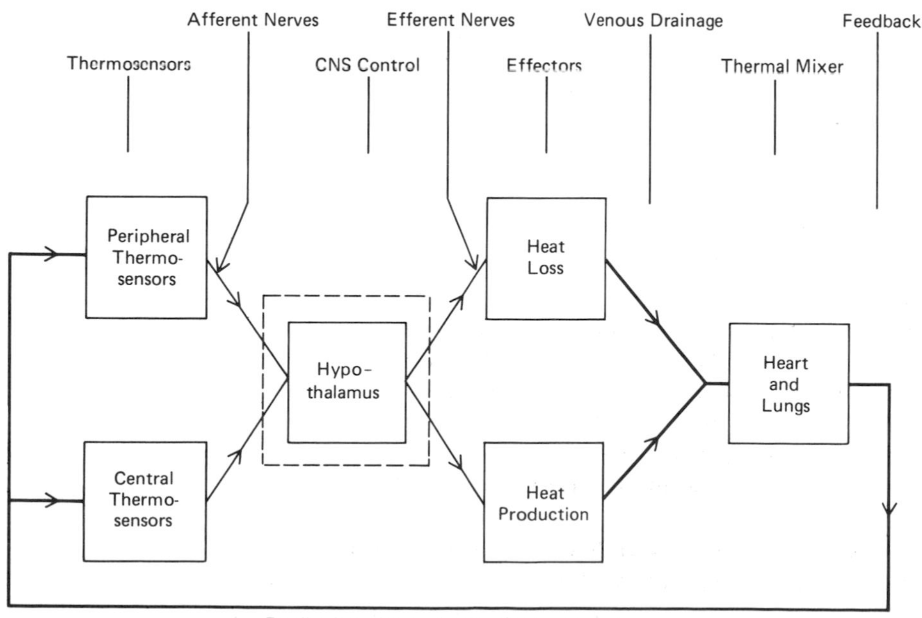

or the bloodstream occurs, and it modifies the input of thermal receptors, resulting in a closed loop (Fig. 19.6).
The operation of the thermoregulatory system is analogous to a system of proportion control. The body is maintained at a given temperature level, the *set point*, and the degree of effector response is proportional to the deviation of actual temperature from the set point. Thus, if core body temperature drops 2°C, the magnitude of increased heat production by shivering is much greater than would be elicited by a drop of 0.5°C in the core temperature. The following sections examine the various elements of the thermoregulatory system.

Temperature Sensors

Temperature sensors are distributed throughout the body surface and deep within the body. Some sensor cells show unusually high temperature sensitivity, but there is no distinct morphologic type of sensory cell associated specifically with thermosensitivity.
The cutaneous thermoreceptors are of two types—cold and warm. Both types are particularly sensitive to the rate of change in temperature. Cold receptors sharply decrease their rate of nervous discharge (firing) in response to increased temperature and increase rates when temperatures fall. The reverse is true of warm receptors. If temperature change is rapid, the change in response of the receptors is much greater than if the same change of temperature occurs gradually.
Temperature sensors are located in various positions within the body. They occur in the deep viscera (Auerbach's plexus), the hypothalamus, the reticular formation and preoptic region of the brain stem, and the spinal cord. Temperature sensors have been found in the respiratory tract, medulla, and motor cortex, and it is likely that further study will reveal more receptors.

Control Center

The activity of many neural centers in the central nervous system (CNS) alters the body's heat production and heat loss. Consequently not only must there be an integrating center for diverse sensory inputs concerned with the body's heat balance, but also a control to regulate various motor responses. This control is one of the functions of the hypothalamus, which is sometimes referred to as the body's thermostat (see Chaps. 10 and 24). If the hypothalamus is surgically separated from the lower brain and spinal cord, ability to regulate body temperature is lost. Separate sites in the hypothalamus are responsible for control of different effectors.

Heat Loss

Motor centers in the *anterior hypothalamus* control the regulation of heat loss from the body, functioning primarily to prevent overheating. If this center is destroyed, physiologic functioning occurs in a cold environment but there is no control of body temperature in a warm environment, and body temperature rises.

Heat Production

A different center in the *posterior hypothalamus* controls the rates of heat production by the body, thereby preventing the body from becoming chilled. Destruction of the posterior hypothalamus of an animal exposed to cold prevents an increase in energy metabolism, and body temperature falls. The centers in the *anterior hypothalamus* regulating vasodilation, sweating, and panting are sensitive to the temperature of the blood that flows through them. In addition, evidence suggests that the Na^+ and Ca^{2+} fluxes are important in determining hypothalamic function—apparently by determining the actual reference temperature or set point. Changing the Na^+-Ca^{2+} flux results in a change in the level of body temperature. The mechanism for this influence and the relation of blood temperature to blood Na^+-Ca^{2+} fluxes remains to be determined. Data suggest that a slight decrease in hypothalamic temperature of small mammals is responsible for increased heat production by shivering and heat conservation by vasoconstriction.

Regulation of Heat Loss by Effectors

Convection of heat away from the body core to the extremities via changes in volume of blood flow is an important means of regulating heat loss (vasomotor control). The extremities tolerate a much wider range of temperatures than does the body core, and they make perfect thermal windows: sites that can lose large or small quantities of heat depending on the magnitude of heat transfer from the body core via blood flow.

Vasomotor tone is controlled by adrenergic sympathetic nerve fibers, which vary the flow rate of blood to the extremity and the temperature of the blood reaching the skin (see Chap. 14).

In response to cold stress, the general vasomotor response reduces blood flow to the periphery by massive vasoconstriction. In man, as blood traverses the major arteries of the arm or leg, the temperature of the blood falls significantly. Cool venous blood returning to the core in vessels lying close to the artery pick up much of the heat lost by the arterial blood (Fig. 19.7). This system is called *countercurrent exchange*. It allows much of the heat to be returned to the core as the blood completes its circuit through the extremity. The net effect of the system is to reduce heat loss.

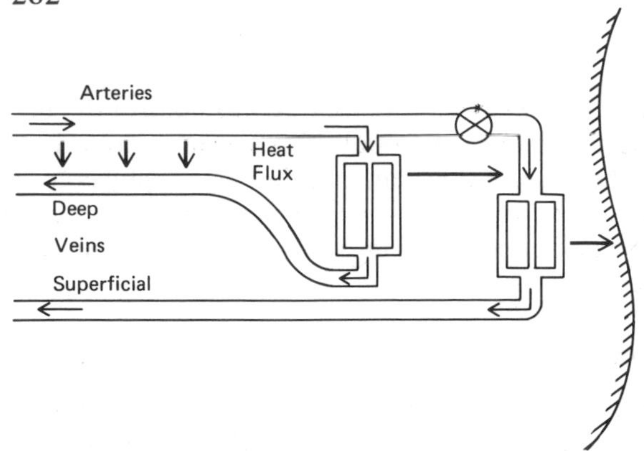

Fig. 19.7. Bloodflow to peripheral tissues, allowing cutaneous vasodilation (valve open), facilitating heat loss or cutaneous vasoconstriction (valve closed), reducing heat loss. Note that blood returns to body core by different vessels.

In subzero weather this heat conservation system is disadvantageous because the temperature of the fingers and toes may fall below freezing (frostbite) as a result of effective heat exchange between arterial and venous blood. Animals adapted to arctic conditions selectively vasodilate to regulate skin surfaces at or above freezing.
When heat stressed, the extremities dissipate excess heat by greatly increasing the blood flow to the skin. Blood returns to the core via veins just under the surface of the skin. By changing the pathway taken by the venous blood, the countercurrent system is bypassed, reducing heat uptake from the descending arterial blood. The proximity of veins to the skin surface maximizes the cooling of the venous blood as it returns to the core.

Evaporative Heat Loss

High environmental temperature imposes a serious thermoregulatory problem for mammals. In most environments mammals are significantly warmer than their surroundings, and the major problem is to reduce rates of heat loss. As air temperature rises, the difference between body and air temperature decreases, thereby reducing the effectiveness of heat exchange by radiation, conduction, and convection. When air temperature exceeds body temperature, the body may actually serve as a heat sink for the environment, increasing the stored heat until a higher equilibrium body temperature is achieved. Under these circumstances evaporation of water represents the major avenue of heat loss from body to environment.

Panting

The form of evaporative heat loss most commonly employed by mammals and birds is panting. The volume of air traversing the moist respiratory passages is dramatically increased by a change in ventilation characteristics, or panting. Respiratory frequency is sharply increased. Dogs, which normally breathe at rates of 30-40 breaths/min routinely attain breathing rates of 300 to 400 breaths/min when they are under heat stress. Some animals pant at frequencies matching the resonant frequency characteristics of their respiratory systems. This is the most efficient frequency at which the respiratory system can be moved and reduces the energy expended. It relies on elastic elements of the system, absorbing some of the energy on inspiration and then releasing it during expiration, causing the lungs to "bounce back."

Panting is only temporarily effective because dehydration becomes a significant factor if it is continued for extended periods, which may upset the body's acid-base balance because excess CO_2 is removed in the lung, resulting in an increased blood pH (see Chap. 17). Although CO_2 loss could be reduced by decreasing the tidal volume at high breathing rates—thereby confining most airflow in the respiratory dead space—most mammals do not utilize such a mechanism effectively and become severely alkalotic during prolonged panting.

Sweating

Man and other mammals, such as horses and cows, dissipate heat mainly by sweat rather than by panting. Sweating is particularly useful for sparsely furred animals. Sweat glands in the skin utilize the entire body surface for evaporation. Sweating does not interfere with normal respiratory function or acid-base balance, and there is little energy cost associated with evaporation by sweating. Prolonged sweating, however, results in dehydration and loss of salts, caus-

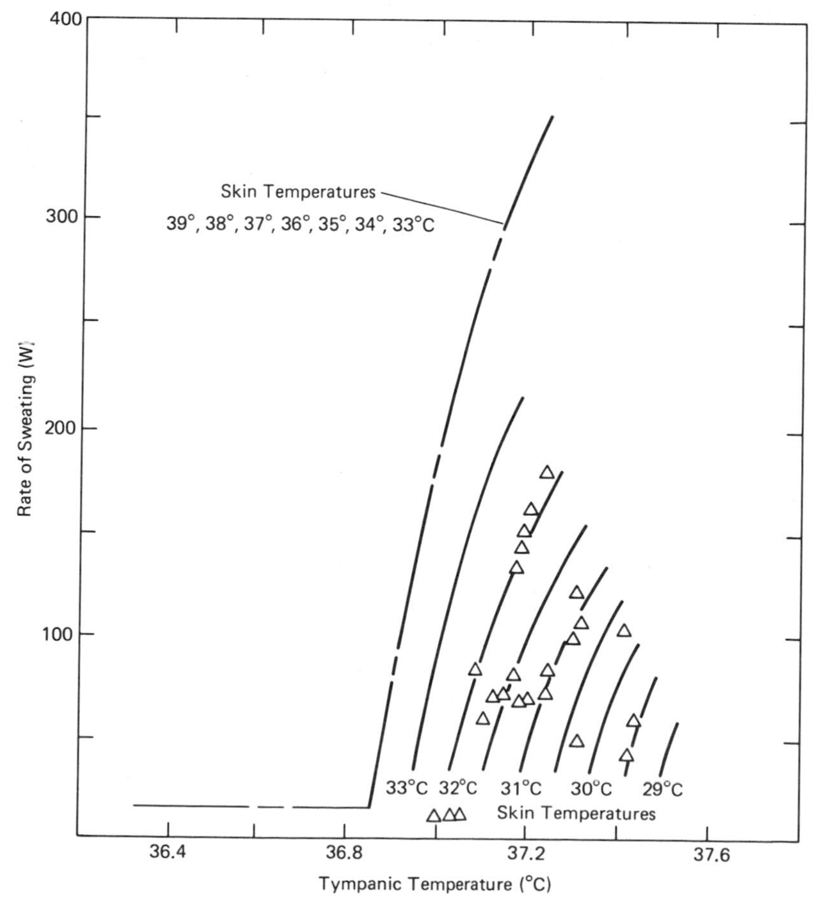

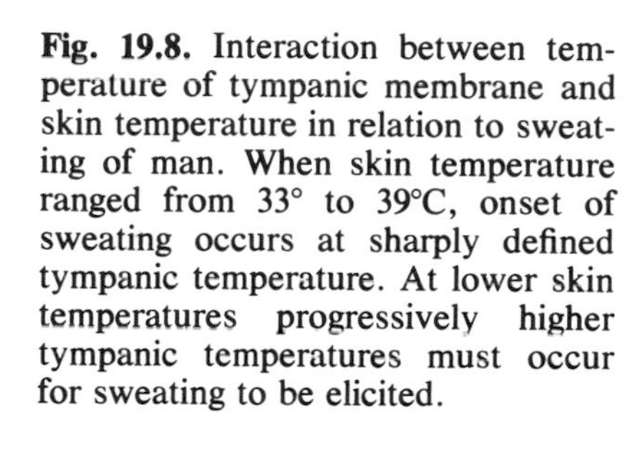
Fig. 19.8. Interaction between temperature of tympanic membrane and skin temperature in relation to sweating of man. When skin temperature ranged from 33° to 39°C, onset of sweating occurs at sharply defined tympanic temperature. At lower skin temperatures progressively higher tympanic temperatures must occur for sweating to be elicited.

ing a disruption of the body's electrolyte balance, which must be corrected by ingestion of salt.
Man's skin contains more than two million sweat glands, which are under the control of sympathetic nerve fibers. They are sensitive to both local skin temperature and central temperature (Fig. 19.8). Maximum sweating occurs at both high skin and high core body temperatures.

Regulation of Heat Production by Effectors

As air temperature falls, the rate of heat production must increase to compensate for an increased rate of heat loss (see "Heat Transfer Coefficient" in this chapter). Increased activity and muscle contraction produce more body heat. This activity is under voluntary control. Another mechanism that increases the body's heat production is *shivering*.
Shivering is an autonomic response to cold, although it can be overridden by voluntary-muscle control. During shivering, both flexor and extensor muscles simultaneously contract at high frequency in a rhythmic fashion. Both the frequency and the force of contraction can be varied. Shivering and activity are not additive; heat is generated by shivering only when the muscles are otherwise inactive. If a man is seated quietly in a cool environment, e.g., he may suddenly begin to shiver, quite without any conscious effort.
Voluntary action, such as walking, causes muscular contraction which overrides the shivering of the muscles. Shivering and walking both produce muscle heat. However, dur-

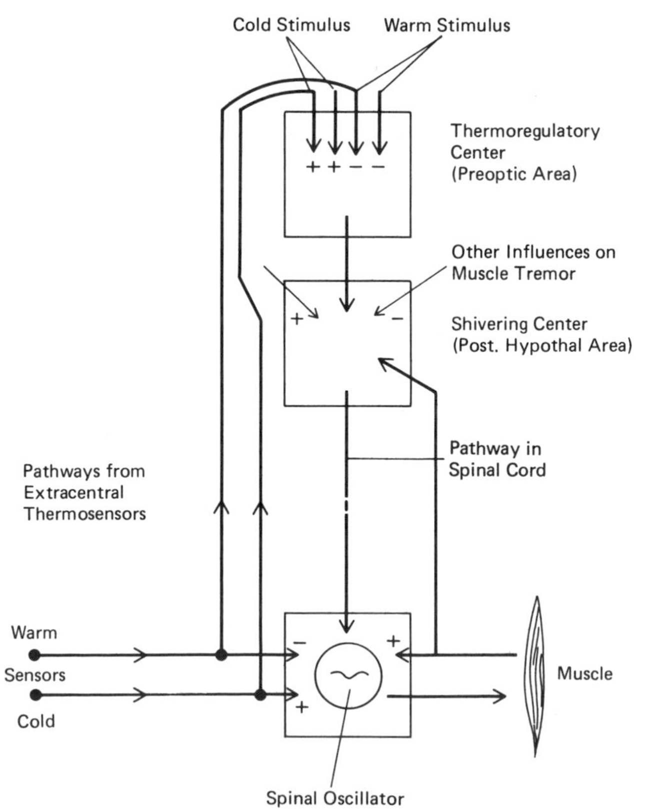

Fig. 19.9. Interaction of posterior hypothalamic shivering center and spinal oscillator controlling shivering. (After Bligh, J, 1973.)

ing activity there is an increased heat loss due to greater convective heat exchange. Reduced muscular movement in shivering increases the effectiveness of thermoregulation by decreasing the amount of heat loss due to convection.

A *shivering center* in the posterior hypothalamus appears to influence the frequency and intensity of muscular contractions during shivering. This center, in turn, receives input from the thermoregulatory center in the anterior hypothalamus and feedback from the muscles. Output from the shivering center passes to all levels of the spinal cord. However, shivering has been demonstrated after the transected spinal cord is cooled, even though signals from the shivering center cannot reach the spinal nerves. This suggests that the rhythmic signals causing shivering in the muscles originate at the level of the spinal cord (Fig. 19.9). The shivering center modulates the rhythmic output of the *oscillator*. Both spinal oscillator and hypothalamic shivering center receive input from thermosensors.

The control of shivering appears to be related to input by both peripheral and deep-body thermosensors. Shivering can be elicited in some species by local cooling of the hypothalamus while skin temperature remains constant. However, in many cases the onset of shivering by an animal placed in a cold environment is often too rapid for any significant change in the core body temperature to take place, thus implicating the importance of peripheral thermosensors. Such a response is highly adaptive, because it antici-

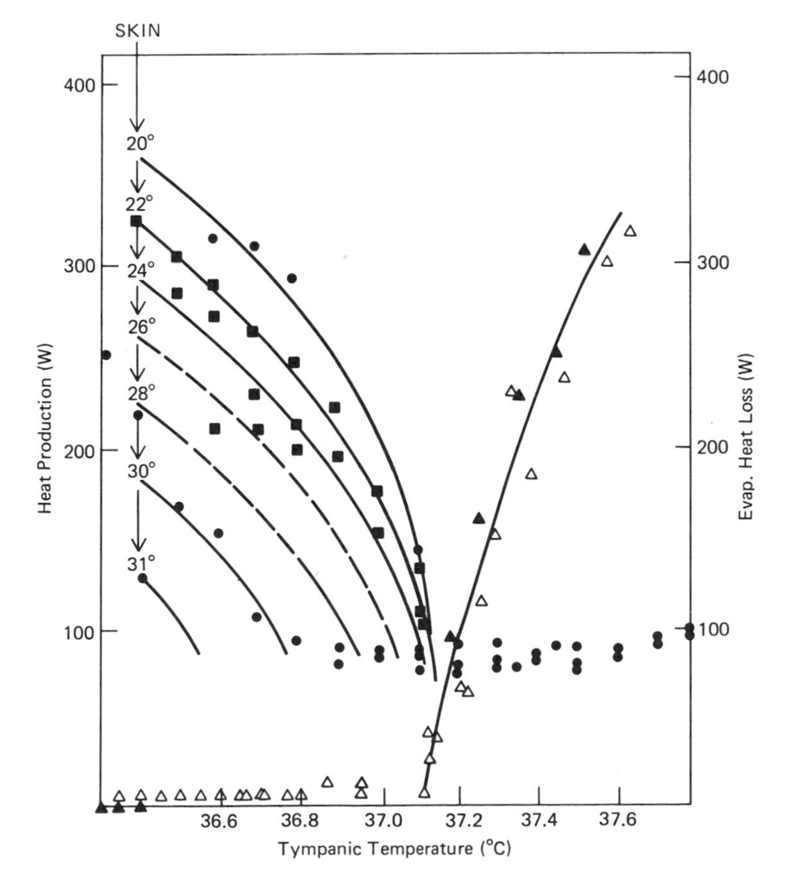

Fig. 19.10. Interaction of tympanic and skin temperatures in relation to onset and intensity of shivering man. As skin temperature increases from 20° to 31°C, a greater drop in tympanic temperature is necessary to elicit shivering (see text). (After Benzinger, TH, 1969.)

pates changes in the core temperature and signals the appropriate thermoregulatory response before core temperature changes.

Both *peripheral and central thermosensors* play a role in determining the onset and intensity of shivering during cold stress. If, e.g., the tympanic temperature of man (an index to brain temperature) is 36.6°C and skin temperature is 28°C, heat production is about 200 Watts (Fig. 19.10). However, if tympanic temperature remains at 36.6°C but skin temperature falls to 22°C, heat production increases by about 50%. The core temperature at which shivering begins and the rate of increase in heat production by shivering decrease as skin temperature increases.

Nonshivering Thermogenesis

It has long been known that increases in heat production can occur that are not attributable to shivering of skeletal muscles. The generalized increase of heat production from various sources in the body is termed nonshivering thermogenesis (NST). It occurs most commonly, but not exclusively, in some animals exposed to cold. Heat produced by NST is additive to the heat produced in the skeletal muscles by shivering or activity. Normally NST occurs before shivering and lowers the temperature at which shivering is elicited. In most species about one-half of the increased heat production resulting from NST occurs in the skeletal muscles.

Brown Fat

Brown adipose tissue—a multilobed tissue located in the subscapular region—is particularly well developed in hibernators. The calorigenic role of brown fat during arousal of hibernators has been repeatedly demonstrated. Brown fat is also well developed in many newborn mammals, including man. During the first few days after birth, nonshivering thermogenesis (via heat production of brown fat) is the major response to cold. Later, shivering constitutes the major response to cold, and this is correlated with the conversion of brown to white fat. White fat represents primarily an energy-storage, rather than a calorigenic, function.

In some species, such as the rat, brown fat persists throughout life and is a source of NST, particularly after exposure to cold. The contribution of brown fat to total heat production by nonshivering means is difficult to assess. Brown fat produces heat in the rat, but it represents 1% or less of the body mass compared with 45% of the body mass as skeletal muscle. Surgical removal of all traces of brown fat in rats results in a reduction but not an elimination of NST. In general, large mammals tend to rely more heavily on shivering in response to cold stress, whereas NST appears to be more important in small mammals.

Thermoregulation and Exercise

During exercise the core body temperature increases, and the magnitude of the rise is proportional to the work rate (internally generated heat) and independent of the environmental temperature, at least in large mammals. In small mammals body temperature varies both in relation to the work rate and the ambient temperature that determines the temperature gradient for heat loss. The coupling of T_b to air temperature in small mammals, but not in large ones, probably reflects the greater rates of heat loss per unit surface in small mammals as a result of higher surface-to-volume ratios and poorer insulation.

The rise in core temperature in man at different work rates suggests that the set point at which body temperature is regulated is changed as a result of exercise. However, a simpler interpretation is that at higher rates of internally generated heat production (after exercise), a larger amount of heat is stored in the body and a new equilibrium between heat production and heat loss is established. If skin temperature is held constant, man's sweating rate is closely related to the core body temperature whether at rest or during exercise (Fig. 19.11). Similar results occur in dogs.

These studies strongly suggest that core temperature is an important determinant of effector function in thermoregulation. The core body temperature rises linearly with work rate (and therefore metabolic rate). The mean skin temperature is a function of the air temperature and is further cooled by evaporation, and it is not closely coupled to

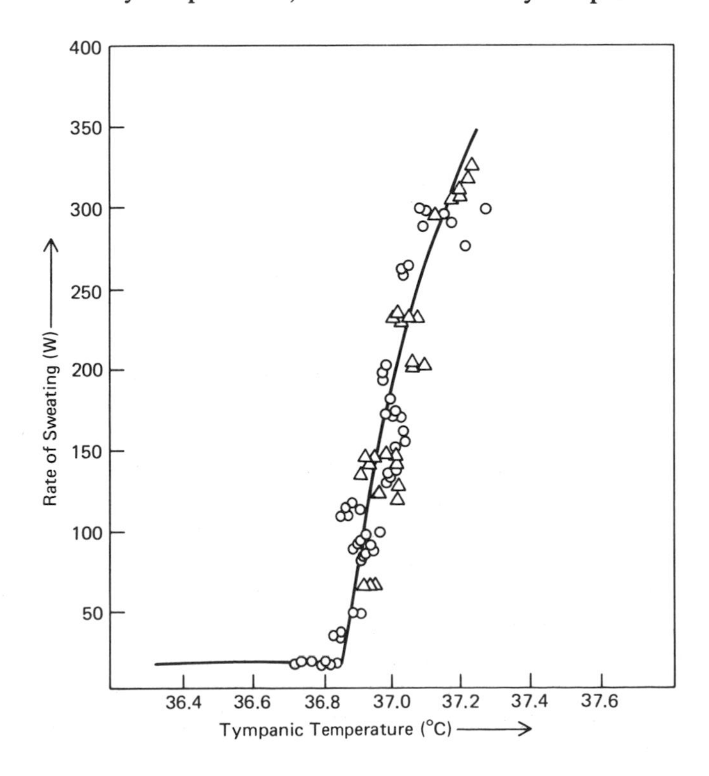

Fig. 19.11. Rate of sweating in man at rest (○) and during exercise (△). Skin temperatures were between 33° and 38°C, and core temperature (tympanic) ranged from 36.7° to 37.3°C. Similar relations at rest and during exercise suggests that set point for regulation has not changed during exercise.

either core temperature or metabolic rate. As mean skin temperature increases at a given core temperature, the rate of sweating also increases. As shown in Fig. 19.8, there can be substantial differences in the rate of sweating at any given core temperature, depending on the level of the mean skin temperature.

Physiologic Adjustment and Extreme Temperature

Normal seasonal changes in climate elicit physiologic changes in animals called acclimatization. In contrast, exposure to prolonged low or high temperatures in the laboratory without changes in other environmental parameters, often results in different physiologic changes called *acclimation*.

Cold

Endotherms exhibit diverse physiologic responses to changes in climate, particularly to cold. Heterothermic birds and mammals shut down their metabolic machinery and allow body temperatures to fall within a degree or so of the environmental temperature. This adaptation reduces the energy expenditure necessary for survival. Heterotherms are not totally at the mercy of the environment, however. If the air temperature drops below freezing, the animals can maintain body temperature above freezing, and therefore some regulation does occur. Impressive increases of insulation during winter as a result of seasonal molting of the fur in some animals reduces heat loss from the body so that heat production is maintained at similar levels in both winter and summer.

Many animals do not show a significant change in body insulation per se and must rely on an increased capacity to generate metabolic heat.

Human beings do not exhibit substantial quantities of brown fat after the first few weeks of life, and increased heat production must occur by shivering or activity. These are only useful on a short-term basis. A further response to cold is a massive vasoconstriction that helps to maintain core temperature by reducing heat transfer to the peripheral areas of the body. As pointed out earlier, such a response is only of limited utility, particularly at very low temperatures, due to the loss of function of hands and feet and the risk of frostbite. Man's major response to cold is behavioral rather than physiologic. The sensation of "feeling cold" elicits behavioral adjustments to control body temperature (heated dwelling, blankets, and clothing among others). Efforts to demonstrate specific physiologic changes associated with acclimatization in man inhabiting cold regions are inconclusive.

Heat

A hot environment is potentially more stressful than a cold one. When the temperature of the surroundings exceeds body temperature, the flow of heat will be reversed, causing the body to act as a heat sink for the environment. Heat is produced by metabolism and is also gained by radiation, conduction, and convection—major avenues of heat loss in a cold environment. If no behavioral means of avoiding heat stress is available, the only way of ridding the body of heat is by evaporation.

Large mammals, such as the camel, that are adapted to hot, dry conditions can tolerate large fluctuations in core body temperature, a process called *facultative hyperthermia*. As heat flows into the body, the camel does not increase its rate of evaporation but instead stores the heat, causing body temperature to rise. This is a viable strategy because of the camel's large bulk and relatively low surface-to-volume ratio. Body temperature rises so slowly that it does not reach lethal levels by the end of the day. At night, when air is cooler, body temperature drops to normal levels again. By abandoning the precise regulation of core temperature, the camel reduces the need for evaporative cooling, which has obvious adaptive value in the desert where free water is scarce.

Although man cannot live in hot, dry environments without access to free water, he shows definite physiologic adjustments to hot environments. On initial exposure to heat stress, rectal temperature, oxygen consumption, heart rate, and cardiac output rise rapidly. The sweat rate is modest and insufficient to maintain core body temperature (Fig. 19.12). After 12 days exposure there are obvious signs of adjustment. Sweat rate has doubled, whereas heart rate and

Fig. 19.12. Rectal temperature, heart rate, sweat rate, and oxygen consumption of young men working for five hours in hot environment (T_a = 36°C); on first day of exposure and after 12 days' exposure.

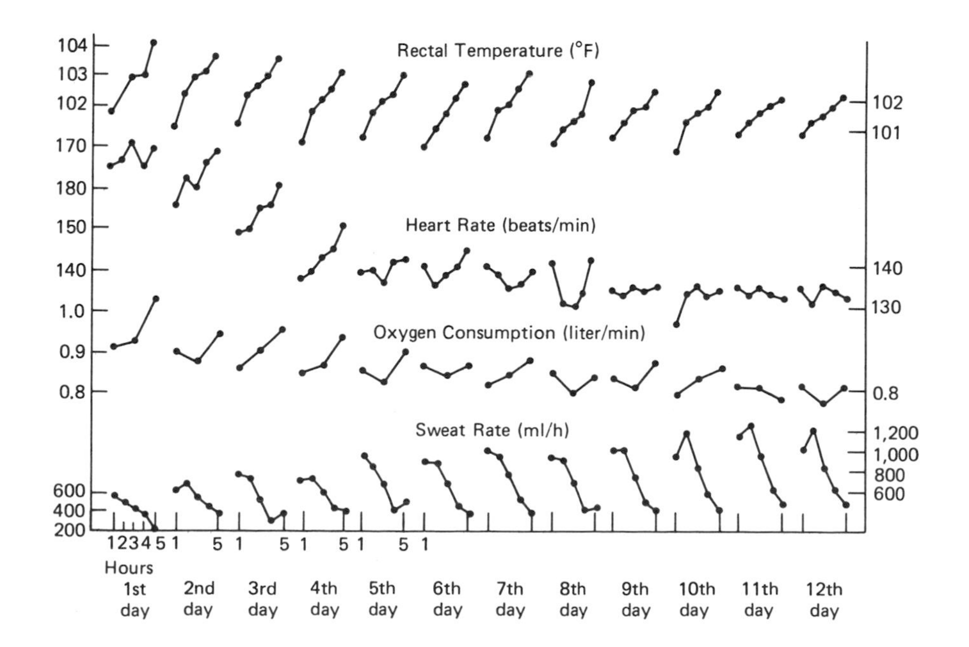

oxygen consumption have dropped about 40%. The increased evaporative component and decreased metabolism reduce the rise in core temperature. Associated with more effective control is a massive vasodilation to deliver heat to the periphery and an increase in the magnitude and onset of sweating. The response is evident after about five days and is soon lost if the subject is not repeatedly exposed to heat stress. *Evaporative cooling* is only a short-term solution because excessive sweating can disrupt the body's fluid balance and lead to circulatory failure from reduced plasma volume.

Heat stroke is a condition in which the temperature of both peripheral and deep-body tissues increases, resulting in unconsciousness; the body does not sweat and evaporative cooling is lacking. The body often becomes flushed, indicating vasodilation in peripheral blood vessels but if heat stress is prolonged, increased temperatures will damage tissues of the brain and nervous system, ultimately leading to death.

Fever

Fever is a physiologic condition characterized by an elevated core body temperature commonly caused by the presence in the blood of certain chemicals called *pyrogens*. One type of pyrogen is derived from microorganisms and is termed *bacterial endotoxin*. Fever resulting from experimental injection of endotoxin usually does not occur in man for at least 30 minutes. The second type, *endogenous pyrogen,* is produced by the body tissues and has been isolated from leukocytes. Onset of fever following injection of endogenous pyrogens is almost immediate.

Pyrogenic substances increase the set point at which the body temperature is regulated. At the onset of fever, the body responds as if to cold stress. Heat loss is reduced by vasoconstriction in the peripheral tissues, and heat production is increased. The sensation of chills occurs and the

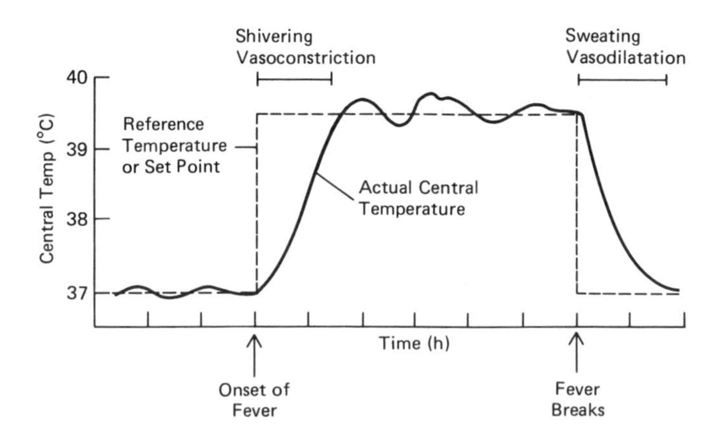

Fig. 19.13. Time course of typical febrile episode. Body temperature lags behind rapid changes in set point. During fever, temperature is regulated at higher level.

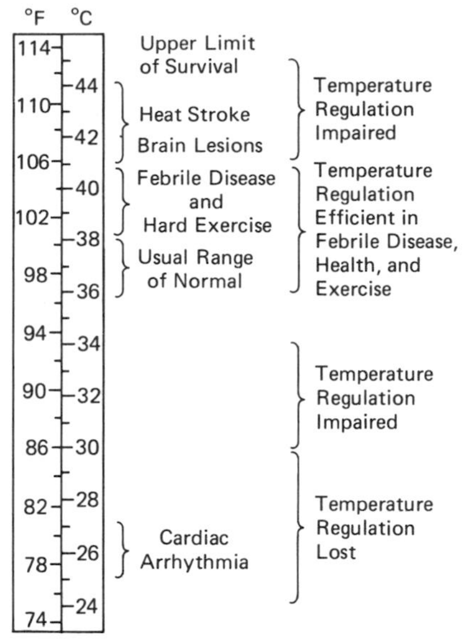

Fig. 19.14. Normal and extremes of body temperature.

body responds by shivering. The core body temperature rises to some new level where it will be regulated as long as fever-producing pyrogen is present (Fig. 19.13). After the pyrogen is removed, the set point in the hypothalamus is lowered and the body responds as if to heat stress. Cutaneous vasodilation and sweating are common during this phase and the body gradually returns to normal levels.

Fever and Infection

It has long been maintained that fever occurs because elevated body temperature is somehow advantageous in fighting the infection produced by invading organisms. Little clinical evidence supports this view. There are obvious disadvantages to high fevers. All metabolic and physiologic processes operate at dangerously high levels, and body temperature is already close to the temperature at which proteins begin to denature (Fig. 19.14). Fever is usually regarded as disadvantageous, and treatment of most illnesses usually includes some method to reduce body temperature, usually by the administration of antipyretic drugs.

Selected Readings

Bligh J (1973) Temperature regulation in mammals and other vertebrates. American Elsevier, New York

Benzinger TH (1969) Heat regulation: homeostasis of central temperature in man. Physiol Rev 49: 671

Dubois EF (1948) Fever and the regulation of body temperature. C C Thomas, Springfield, Ill

Hammel HT (1968) Regulation of internal body temperature. Ann Rev Physiol 30: 641

Hardy JD (1961) Physiology of temperature regulation. Physiol Rev 41: 521

Heller HC, Crawshaw LI, Hammel HT (1978) The thermostat of vertebrate animals. Sci Amer August. 102

Review Questions

1. Of what significance to the heat balance of man is a decline in oxygen consumption during heat acclimation?
2. Describe the advantages and the disadvantages of evaporative heat loss by sweating and panting.
3. Since in order to maintain a constant core temperature, heat production must equal heat loss, and since heat loss is proportional to the difference between body and air temperature, how can heat production be constant over a range of temperatures in the thermoneutral zone?
4. What is the physical basis of wind chill?
5. Assuming that the metabolic rate of a marathon runner is determined by his performance during a race, describe several mechanisms, that can control body temperature.

6. Explain the major difference between heat stress at high environmental temperatures and that incurred during exercise. How do mechanisms of thermoregulation differ in the two circumstances?
7. What are the major avenues of heat gain and heat loss in (a) a man in the desert at midday and (b) a man in the arctic at night?
8. Of what significance is nonshivering thermogenesis?
9. Why does the thermoregulatory system operate with both peripheral and central thermoceptors? Speculate on the interrelationship between peripheral and central thermoceptors at (a) low ambient temperature, (b) high ambient temperature, and (c) exercise.

Chapter 20

Movements of Gastrointestinal Tract

The digestive and absorptive functions of the gastrointestinal tract (GIT) include chewing of food and mixing it with saliva (mastication and salivation), swallowing, and its movement through esophagus and stomach, where digestion begins, to the small intestine (the site of further digestion and absorption). From the small intestine the food mass moves into the large intestine and into the rectum where undigested matter and feces are expelled. The structure and function of these parts must be such as to propel and move food (motility) from one end of the tract to the other.

Anatomic Considerations

The location and general anatomy of the principal parts of the digestive tract or alimentary canal are shown in Fig. 20.1. Further details of anatomy will be presented as each part of the tract is treated. The teeth, tongue, salivary glands, and mouth are the uppermost organs of the alimentary canal where food is masticated, mixed with saliva, and swallowed.

The esophagus begins at the end of the pharynx and extends downward 10 to 12 inches (250-300 mm) to the stomach. The detailed structure (histology) of the wall or lining of the esophagus is essentially the same as the remainder of the digestive tract, except for minor differences in the type and structure of the lining of the lumen (inside of opening) and glands in the mucosa and submucosa. The organization of the layers of tissue making up the esophagus are shown in Fig. 20.2 (see also Chap. 1). From the outer wall to the

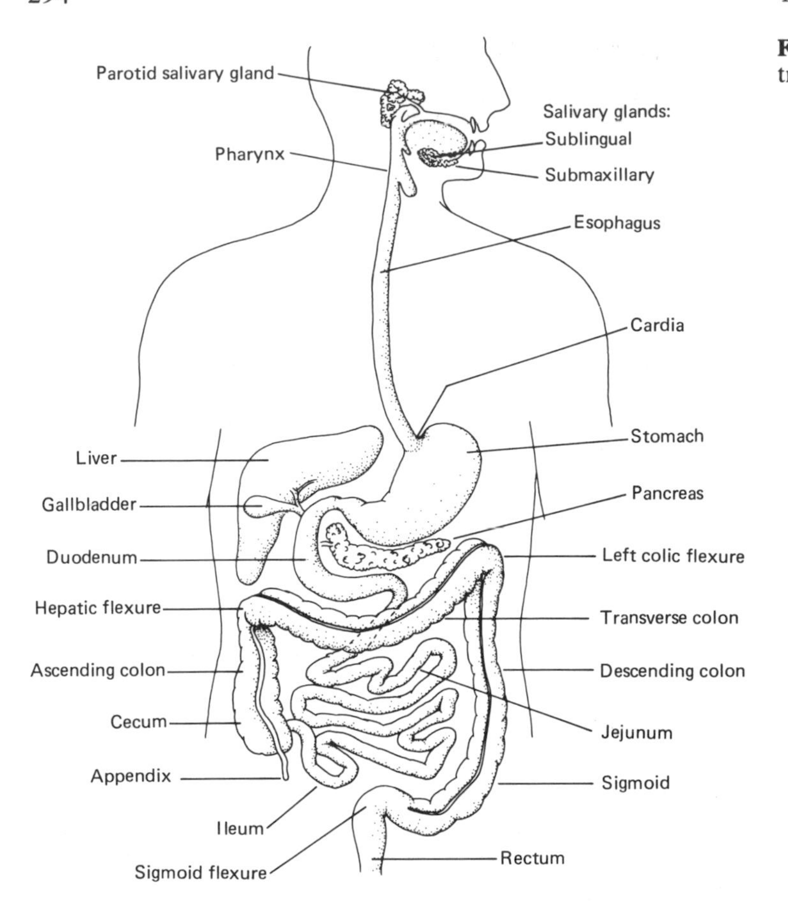

Fig. 20.1. Organs of gastrointestinal tract.

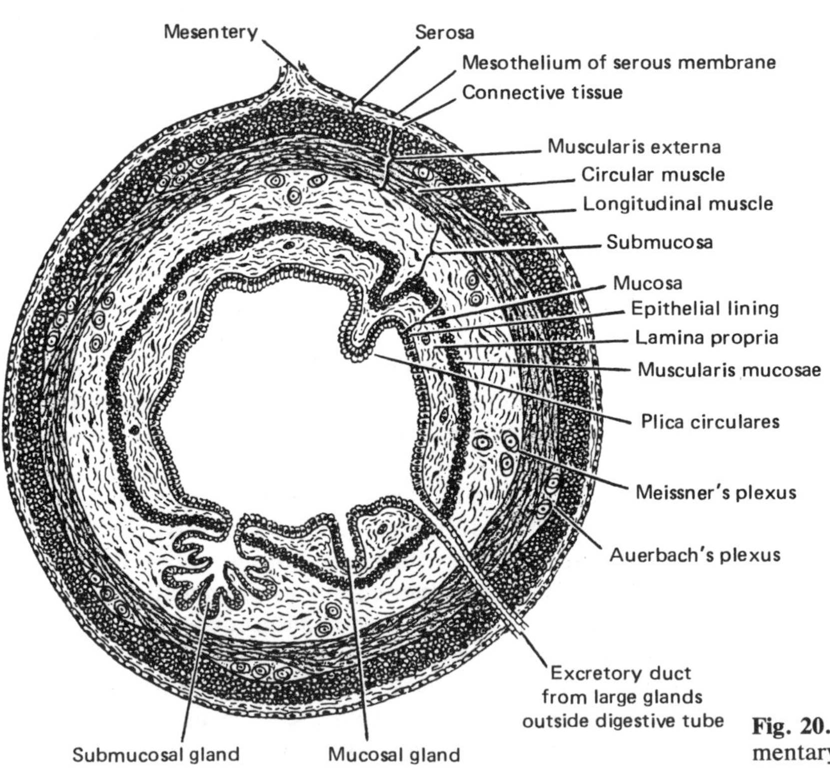

Fig. 20.2. Cross section of wall of alimentary canal.

lumen are connective tissue and muscle of two types (circular and longitudinal), the submucosa, mucosa, and epithelial lining of the inside lumen. Glands are present in the mucosa and submucosa.

Between the circular and longitudinal muscles are specialized nerve cells, aggregated into a network or plexus (Auerbach's plexus, or myenteric plexus). Internal to these cells are other nerve cells making up Meissner's plexus, or submucosal plexus. These are the intrinsic nerves, and they extend throughout the alimentary canal. The plexuses contain nerve cells with processes that originate in receptors in the wall of the gut or mucosa and are responsible for the automatic contractions and movements of the tract, independent of the extrinsic nerves.

Neuromuscular Apparatus

In addition to the intrinsic nervous system, there is an extrinsic nerve supply to most of the smooth muscles of the GIT, which are involved in the stimulation of muscle and subsequent motility. Figure 20.3 A, B, and C show the type of muscle and cells innervated; Table 20.1 summarizes the innervation by area or organ involved (see also Chap. 10). In general, the sympathetic motor nerves have an inhibitory or relaxing effect on gastrointestinal smooth muscle, and the parasympathetic nerves have an excitatory effect and produce most of the motility effects. The preganglionic parasympathetic fibers synapse with the ganglion cells in the myenteric and submucosal plexuses (Fig. 20.3B), and the fibers

Table 20.1. Innervation of gastrointestinal tract by sympathetic (S) and parasympathetic (PS) motor (efferent) nerves only.

Organ Innervated	Nerves Involved[a]	Effect[b]
Stomach	S (Sp1)	−
(muscle)	PS (V)	+
Pyloric sphincter	S (Sp1)	+
	PS (V)	−
Small intestine	S (Spl)	−
(muscle)	PS (V)	−
Ileocolic sphincter	S (Spl)	−
	PS (V)	+
Cecum muscle	S (Spl)	−
	PS (V)	+
Large intestine		
Ascending colon	S (Spl)	−
	PS (V)	+
Descending colon	S (HYP)	−
	PS (Pel)	+
Anal sphincter		
Internal	S (Spl)	+
	PS (Pel)	−
External	Somatic (Pud)	+

[a] Spl = splanchnic nerve plexus; V = vagus nerve; HYP = hypogastric plexus; Pel = pelvic nerve; Pud = pudendal.
[b] −, inhibition; +, excitation.

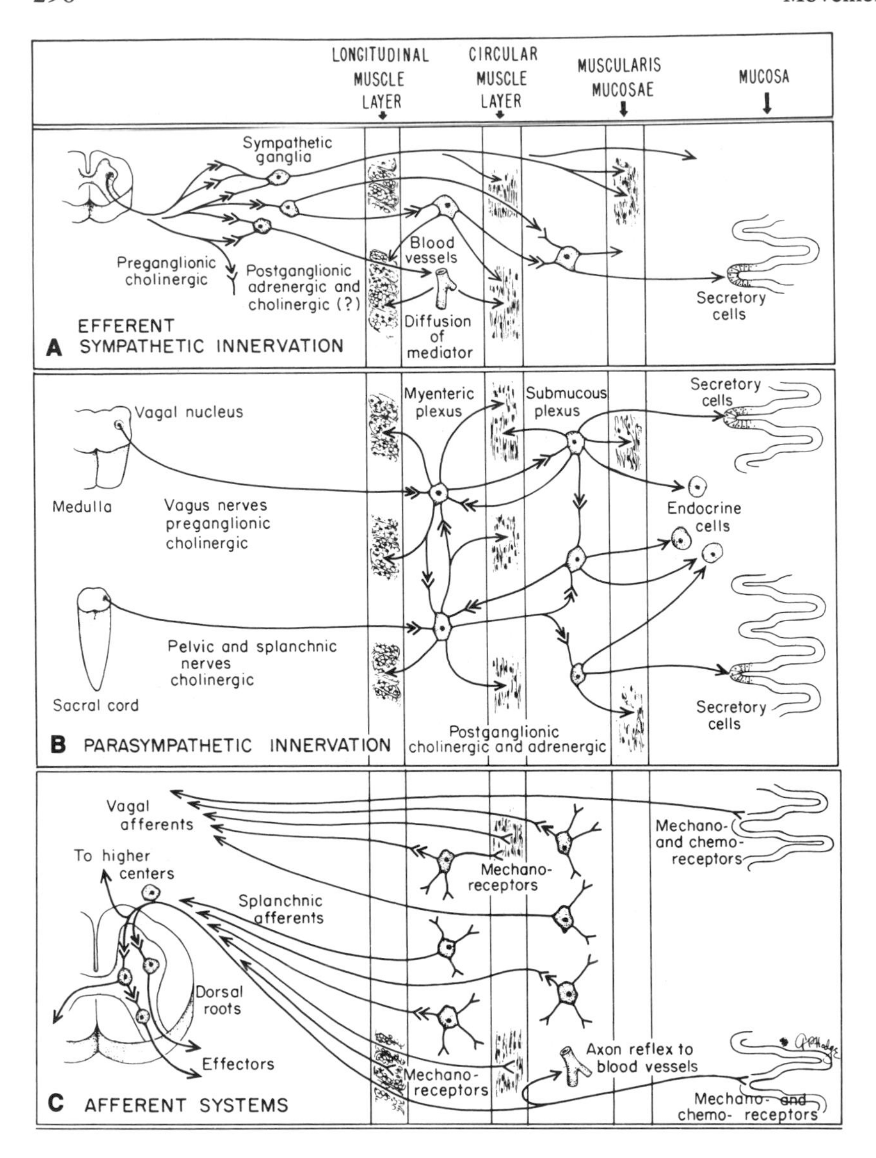

from these nerve ganglia (postganglionic) innervate and stimulate the smooth muscle (longitudinal and circular) by releasing acetylcholine (see Chap. 11). After destruction or denervation of extrinsic nerves, smooth muscle of the stomach and intestine is still capable of rhythmic contractions, although they are weaker ones. This means that such contractions are mediated by the intrinsic nerves (Auerbach's and Meissner's plexuses). Even after abolition of the intrinsic nerve effect (local) by cocaine, smooth muscle is still capable of some automatic movements but not peristaltic movements; the latter require intact intrinsic nerve activity. The afferent nerve system to the tract is shown in Fig. 20.3C. Stimuli to the mucosa, in which are located the receptors (mechano- and chemoreceptors), are relayed through vagal afferent nerve fibers and splanchnic afferents to the central nervous system (CNS), and the response is mediated by the efferent system previously mentioned. These same

Fig. 20.3A–C. Extrinsic and intrinsic nerves to the gastrointestinal tract. **A** sympathetic nerves which are efferent; **B** parasympathetic nerves which are motor and cholinergic; **C** afferent nerve fibers which receive impulses from gastrointestinal tract and relay them to central nervous system. (Reproduced with permission from Davenport, H. W.: Physiology of the digestive tract, 4th ed. Copyright © 1977 by Year Book Medical Publ., Inc., Chicago.)

Fig. 20.4. Major blood vessels of digestive organs.

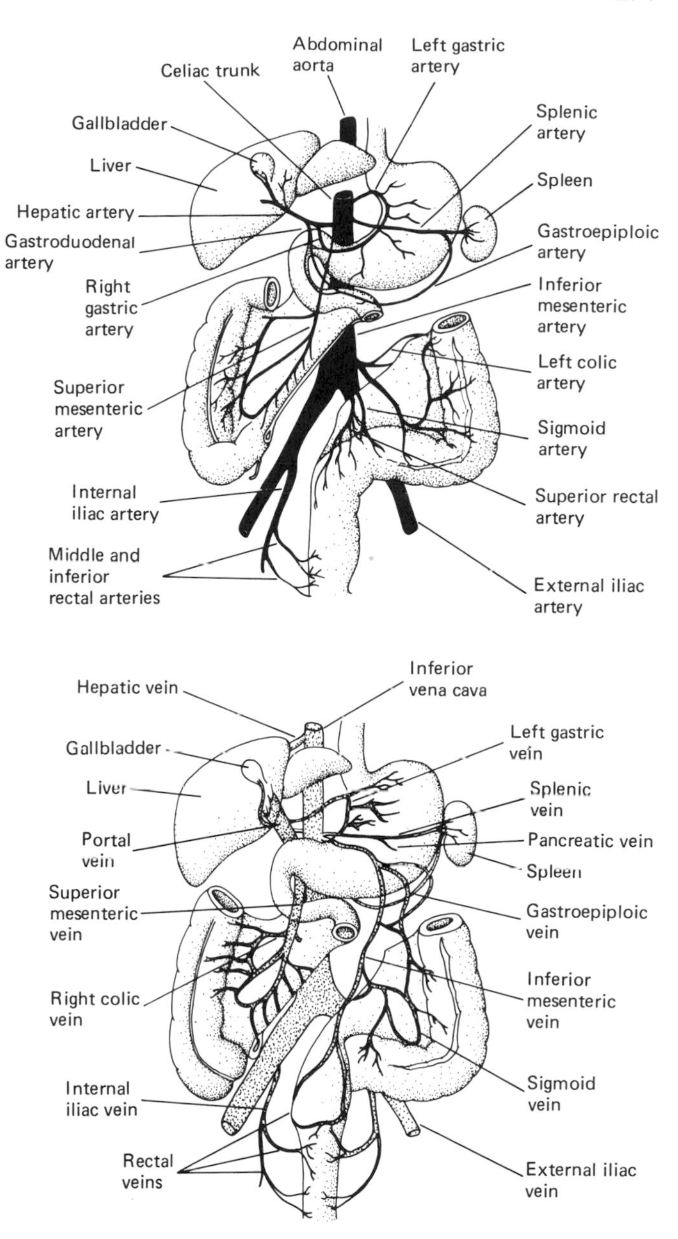

afferent mechano- and chemoreceptors may be involved in a simple reflex whereby the efferent or motor response is mediated via the spinal cord (see Chap. 7).

The major blood vessels to the GIT are shown in Fig. 20.4. Each organ of the trach receives blood from an artery and is drained by a vein (the artery is afferent, the vein efferent). The liver, however, has an artery and two types of veins: a portal vein that carries blood and absorbed nutrients from the intestinal tract to the liver (afferent) and a vein (hepatic) that carries blood from the liver (efferent) to the inferior vena cava.

The blood flowing through the GIT in a man of average size is about 1500 ml/min. The proportion of total blood flow through each blood vessel is shown in Fig. 20.5. The liver

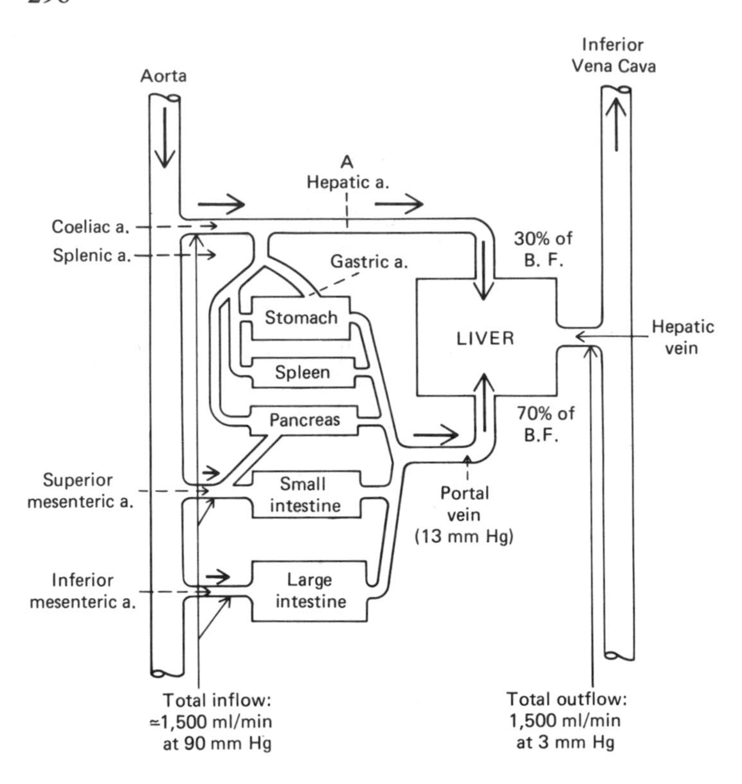

Fig. 20.5. Blood flow (BF) through GIT of man. (After Selkurt, EE, Physiology, 4th ed., Little, Brown, Boston, 1976.)

receives 30% of its blood from the hepatic artery and 70% from the portal vein.

Eating and Drinking

As food enters the mouth and is chewed, it stimulates the taste buds and olfactory senses, which are mainly responsible for the satisfaction of eating. Sight of food and contact with it in the mouth cause reflex secretion of saliva; the latter mixes with the food, softens and lubricates it and when chewed, it forms a bolus, which is swallowed.

Taste and sight of food stimulates receptors in the mouth, which relay the impulse to the brain or lateral hypothalamus, which is the feeding or *appetite center* (see Chap. 10). Stimulation of this area in animals causes voracious eating. Situated ventromedially in the hypothalamus is another area that has an effect the reverse of that of the feeding center. It is the *satiety center,* which tells the person that he or she is satisfied and no longer hungry. Fullness in stomach or distention in any part of the GIT stimulates the satiety center and depresses the feeding center.

Drinking of water is influenced in part by the amount and kind of food eaten and is also regulated by the drinking center in the hypothalamus (see Chap. 10). Other factors affecting fluid intake and outgo from the body are considered in Chaps. 12 and 24.

Chewing and Swallowing

Mastication involves the muscles of the jaws, lips, cheeks, and tongue. Their actions are coordinated by nervous impulses in cranial nerves V, VII, IX, X and XII. Although chewing is a voluntary act, it is also partly reflex in that it occurs when food is in the mouth of an animal whose cerebral cortex has been removed. The crushing force exerted by the teeth is much greater (454–545 kg) than is required for most foods.

Swallowed food passes from the mouth, pharynx, and esophagus to the stomach: The mechanics are as follows:

1. Bolus (bite or ball or food) is passed from mouth and isthmus of fauces to pharynx. The food or water is rolled back of the tongue, and the front of the tongue forces it upward against the hard palate, followed by contraction of mylohyoid muscles, which moves bolus into pharynx.
2. Bolus is moved into the esophagus.

The esophagus is divided into three functional parts, including: (1) the upper pharyngeal sphincter (pharyngoesophageal), (2) the body, and (3) the lower esophageal sphincter (gastroesophageal). All three parts exhibit characteristic contractile activities during rest and swallowing.

Contractile Movements of Esophagus

At rest, the upper sphincter is closed and the body of the esophagus is flaccid. This closure prevents large volumes of air from entering the stomach during respiration. During a swallow, the upper sphincter relaxes and opens and allows food to enter. The sphincter then contracts, and this is followed by a peristaltic contraction wave, which carries through the body of the esophagus toward the stomach (Fig. 20.6). Just before the contraction wave reaches the lower esophageal sphincter, the sphincter relaxes (opens) and

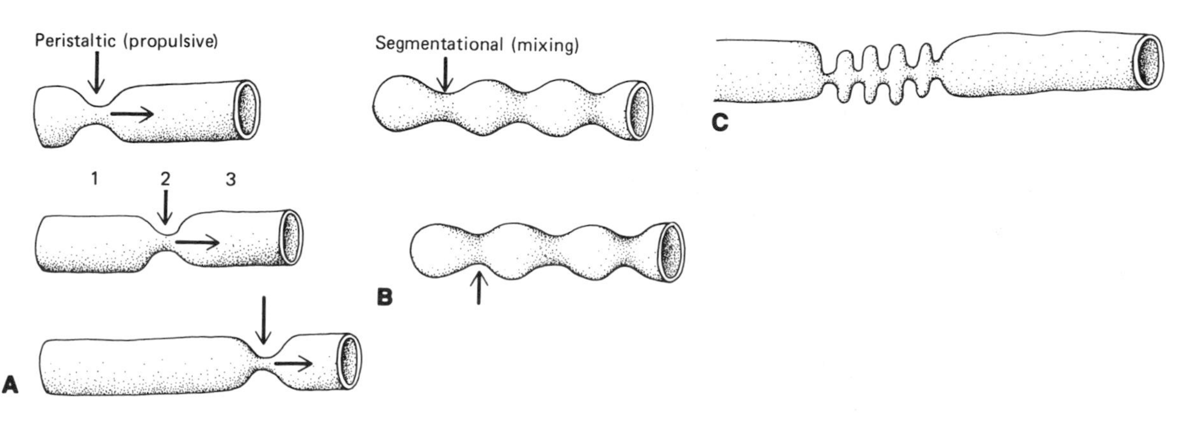

Fig. 20.6A–C. Gastrointestinal contractility. **A** peristaltic waves. **B** segmentation contractions in small intestine. **C** multihaustral contractions in colon. Waves begin at 1 and spread to points 2 and 3. Relaxation follows contraction.

allows food to enter the stomach and then contracts again. If the bolus is watery, it tends to be swallowed and drops to the stomach by gravity. But if it is solid or semisolid, it is propelled to the stomach by peristaltic waves at a rate of 2 to 4 cm/s. The pressure generated by the peristaltic contraction ranges from 30 to 140 mm Hg. Often more than one peristaltic wave is required to sweep the bolus completely into the stomach.

Swallowing is a voluntary act in which the cerebral cortex plays a dominant role; swallowing may also be induced reflexly by a collection of fluid or saliva in the mouth. Evidence exists for a swallowing center in the medulla, which sends motor impulses to the muscles involved in swallowing. The vagus nerve is important in controlling the peristaltic waves of esophagus, because sectioning (cutting) of this nerve abolishes these waves.

Stomach Motility

The stomach is a bean-shaped organ made up of the upper fundus and cardia (near the entry of the esophagus) and the corpus (body), and the lower region, the pylorus which joins the duodenum (Fig. 20.7). The human stomach has a capacity of 50 ml when empty to 750 ml or more when full or distended.

The gastric epithelium is characterized by specialized cells. In mammals the parietal cells produce HCl in the gastric juice, and the chief cells produce pepsinogen granules, but in birds chief cells produce both acid and pepsinogen. (Fig. 20.8).

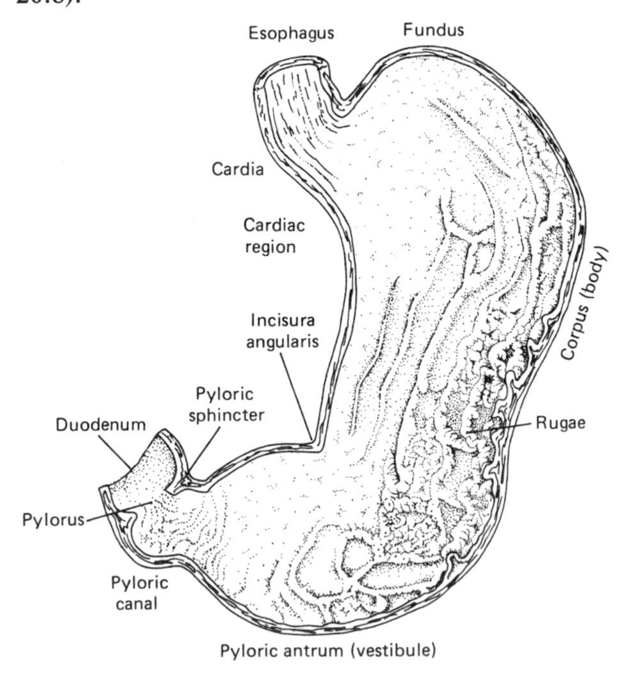

Fig. 20.7. Human stomach and its divisions.

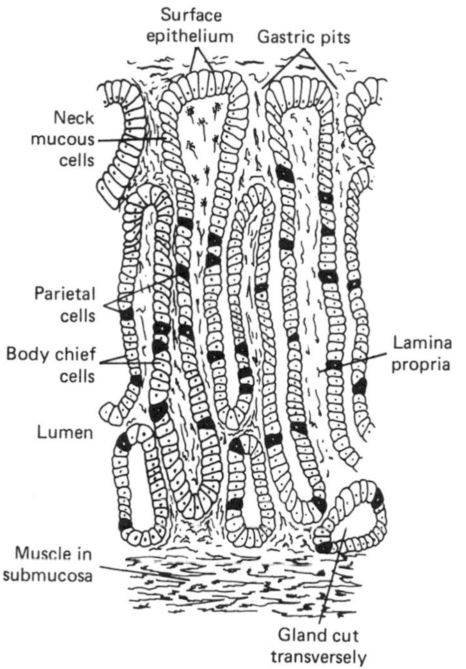

Fig. 20.8. Specialized cells of gastric mucosa. Chief cells produce pepsinogen and parietal or oxyntic cells secrete HCl.

As already mentioned, the upper stomach comprises the fundus and a small part of the corpus, and the lower part consists of corpus and antrum; the two parts exhibit different patterns of contractility. The upper stomach exhibits little activity, but the lower stomach (distal corpus and antrum) is very active.

Contractions begin and usually increase in intensity in the mid-stomach as they move toward the gastroduodenal junction. These mainly peristaltic waves travel at the rate of three per minute. Associated with these contraction waves are pressure waves of varying amplitude and duration. Types I and II are slow rhythmic pressure waves that differ in amplitude. Their duration is from 2 to 20 seconds, and they occur at a maximum frequency of two-to-four per minute. These pressures are probably generated by peristaltic contractions. Type III is a complex pressure wave lasting about one minute.

Ingested solids tend to collect in the upper stomach, with little mixing, and the food is in layers or strata. Ingested liquids tend to flow over these strata to the lower stomach where liquids and solids are mixed.

Gastric Emptying

The rate of passage of the ingested mass (ingesta) from stomach to intestine depends mainly on its physicochemical composition in stomach and duodenum. Carbohydrates leave the stomach most rapidly, proteins less so, and fats remain longest in the stomach.

Consistency of gastric contents also influences emptying time. Large chunks of meat remain in the stomach longer than small ones. Hypotonic solutions remain in the stomach longer than isotonic ones, and those with pHs of 5, 3, or lower retard emptying.

The interaction of stomach and duodenum is responsible for gastric emptying, but the exact mechanisms involved are not known. Several possibilities are suggested, however, such as (1) activity of pyloric sphincter, (2) gastrointestinal hormones, and (3) coordinated cycles of activity of antrum and proximal duodenum. Contraction of antrum is followed by sequential contraction of pylorus and duodenum.

Pressure differences in the pyloric region and duodenum obviously affect the passage of ingesta. If the pressure is lower on the stomach side, there is no passage even though the pyloric sphincter is inactive or absent, as by surgical removal. Thus the pyloric sphincter has little influence on gastric emptying.

The gastrointestinal hormones—gastrin, secretin, and cholecystokinin—inhibit gastric emptying but in uncertain ways. Fat in the intestine tends to inhibit gastric emptying, probably by releasing secretin.

Nervous Control of Motility

As pointed out previously, the principal extrinsic nerve affecting motility of the stomach and initiating the peristaltic waves is the vagus. Irritant stimuli, hypertonic solutions, and duodenal distention inhibit gastric motility by the *enterogastric reflex,* the receptors of which are in the duodenum; they relay impulses afferently by the vagus nerve; the motor or inhibitory response to the stomach is through the efferent fibers of the vagus. Intrinsic nerves may also be involved in this reflex.

Gastric emptying is dealyed when the ileum is full (ileogastric reflex) and when the anus is mechanically stimulated (anogastric reflex). Various emotions, such as anxiety and resentment, increase gastric motility; fear and anger inhibit it.

Vomiting represents emptying of the stomach in the reverse or abnormal direction. The major force is caused mainly by contraction of abdominal, rather than gastric muscles. Curare, a drug that paralyzes striated muscles of the abdominal wall abolishes vomiting. Vomiting is a complex reflex act and is coordinated by a center in the medulla. Afferent impulses arise from stimuli in the pharynx and stomach and from those resulting from motion sickness or pain.

Intestinal Motility

The small intestine in man is comprised of three areas. The first part, the duodenum ("12 fingers") is caudal to and continuous with the stomach, approximately 20 cm long and arranged within a loop (Fig. 20.1). The second part is the je-

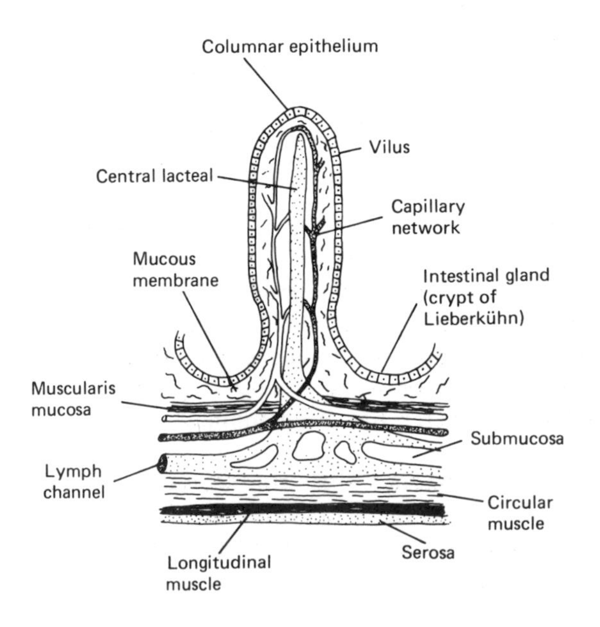

Fig. 20.9. Structure of villus of small intestine.

junum, about 3.5 m long, and the third part is the ileum (4.5 m long). There is no sharp boundary between jejunum and ileum, although there are structural differences in the epithelial lining of the villi (Fig. 20.9), the fingerlike projections or folds of lining that increase the surface and absorptive areas tremendously (about 30 times). Situated on each fold, or villus are smaller folds or projections (microvilli), which further increase surface area. The diameter of the small intestine is about 3-4 cm. The ileum ends at the ileocecal valve, at the entrance to the colon.

The large intestine, or colon, is relatively short (110 cm) and has a diameter of 7-10 cm. It begins at the cecum (Fig. 20.1) and extends to the rectum and anus, the terminal portion of the tract. The colon is divided into (1) the ileocecal region; (2) the main portion including *ascending, transverse,* and *sigmoid colon;* and (3) the anus and rectum (anorectum). There are no villi in the colon.

Movements of Small Intestine

Food, partially digested in the stomach, passes to the small intestine where it is completely digested and where nutrients are absorbed. Waste products and undigested food pass into the colon. These processes are aided by the movements of the small intestine, two types of waves or contractions, including segmentation, also referred to as type I contraction and peristalsis (Fig. 20.6).

Segmentation contractions are ring-like and occur at fairly regular intervals (about 10/min) and serve to mix the chyme, or ingesta. Areas of contraction (Fig. 20.6) are followed by areas of relaxation and vice versa.

Peristaltic waves tend to sweep along the intestine in the aboral direction (away from the mouth). The wave normally is slow and moves at a rate of 1–2 cm/s but may be much faster at rates of 25 cm/s (a peristaltic rush) for an obstructed gut. Peristaltic waves propel the ingesta; segmentation contractions, which are rhythmic, mix the ingesta and move it to and fro but do not propel it.

Control of Motility

Segmentation and peristaltic contractions continue even after abolition of the extrinsic nerves (vagus and sympathetic), although the intensity of contractions is lessened. This suggests intrinsic nerve control of these contractions, and this is true for peristaltic contractions. Abolition of extrinsic and intrinsic nerve activity by cocaine also abolishes peristalsis but not segmentation. Segmentation is initiated by the intestinal smooth muscle itself, which may respond to locally produced mechanical and chemical stimuli. One of these chemicals is serotonin (5HT), which is produced in the gut and which stimulates motility. It is probably a neurotransmitter between sensorimotor neurons in the peristaltic reflex.

Peristalsis involves contractions of longitudinal muscle first, followed by contraction of circular muscle; the latter is normally excited by the intrinsic nerve plexuses, which in turn are influenced by extrinsic nerves. Distention of the gut evokes the *peristaltic reflex*.

To summarize, the contractions of the small intestine are regulated and controlled by: (1) extrinsic and intrinsic nerve activity, (2) activity of smooth muscle alone, and (3) local chemical and mechanical factors.

Ileocecal Sphincter

Usually within $3^1/_2$ hours after evacuation of the human stomach, the contents arrive at the terminus of the ileum, where the valve or sphincter is located which by opening and closing allows or prevents the passage of ingesta into the colon.

The valve relaxes when a peristaltic wave reaches the lower ileum and allows ingesta to move into the colon, after which the valve closes.

One of the factors in closing the valve is the *myenteric reflex from the cecum*. Mechanical stimulation or distention of cecal mucosa causes the valve to contract even after extrinsic nerve abolition. The *gastroileal reflex* is induced and the valve relaxes when food enters the stomach, and gastric secretion and emptying occur.

Motility of Large Intestine

The material discharged from the ileum to colon (about 1500 ml daily) is undigested and resembles watery feces because at this stage some water and electrolytes remain to be absorbed in the colon. About 1.3 liters of water containing electrolytes are absorbed by the colon daily, a relatively small amount but enough to cause the formation of firm feces.

In man the colon is inactive much of the time, but infrequently there are strong contractions forcing material from the upper or proximal colon toward the distal colon. After food leaves the stomach, it requires six to eight hours for it to reach the distal colon, where it is held for varying times before defecation.

The ingesta is propelled within the colon by a combination of three types of movements or contractions, including: segmentation and propulsion, multihaustral propulsion, and peristalsis.

Haustral contraction is the extreme type of segmentation whereby the colonic mucosa is folded into sacs, or haustra (Fig. 20.6).

Nervous control of the colon is similar to the rest of the intestinal tract except that the vagus nerve innervates only part of the colon (probably the upper one-third of transverse colon). The lower colon is innervated by the parasympathetic pelvic nerve (Fig. 20.3). Both myenteric and submucosal plexuses (intrinsic nerves) are in the colon.

Defecation

Defecation is both voluntary and involuntary, and the final act is controlled by the internal and external anal sphincters. The internal sphincter is composed of smooth muscle. It is innervated by hypogastric sympathetic fibers and is under involuntary control and parasympathetic fibers (voluntary) in the pelvic nerve. Stimulation of sympathetic fibers results in contraction (closing) of the valve. Activation of parasympathetic fibers from the pelvic nerve causes relaxation (opening).

The external anal sphincter is made up of skeletal muscle and is innervated by motor fibers from the pudendal nerve (somatic nerve). During periods of no activity the internal and external anal sphincters are tonically active and contracted or closed by impulses from sympathetics and pudendal nerve, respectively. The external sphincter can be contracted or relaxed by voluntary control.

Distension of the rectum initiates reflex contraction of the colon and relaxation of the internal anal sphincters and the desire to defecate. The reflex is mediated by the pelvic parasympathetic fibers, which causes the internal anal sphincter to relax and open and to decrease the impulses in the pudendal nerve that cause contraction of the external anal sphincter. The *defecation reflex* is inhibited by pain or fear due to sympathetic nerve discharge.

Distention of the stomach by overeating may initiate contractions of the rectum and a desire to defecate (*gastrocolic reflex*). This is particularly true in children, in whom defecation after meals occurs frequently.

Constipation and Diarrhea

Many people defecate irregularly, once in two or three days or less often, without untoward symptoms. There may be mild abdominal discomfort and distension, but there is no autointoxication attributable to absorption of toxic substances.

The principal cause of constipation is irregular bowel habits and suppression or inhibition of the normal defecation reflexes. This may lead to weakened reflexes and finally to loss of colonic tone and motility.

Diarrhea occurs when the ability of the small intestine and the colon to absorb water is impaired. If the defect in absorption capacity is mainly in the small intestine, the large volumes of fluid entering the colon overwhelm its absorptive capacity. If the contents of the ileum contain some poorly absorbable substance like magnesium sulfate, this

prevents normal absorption of water in the colon. Inflammation of the small intestine caused by toxins and bacteria impairs absorption of water and produces diarrhea.
Relevant articles and monographs on the topics discussed in this chapter will be found in the "Selected Readings" section at the end of Chap. 21.

Review Questions

1. In a diagram, name and locate the principal parts of the alimentary canal.
2. What is a portal blood supply to an organ like the liver? Explain.
3. What are the feeding and satiety centers?
4. Define and describe peristalsis. Where does it occur?
5. What are the functions of the chief and parietal cells of the stomach?
6. Name factors that inhibit and that induce gastric motility. What hormones inhibit gastric emptying?
7. Name the parts of the intestines (small and large). What are their sizes?
8. What is the name of the main absorptive surface area of the small intestine?
9. What types of waves or contractions are found in the small intestine and what functions do they perform?
10. Peristaltic waves are controlled by intrinsic nerves (true or false) and involve contraction of longitudinal smooth muscle first (true of false).
11. What is the myenteric reflex?
12. What is the gastroileal reflex?
13. What types of movements (waves) are observed in the large intestine?
14. What is the defecation reflex?

Chapter 21

Secretion, Digestion, Absorption

Salivary Glands

The average human adult secretes from 1-2 liters of saliva daily, which helps keep the mouth wet, aids speech, and lubricates food for easier swallowing. Dry mouth results when saliva is scanty. The enzyme in saliva, ptyalin, is secreted in the salivary gland as zymogen granules (Fig. 21.1).

There are three types of salivary glands: (1) parotid, (2) submaxillary, and (3) sublingual. Characteristics of these glands are summarized in Table 21.1. Saliva also contains concentrations of Na^+, K^+, Cl^- and HCO_3^- that are usually higher than those in blood plasma; however, the concentrations may vary considerably, depending on the secretory state of the glands. The pH of saliva approaches 7.0. Saliva is produced in response to impulses arising in the salivary centers in the medulla oblongata. These impulses are initiated by stimulation of receptors in the mouth, nose, eyes, and parts of the gastrointestinal tract (GIT). The response to such impulses reaching the salivary centers is sent to the salivary glands by their sympathetic and parasympathetic innervation. Stimulation of the parasympathetic nerves (VII and

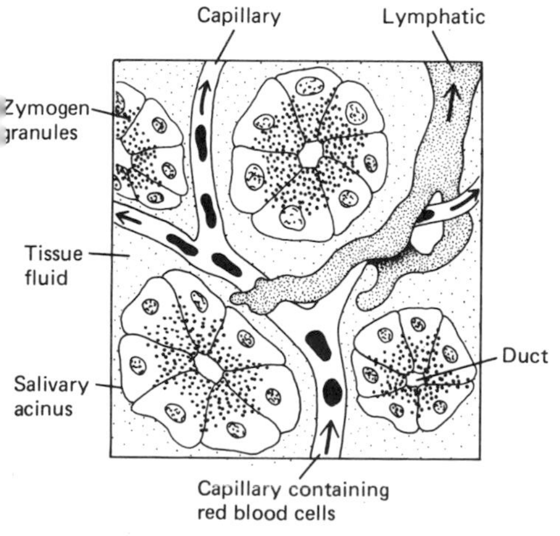

Fig. 21.1. Serous salivary gland, with ducts, blood vessels and lymphatics.

Table 21.1. Salivary glands.

Gland	Nerve Supply (parasympathetic)	Type of Secretion	Total Secretion (%)
Parotid	Glossopharyngeal (IX)	Serous	25
Submaxillary	Facial (VII)	Serous and mucous	70
Sublingual	Facial (VII)	Mucous	5

IX) results in increased serous (watery) secretion and vasodilatation. These cholinergic nerves release acetylcholine, and their effect can be altered by atropine, which blocks the action of the secretion and causes dry mouth.
Sympathetic nerves reach the salivary glands by the superior cervical ganglia. Their stimulation results in a decreased secretion containing much mucus, but the exact mechanism by which this is accomplished is not clear.
Sight of, smell of, and taste of food evoke *reflex secretion* of saliva. Secretion in response to food in the mouth is an unconditioned reflex. Sight and smell of food or sounds associated with food cause the mouth to water and are conditioned reflexes that are strong in dogs, weaker in human subjects. Figure 21.1 shows the typical salivary-gland cells which secrete zymogen granules, and fluid which reaches the salivary ducts and capillaries.

Gastric Secretion

The stomach secretes HCl, pepsinogen granules, water, hormones, mucin, and other substances. Specialized gastric mucosal cells in mammals—parietal or oxyntic (acid-forming) cells—secrete the HCl or (H^+), and the smaller chief cells secrete pepsinogen granules. In certain non-mammalian species (Aves) there are no parietal cells—only chief cells, which secrete acid and pepsinogen granules. Other cells of the mucosa secrete serous and mucus secretions (Fig. 20.7). Secretion is regulated by neurohumoral or hormonal factors, gastrin, histamine, and other substances.
The acid concentration of gastric juice is determined by the rate of secretion and by the buffering or diluting action of food and other substances.
Gastric mucosal area (surface) in man is about 800 cm^2 and is divided into zones, according to the type of glands distributed therein. There are (1) cardiac glands and (2) oxyntic and pyloric glands. Cardiac glands secrete mainly mucus and some electrolytes and little or no HCl or pepsinogen. The *oxyntic area,* made up of the fundus and body of the stomach, represents about 75% of the total gastric mucosal area. Oxyntic glands have three main types of secretory cells, including: (1) oxyntic or parietal, (2) mucous cells, and (3) chief cells. Pyloric glands contain cells that secrete mucus and alkaline juice.
The gastric mucosa secretes acid that is strong enough to digest itself but normally does not. The *gastric mucosal barrier* prevents rapid penetration of acid into the mucosal cells. Certain substances and drugs, such as aspirin (salycylic acid), bile salts, and strong concentrations of alcohol, tend to weaken this barrier by penetrating and destroying

the mucosal cells. The physiologic nature of the resistance of the gastric mucosa to acid penetration and damage is not known but probably resides in the mucosal cells themselves; however, the secretion and covering up of the mucosa with mucin may provide part of the protection.

Composition and Measurement

The gastric juice is composed of an acid component, HCl, and an alkaline one made up of pepsinogen and certain electrolytes, such as Na^+, K^+, and Cl^+. It also contains mucin, water, cells, and certain enzymes.

The (H^+) is a function of secretion rate; as secretion rate increases so does (H^+), but (Na^+) falls. The acid of gastric juice is expressed in milliequivalents per liter or per given time (hour). The maximum (H^+) is about 150 mEq/liter. Basal secretion refers to secretion after fasting, usually during four 15—minute collecting periods. Peak secretion represents the amount of juice collected in the first hour after administration of a secretory stimulus (food, histamine, or gastrin) or vagus nerve stimulation.
Human adults secrete 2-3 liters of gastric juice daily. Basal and peak secretion rates of acid are shown in Table 21.2. The natural stimulus for gastric and acid secretion is food. *Histamine, pentagastrin,* gastrin, insulin, and the vagus nerve also stimulate secretion. As food is seen, smelled, or tasted (cephalic phase), gastric secretion begins and continues for three-to-four hours after eating (gastric phase) (Table 21.2).

Table 21.2 Gastric acid (HCl) secretion in man (mEq/h).

Basal and peak (a) secretion rates		
	Peak Rate	
Basal Rate	After meals	After histamine
1.4	30	34.5

a. Before and after eating	
Hours	Concentration (mEq/h)
0	2.0
0.5	17.0
1.0	27.0
1.5	30.0
2.0	26.5
3.0	20.0
4.0	7.0

Adapted from Davenport HW (1977) Physiology of digestive tract, 4th edn. Yearbook Medical, Chicago.

Histamine and Gastrin

For many years it was believed that histamine produced locally increased acid secretion by stimulating the parietal cells, and that the effects of gastrin in increasing acid secretion resulted from the stimulating effects of histamine-released gastrin. More recent evidence suggests that gastrin stimulates the parietal cells and that histamine may not be the physiologic mediator. Gastrin is many times more potent (on a molar basis) than histamine in stimulating gastric secretion (HCl), but it is a weak stimulator of pepsin secretion.

However, the exact role of gastrin and histamine in gastric secretion is not clear, although evidence indicates that histamine receptors that secrete acid (H_2) can be blocked by pharmacologic agents. These H_2 antagonists block gastric secretion regardless of whether the secretion is induced by gastrin, histamine, or vagal nerve stimulation. Arguments continue as to whether there are separate receptors for each gastric acid stimulus.

Gastrin is produced by cells in the antrum of the stomach. It is a 17-amino acid peptide amide with a low molecular weight. Gastrin not only increases gastric secretion but also stimulates motility of stomach, esophagus, intestines, gallbladder, and uterus.

Regulation of Gastric Secretion

Humoral and neural mechanisms are involved in the regulation of gastric secretion of which there are three phases, including cephalic, gastric, and intestinal.

The *Cephalic phase* represents secretion caused by factors acting in the head, such as sight, taste, smell, and chewing. They are unconditioned reflexes, but conditioned reflexes may also be involved, such as the association of food with indifferent stimuli, or sound (Pavlov's reflex). The cephalic phase of secretion elicits gastric juice that is high in acid and pepsin. The cephalic phase, mediated solely by the vagus nerve, can be subdivided into stimulation by vagally released gastrin, vagal activation of oxyntic glands directly, and sensitization of oxyntic glands to vagal impulses by gastrin. Stimulation of the vagus nerve releases gastrin from the area of the antrum, because surgical removal of the latter abolishes or markedly reduces the secretion. The *gastric phase* occurs when food enters the stomach or makes contact with the gastric mucosa; it lasts for three-to-four hours. This occurs even in an isolated denervated stomach pouch and independently of the cephalic phase. Certain foods like proteins are efficient producers of acid secretion. The gastric phase is probably mediated by cholinergic reflexes, which may be initiated by mechanical (distention) or chemical stimulation of local receptors of: (1) the antrum wall area (where gastrin is produced) or (2) the oxyntic gland area (where acid is produced). This is illustrated in Fig. 21.2. It is postulated that acetylcholine (Ach) is the final stimulator of (1) gastrin release and (2) pepsin and acid release (H^+).

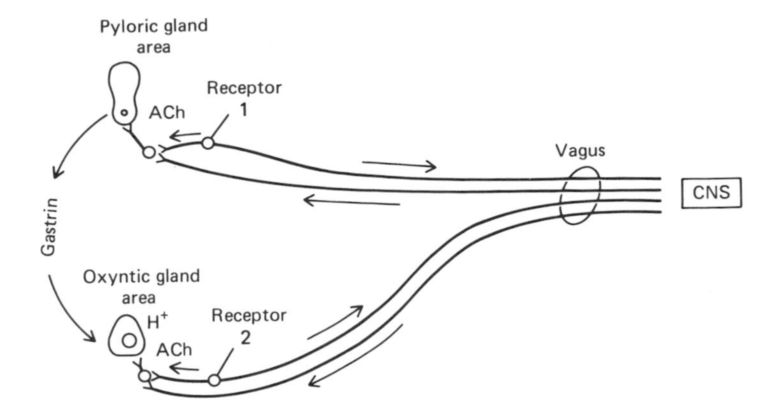

Fig. 21.2. Effects of nerve stimulation and release of gastrin on acid secretion. Stimulation of receptors locally (1) by intrinsic nerve stimulation or by vagal stimulation causes release of acetylcholine (Ach), which releases gastrin from cells of antrum area, which in turn stimulates acid secretion (H^+) in oxyntic gland area. Afferent (→) and efferent (←) pathways of vagus are indicated by arrows. Stimulation of receptor (2) locally or by vagus also releases Ach and causes oxyntic cells to secrete acid (H^+). Response of cephalic stimuli such as sight, smell, and taste can also be relayed over the efferent vagal pathways. (After Grossman, MI, Physiologist 6:349, 1963.)

In the *intestinal phase,* stimulation of intestinal mucosa by ingesta from stomach or from food introduced directly to an intestine separated surgically from the stomach stimulates gastric secretion. This indicates that the intestine releases, directly into the bloodstream, gastrin or some other hormone which when absorbed reaches the stomach and causes it to secrete gastric juice. Cholecystokinin, which is also released from the intestine, exhibits the same physiologic activity as gastrin and may be responsible for part of the intestinal phase of gastric secretion.

Inhibition of Gastric Secretion

Several substances inhibit gastric secretion when they come into contact with the duodenal mucosa, such as fats, acids, and products of protein digestion; acid in contact with antral mucosa inhibits or retards release of gastrin. Inhibition of appetite by emotional influences depresses or abolishes vagal reflex stimulation of the cephalic phase of secretion. *Antacids* (antacid preparations) vary considerably in their ability to neutralize acids. Aluminum phosphate, e.g., is relatively ineffective compared to a preparation containing aluminum hydroxide, calcium carbonate, and magnesium hydroxide.

Secretion of pepsin is stimulated by most factors that stimulate by acid secretion, namely, food, gastrin, histamine, secretin, and the vagus nerve. *Motilin,* a polypeptide containing 22 amino acids stimulates pepsin and acid secretion, and it is found in the mucosa of duodenum and upper jejunum. This putative hormone also stimulates the stomach to contract strongly and may be involved in gastric emptying.

Gastroinhibitory peptide is an inhibitor of gastric secretion and stimulates secretion of intestinal juice and pancreatic hormone. It is released mainly by ingested food, particularly fat and carbohydrates, and is found in the duodenum and jejunum. It was synthesized in 1975. *Secretion of intrinsic factor* by the gastric mucosa occurs in the parietal or oxyntic cells. This factor is necessary for the adsorption of vitamin B_{12} from the ileum.

Pancreatic and Intestinal Secretion

Pancreatic secretions includes those from endocrine and exocrine glands. The exocrine secretion is composed mainly of enzymes, water, and electrolytes. The exocrine portion of the pancreas is made up of compound alveolar glands resembling the salivary glands; its acinar cells likewise secrete zymogen granules that are discharged into ducts, which convey the material into the duodenal lumen (Fig. 21.3). Less is known about how the fluid juice is secreted except that the juice is also assumed to be secreted by the acinar cells.

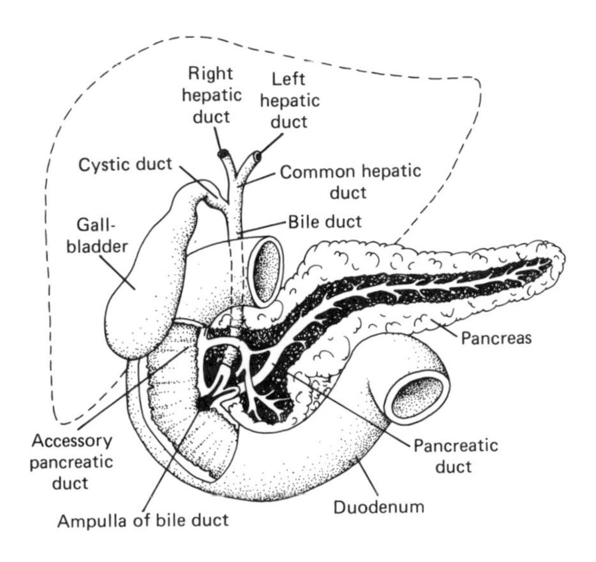

Fig. 21.3. Pancreas, pancreatic ducts, and hepatic and bile ducts and their point of entry into duodenum.

Composition and Regulation

Secretion of pancreatic juice is variable but averages about 2 liters daily in man. It has a pH of 7.6-8.2 (alkaline) and contains a high level of bicarbonates. The enzyme content of the juice is variable, depending on the stimulus and ranges from 0.1%—10%. These enzymes include trypsin, chymotrypsin, carboxypeptidases, (proteolytic enzymes), lipase (fat splitting), amylase (carbohydrate digestion), and others to be discussed under digestive enzymes. Two of them, trypsinogen and chymotrypsinogen, are inactive when secreted and become activated later.

Secretion is under neural and hormonal control, mainly the latter. Stimulation of the vagus nerve causes secretion of small amounts of pancreatic juice, rich in enzymes. The effect can be blocked by administration of atropine. The *gastropancreatic reflex* operates when the stomach is distended, by afferent and efferent fibers of the vagus nerve. Sympathetic stimulation results in a small volume of enzyme-rich fluid.

Hormonal Regulation

Secretin, a hormone produced by the intestinal mucosa, causes copious secretion of pancreatic juice that is low in enzyme content. First discovered in 1902 it was not synthesized until 1966; it contains 27 amino acids. The only effective stimulus for its release is the acidity of the upper intestinal mucosa.

Cholecystokinin, found in intestinal mucosa, has been isolated, purified, and synthesized. Although it has 33 amino acids, all of its activity is determined by the C-terminal eight acids; the last five acids are the same as those in gastrin, and indeed it has some of the same effects as gastrin. Its main effect is to stimulate the gallbladder, but it also induces secretion of enzymes from the pancreas.

Pancreatic polypeptides, first discovered in the avian pancreas, have been isolated and purified from several mammals including man and are all structurally similar. This hormone inhibits gallbladder contraction, gut motility, and pancreatic enzyme output but increases bile-duct tone and pancreatic secretion of water and HCO_3^-.

Secretion in the *small intestine* is governed mainly by the function of Brunner's glands and the glands of Lieberkühn. Brunner's glands of the duodenum secrete a thick alkaline mucus, which probably protects the duodenal mucosa from gastric acid. The glands of Lieberkühn secrete mucus and an isotonic fluid. Vagal stimulation increases the secretion of Brunner's glands.

Hepatic and Biliary Secretion

The liver is the largest single organ in the body and has many functions, including (1) formation of bile, (2) metabolism of many substances and nutrients absorbed from the in-

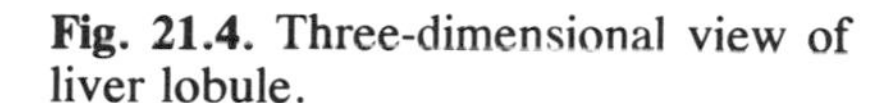

Fig. 21.4. Three-dimensional view of liver lobule.

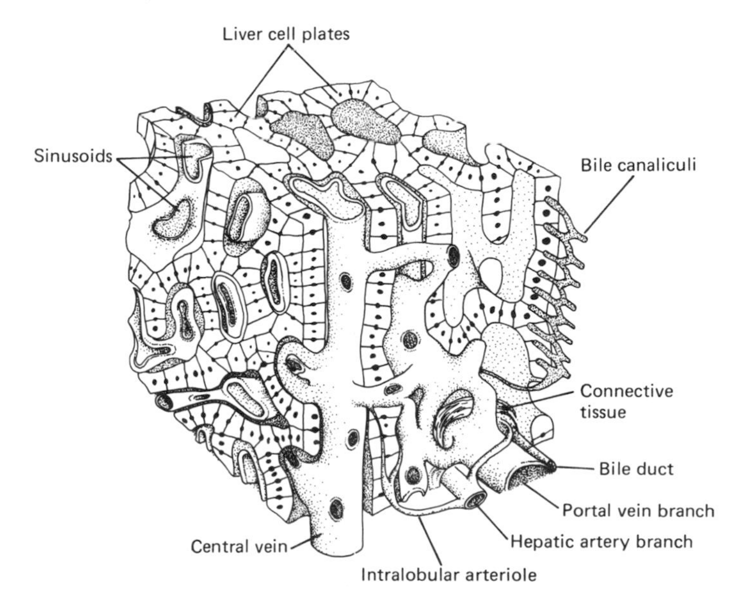

testinal tract, (3) synthesis and storage of certain compounds, and (4) breakdown and detoxification of drugs and other products (see also Chaps. 22 and 23).
The anatomy of the hepatic blood vessels and the location of hepatic and bile ducts are shown in Figs. 20.4, 21.3 and 21.5 and have been discussed previously. The liver is in the abdominal cavity (Fig. 20.1) and has a right and a left lobe (Figs. 21.3 and 21.4), a common bile duct, a cystic duct, and a gallbladder. Some species like the rat, pigeon and horse do not have gallbladders. The bile ducts and gallbladder release bile, which empties into the duodenum (Fig. 21.3).
The liver lobe is divided into lobules, the functional units of which are shown in Fig. 21.4. It has blood vessels and specialized liver cells (Kupffer's) arranged in columns. Between these cells are sinusoids receiving both arterial and afferent and efferent venous blood, and the latter two empty into central veins (efferent).

Composition of Bile

Liver bile contains about 97% water; the remaining constituents and their percentages are: bile salts (0.7), bile pigments (0.2), mineral salts (0.8), fatty acids (0.14), lecithin (0.02), and cholesterol (0.06). Its pH is 7.4. The bile salts are synthesized in the liver from cholesterol.
Gallbladder bile is more concentrated than liver bile, containing less water and more bile salts; this fact suggests that water and some mineral salts are absorbed in the gallbladder. Bile in the gallbladder is more acidic (pH 5-6). Bile acids are important constituents, playing a prominent role in digestion. The principal human bile acids and their percentages are: cholic acid (50), chenodeoxycholic acid (30), deoxycholic acid (15), and lithocholic acid. (5).

Bile Salts

The bile salts combine with fat particles in the intestine to form micelles from which the fats can more easily be transported and absorbed (hydrotropic effect). If bile is prevented from entering the intestine, at least 25% of the fat remains unabsorbed and appears in the feces. More than 90% of the bile salts are absorbed in the ileum (active transport), and the remaining ones pass out in the feces. The absorbed salts are carried to the liver and resecreted. This recycling of bile occurs six to eight times daily.
Bilirubin is a pigment formed from the breakdown of hemoglobin and most of it is bound to albumin in plasma. The free bilirubin goes back to the liver and is reused in synthesis. The color of bile is determined by the pigment: golden yellow if it contains bilirubin and green-to-black if the pigment is biliverdin.
Jaundice, a disorder characterized by yellowness in the skin and mucous membranes, is caused by the accumulation of free or conjugated bilirubin.

Gallbladder

In species with gallbladders, bile is stored and released from time to time; surgical removal of the gallbladder, however, does not impair health; bile, secreted constantly at a slow rate, passes directly to the intestine.

The gallbladder begins to contract soon after ingestion of food, and emptying occurs irregularly and incompletely during several minutes. As it contracts, pressure increases in the bile ducts and finally opens the sphincter in the duct from the gallblader (cystic duct), and bile flows or spurts into the duodenum.

This contraction is under neurohormonal and humoral control. The efferent neural pathways to the gallbladder and the sphincter are in the vagus nerve, and gall-bladder contraction is part of the cephalic phase of digestion and secretion. Afferent impulses arise from the duodenum and other organs and are mediated through the efferent vaga pathways.

Humoral control is mediated by substances in food, such as fat, which causes the release of a hormone from the intestinal mucosa (cholecystokinen); this in turn stimulates the gallbladder to contract. Gastrin also stimulates contraction of the gallbladder.

Gallstones are caused by the excessive accumulation of substances, mainly cholesterol, which settle out to form a ball or stone. This occurs when there is *stasis* (stoppage or diminution) of bile flow, or obstruction in the cystic duct so that water but not cholesterol is absorbed.

Digestion

Digestion involves physicochemical processes by which ingested food is converted into simpler constituents and nutrients that can be absorbed into the bloodstream. It begins in the mouth where saliva contributes the enzyme ptyalin, which plays a minor role in digestion of starch.

Digestion continues in the stomach where pepsin is formed and operates optimally at a pH of 2.0 on native proteins; these latter are split into proteoses and peptones (and more rarely into amino acids).

Pancreatic juice contains the enzymes trypsin, chymotrypsin, and carboxypeptidases (Table 21.3); fat-cleaving enzymes (lipases); and enzymes that act on carbohydrates (amylases) (Table 21.3). The proteolytic enzymes break down complex proteins to amino acids in which form they are absorbed. Lipase also converts compound lipids into absorbable simpler forms—glycerols and fatty acids. Edible carbohydrates are in the form of polysaccharides, or oligosaccharides, disaccharides, and the simpler monosaccharides (see also Chap. 23). Examples of compounds be-

Table 21.3. Enzyme digestion in gastrointestinal tract.

Organ	Principal Digestive Enzyme	Enzyme Substrate	Principal Product(s) of Enzyme
Mouth (saliva)	Ptyalin (amylase)	Starch (not important)	Maltose
Stomach	Pepsin, pH 2-3	Proteins	Proteoses, peptones
Duodenum, pancreatic juice	Trypsin, pH 6-8	Proteins, polypeptides	Amino acids and polypeptides
Duodenum, pancreatic juice	Chymotrypsin, pH 6-8	Proteins, polypeptides	Amino acids and polypeptides
Duodenum, pancreatic juice	Carboxypeptidase, pH 6-8	COO-end of peptides	Amino acids
Duodenum, pancreatic juice	Lipase, pH 8.0	Fats	Fatty acids, glycerol, monoglycerides and diglycerides
Duodenum, pancreatic juice	Amylase	Starch, glycogen hydrolyzes 1, 4 alpha linkages	Maltose, dextrins, and glucose
Mucosa	Ribonuclease, deoxyribonuclease	Ribonucleic and deoxyribonucleic acid	Nucleotides

longing to the poly-, di-, and monosaccharides are starch and glycogen, maltose, and glucose, respectively.

The precursor of *trypsin* is trypsinogen, which is converted to active trypsin in the intestine by enterokinase, an enzyme in intestinal juice. Trypsin digests denatured and partially digested proteins coming from the stomach. Its end products are amino acids and polypeptides. A trypsin inhibitor produced by the pancreas prevents trypsin in large amounts from digesting or injuring the intestinal mucosa.

Chymotrypsin acts on proteins and polypeptides (as does trypsin) and is preformed as chymotrypsinogen, which is converted by trypsin or enterokinase in the intestine to its definitive form. Both of the proteolytic enzymes split linkages in the interior of the molecules (endopeptidases).

Carboxypeptidase splits amino acids from peptides with free carboxyl (COO^-) groups. *Lipase* works on fats to form free fatty acids, glycerol, and glycerides. Liver bile, which is added to the duodenal contents, activates lipase. *Amylase* degrades starch and glycogen to maltose, dextrins, and glucose.

The *intestinal mucosa* produces enzymes with action similar to those of the pancreas, but their role is minor compared to the latter. There are, however, certain specific amylases, such as lactase (works on lactose), sucrase (works on sucrose), dextrinase, and disaccharidase.

Absorption

Substances are moved across the mucosa of the GIT and absorbed in the bloodstream by several mechanisms, including: (1) active transport, (2) diffusion, and (3) pinocytosis (see also Chap. 2).

Carbohydrates

Complex polysaccharides and some disaccharides are ultimately broken down in the intestinal tract into simpler forms, namely, the monosaccharides, which are readily absorbed. Some disaccharides are absorbed unchanged.
Glucose and other hexoses (6-carbon sugars) and pentoses (5-carbon sugars) are readily absorbed from the duodenal and ileal mucosas into the blood capillaries; these latter drain into the hepatic portal vein carrying blood to the liver. Others, together with some pentoses may be absorbed by diffusion; but glucose and galactose, are transported across the mucosa by an active process called carrier-mediated transport (see Chap. 2). The carrier is Na^+, or glucose is carried in combination with Na^+.

Lipids

The more complex lipids (see previous section) are finally broken down into fatty acids, glycerol, and monoglycerides and absorbed with little or no energy expended.
Absorption of fatty acids depends on the size of the compound (number of carbon atoms). Those containing less than 12 carbons are transferred directly across the mucosal cells to the portal blood. Those fats with more than 12 carbons are transported or absorbed in the lymphatics. Actually, most of this fat is carried as triglycerides; in the intestines, however, triglycerides are broken down into monoglycerides, absorbed but then reformed (resynthesized) into triglycerides by the mucosal cells and then enter the lymphatics.
Fat is thought to be absorbed as a finely dispersed emulsion and as micelles, an aggregation of molecules containing fat and bile salts. Absorption of fat is greatest in the duodenum, but some of it is absorbed in the ileum.

Proteins

The digestion products of proteins are free amino acids and some di- and tripeptides containing amino acid residues. Most of the proteins are absorbed as free amino acids by different active carrier-mediated processes involving Na^+. Certain amino acids (L-acids) are absorbed more rapidly than D-acids, and the rate of absorption is more rapid in the duodenum and jejunum than in the ileum.

Water and Electrolytes

Approximately 8 liters of water daily are absorbed in the small intestine, about 1.3 liters in the colon. This water is absorbed mainly by osmosis and most rapidly from hypotonic solutions. The osmotically active particles of digestion are absorbed in the intestine; water also moves passively out (absorbed), together with the osmotically active particles. Electrolytes are absorbed more readily from duodenum and jejunum than from ileum. Monovalent ions like Na^+, K^+, Cl^-, and HCO_3^- are absorbed more rapidly than polyvalent ones, such as Ca^{2+} and Mg^{2+}. Absorption of Na^+, Ca^{2+}, and Fe^{2+} is by an active mechanism. Iron must be in the ferrous state (Fe^{2+}) to be absorbed, and its absorption is increased when the body stores of iron are low and when RBC formation is increased

Selected Readings

Code CF (1968) (ed) Handbook of physiology: alimentary canal, Vol. V. American Physiological Society, Washington

Crane RK (1977) (ed). Gastrointestinal Physiology, 1977 International Review Physiology, Vol. 12, Chaps. 2, 3, 5, 6. University Park Press, Baltimore

Davenport HW (1977) Physiology of digestive tract, 4th edn. Year Book Medical, Chicago

Jacobson ED, Shanbour WL (1974) (eds) Gastrointestinal physiology, Vol. 4, Chaps. 1, 4, 5–10. International Review of Physiology, University Park Press, Baltimore

Ganong WF (1977) Review of medical physiology, 8th edn. Lange Medical, Los Altos, California

Knoebil LK (1977) In: Selkurt EE (ed) Physiology, 4th edn. Little, Brown, Boston

Review Questions

1. Name and locate the salivary glands. Which gland secretes the most juice?
2. What is meant by gastric mucosal barrier?
3. How much gastric juice is secreted daily in man?
4. What are the phases of gastric secretion? Name and describe them.
5. Name three stimulants of gastric secretion.
6. Locate the pancreas and describe the function of the exocrine-acinar cells.
7. Name the types of enzymes produced by the exocrine pancreas, as to the type of substrates on which they act.
8. What hormone causes exocrine pancreatic secretion?
9. What nerve most affects exocrine pancreatic secretion?
10. What are the main functions of the liver?
11. Where is bile produced?
12. What is the function of bile salts?
13. What causes gallstones?
14. Define the following and give the substance on which they act: (a) carboxypeptidases, (b) trypsin, and (c) proteins.
15. At what pH is digestion in the small intestine optimum?
16. What is enterokinase and its action?
17. How are the following absorbed: (a) carbohydrates, (b) lipids, and (c) proteins.
18. What is carrier-mediated absorption? Give an example.

Chapter 22 Intermediary Metabolism

By definition, "intermediary" means "between the most complex (structure) and the least complex", whereas "metabolism" is a term derived from Greek meaning "a change or alteration." In terms of the nutrient needs of vertebrate organisms, then, the building blocks of the three major energy substrates may be arranged (synthesized, anabolized) into larger, more complex structures within the body. Also, very large (complex) organic compounds may be broken down (catabolized) into fundamental, basic building blocks. These basic units of body materials are monosaccharides (for the complex sugars or carbohydrates), amino acids (for the complex polypeptides and proteins), and glycerol and free fatty acids (for the lipid or fat moiety). Thus intermediary metabolism represents the sum total of chemical events (usually catalyzed by enzymatic proteins) occurring simultaneously in both anabolic and catabolic processes.

Anabolism and Catabolism

Anabolism (synthesis), but not catabolism, is usually regarded as being species specific. Each organism may, on the one hand, take the same basic components, amino acids, e.g., and with them build different complex structures, such as polypeptides or proteins. The finished product would be specific for that organism, the tissue machinery involved within that organism, and the relevant enzymatically driven reactions specifically concerned with such chemical interactions.

On the other hand, each organism may take the same polypeptide or protein (in this example) and break it down. However, no matter what type organism does it or how the catabolic procedure is carried out, the identical amino acids result. They were put together differently, but when disassembled the same building blocks are found. Thus catabolism is nonspecific.

Since anabolism and catabolism are governed by chemical reactions within the body, and because the reactions are usually regulated by enzymatic catalysts (some of which are rate-limiting), the simultaneous activity of each process may lead to a dynamic equilibrium, growth, or wasting of body structure. When anabolic processes result in the same amount of tissue accumulation (or energy production) as catabolic processes degrade (or release), a dynamic equilibrium exists with neither net growth nor net body tissue loss experienced. We call this *homeostasis,* i.e., the organism represents a homeostatic norm (Fig. 22.1) If, however, anabolic processes exceed simultaneously occurring catabolic processes, a net accumulation of tissue matter occurs. True growth results. Finally, if the degrading or fragmenting (catabolic) reactions "overwhelm" simultaneously occurring synthetic processes, energy is released and structural tissue will be broken down. Catabolism thus exceeds anabolism and as a result body structure and content diminish; the organism loses true structural mass (Fig. 22.1). Usually in adult organisms the anabolic reactions are balanced by the fragmentation or catabolic events (equilibrium). See Chap. 2 for additional discussion of cellular structure and function.

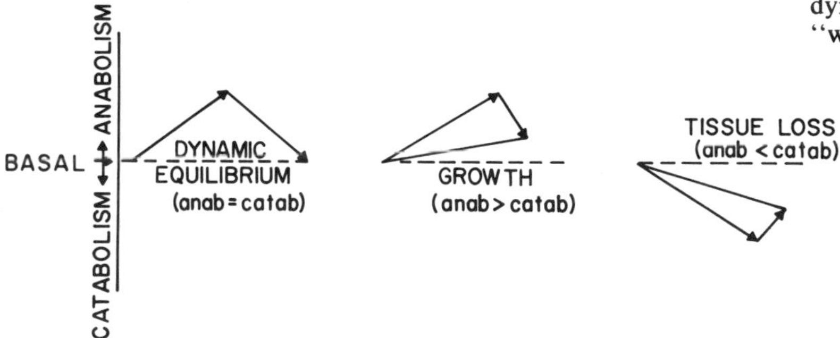

Fig. 22.1. Interrelationships of anabolism and catabolism with respect to dynamic equilibrium, growth, and "wasting" processes.

Source of Energy Production

Energy metabolism of cells (production, transfer, and transformation) occurs mainly in mitochondria. The cytoplasm of animal cells—the "liquid" portion of the cell—serves as a solvent for nutrients and their precursors. The high protein

concentration of the cytoplasm also represents a "bank" of many enzymes playing a critical role in the release of energy stored within the structure of nutrients per se. The precursors are those substances of very small molecular weight (usually less than 50 daltons each) taken in from the environment for synthesizing certain larger (M.W. 50–300 daltons) intermediates; the latter then furnish the basic building blocks (M.W. 200–400 daltons), such as amino and fatty acids, glycerol, and monosaccharides. The initial precursors for this sequence of steps include H_2O, CO_2, NH_3 (nitrogen), and of course these same substances are the endproducts of total catabolism of the aforementioned nutrients.
One other product of catabolism is the energy (stored as potential energy in the bond structure of each nutrient) released either as heat (kinetic energy associated with molecular motion) or as energy transferable to existing substances. The latter incorporate this energy into bond structures of biomolecules. Such "trapped" energy is therefore available for future use (potential energy) and itself may be transformed into kinetic energy, such as in energetics of muscle contraction and locomotion (see Chap. 11). Energy dissipated as heat contributes much to maintain both optimum temperature for enzymatic action and overall body temperature of an organism (see Chap. 18).

Enzymatic Action

The word "enzyme" is derived from the Greek and literally means "in yeast" or "leaven." This derivation is justified because recognition of the very existence of enzymes came from Louis Pasteur's celebrated studies on the role that yeast cells played in the grape fermentation process.
Enzymes are proteins acting like organic catalysts. This was first demonstrated by F.W. Ostwald in 1893. Enzymes then are regulators of chemical reaction rates; usually they speed up reactions even without a change in temperature. They operate best at a *specific pH,* at *certain temperatures,* and only with *specific substrates* (optimum conditions).
Further, like nonproteinaceous catalysts, enzymes complex or interact with the substrate but do not enter into the reaction. Rather they are usually released from the enzyme-substrate complex intact, unharmed, and available to regulate another reaction rate involving the same substrate. Thus:

$$\text{Enzyme} + \text{Substrate} \longrightarrow \text{Enzyme-Substrate Complex} \longrightarrow \text{Product} + \text{Enzyme}$$

$$(E) \quad + \quad (S) \quad \longrightarrow \quad (ES) \quad \longrightarrow \quad (P) \quad + \quad (E)$$

The molecular activity of an enzyme is defined as the number of substrate molecules that can react with a single enzyme molecule per unit of time, usually one minute. Enzymes are specific in the reactions which they regulate because they must interact physically (make contact) with the substrate, much like a lock-and-key relationship (see Chap. 25).

Since enzymes are not destroyed by the reaction that they regulate, their viable existence may extend for weeks, months, or even years. It also follows that because there are no by-products in enzyme-controlled reactions, the product yield is always 100%. The simultaneous operation of hundreds of enzymes (both anabolic and catabolic) results in the diversity of highly specific end products made available to the organism (Fig. 22.2). Enzymes therefore accomplish in seconds or fractions thereof what may otherwise take days or weeks at the biochemist's bench.

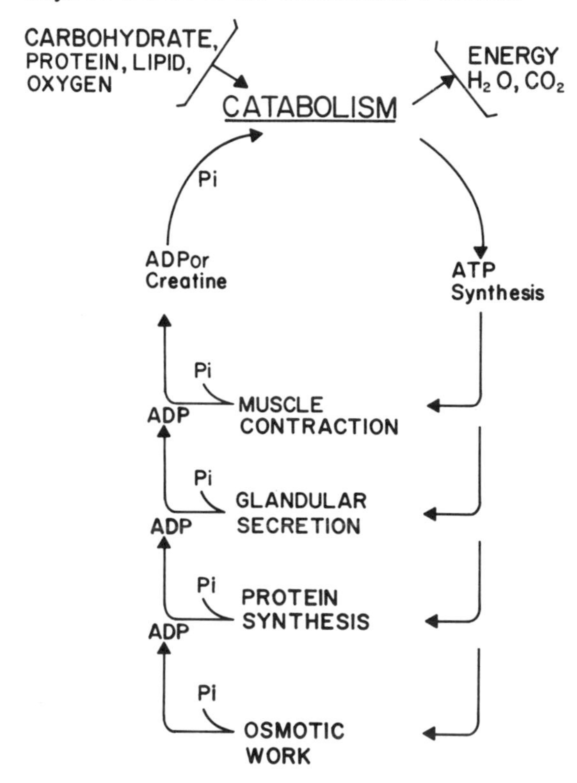

Fig. 22.2. The role of catabolic actions in providing energy for cellular work. P_i = inorganic phosphorus, ATP = adenosine triphosphate, and ADP = adenosine diphosphate.

Enzymatic activity of effectiveness may be altered within an organism by substrate molecules, inhibitors, and hormones. Because the protein enzyme is a fairly large molecule relative to the substrate molecules, the restricted area on the catalyst where physical contact occurs to form the ES complex represents a very small portion of the surface area. This area is the "active site," and it is only at this region that (ES) complexes may be formed. It is obvious that there is a superabundance of substrate, each active site becomes engaged and enzyme activity is maximum. Excess substrate molecules would not be able to interact with enzyme sites until some of the latter are freed of (ES) by elaboration of more product (P). Thus the maximum rate of enzyme-catalyzed reaction occurs when all active sites are occupied. Excess substrate cannot push the reaction further or faster.

Inhibitors

Enzyme kinetics may also be altered by inhibitors. Competitive inhibition is one of the more important forms of inhibition in biologic systems. It is possible that the configuration of the active site on an enzyme is complemented by the configuration on two or more other molecules. The substrate molecule competes with the nonsubstrate substance for contact at the active site because both have the necessary geometry to fit "their keys into the enzyme's locks." Naturally as the nonsubstrate (competitive inhibitor) concentration is increased, there is less statistical chance that the substrate will make favorable contact with the active site. Thus chemical reaction rates can be decreased by inhibitors, those substances that compete for active sites on the enzyme protein. Other forms of inhibition include noncompetitive, uncompetitive, and irreversible ones, and the reader is referred to the review literature at the end of this chapter for a more detailed discussion of the subject.

Energy

Technically speaking, energy is not and cannot be synthesized by the living cell. In accordance with the first law of thermodynamics (conservation of energy), energy can be neither destroyed nor synthesized; rather it is transferred or transformed within the organism. Chemical energy stored within the cell is potential energy, and it is analogous to that energy entrapped within a stretched rubber band. The stretching of the band required energy, which is released as heat or work (kinetic energy) or both when it shortens. Energy is also required to synthesize complex nutrients, or macromolecules, within a cell.

Energy is also released during catabolic degradation of a complex chemical as it is reduced to its intermediate forms, or finally to its fundamental forms of H_2O, CO_2, NH_3 and N_2. Fortunately, enzymatically controlled reactions release bond energy gradually; each bond is broken step-by-step until all energy is released. However, the stepwise release of molecular energy from complex substrates undergoing fragmentation allows coexistent molecules to "trap," or shelve, the energy, packet by packet, and thereby store this energy as high-energy bonds in other chemical forms. The deposition of high energy therefore may be stored within an existing compound without essential alteration of that substance; it may also be employed to form an entirely new substance containing potential high energy.

Anabolism requires energy—most frequently from rupture of high-energy phosphate bonds—to biosynthesize complex molecules from the basic building blocks to form potential energy-laden cellular components. *Catabolism* fragments energy-rich complex molecules, releasing packets of energy step-by-step so that up to 42% may be trapped or deposited in cellular compounds for future transformation (Fig. 22.2).

Potential and Kinetic

Energy "trapped" within a molecule merely by nature of bond structure is considered to be potential energy. This is equivalent to a ball resting at the top of a hill. Rupture of bond-structures releases potential energy (as in pushing the ball downhill) as molecular motion (heat) is transformed to produce work (as in muscle contraction), or is transferred as potential energy to a different family of intracellular chemical compounds.

Formation of Water

When $H_2 + {}^1/_2\, O_2 \rightarrow H_2O$, approximately 68 kcal of energy are released for each mole of H_2O formed (see also Chap. 18). Thus water contains 68,000 fewer calories than do the parent reactants within their initial bond structures. Also when the water molecule is fragmented to hydrogen and oxygen, energy is released. Most intracellular chemical reactions require small amounts of energy to initiate or more often to speed them up. Thus a mixture of oxygen and hydrogen at 20°C may require minutes to form reasonable amounts of water. If the oxygen and hydrogen are heated, however, the molecules join quickly to form water and to release the previously mentioned 68 kcal/mole. The addition of energy to facilitate the synthetic process is called "energy of activation." Basically, the heat speeds up the collision of molecules, making the combination less random and likelier.

High-Energy Phosphate Molecules

The compound adenosine triphosphate (ATP) is composed of an adenine base, a five-C sugar (pentose, ribose), and three phosphate groups linked in series with the sugar (Fig. 22.3). Rupture of the terminal phosphate bond (O—P) releases considerably more energy (7 kcal/mole ATP) than rupture of the usual low-energy bonds of C—O, H—O, or H—H. The 7 kcal/mole released is available immediately for heating the cellular environment, transformation into work or energy, or transfer to recipient molecules (such as ADP or creatine) for subsequent use. Thus ATP, a reservoir of high energy, is preferentially employed in regularly active tissues, such as skeletal muscle. The transfer of energy from ATP to the sliding filaments of skeletal muscle is discussed in Chap. 11.

Since ATP is a depository for energy transferred from other bonds, it requires that at some time in the cell's history a

Fig. 22.3. Structural formulas of adenosine triphosphate, adenosine diphosphate, and cyclic adenosine 3′, 5′-monophosphate.

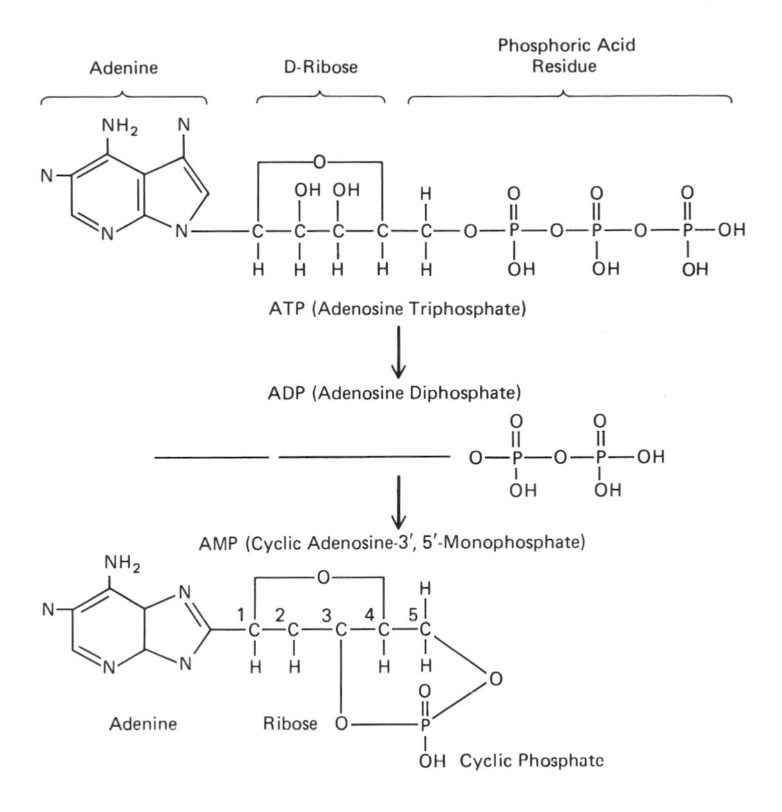

transfer of energy had to be made to a receptive molecule. This molecule then was altered to form ATP. The recipient of choice is adenosine diphosphate (ADP) and under certain conditions in muscle it also may be the creatine (C) molecule. In each case high energy is located in the terminal phosphate group, O—P in ATP and O—N in CP. Hence

$$ADP + P_i + Energy \longrightarrow ATP + H_2O$$

(Adenosine diphosphate) + (Inorganic phosphate + 7 kcal/mole) ⟶ (Adenosine triphosphate) + (Water)

or

$$C + P_i + Energy \longrightarrow CP + H_2O$$

(Creatine) + (Inorganic phosphate) + (10 kcal/mole) ⟶ (Phosphocreatine) + (Water)

and now,

$$CP + ADP \longrightarrow ATP + C$$

(Phosphocreatine) + (Adenosine diphosphate) ⟶ (Adenosine triphosphate) + (Creatine)

In muscle, concentration of CP is five times that of ATP, and creatine acts as a temporary reservoir for energy, transferring it rapidly to an awaiting ADP molecule. This transfer is facilitated by the enzyme creatine kinase, releasing free creatine to be reused in energy transfer. Other metabolic substances may be used as reservoirs in this energy transfer process; this is particularly so in invertebrates.

Energy is constantly being cycled within the cell, and ATP is the common factor supporting energy-required functions, such as active transport mechanisms (see Chap. 2), muscle contraction and locomotion (see Chap. 11), secretory release of hormones (see Chap. 25), and biosynthesis of complex structures like polypeptides and proteins. Although the total amount of energy stored in ATP is not great, the ubiquitous distribution of this substance, as well as the facility with which the energy may be transferred, underscores its importance in cell function. It should also be understood that catabolism of carbohydrate, lipid, and protein provides the energy for resynthesis of ATP. In this manner catabolism fuels the energy-trapping reservoirs, and the latter transfer energy to the common energy carrier, ATP (Fig. 22.2).

Anaerobic and Aerobic Catabolism

Most energy released from the catabolism of nutrients occurs within the mitochondria of cells in the presence of oxygen (aerobic reactions), which acts as a hydrogen acceptor (Fig. 22.2). Energy release, however, does occur during catabolism in the absence of oxygen, such as the initial breakdown of glucose in the cytoplasm of cells. Such reactions are called anaerobic ("without oxygen"). It should be emphasized that a given route between precursor and product (anabolism) and between product and precursor (catabolism) is not the same.

Anabolism is not simply a reversal of catabolic (path) reactions. In most cases these routes are distinct from each other because of differences in the enzyme specificity. *Anabolic* processes require ATP input (producing ADP and inorganic phosphate), whereas *catabolic* reactions release, in small packages, energy that aids the resynthesis of ATP from ADP and phosphate. Oxidative processes (aerobic) yield greater amounts of energy during nutrient catabolism than do anaerobic reactions of degradation.

Anaerobic degradation of nutrients may be best described in the initial steps of glucose catabolism, i.e., $C_6H_{12}O_6 + 6O_2 \rightarrow 6CO_2 + 6H_2O$ + energy. If one mole (180 g) of glucose were catabolized within an instant, the total energy release would be about 673 kcal. Such fragmentation would literally turn the involved cell into a furnace, and the energy evolved would be lost as heat. Instead, through anaerobic means—carefully regulated and slow fragmentation—energy may be released under precisely controlled conditions and transferred to acceptor nucleotides, such as ADP to resynthesize ATP. The explosive release of 673 kcal is thus avoided, even though the degradation of the glucose

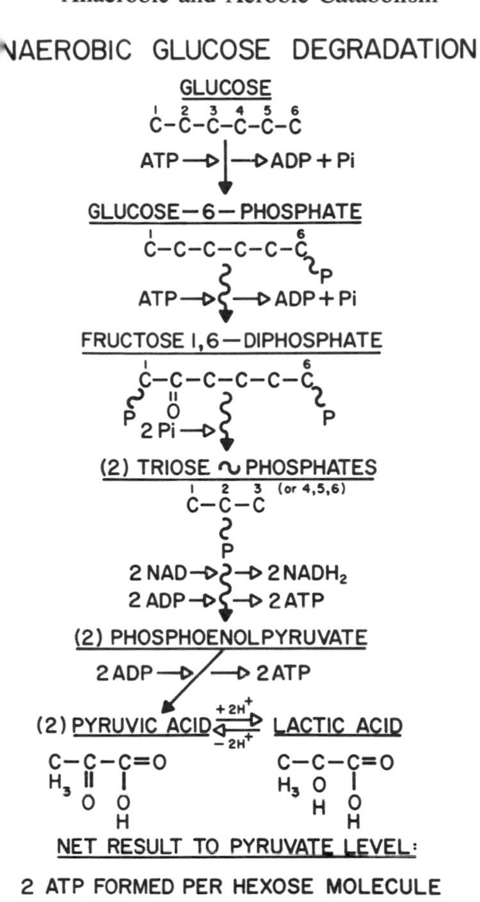

Fig. 22.4. Anaerobic catabolism of glucose. Squiggly lines indicate many reaction steps involved. Numbers on carbon atoms relate to original glucose molecule; ~P is high-energy phosphate.

continues by subsequent aerobic reactions, and eventually all 673 kcal/mole of sugar evolves.

Most commonly, anaerobic degradation occurs within the cytoplasm of animal cells because the enzymes regulating each of the catabolic reactions are restricted to the cytosol compartment.

Catabolism of the 6-C sugar molecule in the relative absence of oxygen results in lactic acid, a 3-C intermediate product (Fig. 22.4). During this transformation, which is normally called *anaerobic glycolysis,* two ATP molecules are required to "charge" the glucose molecule (glucose to glucose-6-phosphate and fructose-6-phosphate to fructose-1,6-diphosphate levels). However, just prior to the formation of pyruvate or lactate, or both, two ATP molecules are regenerated at each of two steps. Consequently the energy yield of anaerobic glycolysis is four ATP molecules less two ATP molecules required initially to prime the glucose molecule for a *net yield* of two ATP molecules per glucose molecule catabolized. Such a yield of energy is low and would not sustain mammalian or avian cellular activity for any length of time. However, it does sustain ATP requirements of fish, amphibia, and rèptiles during physical activity.

Fundamentally *anaerobic glycolysis* is merely a fermentation process, requiring the cooperative sequential action of 11 enzymes. All intermediates are phosphorylated intermediates, and the end product is lactate.

Aerobic degradation of nutrients appears to be the major source of ATP-recovered energy, i.e., energy released, trapped by an acceptor, and subsequently transferred to ADP to form ATP. Protein (from Greek "the first") and lipid (from Greek "fat") catabolisms are initiated under aerobic conditions within the mitochondria where oxidative enzymes are trapped. Some of the intermediates may "leak" to the cytosol but usually are taken back up by the mitochondria, and the final stages of aerobic degradation are carried out therein, with packagable amounts of energy being released at sequential steps (Fig. 22.5). In this way ATP is resynthesized from the energy released and itself may translocate to the cytosol. Owing to the complexity and diversity of protein structures, a definitive number cannot be assigned to the amount of ATP recovered from protein catabolism. However, oxidation of a moderate amount of lipid to CO_2 and H_2O in the mitochondria yields about 48–50 ATP/mole of substrate, whereas the aerobic degradation of lactate to CO_2 and H_2O yields about 30 ATPs. It will require $2^1/_2$ turns through the entire tricarboxylic acid (TCA) cycle to degrade the original hexose completely.

The net yield of ATP from glucose or glycogen (polymerized glucose) under anaerobic conditions is two and three ATPs respectively. Aerobic catabolism initiates both protein and lipid degradation and completes the catabolism of carbohydrate which was initiated by anaerobic glycolysis. Despite the diverse pathways involved in fragmentation of each nutrient type, the paths onverge on a common interme-

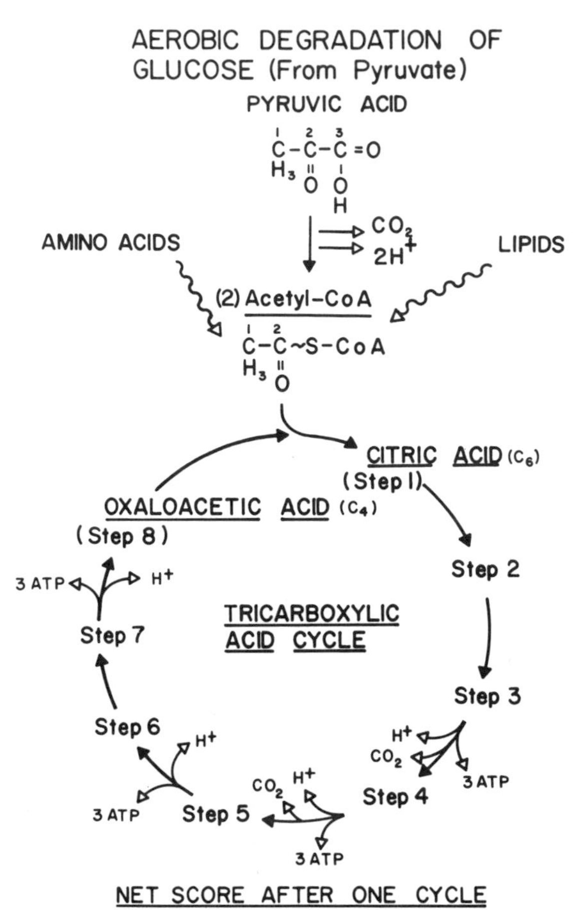

Fig. 22.5. Aerobic degradation of glucose form pyruvic acid to release of energy at ATP and CO_2. Note that ATP resynthesis is associated with hydrogen removal. (Also see Fig. 22.6).

diate form, *acetyl-CoA,* which leads to the final aerobic degradation of each in a common sequence of steps (Fig. 22.5). This common sequence is called the tricarboxylic acid, or Krebs cycle; it is herein that final catabolism of all three substrate types results in CO_2, H_2O, and energy (Figs. 22.5 and 22.6). Energy released in lipid and protein degradation occurs only at this final level of intermediary metabolism. Only carbohydrate degradation releases energy above the level of the TCA cycle, represented by a small amount of ATP recovered as a result of anaerobic reactions.

Metabolism of Carbohydrate, Lipid, and Protein

Intermediary metabolism involves simultaneous anabolism and catabolism of all three nutrient substrates or precursors by coordinated, well-controlled, and enzymically regulated chemical reactions. Carbohydrate catabolism is initiated anaerobically in the cytoplasm, but at the levels of glycerol-3-

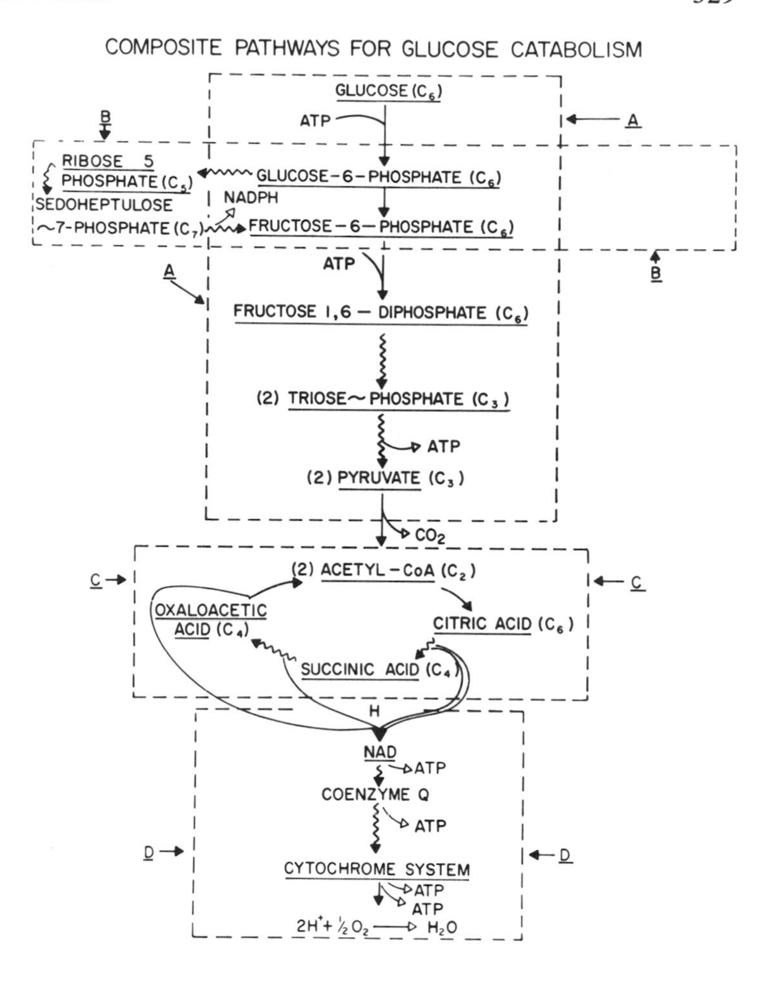

Fig. 22.6A–D. Interrelationships of anaerobic, aerobic, pentose shunt, and cytochrome system in degradation of carbohydrate to release energy, and H_2O. **A** anaerobic glycolytic path; **B** hexose monophosphate shunt; **C** areobic TCA cycle; and **D** oxidative phosphorylation route.

phosphate or pyruvate, degradation is completed in the mitochondria (Figs. 22.5 and 22.6).

Metabolic Pathways

Glycolysis occurs as already described and is shown in Fig. 22.6 by the box labeled A. Through both anaerobic (to pyruvate level) and aerobic mechanisms (from pyruvate to CO_2, H_2O, and E), the immediate energy needs of the cell are met by sugar catabolism. The orginal source of glucose is usually either nonphosphorylated blood glucose or glucose phosphorylated at the 1-C position, which was derived from carbohydrate depots, such as liver or muscle glycogen. In the relative absence of oxygen, lactic acid is formed and acts as a temporary storage of H^+. All these glycolytic reactions are carried out in the cytosol of animal cells and are enzymatically regulated.

Phosphogluconate Pathway

Also known as the hexose (or pentose) monophosphate shunt, this pathway is operative in most vertebrates and in many tissues (Fig. 22.6B). Although not the major pathway of glucose degradation, it affords an alternative path of glucose oxidation, one yielding CO_2 and pentoses. These 5-C sugars are required in the synthesis of nucleic acids (see "Nucleic Acids" in this chapter). Also, this shunt permits the conversion of pentoses into hexoses for subsequent oxidative degradation. Finally (and very important in tissues actively engaged in synthesizing fatty acids and steroid compounds) the *pentose shunt* produces the reduced form of NADP, which is essential for the synthesis already described. All of the critical enzymes associated with this alternative pathway of glucose catabolism are in the cytoplasm of most animal cells, the liver, mammary glands, adrenal glands, and gonads in particular.

The tricarboxylic cycle (Figs. 22.5 and 22.6C) is also known as the citric acid, or Krebs, cycle (after the Nobel laureate Hans Krebs, 1952). The nine enzymatically regulated steps that make up this cycle are those reactions which were already collectively discussed as part of the *aerobic path of glucose catabolism*. The entire process, with energy release as the end product, is carried out within the mitochondria. Actually the oxidative aspects are related to the stepwise removal of H^+ (Fig. 22.6D) from the intermediates.

Except for a pathway in the catabolism of certain amino acids, which bypasses the intermediate acetyl-CoA, the latter acts as a common intermediate for all nutrients, regardless of origin. In this manner the condensation of acetyl-CoA with oxaloacetate allows proteins (as deaminated amino acids), carbohydrates (as decarboxylated pyruvate), and lipids (as 2-C fragments of B-oxidation reactions) to enter that last sequence of controlled reactions leading to CO_2, H_2O, and energy production (Fig. 22.7).

Oxidative Phosphorylation

The role of oxygen in aerobic metabolism is largely linked to the role it plays in removing H^+ from the interior of the mitochondria. As indicated in Fig. 22.6C, CO_2 is removed at two steps and H_3O^+ at least from four steps during the degradation process of the TCA cycle. The hydrogen evolved from breakage of bond structures of the TCA intermediates interacts with the hydrogen acceptor, nicotinamide adenine dinucleotide (NAD) (Fig. 22.6D). As this happens, not only the hydrogen atom but also some of the potential energy from the original molecule is also transferred to NAD. This initial transfer occurs within the mitochondria, on the innermost of the two membranes; all subsequent reactions are carried out here also.

Small amounts of this potential energy are released as electrons from hydrogen atoms transported over a series of "respiratory proteins," the cytochrome system. In fact the electron transport from NADH to oxygen is the immediate energy source for oxidative phosphorylation. Energy is thus transported with electron transport from cytochrome to cytochrome, each one of which differs from the other. Such

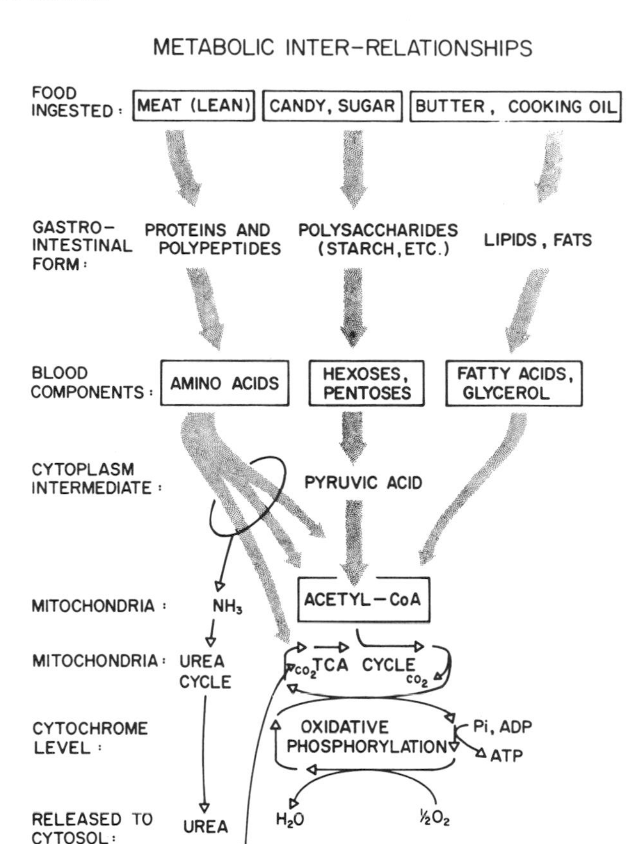

Fig. 22.7 Interrelated events in digestion, absorption, and metabolism of three types of nutrients. Note key role that acetyl-CoA plays; NH_3 is form of nitrogen excretion (as urea in urine).

electron transport not only leaves free NAD and H^+ behind but also releases the energy, step by step. This energy, released in small packets, is that which in the presence of free inorganic phosphate and ADP forms ATP (as already discussed). The H^+ are readily associated with available oxygen within the mitochondria and form H_2O (Fig. 22.6D).

In this way oxidative phosphorylation is responsible for most of the ATP resynthesis and represents a usual production of about *3 molecules of phosphate* transferred *to ADP per oxygen atom utilized*. In this manner roughly *30 ATP molecules* (net) are regenerated by the oxidation of *two pyruvate molecules* to CO_2 and H_2O. Six ATPs are derived from respiratory chain oxidation of NADH produced in anaerobic glycolysis and 2 ATPs (net) from oxidation at the substrate level (outside of the mitochondria) by conversion of glucose to pyruvate. The resulting total of *38 ATP* molecules is "salvaged" from the catabolism of one mole of glucose.

Each terminal high-energy bond of *ATP represents 7–7.5 kcal/mole* when ruptured, and the potential energy released is *266–285 kcal/mole glucose*. Since the intact glucose molecule contains 673 kcal/mole, a net resynthesis of *38 ATP* molecules indicates that the cell (mainly by oxidative phosphorylation in the mitochondria) is (266–285/673 × 100) 39%–42% efficient in recapturing energy from glucose molecules.

Intermediary Metabolism

Intermediary metabolism refers to the step-by-step metabolism of all three families of nutrients. Figure 22.7 shows the interrelationship of three fundamental types of nutrients. Acetyl-CoA plays an important role as a common step to all three paths of catabolism even though only anaerobic glycolysis generates ATP above this level. Most of the energy released and recaptured occurs with the intertwining events of the TCA and oxidative-phosphorylative cycles associated with electron transport. Acetyl-CoA is a necessary intermediate in all lipid—and some protein—synthesis but is not important for polysaccharide formation. The need for energy input (in the form of ATP with release of ADP and P_i) occurs in the initial stages when the precursors (NH_3, CO_2, and H_2O) are joined to form metabolic intermediates; subsequently the latter becomes the fundamental building blocks (amino acids, sugars, fatty acids, and glycerol). Unlike catabolism, the anabolic sequences for all three major pathways (Fig. 22.7) require energy between all levels of super organization above the acetyl-CoA level. It is important to reemphasize that anabolic and catabolic pathways are not merely reversals of the same pathway. Most reactions are independent of the reverse sequence (of precursor and product), and each is independently controlled to a fine degree. The result is coordinated catabolic and anabolic interaction, resulting in liberation of energy and heat, regardless of precursor or nutrient substrate. Usually all nutrients are metabolized simultaneously, emphasizing the need to release energy frequently in small packets, step-by-enzymatically controlled-step.

Nucleic Acids

The metabolism of the cell is largely governed by the activity of the nucleus where the enzymes regulating the controlled steps of catabolism and anabolism are synthesized. Of particular importance is the synthesis of proteins associated with structural elements of cells, or the secretory products of certain cell types. Liver cells, e.g., synthesize only proteins peculiar to the liver cell structure (membrane) or products of liver cell secretion, such as albumin, globulin, and prothrombin.

Differences exist, e.g., between the liver and adrenocortical cell with respect to those "forces" directing both enzyme synthesis and quality of protein end products. The *nucleus,* with its nucleic acid content, is largely responsible for directing protein (and therefore enzyme) synthesis at the level of the cytoplasmic *ribosome,* Without a nucleus, protein synthesis ceases almost immediately and can be restored only by replacement of an otherwise normal nucleus into the

Fig. 22.8. Nuclear direction of cytoplasmic mechanisms. Note "splitting" of DNA so that mRNA synthesis is complementary to one DNA strand only.

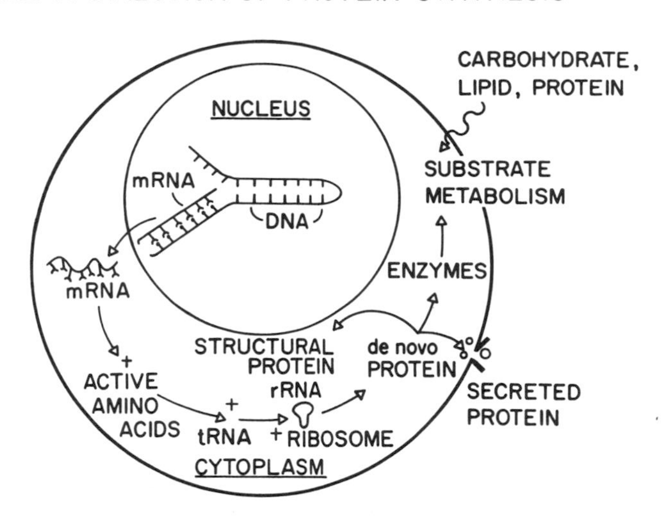

cell. Obviously there must be "contact" between nucleus and ribsome in the form of messenger molecules that carry the directive force from the nucleus to the active site of protein synthesis. This influence of the nucleus is shown in Fig. 22.8.

Deoxyribonucleic Acids

These large nuclear proteins contain coded (blueprint) information within their molecular structures (Fig. 22.9). It is this information that dictates what enzyme and structural protein are synthesized, what proteinaceous product is secreted, and in fact what hereditary trait (features, color, dimensions) is stored or transferred.

Deoxyribonucleic acid (DNA) is a molecule composed of repeating sequences of *5-C sugar* (*pentose*) lacking one oxygen atom (thus *de*oxy), one *molecule of phosphoric acid,* and a heterocyclic derivative of either a *purine* or a *pyrimidine*. The three subunits—pentose, phosphate, and nitrogenous base (purine or pyrimidine)—are called nucleotides and are repeated continually along the long chain of the DNA molecule. The junction of one nucleotide to another is by bond-linkages between the sugar and an oxygen of the phosphate group; DNAs are probably the largest proteins in animal cells and, although concentrated within the cell nucleus, are not restricted to that region. They are very stable and have a relatively long "life."

The characteristics (information) contained within DNA is based collectively upon (1) how long the DNA chain is, (2) the type of nitrogenous base incorporated within each nucleotide, and (3) the sequential position of each nucleotide. Each nucleotide in DNA may contain adenine or guanine (purines) or cytosine or thymine (pyrimidines). The long chain of each DNA molecule is actually made up of two strands (ribbons) intertwined at regular intervals; this is re-

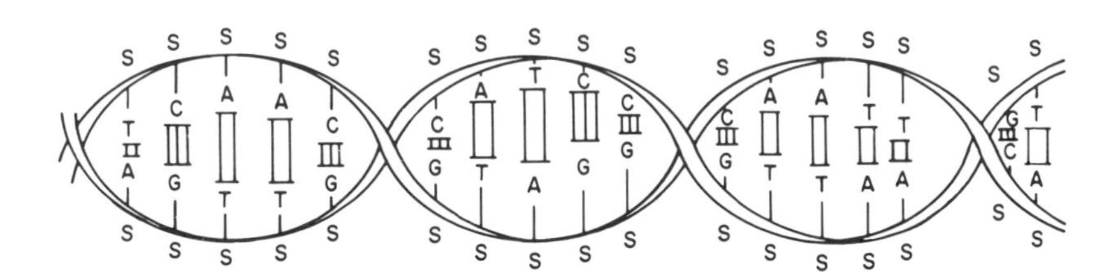

Fig. 22.9. Typical double helical strands of DNA molecule, including base sequences. Note that nitrogenous bases are bonded to the deoxyribose (sugar = S) moiety of DNA strand. Also note that A—T linkages are double hydrogen links and C—G linkages are triple links.

ferred to as a double helix (Fig. 22.9). The binding forces between the two strands are bond-interactions between the purines and the pyrimidines, which are embodied in each nucleotide subunit.

The concentration of purines *equals* that of the pyrimidines in any given DNA molecule; the amount of thymine *equals* that of adenine, and the amount of guanine *equals* that of cytosine. Thus the interaction of purine-pyrimidine between the two strands of DNA are thought to be thymine-adenine and guanine-cytosine, sometimes represented as T—A and G—C (Fig. 22.9).

Even though DNA was isolated as early as 1870, it was James Watson and Francis Crick who were awarded the Nobel Prize (1962) for their elucidation of the molecule's three-dimensional structure. Their work has been confirmed and extended to allow us to understand how genetic information is recorded in the structure of the DNA molecule and consequently how this information is translated to the ribosomes for protein synthesis.

Ribonucleic Acid

Several types of ribonucleic acid (RNA) exist in animal cells, each performing different functions in the process of getting the message from nuclear DNA to the ribosomes engaged in protein synthesis. Approximately 15% of all RNA is located in liver-cell mitochondria, about 25% is in the cytoplasm, 10% is in the nucleus, and 50% is in the ribosomes. The chemical composition of RNA is similar to that of DNA; it contains phosphoric acid, ribose, and any of four nitrogenous bases. However, RNA incorporates a ribose (not deoxyribose as in DNA) in its structure. Also, whereas it may have adenine, guanine, or cytosine in the nucleotide subunit, *uracil* (pyrimidine) may also occur (instead of thymine as in DNA). Another major difference between the two nucleic acids is that RNA molecules are single-stranded structures, tend to be shorter *than DNA chains,* and are therefore of lower molecular weight than are DNA molecules (2.8×10^4 to 1×10^6 daltons versus 1×10^6 to 5×10^9 daltons). Also, RNA is more prevalent in animal cells (2–6 times) than is DNA and appears to have a relatively short life, being catabolized and resynthesized rapidly.

Messenger RNA

This form of ribonucleic acid (mRNA) is synthesized mainly in the cell nucleus (a small amount may originate in the mitochondria) in a process whereby the sequence of the nucleotides of one strand of the chromosomal DNA molecule is transcribed to the single-strand mRNA molecule. Hence, it carries the code of the same base sequence but in complementary form to that of the participating DNA strand (Fig. 22.8). The mRNA then "escapes" by the nuclear pores to the cytoplasm and passes to the ribosomes where it acts as the template or blueprint for the proper amino acid sequence required to synthesize the protein involved. Less then 2% of all RNA is of the mRNA type.

Transfer RNA

This species of RNA (tRNA) was formerly called "soluble RNA" (sRNA) and represents about 15% of all RNA in animal cells. These are small RNA structures, containing less than 90 bases, and their action is highly specific. That is, each tRNA is regarded as a "carrier" of specific amino acids. The tRNAs act to transfer individual amino acids on ribosomes during protein biosynthesis by matching *their anticodons with the mRNA codons.* Since more than 50 tRNAs have been characterized, it is obvious that a given amino acid may have several tRNA associates. Yet each of the 20 amino acids occurring in protein is known to have at least one tRNA as its carrier (Fig. 22.10).

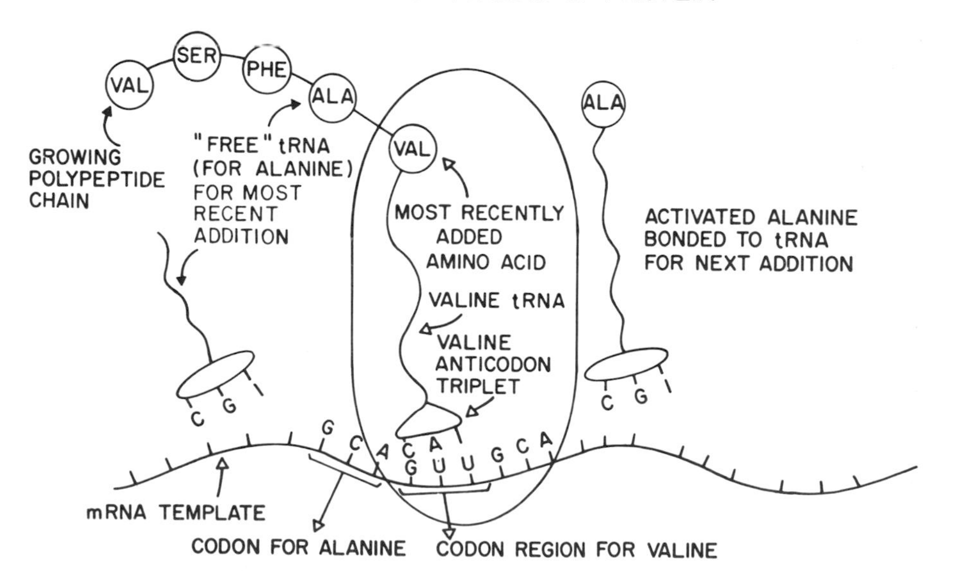

Fig. 22.10. Protein synthesis at ribosomal level. (See text). (A = adenine, C = cytosine, G = guanine, I = indeterminate, U = uracil).

Ribosomal RNA

Although this form of RNA (rRNA) comprises more than 75%–85% of all cellular RNA, its function remains obscure. At least four different forms of rRNA are known, all of which are closely associated with the ribosomal particles. They are probably intimately involved in the final stages of protein synthesis but exactly how is yet to be elucidated. They do not appear to carry coded information; possibly they act as intermediates between tRNA and mRNA so that individual amino acids are aligned in proper sequence as dictated by the "messenger."

Protein Synthesis

In *protein biosynthesis* from amino acids, ATP directly contributes much of the required energy; such is needed to activate each amino acid preparatory to its incorporation in the growing peptide chain at the ribosome. In nucleic acid biosynthesis from nucleotides, much of the energy comes from the potential energy of ATP as well as from UTP, GTP, and CTP. These latter are the high-energy triphosphate forms equivalent to the respective RNA bases. A definite sequence of steps in polypeptide (and protein) biosynthesis is thought to occur, one involving the genetic code and its transcription and translation to ribosomal elements in the cytoplasm.

The Genetic Code

Genetic information is stored within the highly coiled DNA molecule of chromosomes. Each sequence of amino acids making up a particular protein is coded by a given sequence of nucleotides in the DNA molecules. The term *genes* refers to that portion of the DNA chain that specifies one complete *polypeptide chain*. Genes remain within the chromosome, but the message is transcribed (transcription) to form mRNA, which carries the instructions to the biosynthetic ribosomal level for the translation event.

Three contiguous nucleotides in DNA are responsible for coding each of the 20 amino acids employed in protein synthesis. The specific genetic code words for each amino acid was announced in 1961–1965 from three major laboratories, those of M. Nirenberg, S. Ochoa, and H.G. Khorana. Since there are only four different bases in the DNA nucleotides, different sequences (in triplets) of bases are required to govern complete coding of all available amino acids. Coding triplets (of DNA nucleotide bases) passed on to mRNA as complementary genetic instructions are called *codons* and act as the template in the RNA that specifies the exact position and sequence of each amino acid during protein synthesis. (Fig. 22.10).

For the genetic instructions to be passed on from one organism to another, generation after generation, a replication of DNA is required; DNA is the only protein not requiring an "outside" template for its biosynthesis, i.e., for replication. Rather the DNA molecule "reproduces" itself by separating the two double strands, breaking not only the purine-pyrimidine junctions but also the juxtapositioned pyrimidine-purine bonds. In this manner two "halves" of the DNA strands have exposed nucleotide purines and pyrimidines that can associate with the appropriate, freely available *ATP, GTP, CTP, and TTP triphosphorylated bases*. After the pairing off of the appropriate bases (A—T, G—C), the nucleotides are joined and pyrophosphate is released. Thus both strands of DNA act as templates in replication, whereas mRNA synthesis involves only one strand.

The replicated DNA contains an original strand in addition to the newly matched nucleotide strand, and two such identical double helical DNA molecules are formed. It may take 8–12 hours to replicate DNA but as cell division nears, the two duplicated identical threads of DNA become packed, or tightly coiled, and form bodies called *chromosomes*. One copy of DNA is passed to each of the *two daughter cells*—each an exact copy of the genetic code from the original parent cell (see Chap. 26).

Ribosomal Synthesis of Protein

Once the mRNA is synthesized on one of the two DNA strands (as a molecule with a sequence of nucleotides in mirror image of the DNA sequence), it leaves the nucleus and enters the cytoplasm to become associated with the ribosomes. At this point the mRNA, carrying codons for specific sequences of amino acids, becomes associated with a subunit of the rRNA. Other than this "attachment" role of rRNA, not much is known about its function; *rRNA does not carry coded data*. Concomitantly, free amino acids in the cytosol are activated enzymatically by an ATP-dependent reaction that results in an ester linkage with certain tRNAs. The activated amino acid-tRNA complex formed binds to the ribosome in a special way, namely, the base pairs in tRNA (anticodons) line up with their corrresponding mRNA codons (Fig. 22.10). In this manner only the specific amino acid corresponding (dictated) to the code can be added. Enzymes now catalyze peptide bond formation while each is still attached to its respective tRNA.

With one end of the growing peptide chain always attached to a tRNA, amino acids are added, one by one, starting at the N-terminal end of the chain to be synthesized and growing toward the —COOH terminus (Fig. 22.10). Most workers feel that a special codon is required for termination of chain length, but definitive data are lacking. The final synthesized protein structure is now released from the ribosome and may go to the *Golgi area* where it can be packaged and incorporated within *membrane-lined vesicles*. Alternatively, it may be released to the cytosol for further distribution within the cell. The rRNA is neither altered nor destroyed in the process and is available for reuse.

The primary structure of the protein is considered to be its specific amino acid sequence; the secondary structure is the coiled helical arrangement of a peptide chain along a linear axis, and the tertiary structure is the three-dimensional consideration of a folded protein as seen in globular or spherical proteins. The final protein product released from the ribosome, therefore, may be of a randomly oriented coil (helical, coiled, or folded) and frequently with the aid of disulfide bonds (—S—S—). Usually those proteins involved as structural cellular elements are extended, long, and fibrous; they are likely to be sheaths of parallel coils. On the other hand, proteins with a three-dimensional conformation usually are almost spherical and are represented by dynamically active proteins, such as enzymes or hormones.

Some proteins are conjugated with metals or organic reactive groups. The diversity of the biosynthetic process is emphasized by the fact that the range of investigated proteins spans molecules with M.W. between *5000* and *1,000,000 daltons each*. Proteins may also be synthesized in animals on "free" ribosomes (not attached to the endoplasmic reticulum), such as hemoglobin or in mitochondria, although the majority occurs as previously described on the "fixed" ribosome.

Selected Readings

Ganong WF (1979) Review of medical physiology, 8th edn. Lange Medical, Los Gatos, California

Gordon MS (1977) Animal physiology, 3rd edn. MacMillan, New York

Harper HA (1977) Review of physiological chemistry, 16th edn. Lange Medical, Los Gatos, California

Lehninger AL (1975) Biochemistry, 2nd edn. Worth, New York

Vander AJ, Sherman JH, Luciano DS (1980) Human physiology, 3rd edn. McGraw-Hill, New York

Review Questions

1. Specifically what is a gene?
2. In what way(s) does the synthesis of mRNA differ from that of DNA replication?
3. What components make up nucleic acids and of what specific importance are the nitrogenous bases in terms of the genetic code?
4. What specific molecular role does oxygen play in nutrient metabolism?
5. Is anaerobic metabolism efficient enough in terms of energy production to support life processes in mammals? In birds? In other vertebrates?
6. Quantitatively, what is the fundamental advantage of aerobic over anaerobic (or vice versa) metabolism of carbohydrate?
7. What intermediate of nutrient metabolism is central to all forms of substrates and what significance does such an observation carry?
8. How do enzymes interact with the reactants of a biologic reaction? Under what conditions?
9. What relationship does removal of H^+ have to the recapturing of high-energy phosphate?
10. What is the energy source for base bonding in the DNA structure?

Nutrition

Chapter 23

The nutrients are commonly divided into the carbohydrates, fats, proteins, minerals, vitamins, and—last but not least—water. The actual needs of the body, in addition to water, are for energy and certain amino acids, vitamins, and minerals. Theoretically, either carbohydrates, fats, or proteins may serve as energy sources, the last-mentioned especially when consumed in excess of protein needs. There are, however, certain limitations to this freedom of choice. (1) Completely fat-free foods are not pleasing to the palate, and very high levels of fat are not well tolerated by many persons. (2) Protein as the major source of energy has the disadvantage that it requires an efficient mechanism for the removal of excess nitrogen. (3) Nutritionally speaking, small amounts of carbohydrates (about 50 g glucose daily) and of fat for the supply of essential fatty acids (especially linoleic acid) are needed. (4) Carbohydrate is generally the cheapest source of energy. As a result of these considerations, all three nutrients are supplying energy to man, with increased emphasis on carbohydrate in poorer societies and on fat and protein in more prosperous populations. Further details on energy production are discussed in Chaps. 18 and 22.
Energy can also be supplied by alcohol and certain synthetic hydrocarbons. However, the former is believed to be deleterious to health and about the latter little is as yet known.

Fats, Carbohydrates, and Proteins

Carbohydrates and proteins supply roughly 4 kcal/g, fats supply *2¼ times* this amount. Therefore a person obtaining 40% of dietary calories from fat will have consumed a diet containing 23% fat if the noncaloric nutrients (minerals and water) are not considered. Energy needs of adults vary from *1800* to *3000 kcal* under moderate activity. About two-thirds of this amount meet the maintenance (basal) needs of the organism. With heavy work or sports activity, caloric need may increase appreciably.

Fig. 23.1A–G. The chemistry of selected nutrients. Different presentations of glucose (α-D-glucopyranose) according to **a** Fisher; **b** Haworth; **c** "chair," the stereochemically most accurate presentation; **d** sucrose (1-α-D-glucopyranoside-β-D-fructofuranoside); **e** a triglyceride (lineo-oleo-stearin); **f** serine, a hydroxy-substituted aliphatic amino acid; and **g** tryptophan, an aromatic amino acid.

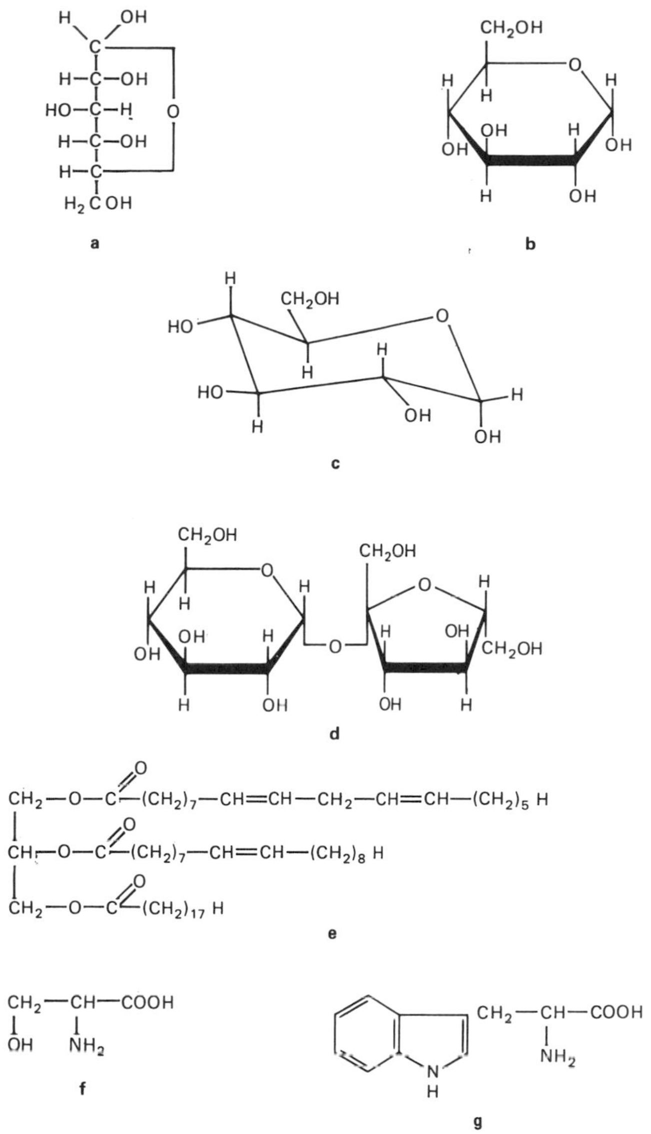

The *major sources* of carbohydrate are the starches of grains, especially of wheat, corn, and rice. Fruits and vegetables contain starch as well as sugar, mostly the disaccharides sucrose (Fig. 23.1d) and maltose and the monosaccharides glucose (Fig. 23.1 a–c) and fructose. Milk is the source of the disaccharide lactose. In some of the economically more advanced countries, as much as *one-third of the carbohydrate* may be ingested as sucrose. Most recently, sugars have surpassed cereal products as the major group of carbohydrates in the American diet (Table 23.1).

Most of the *fats ingested* are triglycerides, consisting of a glycerol molecule linked to three fatty acids (Fig. 23.1). The latter may have different chain lengths and degrees of unsaturation. Phospholipids play a secondary role as a source of lipid calories. Just as with carbohydrates, fats can be synthesized by the organism, with the exception of the polyun-

Table 23.1. Contributions of major food groups to available nutrient supplies (U. S. civilian populations).[a]

Food Groups	Energy		Protein		Fat		CHO	
Period:[b]	I	II	I	II	I	II	I	II
Meat, poultry, fish	18.1	20.0	35.7	42.6	32.9	34.1	0.1	0.1
Eggs	2.6	1.8	6.8	4.8	4.0	2.7	0.1	0.1
Dairy (excl. butter)	13.5	11.1	24.4	22.0	16.6	12.5	7.7	6.7
Fats, oil, butter	16.0	18.1	0.2	0.2	39.5	43.3	[c]	[c]
Fruits	3.3	3.0	1.2	1.1	0.4	0.4	7.1	6.6
Potatoes (incl. sweet)	3.0	2.9	2.5	2.3	0.1	0.1	5.5	5.4
Vegetables	2.7	2.7	3.8	3.6	0.4	0.4	4.7	5.3
Dry beans, peas, nuts	2.9	3.1	5.2	5.4	3.3	3.8	2.3	2.1
Flour, cereal products	21.4	19.2	19.9	17.6	1.6	1.3	37.6	34.7
Sugar, sweeteners	15.8	17.3	[c]	[c]	0	0	34.2	38.5
Miscellaneous	0.8	0.7	0.4	0.4	1.4	1.2	0.7	0.5

[a] Figures are in percentage of total average daily consumption of nutrient. Adapted from National Food Review, USDA, January 1978.
[b] Period I, 1957–59 average; Period II, 1977.
[c] Less than 0.05%.

Table 23.2. Major amino acids occurring in proteins.

Amino Acid	No. of C-atoms	Characteristics
I. Aliphatic Amino Acids		
Glycine	2	
Serine	3	Hydroxylic group (OH)
Alanine	3	
Threonine[a]	4	Hydroxylic group (OH)
Cysteine[b]	3	Contains sulfur (SH)
Aspartic Acid	4	Acidic
Valine[a]	5	Branch chain (CH_3)
Methionine[a]	5	Contains sulfur (S)
Glutamic Acid	5	Acidic
Leucine[a]	6	Branch chain (CH_3)
Isoleucine[a]	6	Branch chain (CH_3)
Arginine	6	Basic
Lysine[a]	6	Basic
Hydroxylysine[b]	6	Basic
II. Nonaliphatic Amino Acids		
Proline	5	Imino acid (no amino group)
Hydroxyproline	5	Imino acid (no amino group)
Histidine[a]	6	Basic
Phenylalanine[a]	9	Aromatic ring
Tyrosine[b]	9	OH-phenylalanine, aromatic ring
Tryptophan[a]	11	Aromatic ring

[a] Essential amino acids, not synthesized by the organism; histidine is probably not an essential amino acid for healthy adults.
[b] The organism can synthesize cysteine only from methionine, hydroxylysine from lysine, and tyrosine from phenylalanine.

saturated linoleic acid, a precursor of arachidonic acid and required for the synthesis of prostaglandins. Linoleic acid should provide 1%–2% of the total calories, an amount easily met by the average American's diet.

The *recommended allowance for protein* can be met by diverse items such as one pound of oatmeal or by one-half pound of lean meat or peanuts. In actuality, the demand is for a small number of essential amino acids that cannot be synthesized by the organism or that may not be synthesized at the required rate (Table 23.2). Protein quality is to a large extent determined by the ratio of these amino acids to total dietary protein in the food. Generally, foods from animal sources—meat, fish, eggs, and milk—have an amino acid composition more closely suited to human needs. However, judicious blending of sources of vegetable protein can result in an amino acid mixture similar to that provided by meat. The successful use of corn-soybean mixtures as the nutritional backbone of the American meat industry and the not inconsiderable number of strict vegetarians attest to the successful utilization of protein from plant sources. Preference for a rib steak to a dish of lentils is due more to the nature of the lipids contained in the steak than to its protein content.

Minerals and Vitamins

The skeleton is the major reservoir of minerals in the body (see Chap. 1). Most of the calcium, phosphorus, and magnesium are found in bone. However, these as well as other minerals also play important roles as cofactors of various enzymatic reactions in the organism or are active in the maintenance of the inner environment (osmotic pressure and acid-base balance) and neuromuscular activity (see Chap. 22). Iron plays a special role as a component of hemoglobin, and iodine as an integral part of the thyroid hormones. The need for some of the trace elements has only recently been recognized, and it is possible that additional elements will be found to be dietary essentials, albeit in very small amounts.

A mixed diet adequate in energy, protein, and vitamins can also be expected to provide the required minerals. Those most likely to be supplied in borderline or insufficient amounts are iron and calcium. In certain regions of the United States, iodine-deficiency symptoms, such as goiter, would be rampant were it not for the supplementation of table salt with iodine. Table 23.3 contains a list of minerals and trace elements known or suspected to be required and one or more of the functions ascribed to them.

As now defined, *vitamins are accessory factors,* necessary for the well-being of some species, that cannot be synthesized in sufficient amounts by this species; as accessory factors they neither provide energy nor become a permanent part of the organism.

Table 23.3. Dietary essentials in human nutrition: Minerals.[a]

Designation	Major Functions	Major Sources	Symptoms of Deficiency[b]
Calcium (Ca)	Bone and teeth, nervous reactions, enzyme cofactor	Dairy products, leafy green vegetables	Calcium tetany, demineralized bones
Phosphorus (P)	Bone and teeth, intermediary metabolism	Dairy products, grains, meat	Demineralized bones
Magnesium (Mg)	Bone, nervous reactions, enzyme cofactor	Whole grains, meat, milk	Anorexia, nausea, neurologic symptoms
Sodium (Na)	Maintenance of osmotic equilibrium and fluid volume	Table salt[c]	Weakness, mental apathy, muscle twitching
Potassium (K)	Cellular enzyme function	Vegetables, meats, dried fruits, nuts	Weakness, lethargy, hyporeflexia
Chlorine (Cl)	Maintenance of fluid and electrolyte balance	Table salt[c]	[d]
Iron (Fe)	Hemoglobin, myoglobin; respiratory enzymes	Meat, liver, beans, nuts, dried fruit	Anemia
Copper (Cu)	Enzyme cofactor (cytochrome-c-oxidase)[e]	Nuts, liver, kidney, dried legumes, raisins	Anemia, neutropenia, skeletal defects
Manganese (Mn)	Enzyme cofactor, bone structure, reproduction	Nuts, whole grains	[d]
Zinc (Zn)	Enzyme cofactor (carbonic anhydrase)[e]	Shellfish, meat, beans, egg yolks	Growth failure, delayed sexual maturation
Iodine (I)	Thyroid hormone synthesis	Iodized table salt, marine foods	Goiter
Molybdenum (Mo)	Enzyme cofactor (xanthine oxidase)[e]	Beef kidney, some cereals and legumes	[d]
Chromium (Cr)	Regulation of CHO metabolism (glucose tolerance factor)	Limited information available	[d]

[a] A human need for the following trace elements is possible but has not been unequivocally established: selenium (Se), fluorine (F), silicon (Si), nickel (Ni), vanadium (V), and tin (Sn). The need for sulfur (S) is satisfied by ingestion of methionine and cystine, and for cobalt (Co) by vitamin B_{12}.
[b] Except for Ca, Fe, and I, dietary deficiency in man is either unlikely or rare.
[c] Many processed foods contain considerable amounts of sodium chloride.
[d] No specific deficiency syndrome described in man.
[e] Examples of activity as enzyme cofactors.

The eight so-called *B-group vitamins* meet this definition completely. They function in anabolic and catabolic enzyme systems, most of which have been well characterized (see Chap. 22). They are water soluble, as are also choline and ascorbic acid. *Choline* actually becomes part of the organism; phosphatidylcholines (also called lecithins), e.g., are found as structural components of body cells, especially of the nervous system. Its inclusion in the list of vitamins is mostly due to the fact that it does not fit into any of the other categories and, where not synthesized by the body, is required in relatively small amounts, albeit larger ones than any of the other vitamins. Although *ascorbic acid* completely fits the just mentioned definition of vitamins and the effects of a deficiency are well known, the exact mode of action has not yet been elucidated.

Table 23.4. Dietary essentials in human nutrition: Vitamins.

Designation	Major Mode of Action	Major Sources[a]	Symptoms of Deficiency[b]
Retinol (A)	Part of visual pigment, maintenance of epithelial tissues	Egg yolk, butter, fish oils; conversion of carotenes[c]	Nightblindness, corneal and skin lesions, reproductive failure
Calciferol (D)	Ca and P absorption, bone and teeth formation	Fish oils, livers; irradiation of sterols[c]	Rickets, osteomalacia
Tocopherols (E)	Antioxidant	Vegetable oils, green leafy vegetables	In animals: muscular degeneration, infertility, brain lesions, edema[d]
Vitamin K	Synthesis of blood coagulation factors	Green leafy vegetables, bacterial synthesis	Slowed blood coagulation
Essential fatty acids	Synthesis of prostaglandins	Unsaturated oils high in linoleic acid	Dermatitis, deranged lipid transport[d]
Thiamine (B_1)	Energy metabolism-decarboxylation	Whole grains, organ meats	Beriberi, polyneuritis
Riboflavin (B_2)	Hydrogen and electron transfer (FMN, FAD)	Whole grains, milk, eggs, liver	Cheilosis, glossitis, photophobia
Nicotinic acid (niacin)	Hydrogen and electron transfer (NAD, NADP)	Yeast, meat, liver[e]	Pellagra
Pyridoxine (B_6)	Amino acid metabolism	Whole grains, yeast, liver	Convulsions, hyperirritability
Pantothenic acid	Acetyl-group transfer (CoA)	Widely distributed	Neuromotor and gastrointestinal disorders
Biotin	CO_2 transfer	Eggs, liver; bacterial synthesis	Seborrheic dermatitis
Folic acid	One-carbon transfer	Leafy green vegetables, meat	Anemia, sprue
Cyanocobalamine (B_{12})	One-carbon synthesis; molecular rearrangement	Animal products, esp. liver; bacterial synthesis	Pernicious anemia
Ascorbic acid (C)	Hydroxylations, collagen synthesis	Citrus, potatoes, peppers	Scurvy
Choline	Fat transport, phospholipid synthesis	Animal products; also synthesized	[d]

[a] Most vitamins, especially of the B-group, occur in a multitude of foodstuffs and in all body cells.
[b] A variety of symptoms occur with certain vitamin deficiencies, vitamin deficiencies are frequently of a multiple nature, and symptoms similar to those described may have their origin in conditions not related to nutrition.
[c] Certain carotenes, found in green and yellow vegetables, are precursors of vitamin A. Certain sterols, including 7-dehydrocholesterol, which is synthesized in the body, are precursors of vitamin D.
[d] No well-defined syndrome is described for man.
[e] Niacin is one of the end products of normal tryptophan metabolism.

Much has been learned during the last 10 years about the function of the four *fat-soluble vitamins*—A, D, E, and K—but the understanding of their exact mode of action on the molecular level will have to await further research. A brief overview of the vitamins can be found in Table 23.4.

Deficiencies and Excesses

The effects of a complete lack or an insufficient amount of a nutrient are generally well appreciated; however, *overdoses* can also be injurious. Excessive intake of energy causes obesity. In certain parts of the world, including the United States and Western Europe, obesity is associated with a negative esthetic effect; more noteworthy, it is shown to be correlated with an increased incidence of various pathologic conditions. A high intake of saturated fats and cholesterol appears also to be positively correlated with cardiovascular disease.

The toxicity of mineral and trace elements depends to a large extent on the organism's capacity for their elimination. Among the trace elements required for the well-being of the human organism, *copper, cobalt,* and *fluorine* are considered to be among the more toxic. Hypervitaminotic syndromes are described for vitamins A and D and, to a smaller extent, nicotinic acid, although specific symptoms following large overdoses are also recognized for some other vitamins. Although caloric overconsumption is commonplace, it is not likely that a person eating customary foods in reasonable or even slightly exaggerated quantities will suffer from the consequences of an overdose of minerals or vitamins. Such overdoses generally result from metabolic disorders, unusual modifications of foods, or purchase and consumption of concentrated sources of vitamins or minerals in multiples of the required dosage. Examples are the toxicity of normal doses of *vitamin D* due to a person's inability to catabolize the vitamin; beer-drinker's myopathy as a result of the use of a cobalt-containing foaming agent; and the intake of high doses of vitamin A, often as an antidote to acne vulgaris.

Textbooks of nutrition generally describe the symptoms of deficiencies, especially of the vitamins, in some detail. The major symptoms of a *thiamin deficiency* in rats, e.g., are described as loss of appetite, weight loss, convulsions, slowing of the heart rate, and lowering of body temperature; and of a *pantothenic acid deficiency* in man as vomiting, malaise, abdominal distress, and burning cramps, followed by tenderness in the heels, fatigue, and insomnia. Clearly, most if not all of these symptoms might be due to a multitude of causes other than vitamin deficiencies. To obtain the symptoms described for pantothenic acid deficiency in man as the actual consequence of a deficiency of this vitamin, it is necessary to use extreme means, such as the consumption of *antivitamins*.

These facts notwithstanding, the descriptions of deficiency symptoms can, at least to the lay public, create the impression that such symptoms—whatever their cause—can be relieved by an increased intake of the appropriate vitamins. A positive reaction to such medication, in the absence of an actual deficiency, would likelier be due to a placebo (imagined) effect than to the actual elimination of a *hypovitaminosis*. Nevertheless, a sizable industry is based on the over-the-counter supply of vitamin and mineral supplements in excess of demonstrated need.

Nutritional Status

The first comprehensive National Nutrition Survey in the United States was initiated in 1967 and covered 70,000 people in low-income areas of 10 states. Anemia, dental problems, retarded growth, and cases of vitamin A, vitamin C, vitamin D, and protein deficiency were discovered. *Anemia* may have its roots in various nutrient deficiencies; in practice, iron deficiency is the most frequent cause. The symptoms of the observed vitamin deficiencies were, in the great majority of cases, restricted to low blood levels of the appropriate vitamin. Similarly, low serum-protein levels were common whereas clinical symptoms of protein malnutrition were relatively rare. Other surveys show that the intake of *calcium* is often below the recommended level. However, some investigators consider the recommendation for dietary calcium to be too high, and—except in extreme cases—it is difficult to relate lower intake to an actual deficiency syndrome.

It can be assumed that the same deficiencies found in low-income areas also occur, albeit less frequently, in other segments of the United States' population.

On a global scale, *undernutrition is widespread* and clinical deficiency symptoms are frequently observed. The greatest problem is believed to be protein-calorie malnutrition—in the most severe form called *kwashiorkor* and *marasmus*. *Anemia and blindness,* the latter a result of vitamin A deficiency, are other frequently encountered problems. Cases of *rickets* (vitamin D deficiency), *scurvy* (vitamin C deficiency), *beriberi* and *pellagra* (thiamin and niacin deficiencies), and symptoms of *riboflavin deficiency* are also seen. The problem can be summarized as one of insufficient food, choice, and information. Thus poverty and ignorance are seen to lead to both malnutrition and early death.

While the populations of the United States and other developed countries do not suffer from widespread nutritional deficiencies, they are prone to symptoms of *overnutrition,* believed to give rise to various chronic ailments. Although individual cases of *toxic doses* of certain vitamins are infrequent, excessive intake of calories and of salt is widespread. *Cardiovascular disease, diabetes mellitus,* and disorders of the *gallbladder* are some of the chronic diseases believed to be related to overnutrition. Excessive *salt intake* appears to favor the development of *hypertension*.

Nutrient Requirements and Recommended Daily Allowances

Individual animals in a group receiving adequate amounts of a feed that satisfies all nutrient needs may still grow at different rates. This can be explained by differences of individuals in their *genetic ability to grow*. When the feed is so modified that one or more nutrients become slightly limiting, the members will still vary in *growth*, indicating a difference in their ability to utilize a given nutrient and, thus, in their requirement for a given nutrient for optimum performance. If one were to feed nutrients at the *average requirement* level, one-half the animals would receive sufficient nutrients whereas the other half would get less than needed and would therefore not be able to perform optimally.

The *Recommended Dietary Allowances* (RDA) are not average requirements for man. They are the levels of intake of essential nutrients considered by the Food and Nutrition Board of the National Research Council to be adequate, on the basis of available scientific knowledge, to meet the known nutritional needs of practically all healthy people. Clearly the RDA for the various nutrients must be significantly higher than the actual average nutrient requirements for people of a given age, weight, and sex. The only exception is the *recommended energy allowance*, which is established at the lowest value thought to be consonant with the health of average persons in each age group. Table 23.5

Table 23.5. Recommended daily dietary allowances for selected nutrients.[a]

Nutrient	Units	Male[b]	Female[b]
Energy	kcal	2900	2100
Protein	g	56	44
Vitamin A	mcg R.E.[c]	1000	800
Vitamin D	mcg[d]	7.5	7.5
Vitamin E	mg α T.E.[e]	10	8
Ascorbic Acid	mg	60	60
Thiamin (B_1)	mg	1.5	1.1
Riboflavin (B_2)	mg	1.7	1.3
Pyridoxine (B_6)	mg	2.2	2.0
Niacin	mg N.E.[f]	19	14
Folacin	mcg	400	400
Cobalamin (B_{12})	mcg	3	3
Calcium	mg	800	800
Phosphorus	mg	800	800
Magnesium	mg	350	300
Iron	mg	10	18
Zinc	mg	15	15
Iodine	mcg	150	150

[a] Nutrients for which a recommended daily allowance (RDA) has been determined by the Food and Nutrition Board, National Academy of Sciences-National Research Council; for a list of estimated safe and adequate daily dietary intakes of additional selected vitamins and minerals, see "Recommended Daily Allowances," National Academy of Sciences, 9th ed., 1980.

[b] Male-age 19 to 22, 70 kg (154 lbs.), 177 cm (70 in) tall; female-age 19 to 22, 55 kg (120 lbs), 163 cm (64 in) tall; for other ages, see "Recommended Daily Allowances," National Academy of Sciences, 9th ed., 1980.

[c] R.E.-retinol equivalent; 1 R.E. = 1 mcg retinol (Vitamin A alcohol) or 6 mcg beta-carotene.

[d] As cholecalciferol; 10 mcg cholecalciferol = 400 I U vitamin D.

[e] Alpha-tocopherol equivalents; 1 mg d-α-tocopherol = 1αT.E.

[f] 1 N.E. (niacin equivalent) is equal to 1 mg of niacin or 60 mg of dietary tryptophan.

shows the RDAs of selected nutrients for young adult females and males. For other ages, consult the National Academy of Sciences publication listed in the "Selected Readings" at the end of this chapter.

Outlook

Table 23.6. Average daily consumption of major nutrients by the U. S. civilian population.[a]

Nutrient	1957–59 Average	1977
Energy (kcal)	3140	3380
Protein (g)	95 (12)[b]	103 (12)
Fat (g)	143 (41)	159 (42)
CHO (g)	375 (47)	391 (46)

[a] Adapted from National Food Review, USDA, January 1978.
[b] Figures in parentheses are percentage contribution of nutrient to total caloric intake.

In 1977, the Senate Select Committee on Nutrition and Human Needs published dietary goals for the United States. The Committee recommended that Americans consume less *fat*, particularly saturated fats, less *cholesterol*, less *sugar*, and less *salt*. It was also recommended that there be an increase in the consumption of *vegetables*, *fruits*, *grain products*, and *unsaturated oils*.

At present Americans obtain 42% of their calories from fats and 20% from sugar, and total caloric intake is on the increase (Table 23.6). The average intake of cholesterol exceeds 500 mg, and salt is being consumed at the rate of 6 to 18 g/day. However, the nature of the fats consumed has changed in recent years: butter and lard consumption has decreased and the use of the more unsaturated salad and cooking oils has climbed steeply. This qualitative change in fat consumption has also brought about a significant decrease in cholesterol intake in the last 40 years. Some scientists feel that the decline in coronary heart disease mortality, which began in the late 1960s may be related to these changes in fat consumption and that further improvements are possible. Others, however, criticize the "dietary goals" for having been developed without adequate supporting evidence and feel that they are not justified based on current knowledge. Another criticism centers around the lack of proof that the adaptation of these goals will actually benefit the public. Although these criticisms cannot be completely discounted, the views expressed in the "dietary goals" appear to be the best advice possible at present. As with all dietary recommendations, this advice will be updated and modified when new information becomes available.

Selected Readings

Baker H, Frank O (1968) Clinical vitaminology. Wiley, New York

Goodhart RS, Shils ME (eds) (1979) Modern nutrition in health and disease, 6th edn. Lea and Febiger, Philadelphia

Harper AE (1978) Dietary goals—a skeptical view. Amer J Clin Nutr 31: 310

Hegsted DM (1978) Dietary goals—a progressive view. Amer J Clin Nutr 31: 1504

Mitchell HS, Rynbergen HJ, Anderson L, Dibble MV (1976) Nutrition in health and disease, 16th edn. Lippincott, Philadelphia

The Nutrition Foundation, Inc. (1976) Present knowledge in nutrition. The Nutrition Foundation, Inc., Washington
Recommended Dietary Allowances. (1980) 9th revised. Food and Nutrition Board, National Research Council, National Academy of Sciences, Washington
Select Committee on Nutrition and Human Needs (1977) U. S. Senate. Dietary goals for the United States, 2nd edn. U. S. Government Printing Office, Washington
Underwood EJ (1977) Trace elements in human and animal nutrition, 4th edn. Academic, New York

Review Questions

1. Energy can be provided by carbohydrates, proteins, and fats. What are the limitations to the use of these nutrients as energy sources for man?
2. What are the nutritionally important mono- and disaccharides?
3. Which amino acids are dietary essentials for man?
4. Which amino acids can be synthesized only from a certain essential amino acid but not from the nonessential amino acid pool?
5. Name the minerals shown to be essential for man, and note one major function of each.
6. Which conditions are likely to cause a hypervitaminosis in man?
7. What is the danger in describing the symptoms of nutrient deficiencies in publications read by people devoid of a scientific background?
8. What are the major nutrient deficiencies in the United States, and what are those observed on a global scale?
9. What are RDAs and how do they differ from actual requirements?
10. What are the major recommendations of the Senate Select Committee on Nutrition and Human Needs?
11. What are some of the changes in American nutrient intake over the last 20 years?
12. If 40% of the calories consumed by a given population group were supplied by fats, what would the percentage fat intake of this group be (omitting consideration of noncaloric nutrients, such as minerals and water)?

Chapter 24 The Kidney

To clarify and emphasize the role of the kidneys as vital organs, several points can be made. The first is that they receive about 20%–25% of the total cardiac output, more blood per unit weight than any other major organ in the body. Through the formation of urine, the kidneys remove (1) wastes or metabolic by-products from the plasma, such as urea; (2) control total body and plasma levels of various electrolytes, such as sodium, potassium, chloride, calcium, and magnesium; and (3) assist in the regulation of body pH by the adjustment of plasma bicarbonate and the excretion of an acidic urine. They also control the amount of water in the plasma and all other body compartments to maintain constancy in the internal fluid environment. In addition the kidneys release two hormones, renin and prostaglandins, which have the capacity to influence cells and alter physiologic processes throughout the body.

Functional Anatomy

The kidneys are a pair of bean-shaped organs in the abdominal cavity. (Fig. 24.1A). In man each weighs approximately

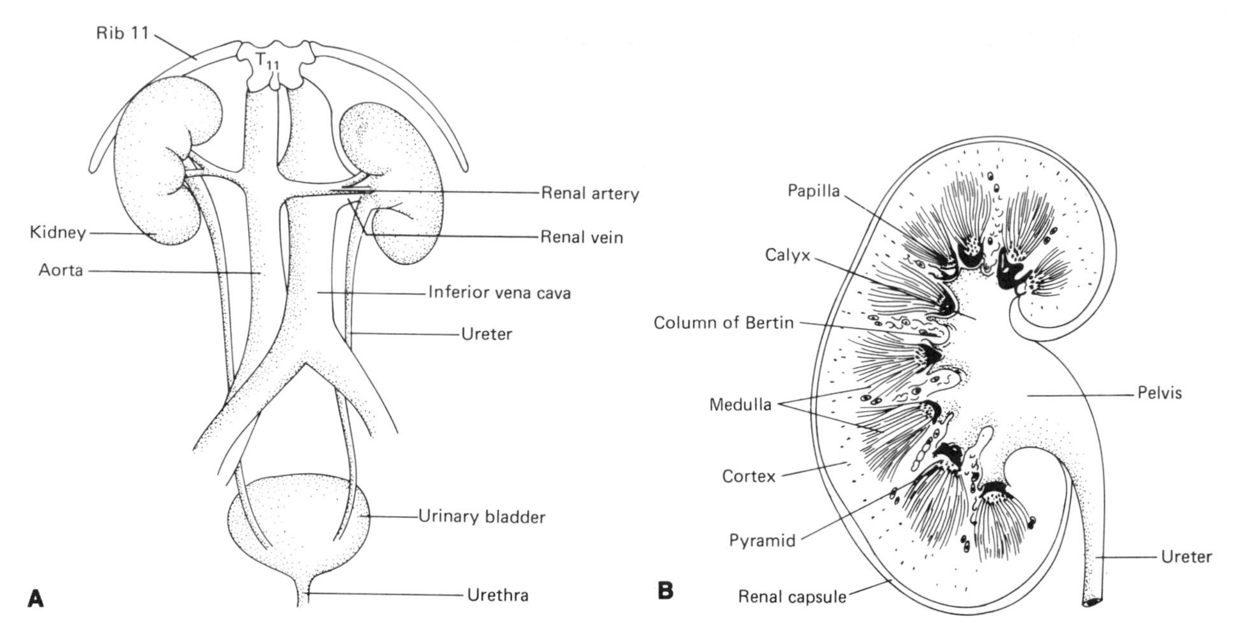

Fig. 24.1. **A** anatomic relationship of kidneys, ureters, and bladder within abdominal cavity; **B** cross section of kidney.

150 g. They are surrounded and supported within the abdominal cavity by connective tissue called renal fascia and adipose tissue. If one examines the kidneys in cross section (Figure 24.1B), the major anatomic landmarks can be noted. Along the medial borders of the kidney there is an indentation called the renal hilus. Through the hilus pass the major blood vessels, the renal nerves, the lymphatics, and the ureter. The outer aspect of the kidney, which is smooth and reddish-brown, is the cortex. Cortical tissue differs markedly in appearance from the whitish-gray medullary tissue

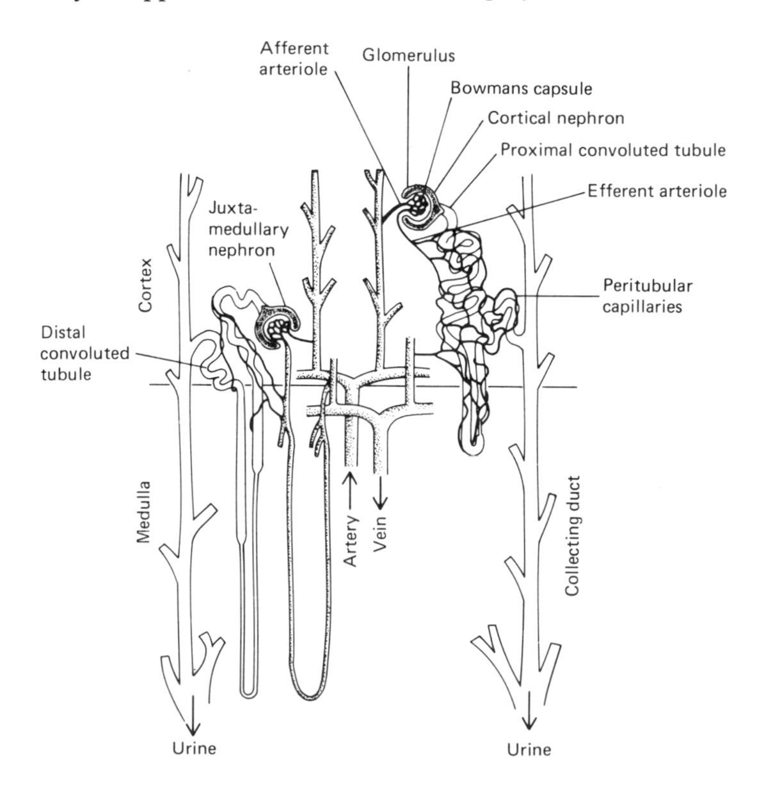

Fig. 24.2. The nephron. Glomerular capillaries within the Bowman's capsule and peritubular capillaries are shown. (Modified after Smith, H (1951) The kidney, Oxford University Press.)

extending from the inner aspect of the cortex to the renal hilus. These two areas, the cortex and medulla, have major functional differences.

Within the medullary region are apical structures called renal pyramids. The apexes of the renal pyramids, or the papillae, which point toward the hilus, are encapsulated by the minor calix. These minor calixes merge to form major calixes, which in turn form the renal pelvis. The ureter from each kidney originates in the renal pelvis, leaves the kidney through the hilus, and proceeds through the abdominal cavity to the bladder. Columns of reddish-brown cortical tissue, located between the renal pyramids and which penetrate the medullary region, are called the columns of Bertin.

The microscopic anatomy of the kidney reveals that each kidney consists of about 1.25 million individual units called nephrons. The combined function or actions of all nephrons represent total kidney organ function. We now focus on the processes that occur within a single nephron. (Fig. 24.2)

The Nephron

Each nephron consists of two major parts: a blood supply and a renal tubule. It is through the renal tubule that fluid passes that is destined to become urine. The first structure of the nephron is the glomerulus, a small network or tuft of capillaries branching off from the afferent arteriole. Almost totally surrounding each glomerulus is a structure known as the Bowman's capsule, which is generally described as the invaginated end of the renal tubule and whose function is (1) to collect fluid and molecules passing through the glomerular capillaries and (2) to direct the fluid into a system called the renal tubule. The combination of the glomerulus and the surrounding Bowman's capsule is the renal corpuscle.

The *renal tubule* has several major features. Shortly after its beginning at the renal corpuscle, it makes a series of bends and loops. This region is called the proximal convoluted tubule (PCT). After the PCT, it straightens and proceeds as a U-shaped structure known as the loop of Henle. The portion of the tubule from the PCT to the bottom of the loop is the descending limb; the ascending limb of the loop of Henle is that segment that continues after the bottom of the "U" and parallels the descending limb. Subsequently another series of twists and turns is called the distal convoluted tubule (DCT). Finally the DCTs from several nephrons join to form collecting tubules, which unite to form even larger collecting ducts. The renal tubule is a continuous structure extending from Bowman's capsule to the collecting ducts.

To understand the functional roles of the various parts of the nephron, it is crucial to orient the nephron's location within the total kidney. The glomerulus, Bowman's capsule, PCT, DCT, and secondary capillary network are all located predominantly in the outer cortical regions of the kidney (Figs. 24.2 and 24.3). The loops of Henle and the collecting

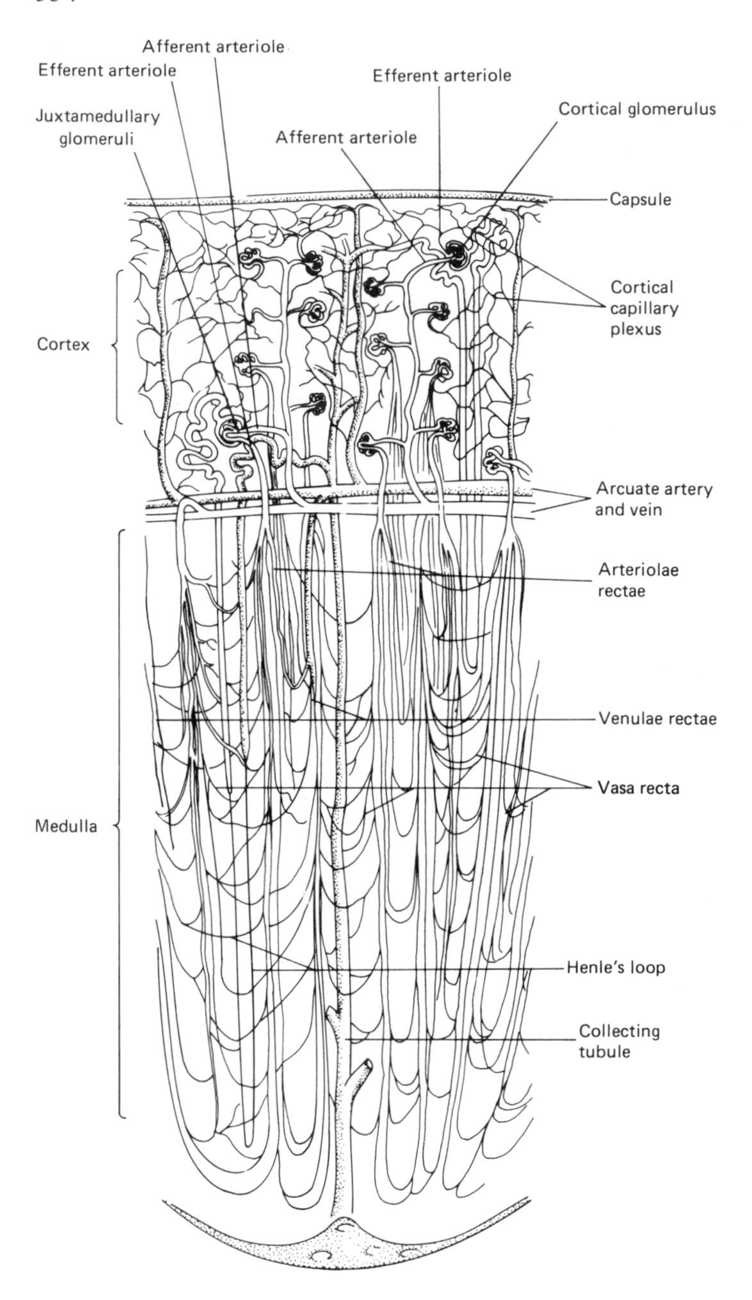

Fig. 24.3. Vasa recta, the capillaries that penetrate medulla and surround loop of Henle and collecting tubule.

ducts are in the medulla. A small fraction of the nephron population lies deeper within the cortex. These nephrons, because of their location, are referred to as juxtamedullary nephrons. Various parts of the cortical nephron, including the afferent and efferent arteriole, the PCT, and the DCT are innervated by sympathetic nerve fibers that also gain access to the cortical regions by the columns of Bertin.

Blood Supply

In the *cortex* there is also a vascular component to each nephron. After the glomerular capillary network these small

Fig. 24.4. The nephron, showing processes of filtration, reabsorption, and secretion.

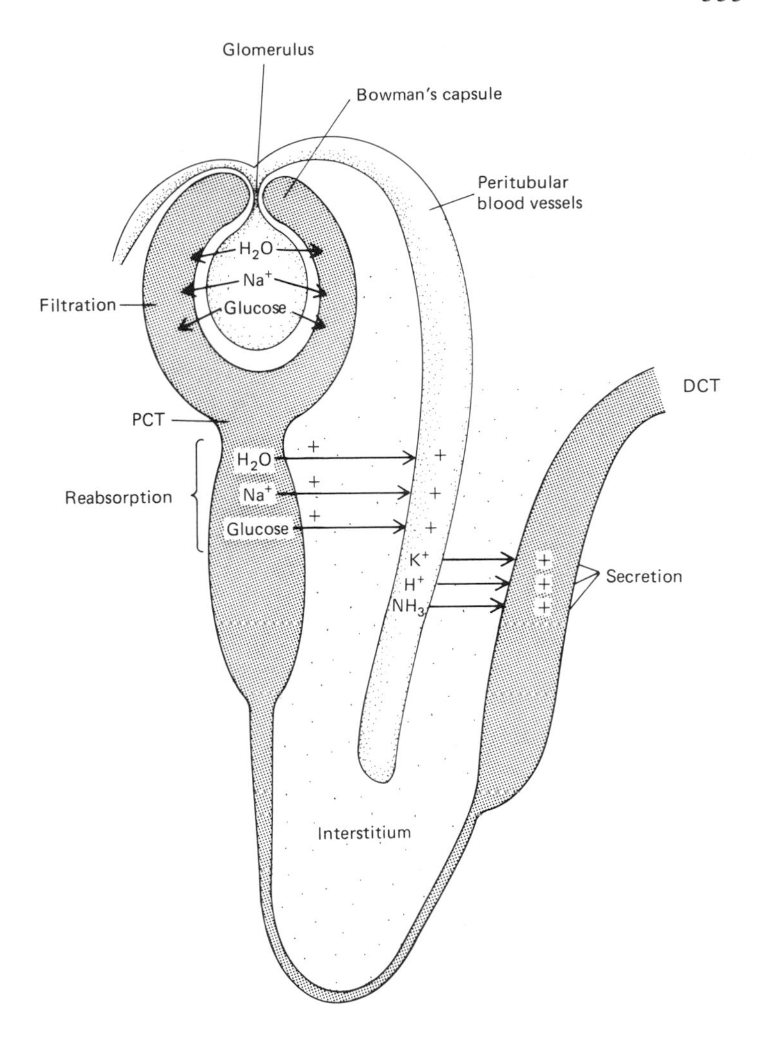

vessels recombine to form the efferent arteriole, which exits from the renal corpuscle (Figs. 24.2 and 24.3). The efferent arteriole subsequently branches to form a secondary capillary network, or peritubular capillary system, surrounding the PCT and DCT. The kidney is one of the few structures in the body in which there are two capillary networks located in series. The peritubular capillary system reunites to form veins, which eventually lead into the renal vein.

Blood flow to the *medulla* occurs through a specialized group of blood vessels called the *vasa recta* (Fig. 24.3). They are long capillary loops that originate as branches from cortical blood vessels, descend into the deep medullary region, and then return to the cortex to recombine and form veins. Even though the cortical and medullary regions contain blood vessels, more than 90% of total renal blood flow goes to the cortex.

The three basic processes that occur within the kidney or each nephron are glomerular filtration, tubular reabsorption, and tubular secretion (Fig. 24.4).

Glomerular Filtration

Glomerular filtration is the process by which fluid and molecules pass from the circulatory system or glomerular capillaries into the Bowman's capsule or renal tubule (Figs. 24.2 and 24.4). The glomerular capillaries may be described as ultrafilters, allowing the passage of plasma water and small molecules into the renal tubule. They functionally behave as though they contained cylindrical pores, 75–100 Å in diameter, even though anatomically pores have not been identified and probably do not exist. The major features of the glomerular capillary membrane are epithelial cells adjacent to the lumen of the capillary basement membrane and endothelial cells adjacent to the space in Bowman's capsule. Water, electrolytes, and small molecules are freely permeable through the glomerular capillary.

Larger molecules, such as proteins, do not pass through freely. Very large molecules, such as complex proteins or blood cells do not normally traverse the glomerular membrane. The major restrictive sites for the passage of intermediate molecules are at the basement membrane and the spaces between adjacent endothelial cells called slit pores. The electronic charge of larger molecules influences their permeability. The permeability of the glomerular capillary is not fixed and may be altered by circulating hormones and other agents.

Glomerular filtration rate (GFR) in man—or the amount of filtrate entering the renal tubule for both kidneys—is about 125 ml/min; both kidneys receive about 650 ml/min of renal plasma flow (RPF).

Only RPF and not blood cells pass through the glomeruli. A term called filtration fraction is defined as the quotient of GFR and RPF $\left(\frac{125}{650} = 0.20\right)$. This indicates that of all the plasma passing through the glomerular capillaries, only 20% is filtered through the glomeruli and enters the renal tubule. The other 80%, which is not filtered, continues out the efferent arteriole, passes through the secondary capillary network (or peritubular capillaries), and returns to the general circulation by the venous system.

Clearance Techniques

The measurement of GFR, or the quantification of the amount of fluid passing through the glomerular membrane, was a difficult task for many years. The solution to the problem revolved around the use of what is called clearance techniques. Clearance refers to the amount of a substance removed from the plasma per unit time.

A simple equation or relationship may be formulated for the clearance of a substance (compound A) as follows:

1. Plasma concentration of A × Amount of plasma filtered into renal tubule = Urine concentration of A × Urine volume

This simply indicates that the product of plasma concentration of "A" and the rate at which it is filtered into the renal tubule is equal to the amount excreted in the urine, or urine volume × urine A concentration. The obvious assumption is that "A" is neither added to nor removed from the urine by the kidney. The equation can be rearranged to read:

2. $$\text{Amount of plasma filtered into renal tubule} = \frac{\text{Urine concentration of A} \times \text{Urine volume}}{\text{Plasma concentration of A}}$$

Since the amount of plasma filtered is equal to GFR, the final equation is:

3. $$\text{GFR (ml/min)} = \frac{\text{Urine A (mg/ml)} \times \text{Urine volume (ml/min)}}{\text{Plasma A (mg/ml)}}$$

or

$$\text{GFR} = \frac{U_A \times V}{P_A}.$$

Inulin Clearance

The substance chosen or selected to measure clearance must meet specific criteria. First, the substance must be freely filterable through the glomerular membrane. Secondly, it is necessary that "A" be neither added to nor removed from the renal tubule. Last, the substance chosen should be relatively inert with no other physiologic effects, and it should be something that is easily measurable in the plasma and urine. A substance appearing to fit all these criteria is a small sugar molecule called inulin, which is normally not found in the plasma. To measure GFR, inulin must be infused intravenously; its clearance is synonymous with GFR.

There is another substance, creatinine, which is normally found in the plasma and is filtered and excreted, whose clearance approximates that of inulin of GFR. The clearance of urea may also be utilized to assess GFR although this procedure is of questionable validity because of extremely variable blood urea values.

Renal Plasma Flow

The clearance principle may also be used to measure renal plasma flow (RPF). This is an adaptation of the Fick principle (see also Chaps. 2 and 14). There is a relationship which states:

Renal plasma flow (ml/min) (RPF)

$$= \frac{\text{Urine concentration of A (mg/ml)} \times \text{Urine volume (ml/min)}}{\text{Renal artery concentration of A (mg/ml)} - \text{Renal vein concentration of A (mg/ml)}}$$

The choice of A in this case may be anything that is excreted with the sole restriction that it be neither removed from nor added to the urine by the kidney itself. Paraaminohippurate (PAH) appears to be an ideal substance because it is freely filterable, and in man more than 90% of the PAH may be completely removed from the renal plasma so that its renal venous concentration is considered to be negligible. This enables a simplification of the equation to:

$$\text{RPF (ml/min)} = \frac{\text{Urine PAH (mg/ml)} \times \text{Urine volume (ml/min)}}{\text{Renal arterial PAH (mg/ml)}}$$

Also PAH, which must be added intravenously to the circulation, is not metabolized by the cells of the body; therefore arterial and venous plasma levels are equal. This fact enables the PAH to be measured in a peripheral venous blood sample, obviating the difficulty of obtaining an arterial blood sample.

A value for whole blood containing plasma and red cells (RBF) is calculated from RPF as follows:

$$\text{RBF} = \text{RPF} \frac{1}{(1 - \text{Hematocrit})}$$

Since filtration fraction (FF) $= \frac{\text{GFR}}{\text{RPF}}$, then

$$\text{FF} = \frac{\text{Clearance of inulin}}{\text{Clearance of PAH}}$$

Problems with the clearance measurements of inulin and PAH are that: 1) they must be added to the circulation, 2) timed urine collections of at least 20–30 minutes in duration are required and 3) the resultant clearance values do not represent instantaneous measurements of GFR or RPF but average measurements over the time interval that the urine was collected.

Renal plasma flow may also be determined directly by using an *electromagnetic flowmeter*. An electromagnetic flowprobe, surgically implanted around the renal artery, records blood velocity from which volume of blood flow may be determined (see Chap. 14). The value of this device is that it provides an instantaneous measure of renal plasma flow without having to infuse PAH or measure it, in both the plasma or urine. Since the flow probe must be surgically implanted on the blood vessel, its use is primarily restricted to animal research and is not for general clinical applications.

Physical and Electrochemical Influences

Glomerular filtration rate is controlled by hydrostatic forces and the physical or electrochemical constraints of the glomerular membrane (permeability). The physical forces are the same as those responsible for filtration or absorption at any capillary (see Chap. 12). The following equation illustrates the factors.

$$\text{GFR} = (\text{Filtration coefficient}) \times (\text{Effective filtration pressure})$$

The filtration coefficient (Kf) is simply a term describing the permeability characteristics of the glomerular membrane. Effective filtration pressure (EFP) is the driving force resulting from the differences in existing hydrostatic (P) and osmotic (π) pressure of plasma and renal tubule

$$\text{GFR} = (K_f) + (P_{\text{glomerulus}} - P_{\text{tubule}} + {}^{\pi}\text{tubule fluid} - {}^{\pi}\text{plasma}).$$

Effective filtration pressure is responsible for the movement of water and solutes through the glomerular membrane into the renal tubule. If we place representative values for the appropriate pressures (mm Hg) an effective pressure may be calculated. A plus sign (+) = positive pressure forcing fluid out of blood vessel into the tubule; a minus sign (−) = negative pressure forcing fluid out of the tubule into the glomerular capillary.

$$\begin{aligned} \text{GFR} &= (\text{Kf}) \times (+45 - 10 + 0 - 25) \\ &= (\text{Kf}) \times (10 \text{ mm Hg EFP}) \end{aligned}$$

This calculation demonstrates that at any given state of glomerular permeability, or K_f, there will be a 10 mm Hg pressure favoring filtration. It is important to understand that GFR will be influenced by changing any of the parameters shown.

The *filtration coefficient* is traditionally regarded as being relatively constant, but recent experiments indicate that in certain diseases or under the influence of specific hormones it may vary appreciably. A significant means of controlling GFR, however, is by manipulating or altering the hydrostatic pressure within the glomerular capillary. The afferent and efferent arteriole of the glomeruli contain smooth muscle, which may contract in response to either sympathetic nerve activity or hormonal influences. Constriction of afferent arterioles decreases glomerular capillary hydrostatic pressure and causes a fall in GFR. Conversely constriction of the efferent arteriole elevates glomerular capillary hydrostatic pressure and increases GFR. This mechanism of control is of primary importance for the control of renal fluid and electrolyte excretion.

Autoregulation of GFR and RBF

Based on the previous discussion, one would expect GFR to vary proportionally with systemic blood pressure or renal perfusion pressure. Since blood flow is also a function of pressure, renal blood flow (RBF) would be expected to increase and decrease as pressure is raised or lowered. In actuality, this does not occur. There is a wide range of renal perfusion pressure (80–180 mm Hg) over which both GFR and RBF are relatively constant (Fig. 24.5). This means that the kidney's resistance to flow, degree of constriction, or dilatation is being actively adjusted. This phenomenon is called autoregulation.

The factors responsible for autoregulation remain to be defined. Autoregulation will occur in a pump-perfused kidney that has been isolated from the body, indicating that neither the nervous system nor circulating hormones mediate the autoregulatory response. Some data implicate the hormone *renin* (produced within the kidney) as being responsible for autoregulation. However, the predominant belief is that autoregulation results from reflex contraction of vascular smooth muscle in response to alterations in transmural pressure. As renal perfusion pressure is increased, the arterioles reflexly contract, increasing resistance so that flow is held constant. If pressure is decreased, the stimulus for contraction is lessened, and the vasculature dilates and decreases resistance.

Although the kidney has the capacity to autoregulate GFR and RBF many circumstances override autoregulation and drastically influence both parameters. If, e.g., a person suffers hemorrhage or blood loss, there is an increase in sympathetically mediated renal vasoconstrictor activity, which results in a decline of GFR and RBF. If the circulating levels of norepinephrine or epinephrine increase, RBF and GFR are reduced. During strenuous exercise in man, both are decreased by as much as 75%–80%. This intense renal vasoconstriction enables blood flow to be redistributed to contracting skeletal muscle and also conserves body fluids.

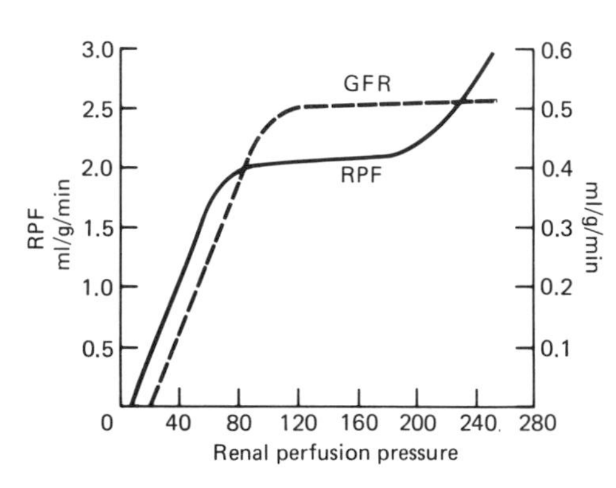

Fig. 24.5. Autoregulation of renal plasma flow (RPF) and glomerular filtration rate (GFR). RPF and GFR increases with renal perfusion pressure up to a point and then level off as pressure continues to rise (autoregulation).

Tubular Reabsorption

In man GFR is normally 125 ml/min or 180 liters in 24 hours (1440 min). Since man produces roughly 1.5 liters/day of urine, he excretes only 1.5/180 (about 1%); 99% of the fluid filtered through the glomerular capillaries is reabsorbed and returned to the circulation (Fig. 24.4). Vast quantities of water, electrolytes, glucose, amino acids, and other vital substances are returned to the circulation via reabsorption.

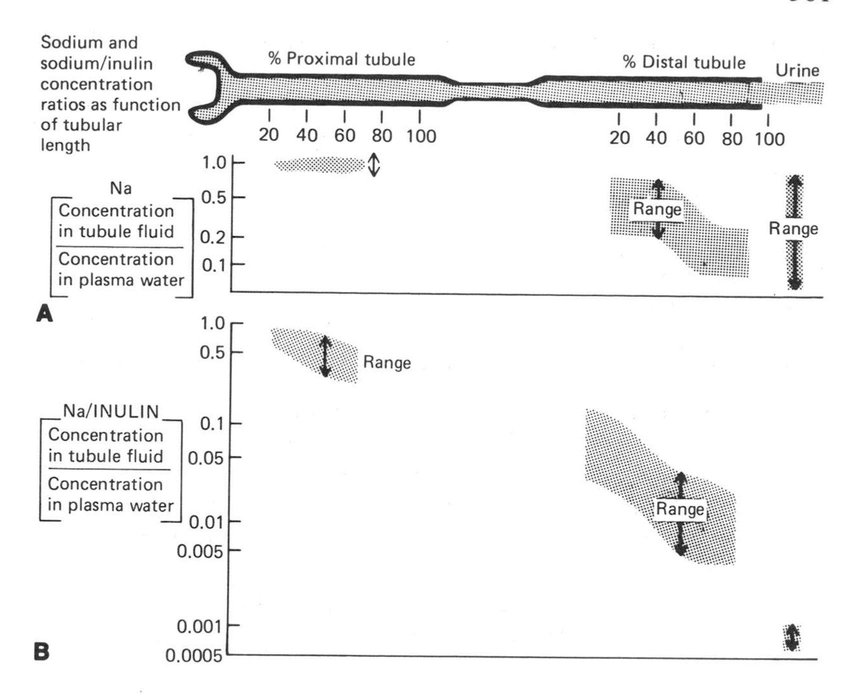

Fig. 24.6A and B. The TF/P ratios for Na^+ and Na^+/inulin. **A** Nephron illustrates sites at which ratios were determined. When no correction is made for water reabsorption in PCT, there is no change in the TF/P ratio for Na^+. **B** At any site along PCT or DCT, when correction is made for water reabsorption by dividing TF/P Na by TF/P inulin, there is progressive decrease in ratio, which indicates that Na^+ is being reabsorbed. (After Netter FH (1973) Kidneys, ureters and urinary bladder. Ciba Medical Illustrations Vol. 6.)

Tubular Fluid/Plasma Ratios

To examine the reabsorption of molecules, a useful term called *tubular fluid/plasma (TF/P) ratio* has been defined. It is the quotient or ratio of the concentrations of any substance measured in the plasma to the same substance measured in the tubular fluid at a specific site within the renal tubule. The TF/P ratio of Na^+ is illustrated in Fig. 24.6.

The plasma sodium concentration (Na^+) is relatively constant. Since Na^+ is freely filterable, the (Na^+) of the TF just inside Bowman's capsule is the same as the plasma (Na^+) (TF/P = 1). It is also observed that at points distal to Bowman's capsule—such as at the end of the PCT—the TF (Na^+) is decreased, and the TF/P sodium ratio is less than 1.0. Consequently, it can be said that if a substance has a TF/P ratio that is less than 1.0, then the substance must have been reabsorbed. As one continues down the renal tubule, if more and more of a substance is reabsorbed, the TF/P ratios determined at specific sites will progressively decrease.

The process by which a substance is moved from the general circulation to the renal tubule at a site other than the glomerulus, is called *secretion*. Consequently, if the TF/P ratio for a substance is greater than 1.0, this indicates that the substance is being secreted into the renal tubule. TF/P ratios can be computed for any filtered electrolyte or substance to quantify the degree of tubular reabsorption or secretion.

A problem with the use of TF/P ratios involves the removal of water. If a fixed amount of substance, e.g. Na^+, is in the renal tubule and water alone is reabsorbed, TF (Na^+) will

Table 24.1. TF/P inulin ratios observed at various points in renal tubule.[a]

Location	Ratio of ^{14}C Inulin, Tubular Fluid/Plasma	Glomerular Filtrate Remaining (%)	Filtered Water Reabsorbed in Segment (%)
Bowman's capsule	1	100	
Junction middle and distal third of proximal tubule	3	33	75 in proximal tubule
End proximal tubule (calculated)	4	25	
Start distal tubule	5	20	5 in loop
End distal tubule	20	5	15 in distal tubule
Ureter	690	0.14	4.86 in collecting ducts

[a] Adapted from Gottschalk (1961) Micropuncture studies of tubular function in the mammalian kidney. Physiol 4:35.

increase, resulting in a TF/P sodium ratio in excess of 1.0. This would indicate (falsely) that Na^+ was being secreted. To correct for the effects of the removal of water from the renal tubule, the TF/P ratio for inulin is calculated. Since inulin is neither reabsorbed nor secreted, its TF concentration will be affected solely by the removal of water. Table 24.1 lists the TF/P ratios for inulin at various points within the tubule and the amount of filtered water that has been reabsorbed.

By dividing a substance's TF/P ratio by the TF/P ratio for inulin, one corrects for the alterations due to the removal of water. An example of this can be seen for Na^+ in Fig. 24.6. To measure the TF/P ratio for inulin or any substance, it is necessary to obtain a sample of the fluid contained at specific sites within the renal tubule by micropuncture. Micropuncture involves the insertion of a very small fine-tipped micropipette through the tubular wall into the lumen of some species (rat) so that a samplc of tubular fluid may be drawn through the pipette.

Mechanisms of Reabsorption

Reabsorption is influenced greatly by the existing electrochemical gradients in the renal tubule (Fig. 24.7). This figure

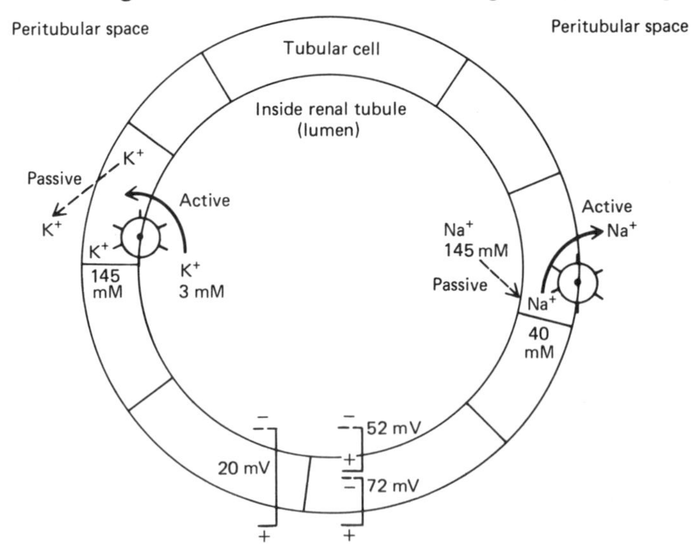

Fig. 24.7. Conditions existing in proximal tubule and active and passive reabsorptive steps for Na^+ and K^+.

shows the inside of the renal tubule, or lumen, the tubular cell surrounding the lumen, and the peritubular fluid and space adjacent to the tubular cell. Reabsorption involves the movement of substances from the lumen to the peritubular space. The membranes, because of their permeabilities and transport capacities, are responsible for the electrochemical and concentration gradients that exist.

Electrochemically, the interior of the tubular cell is negative with respect to both the tubular lumen (−52 mV) and the peritubular fluid (−72 mV). Consequently the lumen is slightly negative when compared to the peritubular fluid (−20 mV). The membrane separating the tubular cell and the lumen in general is selectively permeable to Na^+, whereas the peritubular membrane is selectively permeable to K^+. The chemical concentration gradients are such that (K^+) of the tubular cell is high, and intracellular (Na^+) and (Cl^-) are low.

Transport of Electrolytes

The manner in which Na^+ is reabsorbed is shown in Fig. 24.7. Sodium passively moves down a concentration and electrical gradient from the lumen into the tubular cell. Intracellular Na^+ is actively transported out of the cell into the peritubular fluid against a concentration and electrical gradient. This second step is an energy-requiring process. The energy is derived from the metabolism and subsequent hydrolysis of ATP. The oxygen requirements or consumption of the kidney is proportional to the amount of Na^+ actively being transported. Some of the reabsorbed Na^+ may "backleak" into the tubule through intercellular spaces. This is especially true in the PCT where tubular cell junctions tend to be "loose." In the DCT, however, cells are connected by what is termed, "tight junctions," which prevent a large amount of backleak.

The rate-limiting step for the transport of Na^+ appears to be the passive entry of Na^+ into the tubular cell and not the active transport step, or *sodium pump.* Fig. 27.6 demonstrates, by the use of TF/P ratios, where the filtered Na^+ is being reabsorbed.

Potassium Reabsorption in PCT

Even though the passage of K^+ into the tubular cell is down an electrical gradient (Fig. 24.7), the very large opposing concentration gradient requires that active transport occur; K^+ then passively diffuses out of the tubular cell against an electrical gradient but with a favorable concentration gradient. If total body K^+ is high, it may be actively secreted by the DCT.

Chloride transport from the tubule lumen to the peritubular fluid would be predicted to occur down an electrochemical concentration gradient. However, entrance of Cl^- into the tubular cell is against a strong electrical gradient. (−52 mV). To circumvent this problem two possibilities exist. The first is the coupling of Cl^- with another molecule, such as sodium, to form a neutral molecule. The second alternative is the movement of Cl^- through the intercellular junctions. The majority of Cl^- transport occurs passively, although it has been irrefutably shown that in certain segments of the tubule, such as the ascending limb of the loop of Henle, active transport of Cl^- occurs.

Reabsorption of Glucose

Glucose, which is freely filterable at the glomerulus, is found in the tubular fluid (Fig. 24.8). That under normal circumstances glucose is not present in the urine indicates that all of the filtered glucose is reabsorbed. Glucose reabsorption is an active energy-requiring process. There appears to be a common transport mechanism, or carrier molecule, for

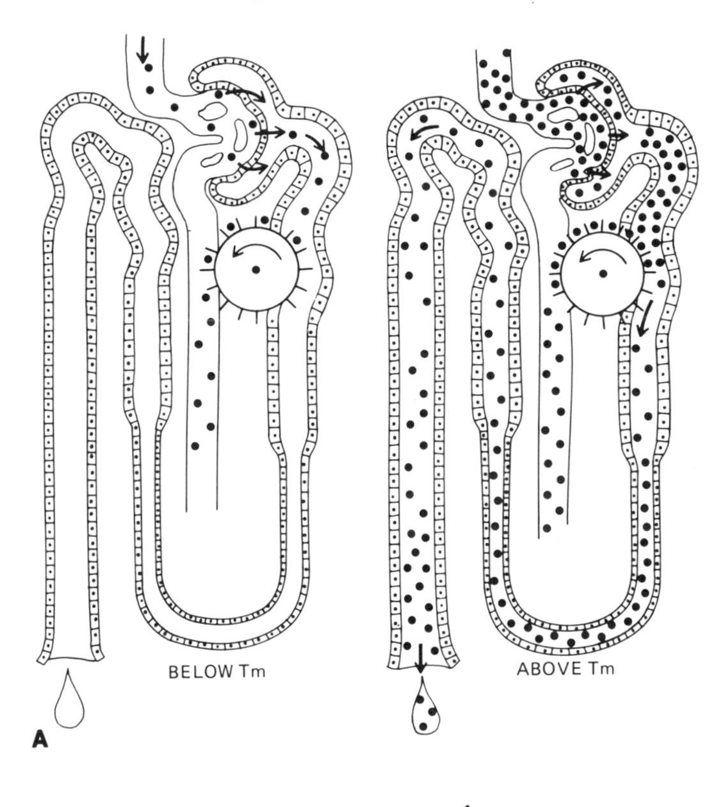

Fig. 24.8. A Events that occur when nephron is operating below or above TM are shown for glucose; **B** Net results are shown graphically. If substance is filtered in an amount below the TM, none will be excreted. When filtered load exceeds the TM, the substance will be present in urine. (After Netter FH (1973) Kidneys, ureters and urinary bladder. Ciba Medical Illustrations Vol. 6.)

glucose, galactose, and fructose. The reabsorptive system for glucose demonstrates what is known as a *transport maximum* (TM). This means that the transport system has a finite capacity to reabsorb glucose.

At normal blood glucose levels the system functions at well below its TM, reabsorbing all glucose, with none appearing in the urine. If blood glucose levels increase, the system reaches its TM, or point of saturation. If plasma glucose is elevated above the TM, then the amount of glucose filtered exceeds the amount that can be reabsorbed. The net effect is glucose in the urine. An example of this is the presence of glucose in the urine, or glucosuria, in an untreated diabetic whose blood glucose values are in excess of the TM value, which is approximately 200 mg/100 ml.

The kidney has the capacity to reabsorb many additional substances. Active reabsorption of PO_4, creatine, sulfate, uric acid, ascorbic acid, calcium, and magnesium occurs. Transport mechanisms for the reabsorption of amino acids and small filterable proteins also exist.

Tubular Secretion

Tubular secretion is the process whereby substances are transported from the peritubular fluid into the renal tubule (Fig. 24.4). Secretion is similar to filtration in that both processes result in substances gaining access to the renal tubule. However, filtration occurs only at the glomerulus, whereas secretion occurs at all parts of the nephron distal to the glomerulus. Secretion may be either an energy-requiring active process, or it can be passive. Most of the active secretory mechanisms exhibit a limited transport capacity, i.e., they demonstrate a TM.

There exists a general common pathway for the secretion of organic acids. Compounds removed from the circulation by this mechanism include phenol red, PAH, penicillin, and glucuronides. A test to evaluate secretory capacity is the measurement of the excretion of phenol red, a compound that must be added to the circulation.

A second secretory mechanism transports strong organic bases. Such compounds include guanidine, thiamine, choline, histamine, and tetraethylammonium. *Passive secretion* involves the movement of substances into the renal tubule down an electrochemical gradient. Materials like weak bases and weak acids are transported in this manner. In addition, K^+ may be secreted passively down an electrochemical gradient in the distal tubule.

Control of Electrolytic Excretion

The kidney, acting like a homeostatic organ, is responsible for maintaining plasma fluid and total body electrolyte con-

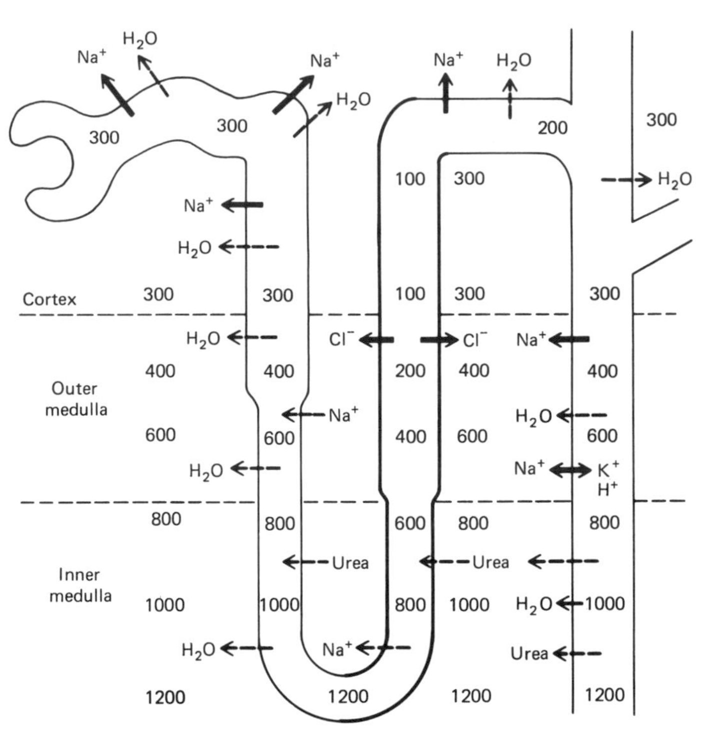

Fig. 24.9. Summary of exchanges of water, ions, and urea in kidney. Numerals give osmolality (mosmol) of tubular urine and peritubular fluid. Note increase in peritubular osmolality as one proceeds into deep medullary region. (300 cortex–1200 medulla). Solid arrows indicate active transport; dashed arrows passive transport. Heavy outline along ascending limb of loop of Henle indicates that this segment is relatively water-impermeable. (Adapted from Gottschalk CW, Mylle M (1959) Am J Physiol 196: 927)

centrations within narrow limits. Sodium deprivation or a decline in total body Na^+ activates the kidney to conserve Na^+ by minimizing the amount excreted in the urine. The kidney reduces Na^+ excretion most effectively by decreasing filtration rate. If a substance is not filtered or secreted, it will not appear in the urine. Because we have the capacity to alter GFR, the filtered load of water, electrolytes, or any substance can be drastically altered. The other option is to increase the reabsorption of Na^+ (Fig. 24.9). Quantitatively, however, it is a reduction of GFR that is responsible for most of the decrease in Na^+ excretion. Even though an increase in Na^+ reabsorption is beneficial and occurs, it is an active energy-requiring process. It is less efficient than simply decreasing Na^+ excretion by decreasing the amount of Na^+ filtered.

Sodium Reabsorption

Several mechanisms facilitate Na^+ reabsorption. One of them is the action of the hormone *aldosterone,* which is released from the adrenal cortex. This hormone has mineralocorticoid activity or the ability to increase the reabsorption of Na^+, primarily in the DCT. Recent studies suggest that aldosterone may increase the production of a *"sodium permease,"* which facilitates the passive entry of Na^+ into the tubular cell, which is rate limiting for Na^+ reabsorption.
Aldosterone is responsible for controlling only a small fraction (2%–3%) of the total Na^+ being reabsorbed. Consequently it is regarded as having a "fine-tuning effect" on renal handling of Na^+. Its influence is extremely important because of the enormous amount of Na^+ transported over the course of several days.

Renal Nerves and Sodium Reabsorption

It is widely known that stimulation of renal sympathetic nerves markedly decreases Na^+ excretion. This effect had been attributed solely to hemodynamic alterations, i.e., decreased RBF and a decrease in GFR or the filtered Na^+ load. Very recent studies, however, show that the renal nerves come into direct contact with the basement membrane of both the PCT and DCT. Stimulation of renal nerves, at a low level that does not alter GFR or the filtered Na^+ load, increases the tubular reabsorption of Na^+. Consequently a direct neural mechanism of control exists over Na^+ reabsorption.

Reabsorption of Water

Of the 180 liters of water filtered daily by the glomerulus 99% is reabsorbed by the renal tubule. The reabsorption of water is a passive process accomplished by osmotic pressure and by an osmotic gradient in the kidney. Figure 24.9 shows the osmolarity of interstitial or peritubular fluid surrounding the renal tubule; it increases progressively from 300 mosmol/liter in the outer cortex to roughly 1200 mosmol/liter in the inner medulla near the papillae.
The values for TF/P inulin (Table 24.1) demonstrate that water is being reabsorbed from both the PCT and DCT. Reabsorption is isosmotic, which means that the fluid reabsorbed and remaining in the tubule has the same osmotic pressure. This indicates that ions and water are being removed at the same rate. The removal of ions, such as Na^+ is active, with water following as a consequence of the osmotic force generated. About 80% of the water filtered is reabsorbed by this mechanism in the PCT, and the remainder is reabsorbed from the more distal segments of the nephron.

Concentration of Urine

The *osmolarity of plasma* is approximately 300 mosmol/liter. The osmolarity of normal urine is 600–800 mosmol/liter and is therefore hypertonic to plasma. How the kidney accomplishes this is explained as follows. The filtrate that passes through the glomerular capillaries has the same osmolarity as the plasma (Fig. 24.9). As fluid descends the loop of Henle, an equilibration occurs between the ever-increasing osmotic environment around the tubule and the tubular contents. It happens because the descending limb is permeable to both water and electrolytes. At the tip of the loop, tubular fluid osmolarity will be about 1200 mosmol/liter. The movement of fluid through the ascending limb is accompanied by a decrease in osmolarity. This is because the ascending limb is impermeable to water and electrolytes, (i.e., Na^+ and Cl^-), are actively being transported out of the tubule (Fig. 24.9). The result is that fluid reaching the DCT is actually hypotonic with respect to plasma with an osmolarity of about 100 mosmol/liter. At first glance this appears to be counterproductive since the problem is to form a hypertonic urine. A crucial point, however, is that the tubular fluid must descend in the collecting tubule back through the medullary osmotic environment. The events that occur during this final descent determine the water content or osmolarity of the urine excreted.

Two choices or circumstances can develop. (1) If the collecting tubule is permeable to water, osmotic equilibration between the surrounding interstitium and the tubular contents occurs. This promotes the formation of urine with the same hypertonicity as fluid found in the deep regions of the medulla (1200 mosmol/liter). (2) If the collecting tubule is *impermeable* to water, no osmotic equilibration occurs. In this instance the hypotonic fluid in the DCT will not be concentrated, and a hypotonic urine is excreted.

Antidiuretic Hormone

The collecting tubule is normally impermeable to water. The action of a hormone called *vasopressin,* or *antidiuretic hormone* (ADH) causes the collecting duct to become water permeable and to form a concentrated urine. ADH is released from the neurohypophysics (see Chap. 25). An increase in osmolarity, indicative of a water deficit, causes an increase in ADH, which results in a decrease of water excreted. Certain conditions associated with a decrease in urine volume, such as dehydration or exercise, have increased levels of ADH. Patients with diabetes insipidus either have low levels of ADH or fail to respond to the hormone. Their urine production may be as high as 5–10 liters/day.

Countercurrent Multiplier

Although questions remain as to the exact mechanism(s) responsible for the creation of the corticomedullary osmotic gradient, this feature involves a countercurrent multiplication system (Fig. 29.9). The high osmolarity in the medullary region is due to increased concentrations of Na^+ and urea. The development of this osmolarity is based on the assumptions that a small 200-mosmol/liter osmotic gradient can be generated between the ascending and descending limbs of the loop of Henle and that the permeability characteristics of various segments of the renal tubule differ. It is the effect of more concentrated fluid turning the hairpin turn

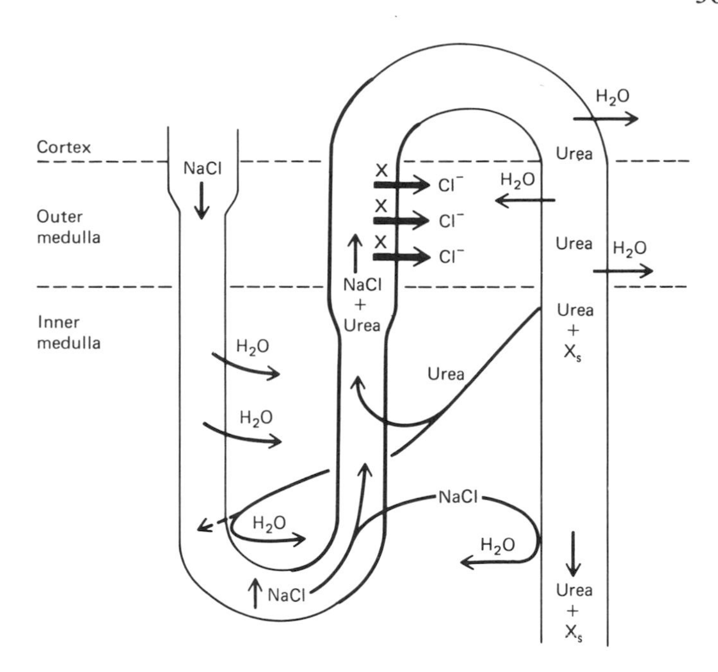

Fig. 24.10. Model of Kokko and Rector. Dark arrows indicate active chloride transport and Xs represents non-reabsorbable solute. In this model a large fraction of the osmolarity in inner medulla is due to urea (Adapted from Kokko, JP, et al. (1974) Fifth International Congress of Nephrology.)

of the loop that causes an increase in the osmolarity of the initial ascending segment. This process continues in the descending limb, resulting in an osmolarity far in excess of plasma's 300 mosmol/liter. Water and electrolytes in the tubule then equilibrate with the peritubular fluid to form the osmotic environment around the tubule.

A formal model utilizing the countercurrent multiplication principle is proposed by Kokko and Rector. They suggest that urea, which is very high in concentration in the distal tubule, diffuses from the collecting tubule and is largely responsible for the extraction of water from the descending limb. Since a large fraction of the osmolarity in the medullary interstitium is due to urea, a Na^+ gradient exists that favors the passive movement of Na^+ out of the initial segment of the ascending limb. Urea then diffuses into the vacated area and is recycled into the medullary interstitium via its exit from the collecting duct (Fig. 24.10).

Countercurrent Exchange

A requirement for the system of concentrating the urine, involving a hypertonic medullary region, is that both the Na^+ and water that are reabsorbed from the tubule be returned to the circulation but that their removal not diminish the high medullary solute concentration. This is accomplished first by having an extremely slow rate of medullary blood flow through the vasa recta (Fig. 24.3). Second, a process known as counterexchange is operative whereby large amounts of solute are not removed from the medulla. There is an exchange of solute in the medulla between the two limbs. The end result is that the plasma leaving the medulla is only slightly more concentrated than the plasma that entered,

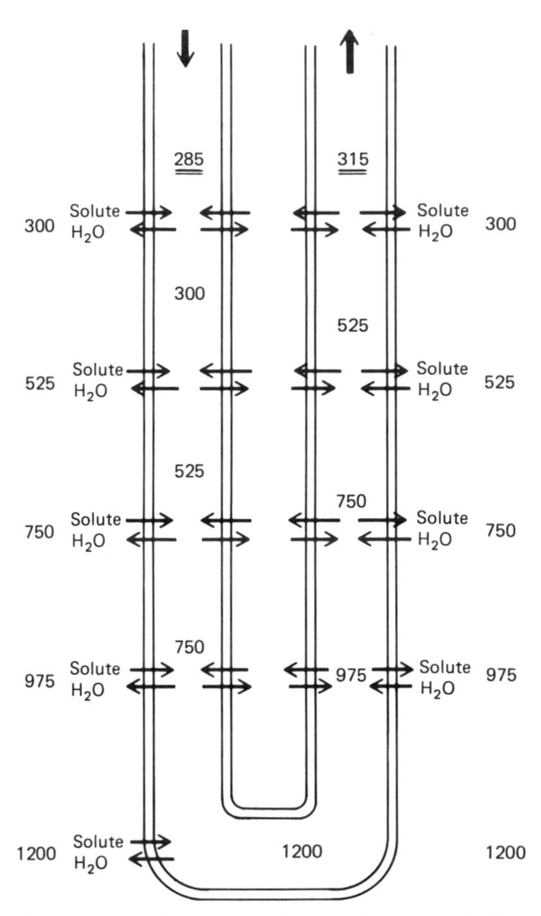

Fig. 24.11. Countercurrent exchange. Blood entering medulla has an osmolarity of 285 mosmol/liter, whereas blood leaving osmolarity of 315 mos-mol/liter. This process whereby blood leaving does not contain large amounts of solute prevents elimination of hypertonicity of medullary interstium. (After Netter FH (1973) Kidneys, ureters and urinary bladder. Ciba Medical Illustrations Vol. 6.)

thus preventing a washout of the medullary solute concentrations (Fig. 24.11).

Renal Hormones

There are two major hormone systems in the kidney. The first is the *renin-angiotensin system,* the major components of which are described in Chap. 29. Renin is a hormone produced by and released from the kidney. It is synthesized in a specialized group of epithelioid cells known as the juxtaglomerular cells (JG cells). These cells line the afferent arteriole and are situated between the afferent arteriole and a specialized segment of the distal tubule called the macula densa (Fig. 24.12). Various stimuli, such as a decrease in renal perfusion pressure, increased sympathetic nerve activity, increased circulating catecholamines, decrease in total body extracellular fluid volume, and changes in the electrolytic composition of distal tubular fluid as sensed by the macula densa increase renin release.

Renin acts on renin substrate, a plasma globulin synthesized in the liver, to produce a decapeptide, angiotensin I. In the presence of a converting enzyme found in both lung and kidney, angiotensin I is cleaved to an octapeptide, angiotensin

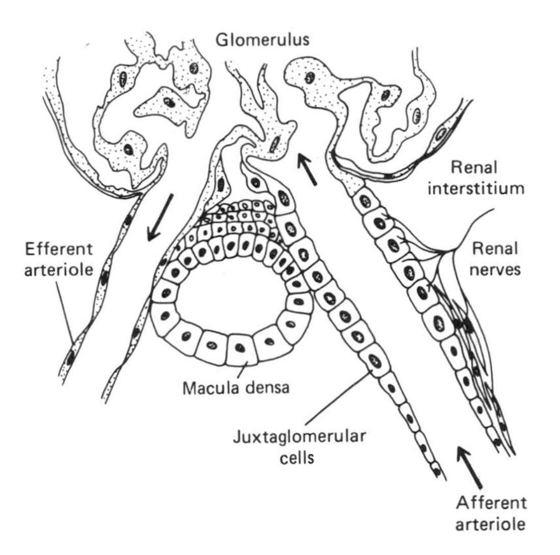

Fig. 24.12. Juxtaglomerular apparatus consisting of macula densa and juxtaglomerular cells (Adapted from Davis I (1971) Circ Res 28: 301.)

II (AII). The biologically active AII is a potent vasoconstrictor (see Chap. 14). It also increases the release of aldosterone, decreases the release of renin, and may directly enhance Na^+ reabsorption. The renin-angiotensin system is implicated in the initiation or maintenance of certain types of hypertension.

The second major endocrine complex within the kidney is the *prostaglandin* (PG) *hormone-like system*. PG's are synthesized from phospholipids and arachidonic acid by an enzyme complex known as PG synthetase, or cyclooxygenase. An entire family of PG hormones with many different physiologic activities exists.

Although PG's may be synthesized throughout the body, the medullary regions of the kidney are able to produce large amounts of PG; the major PGs being PGE_2 and $PGF_{2\alpha}$. In addition to their effects on the cardiovascular system, PGs are able to modify the effects of sympathetic nerve activity, increase renin release, decrease the constrictor effects of AII, and inhibit ADH. Prostaglandins are implicated in many diseases. It is interesting to note that aspirin, a potent PG synthetase inhibitor, decreases the formation of all PG.

The Kidneys and Acid-Base Balance

There are several means by which the kidney assists in the regulation of plasma pH or total body acid-base balance. They include: (1) control of plasma HCO_3^- concentrations, (2) regeneration of HCO_3^-, and (3) secretion of H^+ into the urine.

Control of Plasma-HCO_3^- Concentrations

As discussed in Chaps. 2 and 17, the role of the kidneys is to control the concentration of HCO_3^-. When plasma HCO_3^- is 28 mM/liter or less, all of the filtered HCO_3^- is reabsorbed. When plasma HCO_3^- exceeds 28 mM/liter, the concentration is referred to as the *renal threshold,* HCO_3^- begins to appear in the urine. Bicarbonate reabsorption occurs in both proximal and distal tubules.

Most of the filtered HCO_3^-, approximately 90%, is reabsorbed in the proximal tubule. Figure 24.13 illustrates the HCO_3^- that occurs in a proximal tubular cell. Sodium moves into the cell down an electrochemical gradient. Hydrogen ions, actively secreted in the tubule lumen react with HCO_3^- to form H_2CO_3, which dissociates, forming CO_2 and H_2O. The CO_2 moves into the tubular cell, and the reaction of CO_2 and H_2O to form H_2CO_3 is catalyzed by the enzyme carbonic anhydrase. The newly formed H_2CO_3 dissociates, making available a H^+ for secretion and the HCO_3^- is actively reabsorbed.

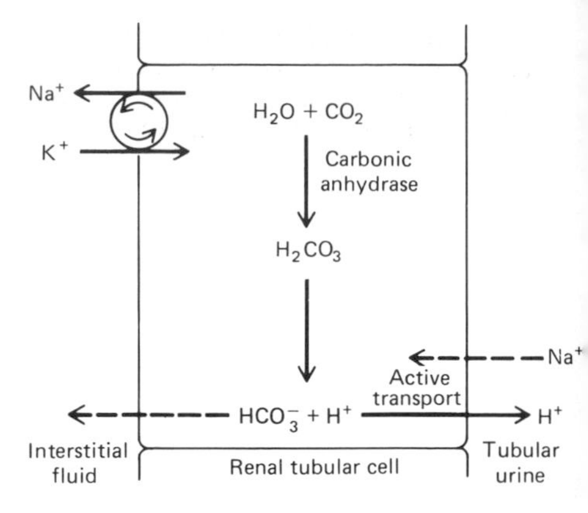

Fig. 24.13. Mechanism involved in reabsorption of HCO_3^- in kidney.

Besides reabsorbing HCO_3^-, the tubules also regenerate HCO_3^- to replace that which is depleted during the buffering of strong acids. How strong acids are converted to a neutral salt and dissociable carbonic acid by HCO_3^- is shown by the equation:

$$\underset{\text{Strong acid}}{H_3PO_4 + 2NaHCO_3} \longrightarrow \underset{\text{Neutral salt}}{Na_2HPO_4} + 2H_2O + 2CO_2$$

As shown in Fig. 24.14, the disodium phosphate (Na_2HPO_4) is converted to the monosodium phosphate (NaH_2PO_4) the net effect of which is the return of Na^+ and HCO_3^- to the circulation.

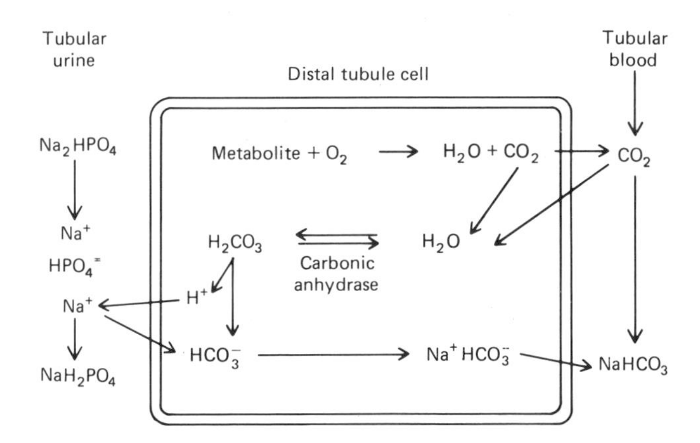

Fig. 24.14. Mechanisms involving conversion of disodium to monosodium phosphate and subsequent reabsorption of sodium bicarbonate.

Secretion of Hydrogen Ions

Hydrogen ions may be actively secreted into the urine. The amount of acid secreted in this manner depends on the urinary pH. The maximum gradient for H^+ secretion is reached when the urine pH reaches 4.5. Hydrogen ions may be combined with HCO_3^- to form CO_2 and H_2O and with HPO_4^{2-} to form $H_2PO_4^-$. An additional method to decrease the number of free H^+ in the tubular lumen is through their reaction with NH_3 to form NH_4^+. Ammonia is secreted by the renal tubule. A small quantity of it may be derived from arterial blood. Being lipid soluble NH_3 can diffuse through the tubular membrane into the tubule lumen. Once there, it reacts with secreted free H^+ to form NH_4^+, which is lipid insoluble, remains in the renal tubule, and is excreted.

Urine Formation

The kidneys are extremely important in attempting to maintain homeostasis. The chemical composition of urine and the amount excreted reflect the fluid and electrolyte status. Alterations in the urinary profile associated with dehydration are shown in Table 24.2.

Table 24.2. Representative urinary profiles of normal and dehydrated subjects.[a]

State	Urine Volume (ml/24 h)	Specific Gravity	Osmolarity (mosmol/liter)	pH	Urine Na^+ (mEq/liter)	Urine K^+ (mEq/liter)
Normal	1500	1.022	800	6.2	138	40
Dehydrated[b]	300	1.028	1100	5.9	110	120

[a] Adapted from Zambraski et al., (1975) Med Sci Sports 7:217.
[b] During dehydration, less urine is produced; the concentrated and acidic urine excreted reflects increased reabsorption of H_2O and Na^+.

Micturition

The urine that leaves the renal tubule via the collecting duct enters the renal pelvis from which the ureters transport the urine to the bladders (Fig. 24.1). The passage of urine through the ureters is facilitated by rhythmic peristaltic contractions, occurring at a rate of 1–5 times/min.

The urinary bladder is a hollow structure composed mainly of smooth muscle. The two major parts of the bladder that make up the body are the detrusor muscle and a small triangular area called the trigone. The right and left ureters enter and the urethra exits from the trigone. The bladder is innervated by afferent and efferent nerve fibers. Sympathetic fibers emerge via the inferior mesenteric ganglia, and parasympathetic fibers from the pelvic nerve innervate the body of the bladder and the internal sphincter. Nerve fibers to the external sphincter originate from the pudenal nerves. Afferent fibers arise from the ureters as well as from the bladder.

The bladder is a distensible organ. In man it is not until bladder volume reaches about 400 ml that the wall stretches almost to its maximum, and the pressure within is elevated drastically.

Micturition is the passage of urine from the bladder. It involves the contraction of detrusor and abdominal muscles

and the relaxation of the internal and external ureteral sphincters. These processes involve both autonomic and voluntary control. The initial phase of micturition involves progressive filling until bladder-wall tension increases to a level that initiates an afferent response. In man, a volume of about 150 ml results in the desire to void. When the afferent responses reach a critical level, a reflex is initiated that contracts the detrusor muscle and relaxes both sphincters. Micturition itself intensifies the efferent response.
Higher-level neural inputs from the brain stem have major inhibitory and facilitative influences. A general inhibitory effect with a concomitant tonic contraction of the external sphincter persists until one desires to void. The desire to urinate evokes the micturition reflex, and there is an inhibition of external sphincter contraction.

Selected Readings

Brenner BM, Rector FC (1976) The kidney. Saunders, Philadelphia
Handbook of Physiology (1973) Renal physiology. American Physiological Society, Washington
MTP International Review of Science (1974) Kidney and urinary tract physiology, Vol. 6. University Park Press, Baltimore

Review Questions

1. List the major anatomic features of the nephron and relate specific structures with their functional role.
2. Define the major processes of filtration, reabsorption, and secretion.
3. Define clearance and indicate how it is used to quantitate glomerular filtration rate and renal plasma flow.
4. List the factors that control GFR.
5. Define the TF/P ratio and indicate why it is useful in studying nephron function.
6. Describe the concentration and electrochemical conditions in the PCT and the steps in the reabsorption of Na^+ and K^+.
7. What are the various mechanisms or control factors that may regulate electrolyte excretion?
8. Describe the events responsible for the reabsorption of 98% of the water filtered at the glomerulus.
9. Describe the renin-angiotensin and prostaglandin hormonal systems.
10. How would the kidneys assist in restoring normal plasma pH if the body were stressed with an acidotic condition?
11. List the events in micturition.

Chapter 25

Endocrinology: Functions of Hypophysis and Hypothalamus

Endocrinology is the study of secretions (hormones) from endocrine (ductless) glands; each secretory cell has a surface close to a venous sinusoid or capillary, and this arrangement facilitates rapid absorption into the circulation. Hormones are usually secreted in small amounts and transported in the bloodstream to specific sites to promote and regulate reactions of the organ or organism. Secretion of a hormone is first discovered by its lack of effect or absence, which occurs after removal (ablation) of the gland or in a diseased gland. The suffix *-ectomy* means "ablation of." Thus thyroidectomy means removal of the thyroid.

Hormones extracted from glands are occasionally injected into an animal whose endocrine gland has been ablated to restore the animal or organ to its normal function; this is known as replacement therapy. The crude extract may contain several hormones, the latter can be separated and purified, chemically identified, and later synthesized.

Mechanism of Hormonal Action

When a hormone is absorbed and transported to its site of action, it may activate the cells of the organ or tissue by at least two general mechanisms. It can: (1) stimulate production of cyclic adenosine monophosphate (cAMP) which in turn elicits specific cellular functions, or (2) activate genes to cause the synthesis of intracellular proteins, which initiate specific cellular reactions (see Chap. 22).

The hormone produces its effect presumably by its combination and action with a specific chemical moiety or part of the cell known as a *receptor*. The nature of this receptor is not fully understood, but it has been isolated and its specificity can be demonstrated by appropriate immunochemical techniques.

Role of cAMP

This compound is believed to play a definite role in endocrine action. It is formed from ATP as a result of the action of the enzyme adenyl cyclase (AC) found on cell membranes, as follows: ATP + AC → cAMP. The nature of the relationship between the cell receptor and the final reaction is not fully understood, but cAMP is proposed by some workers as the second messenger for some hormonal actions. The first messenger is the hormone itself (Fig. 25.1). The cAMP reaction just mentioned is a common one that also occurs in several nonhormonal receptors and reactions. From the time a hormone is synthesized in the cell until it is used or inactivated, it undergoes numerous stages or processes (Fig. 25.2).

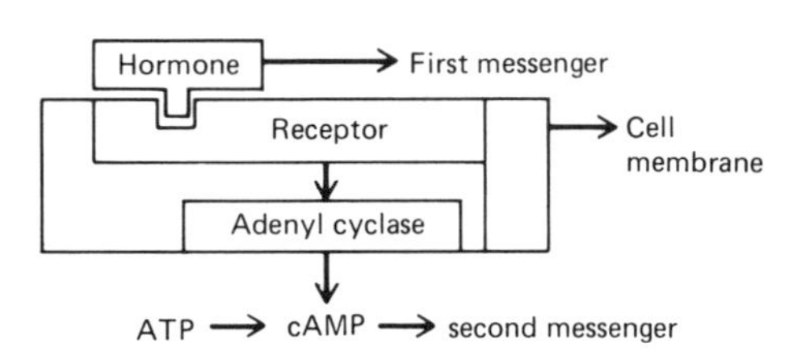

Fig. 25.1. Role of cell receptors and cAMP in hormonal activity (see text).

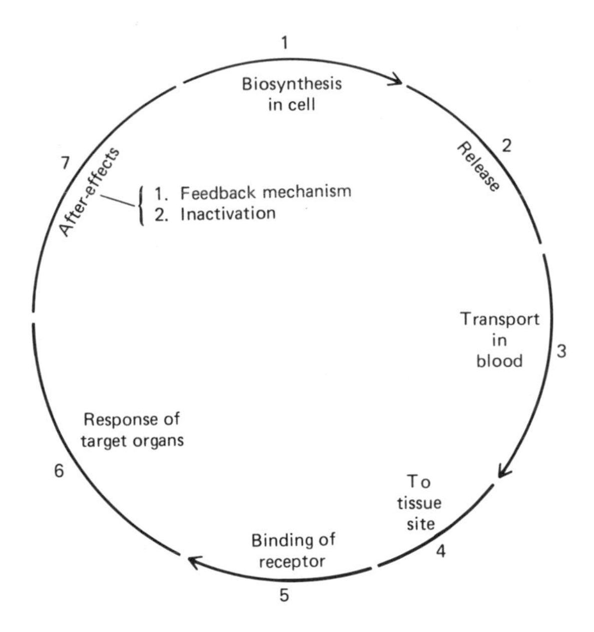

Fig. 25.2. Life history of hormone.

Determining Hormonal Concentration

A hormone may be extracted from the organ producing it to form a crude extract, which after refinement can be injected. Concentration in the bloodstream is determined by biologic assay, appropriate chemical reaction, or immunoassays.

Biologic assays depend on the response of a tissue or organ to the specific hormone to be tested. Follicle-stimulating hormone (FSH), e.g., may be assayed by the effect of its administration on testicular weights of sexually immature animals. The increase in the weight of the testes when compared to testicular weights of nontreated animals gives a measure of the hormone's concentration. Biologic assays are relatively insensitive and imprecise.

If the chemistry of the hormone is known and chemical methods are available for its determination, the *chemical assay* is a moderately efficient method. Many of the methods, however, are not sensitive enough to determine small amounts of the hormone, and certain hormones (peptides and proteins) are virtually impossible to determine by these methods.

Radioimmunoassay is the method of choice for determination of peptides, proteins, and steriod hormones when great sensitivity is desired. This method is based on the immunogenic properties of these hormones, or the antigen-antibody reaction:

1. First a hormone like the one produced by the endocrine gland is purified and injected into an animal for a given time. After repeated injections, the animal develops antibodies in the blood to the protein.
2. A small amount of purified hormone is labeled with a radioisotope (usually radioactive iodine) and a known amount placed in a test tube to which has been added a small measured amount of plasma containing antibodies.
3. Then a known amount of blood plasma from the unknown animal, whose blood is to be assayed is added to the tube. The antibody reacts (binds), with the antigen both labeled and unlabeled. Since the binding capacity of the antibody is limited, the ratio of the bound-labeled to unbound-labeled antigen (free) varies with the amount or concentration of unlabeled antigen in the plasma of the animal tested.

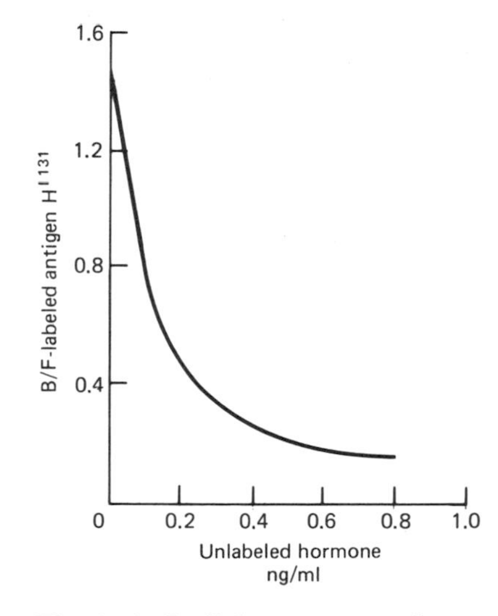

Fig. 25.3. Radioimmunoassay for unknown hormones. Concentration of labeled hormone (H^{I131}) and B/F plotted against concentration of unlabeled hormone (antigen) (see text).

Therefore the proportion of labeled hormone bound to antibody decreases as more unlabeled hormone (antigen) is added, which competes for binding sites on the antibody. Plasma levels of unknown hormone can then be assayed by extrapolating from a plot of bound/free (B/F)-labeled hormone to the concentration of unlabeled hormone (antigen) (Fig. 25.3). Radioimmunoassays have been developed for nearly all hormones, including TSH, FSH, LS, MSH, GH, testosterone, estrogens, and prostaglandins.

Endocrines discussed in this chapter are grouped into the following categories:

1. Hypothalamic releasing factors
2. Tropic hormones from adenohypophysis and their effects on: (a) target organ hormones from thyroids, adrenal cortex and medulla, and gonads and (b) organs or tissues directly (mammary glands, cells [growth])

3. Hormones not stimulated by adenohypophysis, including: (a) pancreas (insulin and glucagon), (b) parathyroid hormone and calcitonin, and (c) others (see Chap. 20).
4. Hormones from neurohypophysis (oxytocin, vasopressin)

Hypothalamic and Pituitary Hormones

The hypothalamus produces numerous hormones that cause the release of tropic hormones from the anterior lobe (adenohypophysis, or pars distalis) and from the posterior lobe (neurohypophysis or pars nervosa) (Figs. 25.4 and

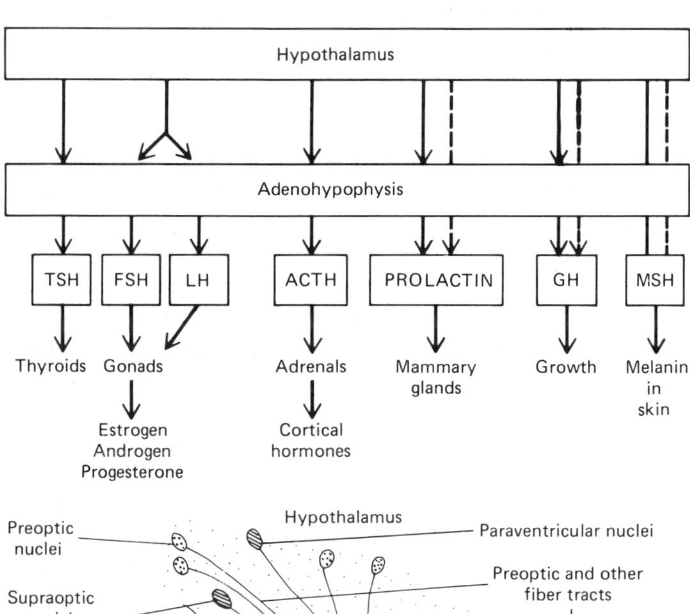

Fig. 25.4. Relationship of hypothalamic-releasing (-) or inhibiting (---) factors on tropic hormones of adenohypophysis (see text).

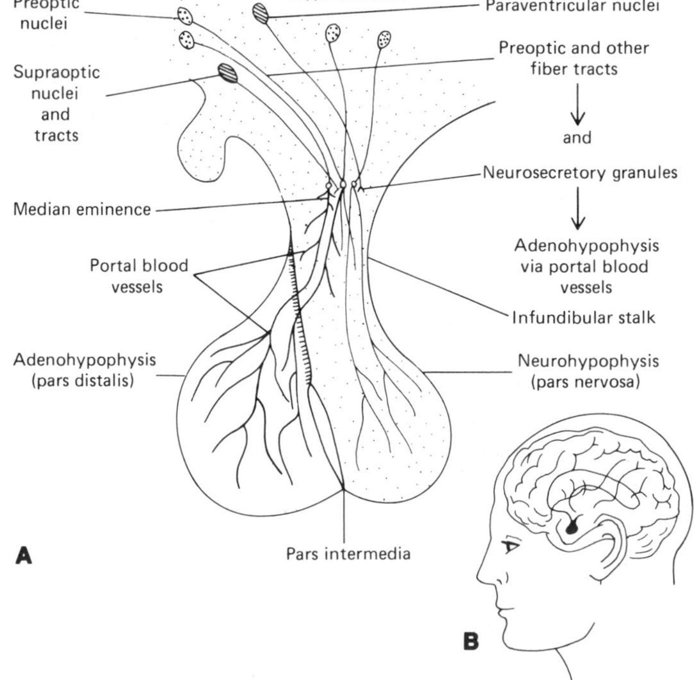

Fig. 25.5A and B. Hypothalamic control of pituitary. **A** Paraventricular and supraoptic nuclei and their fiber tracts transport oxytocin and vasopressin to posterior lobe. Neural stimuli from other fiber tracts, and nuclei from hypothalamus transport neurosecretory material and releasing factors to adenohypophysis by portal blood vessels (veins). **B** Location of pituitary (black).

25.5). These releasing hormones from hypothalamus cause the adenohypophysis to release: (1) adrenocorticotrophic hormone (ACTH), (2) follicle-stimulating hormone (FSH), (3) LH-luteinizing hormone (LH), (4) thyrotropic hormone (TSH), (5) prolactin, (6) growth hormone (GH), and (7) melanocyte-stimulating hormone (MSH).

The hypothalamus also stimulates the *posterior lobe to release:* (1) antidiuretic hormone (ADH) in man (arginine vasopressin) and in birds (arginine vasotocin), and (2) oxytocin.

Details concerning the release and effects of tropic hormones will be considered in the appropriate sections on the tissue and target organ hormones.

Hypothalamic-Releasing Factors

There is at least one hypothalamic-releasing factor for each of the hypophyseal tropic factors (Fig. 25.4) except for FSH and LH. In the latter cases there is only one that has so far been isolated, which causes the release of both FSH and LH. Three of these factors have been isolated and purified from sheep and pigs and their chemical structure is definitely known (Table 25.1).

Table 25.1. Hypothalamic releasing factors.

Releasing Factors	Description	Amino Acid Sequence (not circled)
TSH-RH(TRH)	Tripeptide	(Pyro) -Glu-His-Pro-(NH$_2$)
FSH, LH >RH	Decapeptide	(Pyro) -Glu-His-TrP-Ser-TYR-Gly-Lev-Arg-Pro-Gly-(NH$_2$)
GH-RH (GRH)		Structure indefinite
GH-R (GIF) (somatostatin)	Tetradecapeptide or growth-inhibiting factor	(H)-Ala-Gly-Cys-Lys-ASN, Phe, Phe, TRP,-Lys-Thr-Phe, Thr, Ser, Cys---- (OH)
ACTH-RH[a] (CRH)	Partially purified	Structure indefinite
Prolactin (PRH)		Structure indefinite
Prolactin-PRTH[b] (inhibiting)		Structure indefinite
MSH[c] (RH)	Tripeptide	Pro-Leu-Gly-(NH$_2$)
MSH (RIH)[d]	Pentapeptide	(H) -Cys-Tyr-Ile-Gln-Asn- (OH)

[a] RH, Releasing hormone.
[b] Definite evidence that neurotransmitter dopamine from hypothalamus also has prolactin-inhibiting power.
[c] MSH, Melanocyte-stimulating hormone.
[d] RIH, Inhibiting hormone.

Neurosecretoion

The components of the hypothalmic-hypophyseal system include: (1) the adenohypophysis and (2) nuclei and fiber tracts from the hypothalamus and median eminence, which relay impulses to the neurohypophysis, which releases oxytocin and vasopressin and indirectly to the adenohypophysis, which is stimulated to release various hormones (Figs. 25.4 and 25.5).

The nerve connections from the hypothalamus extend into the neurohypophysis (Fig. 25.5), but there are no neural connections between the hypothalamus and adenohypophysis. There is abundant evidence that neurosecretory material (hormones) from the hypothalamus and median eminence finds its way to the adenohypophysis by the portal blood vessels (veins) in the infundibular stalk (Fig. 25.5). Interruption of this vascular pathway by transection and blockage of blood flow in the infundibular stalk and median eminence prevents the movement of hypothalamic-releasing factors into the adenohypophysis.

Feedback Regulation of Neurosecretion

It is well known that circulating target organ hormones may act either back on the pituitary to decrease or inhibit release of tropic hormones or on the hypothalamus (Fig. 25.6) to decrease release. This is called negative feedback and tends to control the level of hormones required. Steroid target organ hormones (estrogen, androgen, and progesterone) mainly inhibit at the level of the hypothalamus, and thyroid hormones inhibit at the level of the hypophysis. Certain tropic hormones of the pituitary (ACTH, GH, FSH, and LH) inhibit at the level of the hypothalamus and probably reach the gland by back diffusion (short loop).

Hormones reaching the hypothalamus by the bloodstream do so by long-loop feedback. In some instances a target organ hormone stimulates the release of its tropic hormone or the releasing hormone. This is known as positive feedback.

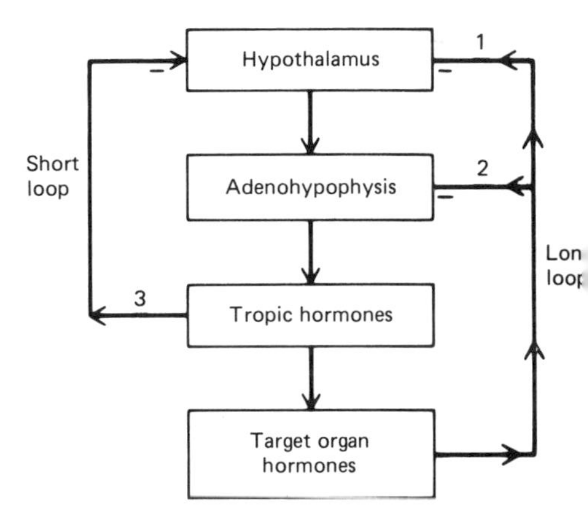

Fig. 25.6. Short (3) and long (1 and 2) feedback (inhibiting -) effect on production and release of hormones from adenohypophysis and hypothalamus by target organ hormones and tropic hormones of adenophypophysis. Steroid hormones (androgen and estrogen) inhibit mainly at level 1 and thyroid hormones at level 2 (see text).

Adenohypophyseal Hormones

The adenohypophysis is located in a saddle-like depression of the sphenoid bone (sella turcica) and weighs about 500 mgs. It is roughly 1.4 cm at its greatest diameter (Fig. 25.5). There are three major cell types of the adenohypophysis as revealed microscopically, based mainly on their staining reaction. They are: (1) chromophobes, whose nuclei stain blue (basic) and which make up about one-half of the total cells in the gland; (2) acidophils, whose cytoplasmic granules take an acid stain (pink to orange granules) and make up about 35% of the total cells; and (3) basophils, whose cytoplasmic granules stain blue and which represent about 10% of the total cells. The exact numbers vary with age and other factors.

There is variation in the size and shape of these cells and in the cytoplasmic granules that may reflect their secretory state. There is fair evidence that the acidophils secrete GH and prolactin and that basophils—varying in type and shape of cytoplasmic granules—secrete most of the other hormones of the adenohypophysis. The best test for localizing cell secretory activity is the immunocytologic technique. The chromophobes are believed to secrete MSH.

All of the hormones from the adenohypophysis have been isolated and purified and some of them have been synthesized. The hormones TSH, LH, and FSH are glycoproteins with molecular weights of about 30,000; they are made up of carbohydrates attached to two polypeptide chains. Each hormone is composed of two glycoprotein subunits of amino acids arranged in sequence (α and β). The subunits of α-TSH and β-LH are almost identical and each contains 96 amino acids. The biologic activity is determined by the β subunits present; α-LH contains 120 amino acids; β-TSH alone has 113 amino acids. The latter are inactive but become active when joined together. Examples are:

α-LH + β-LH = Active LH

α-TSH + β-TSH = Active TSH

α-TSH + β-LH = Active LH

α-LH + β-TSH = Active TSH

Prolactin and GH are proteins with a high molecular weight (about 22,000) and contain about 190 amino acids in a definite sequence. The structural formula for human GH is shown in Fig. 25.7 and shows the sequence and number (188) of amino acids. Human GH is structurally similar to ovine prolactin. Prolactin exhibits considerable growth hormone-like activity, and GH has some prolactin activity.

ACTH and MSH are polypeptides that have some features in common. They possess a common core of seven amino acids with similar adjoining ones; ACTH has a molecular weight of about 4600 and contains a chain of 39 amino acids. The structure of ACTH varies slightly among species as to the sequence of certain amino acids. The first 23 amino acids in the chain are the same for all species and represent the core and most active part of its molecule; amino acids 24–29 vary among the different species.

Activity

The general function of the pituitary hormones will be given here and further details will be considered in the appropriate sections on target organ hormones.

Thyrotropic hormone (TSH): stimulates growth and development of thyroid cells and *initiates* the synthesis and release of the thyroid hormones.

ACTH: stimulates growth and development of adrenal cortex and synthesis and release of cortical hormones.

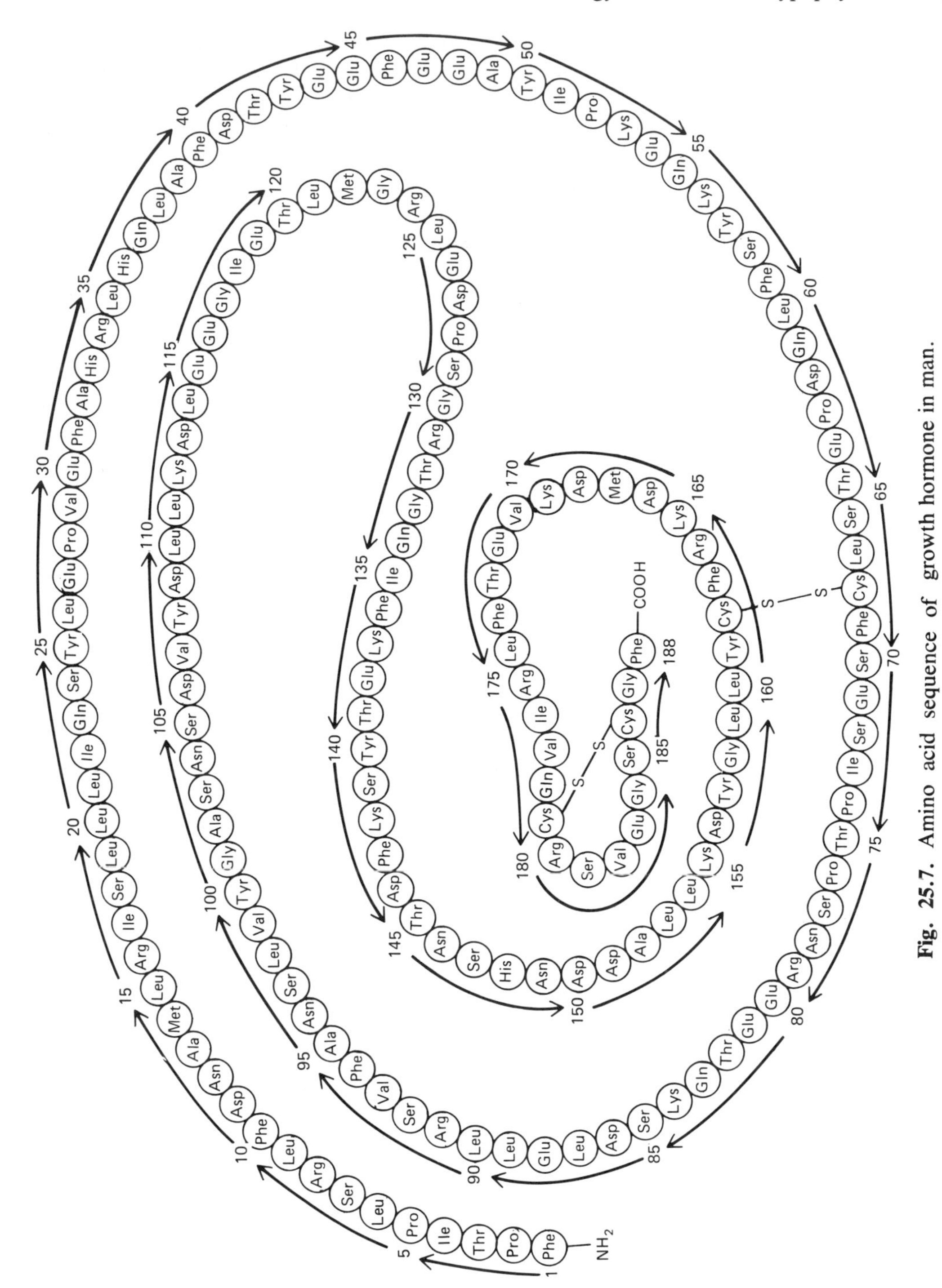

Fig. 25.7. Amino acid sequence of growth hormone in man.

FSH: stimulates ovarian follicle growth and development and subsequent release of estrogen from follicle; also stimulates testicular growth and spermatogenesis.

Luteinizing hormone (LH): causes cyclic release of ova from ovary (ovulation) and development of corpora lutea of ovulated ova, which produce progesterone; also stimulates growth and development of interstitial cells of testis (Leydig's cells) to produce androgen.

Growth hormone (GH) or somatotropin: promotes growth and development of bone and other tissues in the young; its action is mediated by low molecular weight polypeptides called *somatomedins,* which are produced as a result of GH action on the liver; human beings are unresponsive to GH from other species.

MSH: stimulates dispersal of melanin in skin and is present in intermediate lobe of mammals but in the anterior lobe of birds.

Prolactin: helps to maintain corpus luteum in rat and mouse (luteotropic) but not in pig, cow, sheep, or man; its main function is to promote and maintain lactation in mammals and to induce broodiness and crop sac secretion in birds; it may promote growth in some species.

Prolactin-inhibiting factor: inhibits or decreases prolactin secretion in mammals, but there is no inhibiting factor in avian species.

Levels of Hypophyseal Hormones

Blood and pituitary levels of tropic hormones are a reflection of their release and utilization. As the hormones are released, the glandular level usually falls and increases in the blood and vice versa. However, this is not always the case because increased release may increase synthesis with no appreciable change in pituitary level. Removal of or injury to the hypophysis or hypothalamus decreases or prevents release of hormones, and blood levels drop drastically.

Growth Hormone

Plasma levels of GH in human adults are about 3–4 ng/ml; GH has a half-life of about 30 minutes and its calculated daily secretion rate is from 1–4 mg. Recently reported levels, based on mean integrated 24-hour blood samples in man, are: prepubertal males (5.10 ng/ml), pubertal males (7.37 ng/ml), and young adults (3.47 ng/ml).

As would be expected, the levels are higher in growing individuals. There is a tendency toward lower levels in the forenoon and increasing levels toward evening. The levels tend to be higher in females. *The human adult hypophysis* contains variable amounts of GH, ranging from 4–15 mgs. *Starvation* and depletion of body stores of protein (kwashiorkor) increase GH levels in the blood to as high as 40–50 ng/ml; GH-regulating factors are summarized in Table 25.2.

Table 25.2. Growth hormone (GH) regulating factors.

Increasing GH	Decreasing GH
Going to sleep	REM sleep
Exercise	Free fatty acids
Apprehension and stress	Glucose infusion
Hypoglycemia and insulin	GH hormone infusion
Fasting	Somatostatin
Certain amino acids	
Tumors of hypophysis	
L-Dopa	
Infusions of calcium	

Abnormalities of Growth-Hormone Secretion. Dwarfism results from *decreased secretion* of GH during infancy and childhood when plasma levels of GH are low (1–3 ng/ml). Dwarfism from GH deficiency is characterized by a proportionate diminution in size of all organs, whereas that which is caused by thyroid deficiency tends to be disproportionate. The pituitary dwarf experiences deficiencies in other hormones and does not obtain normal adult sexual development.

Increased secretion of GH in children causes abnormal growth and *gigantism,* characterized by extreme heights of 8–9 feet; all tissues grow, particularly the long bones of the body. Gigantism is usually caused by a tumor of the pituitary.

Acromegaly is a condition produced by excess GH secretion in adults (with plasma levels ranging from 20–100 ng/ml) after the long bones have reached their definitive size. The bones can no longer lengthen under the effects of GH but they can thicken, and the soft tissues can grow. The hands, feet, and lower jaw are particularly enlarged as are those of nose and forehead; the lower jaw tends to protrude (Fig. 25.8).

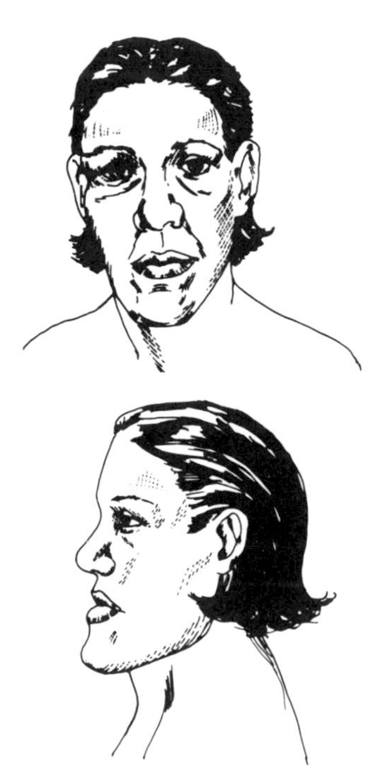

Fig. 25.8. Acromegaly. Note protrusion of lower jaw and enlarged nose.

Action of GH. The basic action or effect of GH, which is on the metabolic processes of the body, is mediated by somatomedins (polypeptides produced as a result of GH action on the liver). These effects include: (1) increased rate of protein synthesis and decreased breakdown of proteins, (2) increased mobilization of fats and use of fats as a source of energy, and (3) decreased carbohydrate utilization.

Levels and detailed actions of TSH, ACTH, FSH, LH, and prolactin will be discussed under target organs concerned, namely, thyroids, adrenals, gonads, and mammary glands. See relevant chapters.

Posterior Pituitary Hormones

The principal posterior lobe hormones of most mammals and man are oxytocin and arginine vasopressin; in birds and

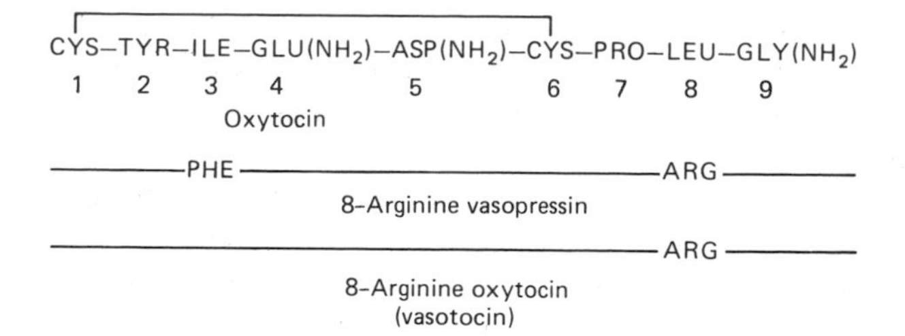

Fig. 25.9. Structure of neurohypophyseal hormones of mammals (oxytocin vasopressin) but vasotocin instead of vasopressin in birds and lower vertebrates. Blank lines indicate amino acids in common with oxytocin.

other lower forms, however, they are oxytocin and arginine vasotocin. Lysine vasopressin is found in pigs.

Each of these hormones is made up of a chain of nine amino acids arranged in a definite sequence attached to a side chain (Fig. 25.9). These hormones are stored in the posterior lobe, and their concentration is high, approximating 25 μg of arginine vasopressin, which is usually 1.5 times that for oxytocin. These hormones are produced in the supraoptic and paraventricular nuclei of the hypothalamus, and the neurosecretory granules from these nuclei migrate down the fiber tracts to the neurohypophysis (Fig. 25.5) where they are stored in association with the carrier proteins—neurophysin I for oxytocin and neurophysin II for vasopressin.

Recent evidence indicates that most of the oxytocin is produced by the paraventricular nuclei and most of the vasopressin by supraoptic nuclei in a number of mammalian species.

Activity

Arginine vasopressin is the antidiuretic hormone (ADH) of mammals and vasotocin is the ADH of avian species. It acts on the distal tubules of the kidneys and collecting ducts to increase water resorption and decrease excretion. It is most important and active in those species (camel, kangaroo, rat) that live in the desert where water is scarce and its conservation most important.

Secretion of ADH is influenced by water and electrolyte levels in blood and tissues. Dehydration and a decrease in body water increase the secretion of ADH. The level of ADH in blood, which reflects secretion rate, varies according to state of hydration (Table 25.3). Accordingly, a decrease in water level and blood volume increases ADH se-

Table 25.3. Hydration and secretion rates and levels of ADH in man.

Man	Levels in Plasma (μunits/ml)	Secretion Rate (ml/h)
Normal man	2	10
Dehydrated man	6–8	20–40
Hemorrhage (severe)	900	—
Water loading (hydrated)	<1	—

—, No data.

cretion greatly, and hydration decreases it. Changes in plasma osmolality (in electrolytes), particularly an increase, is a powerful stimulant to ADH release in mammals but not in birds.

Stimuli other than those already mentioned may induce ADH secretion. They include stimuli from areas of the central nervous system (CNS) above the hypothalamus, such as pain, anxiety, surgical stress, and certain drugs. Increased alcohol intake decreases ADH secretion and results in increased urine output and dehydration.

An abnormal decrease in ADH secretion causes diabetes insipidus, which is characterized by increased consumption of water and urine formation. It may be controlled by injections of arginine vasopressin.

Oxytocin

The plasma level of oxytocin in man is 1 to 5 μunits/ml. Its half-life is short (1–4 min), and it is metabolized rapidly and excreted in the urine. The main stimulis for release of oxytocin in mammals is suckling and distention of cervix and vagina. The suckling reflex involves receptors in the mammary gland that relay the stimulus to the brain and hypothalamus and cause the release from the latter to the bloodstream (Fig. 25.5). Oxytocin is released in short bursts, or episodes of secretion, and in variable amounts.

Oxytocin is carried to the mammary gland and causes contraction of the myoepithelial cells present in the alveoli and milk ducts and results in milk "letdown." It causes milk letdown only (milk already present in milk ducts) and does not affect lactation or milk secretion (see Chap. 26).

The suckling reflex is important in milk cows at milking time. Several stimuli other than suckling itself evokes the suckling reflex, such as pleasant surroundings, sight of milk pail or milking machine, and sight of calf. Fright and excitement inhibit milk letdown.

Delivery of Fetus

The role of oxytocin in contraction of uterine musculature at term and in normal delivery of fetus is not clear. It is known that human blood levels of oxytocin increase during labor to a maximum of 200 μunits/ml. Thus the increased levels of oxytocin may increase contractile movements of uterus and facilitate labor, particularly after labor has begun. Distention of uterus may play a more important role in labor. Actually hypophysectomized women with no posterior lobe hormones can initiate labor.

Vasotocin has the same contractile effect in chicken uterine tissue as oxytocin has in mammals, and its level in the blood increases enormously just before the egg is laid.

Selected Readings

Bentley PJ (1976) Comparative vertebrate endocrinology. Cambridge University Press, London and New York
Catt KJ (1971) An ABC of endocrinology. Little, Brown, Boston
Ganong WF (1977) Review of medical physiology, 8th edn., Lange Medical, Los Altos, California
Greep RO, Koblinsky MA (1977) Frontiers in reproduction and fertility control. MIT Press, Cambridge
Greep RO (1977) (ed): Reproductive physiology, International Review of Physiology, Vol. 13, Chaps. 1 and 8. University Park Press, Baltimore
Ingbar SH (1976) (ed): The year in endocrinology 1975–1976. Plenum, New York
McCann SM (ed) (1977) Endocrine physiology II, International Review of Physiology, Vol. 16, Chapter 1–3. University Park Press, Baltimore
Sturkie PD (1976) Avian physiology, 3rd edn, Chap. 15. Springer Verlag, New York
Tepperman J (1973) Metabolic and endocrine physiology, 3rd edn., Year Book Medical, Chicago

Review Questions

1. Define endocrine and exocrine glands.
2. What are the mechanisms of hormonal action?
3. What are the hormones of the anterior pituitary (adenohypophysis)? Name them and give the principal function of each.
4. What factors influence the release and secretion of these anterior lobe hormones?
5. What is the neurosecretory system? How does it operate?
6. How is the secretory function of the adenohypophysis related to cell types of the gland?
7. What does an excess of growth hormone production cause in the adult?
8. What does a deficiency of FSH production cause?
9. What hormones are produced by the posterior lobe of the pituitary? What are their functions?

Reproduction

Chapter 26

The male reproductive organs (Fig. 26.1) include the testes, penis, and accessory ducts and glands by which secretions of the testes, prostate, and seminal vesicles as well as urine reach the penis.

The testes are paired and consist of seminiferous tubules where spermatogenesis occurs, and efferent ducts by which sperm are conveyed to the epididymis. The sperms are stored there and then carried to the penis either by the ductus or vas deferens. The urethra is the common opening in the penis by which urine from the bladder and secretions from glands and testes make their exit from the body.

The penis is made up of a sensitive area (the glans) and much spongy tissue (the corpus cavernosum and spongiosum), which becomes congested with blood to cause an erection (Fig. 26.1B). The length and diameter of the penis increase greatly on erection.

Erection, Ejaculation, and Emission

Erection is caused by the dilation of the arterioles of the penis. The corpus cavernosum and spongiosum, the spongy portions of the penis, become filled with blood; this congestion compresses the veins, prevents outflow of blood, and serves to maintain the erection. The efferent or motor

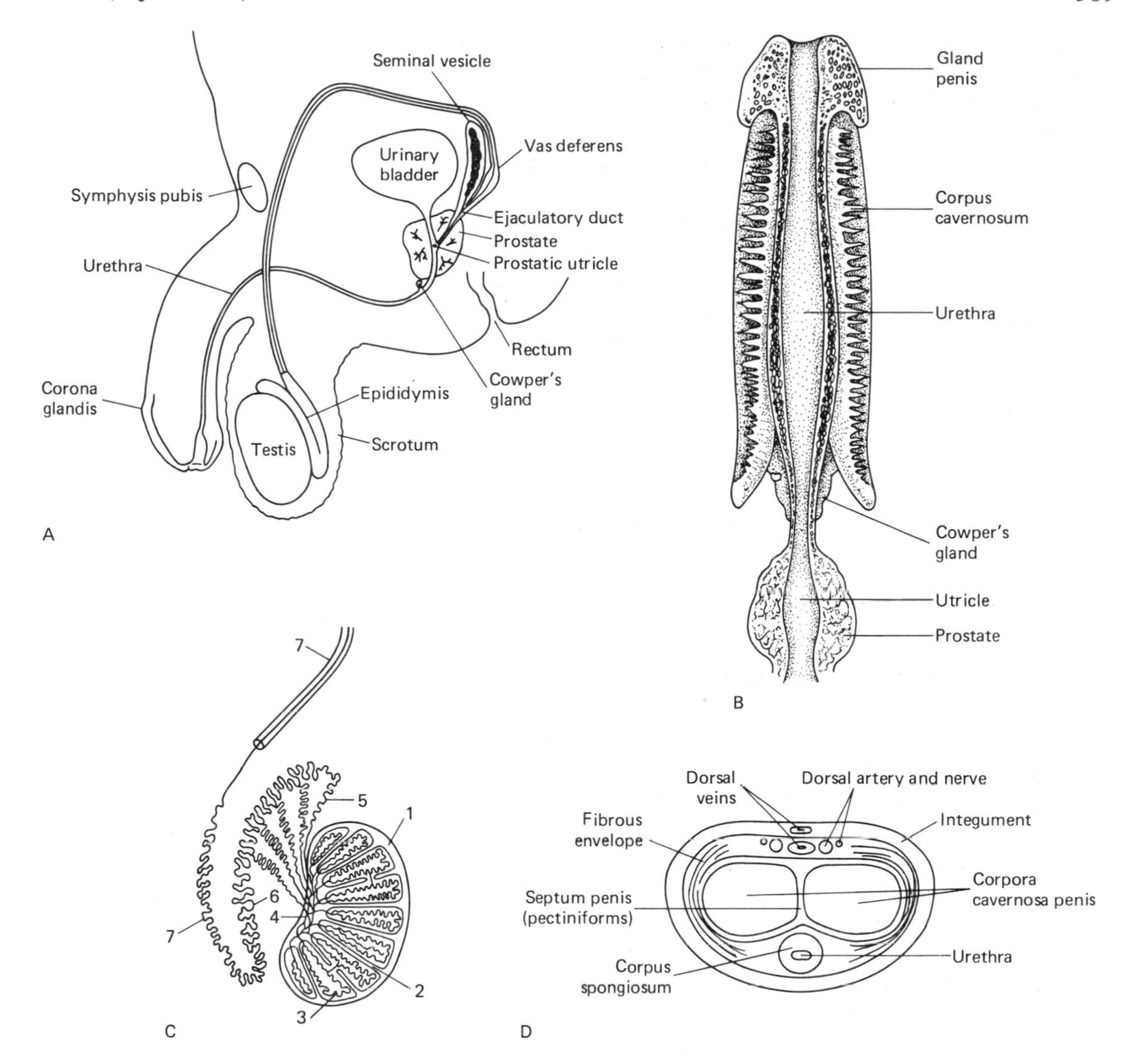

Fig. 26.1. **A** section of pelvic area showing reproductive organs of male; **B** longitudinal section of penis showing detailed structure; **C** section of testes showing (1) tunica albuginea, (2) septum of testes, (3) seminiferous tubule, (4) rete testes, (5) efferent ducts, (6) epididymis, (7) vas deferns; and **D** cross section of penis.

nerves involved in the vasodilatation are the sacral parasympathetic nerves (nervi erigentes).

Erection is produced by physical and psychic stimuli. Physical stimuli, such as touch and massage of the penis, relay the afferent impulse to the pudendal nerve, to the spinal cord reflex center, and to higher brain centers. Other afferent stimuli, such as anal and scrotal stimulation, reinforce the response. Distention of the bladder and irritation of the prostate gland or urethra may also stimulate sexual desire. *Sexual behavior* as influenced by the hypothalamus is discussed in Chap. 10.

Psychic stimulation, such as thoughts of sex, the sight of nudity, and erotic dreams causes sexual arousal and erection. Wet dreams (nocturnal emissions) are common in young males.

Emission begins with contraction of the epididymis and vas deferens, which forces sperm into the urethra where it is propelled further by the contraction and secretions of both the seminal vesicles and the prostate gland. All these secre-

tions, including those from Cowper's gland constitute semen. *Ejaculation* represents the culmination of the act of expelling semen from the urethra and penis.
The afferent nerve pathways involved are mainly from touch receptors in the glans penis that reach the spinal cord through the internal pudendal nerve. Emission is effected by motor impulses orginating in L_1 and L_2 of the spinal cord that pass to the penis (hypogastric nerves). Ejaculation is caused by the rhythmic contraction of the skeletal muscle encasing the erectile tissue of the penis (Fig. 26.1). The spinal reflex centers for this act are located in the upper sacral and lower lumbar segments of the cord; the motor pathways are via the first to the third sacral roots and the internal pudendal nerves.
The volume of ejaculate usually is from 2–6 ml (average 3.5 ml). It is slightly alkaline with a pH of 7.0–7.5. Semen contains large amounts of prostaglandins (hormones) produced by the seminal vesicles.

Spermatogenesis and Spermatozoa

Formation of spermatozoa in the testes is illustrated in Fig. 26.2. The mature sperm (Fig. 26.3), or spermatozoa, develop from: (1) spermatogonia, (2) primary spermatocytes, (3) secondary spermatocytes, (4) spermatids (immature sperm), and (5) spermatozoa.
Sertoli cells may be involved in maturation of the spermatids. About 70 days are required for the formation of mature sperm from the primitive spermatogonial cells.
The spermatozoa vary in size and shape among species but are structurally similar in all (Fig. 26.3). The sperm has a head piece, containing the nucleus; a midpiece; and a long, whip-like tail piece. The head piece penetrates the ovum in fertilization.

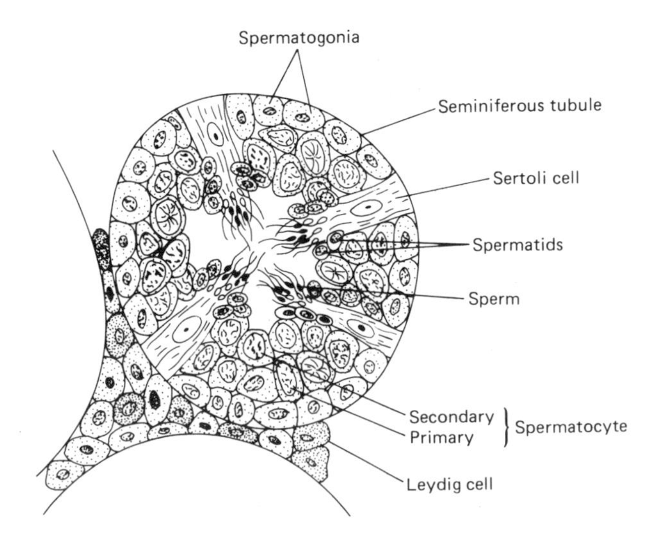

26.2

Fig. 26.2. Cross section of testes showing seminiferous tubules and formation of sperm from spermatogonia, primary spermatocytes, secondary spermatocytes, spermatids, and Leydig cells where testosterone is synthesized.

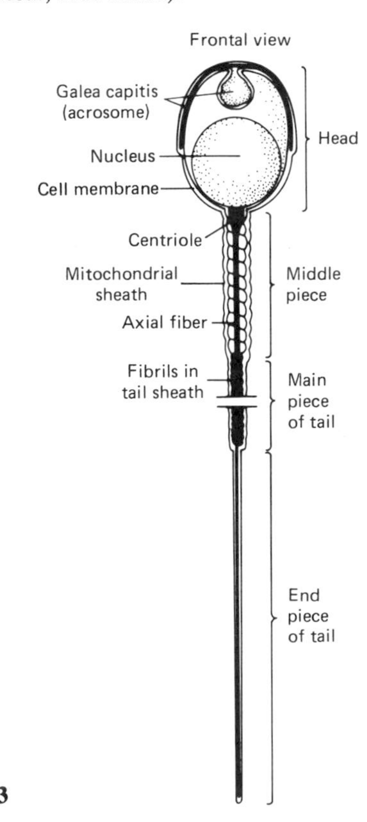

26.3

Fig. 26.3. Diagram of human spermatozoon. (After Ganong WF (1977) Medical Physiology 8th ed. Lange Medical, Los Altos).

Fertility

The ovum is capable of being fertilized for about 24–36 hours after it is expelled from the ovary. The life of most sperm in the female tract is 24–48 hours, although some few remain alive for 72 hours. Thus, fertilization time is limited to a few days at most.
A normal ejaculation of 3.5 ml of semen contains on the average 120 million sperms. The value, however, may range from a low of 40 million to more than 400 million. Only 100 or more of these spermtozoa find their way to the fallopian tubes and only one fertilizes the ovum. The remainder die in the female tract. Since only one sperm is required for fertilization, the total number of sperm in semen would appear not to be important, but it is. When the sperm count is less than 20 million/ml of semen, fertility is poor or greatly reduced. Accordingly, the high numbers of sperm reflect some indication of vitality and fertilizing capacity.

Capacitation and Inhibition

In some species the sperm, after leaving the epididymis, must spend some time in the female body before they are capable of fertilizing ova (capacitation). This time is about six hours in rabbits, but capacitation may not be required in human beings.
The pH of semen is alkaline, that of the vagina acid (pH 3–4). Sperms are not optimally active until pH is about 6.5. Seminal sperms live 24–72 hours at normal body temperature but considerably longer in the epididymis. If the semen is quick frozen, the sperm may be stored and preserved for many months and used for artificial insemination.
Fertility in the male can be decreased or inhibited by agents that interfere with or prevent normal maturation of sperm, but their use clinically is not generally recommended.

Male Hormones

The principal male testicular hormone (androgen) is *testosterone,* which is synthesized in the interstitial cells, or Leydig cells of the testes. The main biosynthetic pathways in man are: cholesterol

↓

pregnenolone, which is acted on by
17 α-hydroxylase to form,

↓

17-hydroxypregnenolone

↓

dehydroepiandrosterone

↓

androstenediol

↓

testosterone

Testosterone is also formed by another pathway, namely:

progesterone
↓
17 α-hydroxyprogesterone
↓
androstenedione
↓
testosterone

but this pathway is less prominent in men. The structures of certain androgens are shown in Fig. 26.6.
More than one-half of the blood level concentration of testosterone is bound to plasma proteins, albumin, and globulin. Most of the circulating testosterone is converted in the liver to 17-ketosteroids and excreted in the urine.

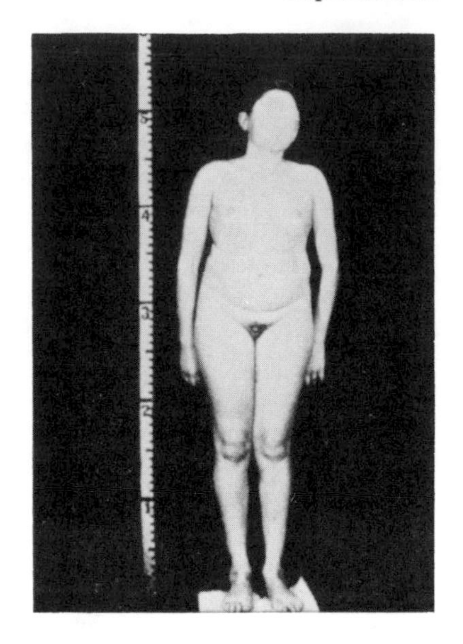

A

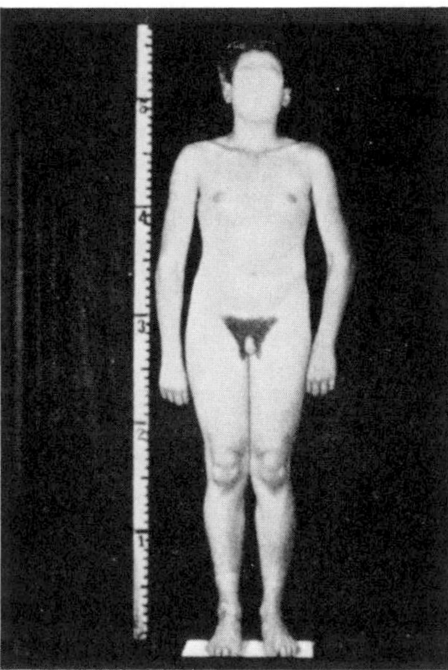

B

Fig. 26.4A and B. Eunchoid male (age 16$^{1}/_{2}$ years) whose tests had not descended. **A** before treatment; **B** after treatment, which included administration of androgen for 6 months.

Action

Testosterone is responsible for the male secondary sex characteristics that develop at puberty, such as deepening of the voice, emergence of pubic and axillary hair, growth of the beard, and increase in size of the external genitalia. Changes also occur in body conformation, such as broadening of the shoulders, enlargment of muscles, and change in hairline. Aggressive behavior is certainly attributable to androgen in experimental animals but to a lesser extent in human males.
Secretion of androgen is involved in male pattern baldness if there is a genetic predilection to baldness. Males castrated before puberty are never bald.
Androgens increase the synthesis, but decrease the breakdown of proteins in the body (*anabolic effect*). Doses of exogenous androgen may increase sexual desire (libido).
A cryptorchid is one whose testicles do not descend from the body cavity; the testes operate at a higher body temperature than that which prevails in the scrotum. This higher temperature prevents normal spermatogenesis, and cryptorchids are sterile, although some of them may produce testosterone, depending on the stage of gonadal development. Administration of gonadotropic hormone may cause the testes to descend.
Eunuchoidism is a condition caused by impairment or lack of Leydig cells either congenitally or resulting from prepubertal castration. A eunuch has small and narrow shoulders, small muscles, and the general appearance of a female. The genitalia are small, the voice is high pitched, and pubic hair sparse or absent. Administration of androgen for three–six months may initiate growth and development of the genitals and pubic hair (Fig. 26.4).

Reproduction in the Female

Reproductive capacity in the human female begins at about age 14–16 years and persists until menopause at 40–50 years. Puberty represents changes occurring at the onset of sexual maturity, whereas prepubertal refers to conditions before puberty. Puberty is characterized in females by changes in size and contours of the body (hips) and in the primary and secondary sex characteristics, such as the emergence of adult-sized genitalia (Fig. 26.5), growth of pubic hair, and development of the breasts (mammary glands).

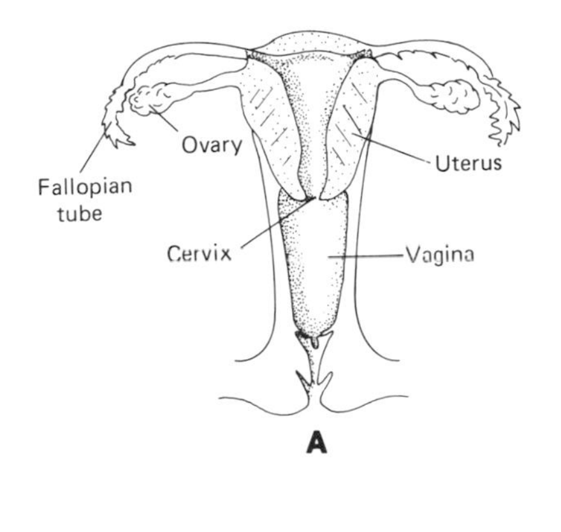

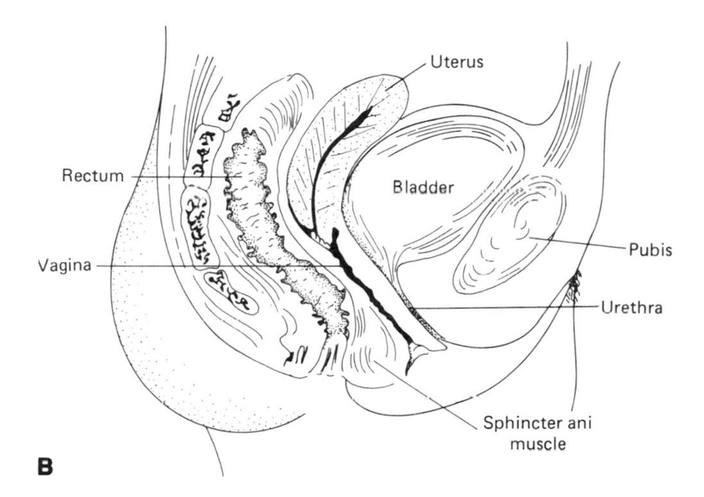

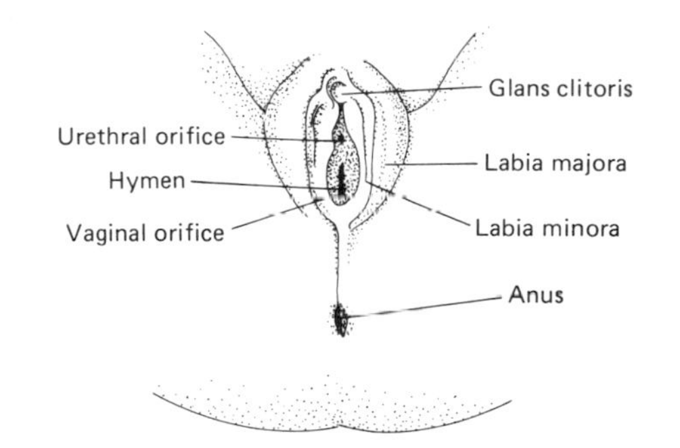

Fig. 26.5A—C. Female genitalia. **A** Upper portion shows ovary, fallopian tubes, uterus, and vagina. **B** Lower portion of pelvic area shows reproductive organs in relation to other organs. **C** External genitals, showing clitoris and vulva, which includes labia (lips) majora and labia minora.

Internally, changes also occur in the ovaries (Fig. 26.7 and 26.8), uterus, and vagina, and this signals the onset of menstruation (menarche). These changes reflect the secretion and action of the pituitary gonadotropic hormones (FSH and LH) and the secretion and action of the ovarian hormones (estrogens, androgens, and progesterone).

Before puberty the secretion of FSH, LH, and ovarian hormones is low. After puberty and during the reproductive cycle, the secretion of these hormones is high and cyclical. After menopause, however, the response of the ovaries to FSH and LH decreases so that secretion of the ovarian hormones (estrogen and progesterone) is low or absent, but secretion of FSH and LH is even higher than it was before

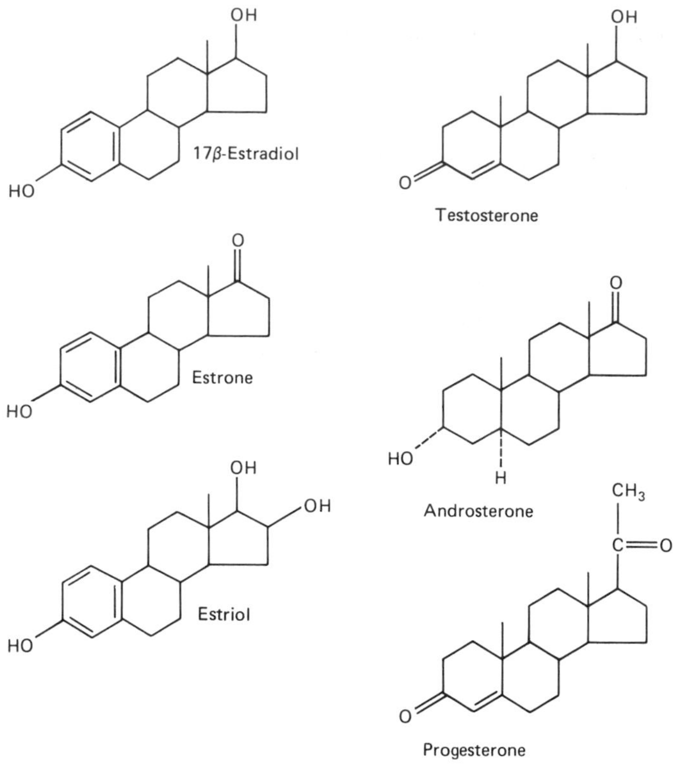

17β-Estradiol
Testosterone
Estrone
Androsterone
Estriol
Progesterone

Fig. 26.6. Structural formulas of certain gonadal hormones (see text).

menopause. This means that the normal inhibition of pituitary secretion of FSH and LH by estrogen (negative feedback) is depressed.

From the beginning of sexual reproduction to its cessation several phases are involved, including: (1) oogenesis and fertilization, (2) ovulation and menstration, and (3) pregnancy and gestation.

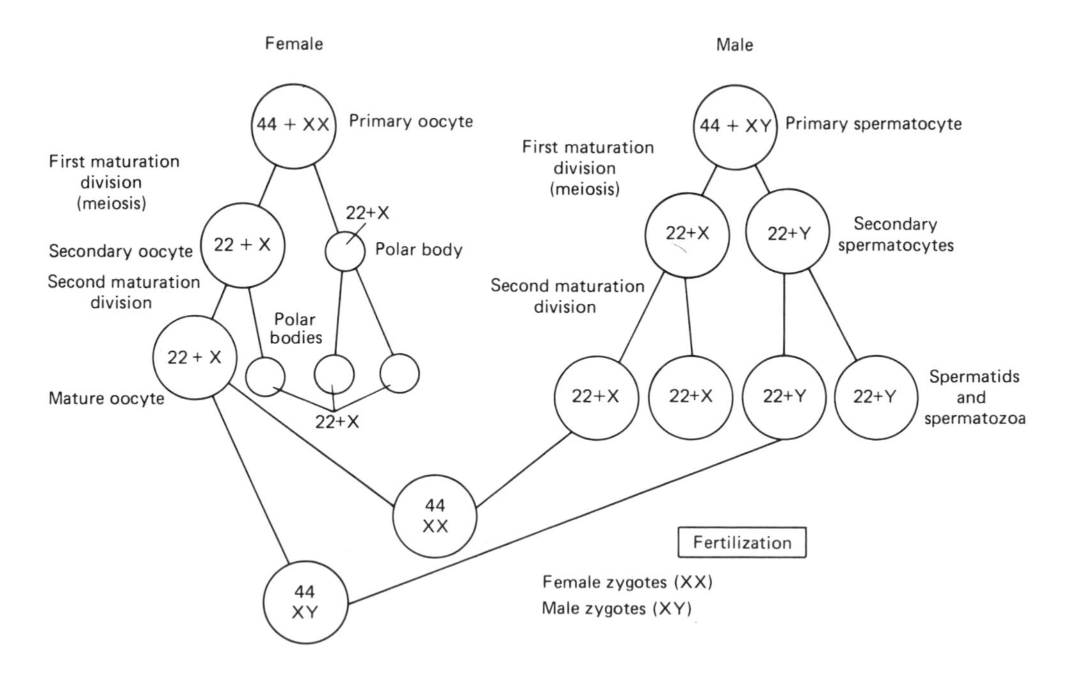

Fig. 26.7. Oogenesis and spermatogenesis (see text).

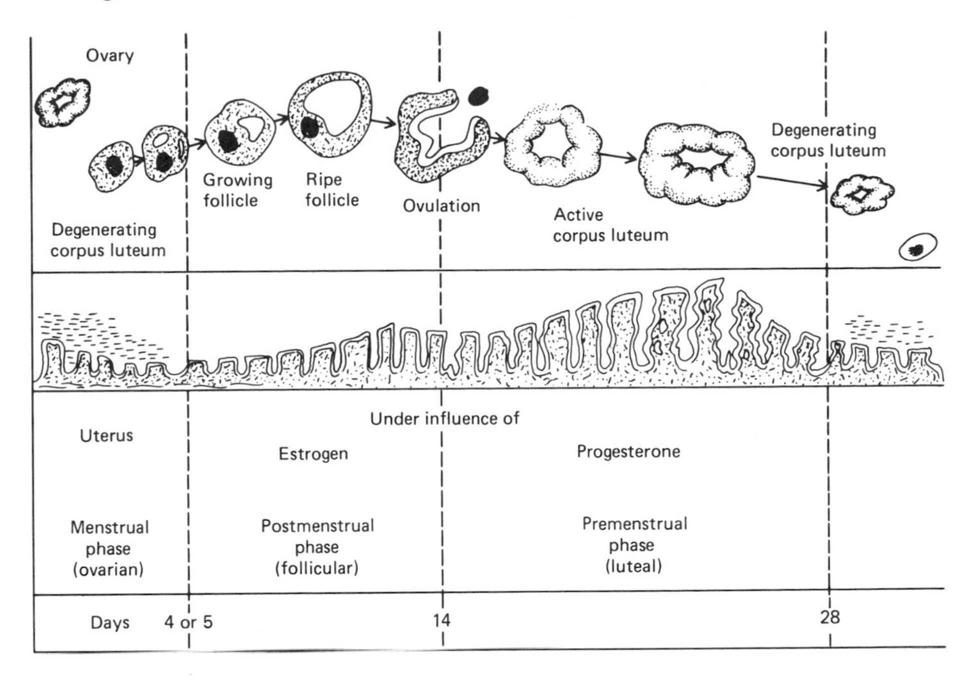

Fig. 26.8. Changes in ovarian growth, ovulation, and endometrium of uterus during menstrual cycle. (After Schottelius BA Schottelius DA (1978) Textbook of physiology Mosby, 81. Conis)

Anatomy of Reproductive Organs

The reproductive organs of the adult female include the internal organs (ovaries, fallopian tubes, uterus, and vagina) and the external genitalia, comprising the vulva (labia majora and minora), clitoris, and orifice to the vagina, which may be covered over by the hymen in the virgin (Fig. 26.6A and B).

Ovary

There are two ovaries in the human female located on each side of the body and uterus, low in the pelvic region. Each mature ovary is about 3.5 × 2 cm × length and weighs 4 g. Both ovaries before birth have about 500,000 ova of varying size, most of which shrink (become atretic) and never ripen or ovulate. Only 300–400 of these ova are ovulated during the normal reproductive life of women (one monthly for 30–35 years). As the ripened ova are shed (ovulated), they are picked up by the adjacent funnel shaped fallopian tubes and transported to the uterus (Fig. 26.5). Such ova are fertilized by spermatozoa from the male in the fallopian tubes, before reaching the uterus.

Growth of Ova

A cross section of the adult ovary reveals ova in various stages of development (Fig. 26.9), including: germinal epithelium, primary follicles, maturing follicles, preovulatory graffian follicle (with ovum), ovulated or released ovum,

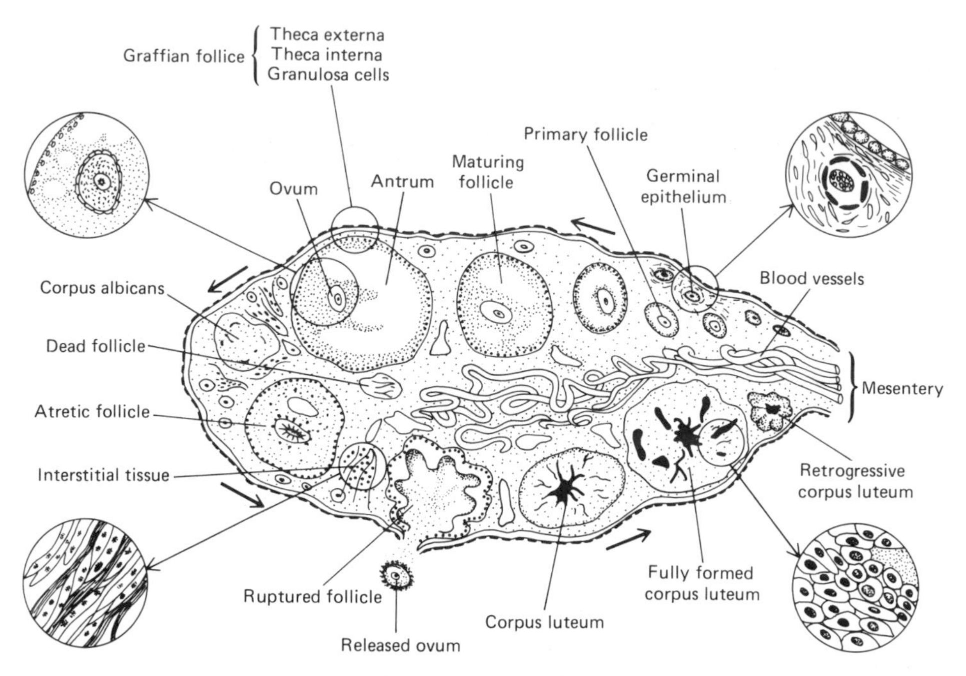

Fig. 26.9. Cross section of ovary showing various stages of develoment of ova to ovulation and formation of corpus luteum. Note enlarged sections showing cells of corpus luteum in different stages. Also a section through mature follicle showing types of cells found (from outside to inside).

atretic follicles, and corpus luteum (formed from ovulated follicles). Some of the corpora lutea are active and some are regressive.

The mature graffian follicle 26.9 shows a large ova (one cell) encircled by the follicle and a cavity—the antrum—containing fluid. The follicle consists of a layer of cells (from outside to inside to antrum) called: theca externa, theca interna, and granulosa cells extending to antrum. During ovulation, which occurs once each month, there is a rupture in these follicular cells, allowing release of ovum.

Uterus

The uterus (womb) is a muscular organ between the urinary bladder and the rectum (Fig. 26.5). Triangular, in adults it is about 7–8 cm long, 3 cm thick, and 3 cm wide at the isthmus (the narrowest part). The opening from the vagina to the uterus is the cervix, which becomes closed by a plug of mucus during pregnancy.

The interior of the uterus is lined with a mucosal surface, the endometrium. Below this surface is the muscular body of the uterus, or myometrium. The endometrium is very vascular and changes greatly during the mestrual cycle. During the period of bleeding (Fig. 26.8), the endometrial lining is shed along with the blood vessels near the surface; later the endometrium regenerates or builds up (proliferates) under the influence of estrogen and continues under the influence of progesterone produced by the active corpus luteum. The latter eventually degenerates and the endometrium is shed again (bleeding).

If pregnancy ensues, the corpus luteum persists under the influence of human chorionic gonadotropin (HCG). This hormone stimulates the secretion of progesterone and corpus luteum; progesterone and estrogen *prepare* the endometrium for implantation of the fertilized ovum and development of the embryo or fetus.

Vagina

The vagina is made up of epithelium (lining the interior), the middle muscular layer, and its outer connective tissue layer. The vaginal epithelium responds to ovarian hormones and undergoes proliferation, differentiation, and desquamation (shedding) of cells. During the prepubertal and postmenopausal stages when little estrogen is produced, the cells become atropic and flattened. Estrogen induces proliferation and an increase in epithelial thickness. Progesterone tends to increase the shedding or desquamation of the surface cells.

Oogenesis and Spermatogenesis

These are the processes by which the chromosome numbers in succeeding generations remain constant at 44 autosomes and XX (2 sex chromosomes) for the female and 44 + XY for the male.

As shown in Fig. 26.7 the cells of the germinal epithelium of the ovary undergo division and growth (oogenesis). These are the oogonial cells and they contain 44 + XX chromosomes. Some of them develop into primary oocytes. Now this cell divides by meiosis (reduction division) so that the resulting cells (secondary oocytes) have one-half the chromosomal number (22 autosomes and 1 X chromosome). This occurs as the follicle has enlarged and just before ovulation. The secondary oocytes then divide equationally (mitosis), and the resulting mature ovum is ready for ovulation and fertilization. The polar bodies that were formed degenerate.

A similar process, *spermatogenesis,* takes place in the testes (Fig. 26.2). Spermatogonial cells begin to grow, divide, and form primary spermatocytes (44 + XY); this is a continuous process. These then divide (reduction division) to form secondary spermatocytes, which contain the haploid (N) chromosome numbers but are of two types with respect to sex chromosome, such as 22 + X and 22 + Y. These develop first into spermatids and then into mature spermatozoa called gametes (Fig. 26.7).

Fertilization occurs when the sperm of the male unites or penetrates the ovum of the female, usually in the fallopian tubes. The female gamete 22 + X unites with the male gamete 22 + X to form a female zygote of 44 + XX. The other male gamete 22 + Y unites with the female 22 + X to form a male zygote of 44 + XY.

Menstruation and Ovulation

Menstruation and ovulation in women are cyclic phenomena (Fig. 26.8). Because they are intimately related, they will be considered together. The average normal menstrual cycle is about 28 days but may vary. If menstruation is absent or has never occurred, it is referred to as primarv amenorrhea. Cessation of cycles in a previously menstruating woman is secondary amenorrhea. Dysmenorrhea refers to painful menstruation.

During the first four to five days of the *menstrual cycle* when bleeding occurs (menstrual phase), the uterine epithelium is shed and passes out with the blood from the small ruptured blood vessels. Menstruation is preceded and caused by cessation of production of progesterone, resulting from degeneration of corpus luteum.

In the *postmenstrual* or *follicular* phase, the ovarian follicles grow, ripen, and begin to secrete estrogen, which reaches a peak preceding ovulation. This coincides with proliferation of the endometrium. In *late postmenstrual* phase ovulation occurs at about midcycle, or day 14, and the corpus luteum is formed from the previously ruptured follicle (premenstrual or luteal phase) and begins secreting progesterone, which maintains and increases the height and integrity of the endometrium. This phase lasts until the end of day 28 and the cycle is repeated.

Ovulation is preceded and followed by the secretion of certain hormones (Figs. 26.8 and 26.10). About three days before ovulation, LH in the blood begins to increase and reaches a peak concentration about 16–24 hours before ovulation; this peak is preceded about 24 hours by a peak concentration of estrogen. In fact, the estrogen peak is thought to trigger the LH release peak, which causes ovulation; FHS also begins to increase when LH does and reaches its peak at about the same time as LH.

Three to four days after peak FSH and LH are observed, the levels drop drastically and remain low while the corpus lutuem is active (premenstrual or luteal phase).

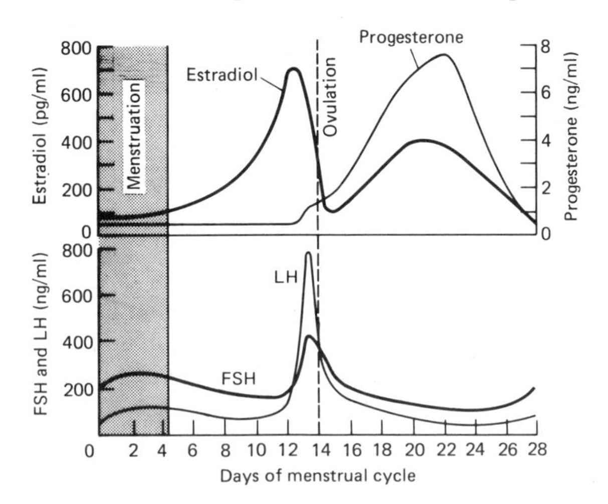

Fig. 26.10. Plasma levels of pituitary hormones (LH and FSH), estradiol, and progesterone as related to menstrual cycle, ovulation, and days before and after ovulation (day 14) (see text). (After Guyton AC (1977) Basic Human Physiology Saunders, Philadelphia)

Immediately after ovulation, both estrogen and progesterone begin to increase and reach peaks at about day 21 to 22 of the menstrual cycle. Four to five days later, estrogen and progesterone levels return to their cyclic low levels just preceding the next menstruation.

Estrogens are *secreted* mainly by the theca interna cells of the ovarian follicles and include 17 β-estradiol, estrone, and estriol. Estradiol is the most active and estriol the least active of the estrogens. *Estradiol* is also secreted at the greatest rate and varies according to the stage of follicular development as follows: (1) early follicular phase (50 μunits daily), (2) just before ovulation (400 μunits daily), and (3) after ovulation (falls drastically).

Males also secrete some estrogens, but most of the male-circulating estrogens (40 μunits daily) are formed from androgens.

Summary of Sequence of Events
in Menstrual Cycle

1. FSH and LH are released durng menstrual bleeding and beginning of growth of ovarian follicle.
2. Developing ovarian follicles secrete estrogen, which causes proliferation of uterine endometrium (postmenstrual).
3. Ovarian follicles undergo oogenesis.
4. Level of estrogen reaches peak value about two days before *ovulation* and probably triggers peak release of ovulating hormone (LH).
5. Levels of FSH and LH reach peak values at 16–24 hours before ovulation.
6. Ovulation occurs 16–24 hours later.
7. After ovulation, corpus luteum is formed, which secretes progesterone (luteal phase) and helps to maintain height and integrity of endometrium.
8. Progesterone and estrogen (estradiol) reach peak secretion midway of luteal phase, or day 21 of menstrual cycle.
9. Progesterone and estrogen return to lowest levels just before next menstrual period.
10. Menstrual bleeding occurs again with shedding of endometrial epithelium and surface blood vessels.

Menstrual and Estrus Cycles

Estrus cycles refer to heat cycles or periods when mammals other than primates mate and reproduce. The female is receptive to the male only during the heat or estrus cycle or near the day of ovulation (day 1), but in the human female, ovulation occurs at day 14 of the cycle and menstration at day 1.

The period after estrus is called metestrus; its length varies, depending on the species and is followed by diestrus, comparable to menstruation, then proestrus (proliferation phase), and then estrus.

Feedback Control of FSH and LH Secretion

The secretion and circulating levels of FSH and LH tend to control the release of these from the pituitary by the short-loop mechanism (see earlier section) by inhibiting or decreasing the release of the hypothalamic releasing factors for FSH and LH (negative feedback).
Estrogen and progesterone affect secretion of FSH and LH by the long-loop mechanism by affecting the release of hypophyseal and hypothalamic factors (see Chap. 25). The low level of estrogen secretion tends to inhibit LH secretion during the early part of the follicular phase of cycle, but high secretion of estrogen is associated with increased release of LH and ovulation, and this represents a positive feedback. Following ovulation (luteal phase) estrogen secretion decreases and progesterone secretion increases, and the latter is probably the main factor in the inhibition of FSH and LH release.

Gonadal Hormones

The structural formulas of the estrogens, testosterone, and progesterone are shown in Fig. 26.6. The naturally occurring *estrogens* in human beings include 17-β estradiol (E_2), estriol (E_3), and estrone (E_1). The latter two are metabolic products of estradiol, which is the most active estrogen; most of it is secreted by the ovarian follicles. Some of the estrone is converted to estriol in the liver. Estriol is believed to be secreted by the placenta from dehydroepiandrosterone. Natural estrogens are relatively inactive when taken orally because they are inactivated in the liver. The estrogen, 17β-estradiol, is synthesized as follows:

Cholesterol ⟶ Pregnenolone ⟶ Progesterone ⟶
17-Hydroxyprogesterone ⟶ Testosterone ⟶
19-Nortestosterone ⟶ 17-β-Estradiol.

This process occurs mainly in the ovarian follicle (theca cells), but it can occur in the Leydig cells of testis, corpus luteum cell, or cells of the adrenal cortex.
A derivative of estradiol, ethenyl estradiol (synthetic) is a potent estrogen when administered orally as is diethylstilbestrol (nonsteroid), another synthetic substance with estrogenic activity; the latter has been often recommended as the "morning-after pill" for contraception.
Progesterone is an important intermediate in steroid biosynthesis in many organs in addition to the ovarian follicle (corpus luteum) where it is secreted in large amounts. It has a short half-life, is converted in the liver to pregnanediol, and is later conjugated and excreted in the urine.

Action of Estrogens

Estrogens are produced by the developing ova and play a prominent role in the menstrual cycle and implantation, as previously discussed. They also increase blood flow to the uterus and increase the activity and contractility of uterine muscle; they increase the binding of calcium in the muscle and cause some degree of salt and water retention in the tissues. Estrogens also influence growth of mammary glands (see "Lactation" in this chapter).

Secondary sexual characteristics that develop at puberty (see previous section) are attributable to the action of estrogens. The wide hips and narrow shoulders represent prominent effects of estrogens on body conformation, and they also affect the distribution and accumulation of fat in the breasts and buttocks.

Pregnancy and Parturition

Fertilization of the ovum occurs after ovulation in the fallopian tubes, and the ovum is penetrated by only one sperm, and the nuclei of the sperm and ovum fuse to become a zygote. The zygote now migrates down the tube to the uterus and becomes implanted in the endometrium within $7^1/_2$ days after fertilization, and after having reached the 200-cell stage of division (blastocyst). The ovum will not implant unless the endometrium has been properly prepared by estrogen and progesterone. The blastocyst invades the endometrial mucosa while the latter grows around it and is now called a fetus or embryo. A placenta then develops followed by the amnion and chorion (Fig. 26.11).

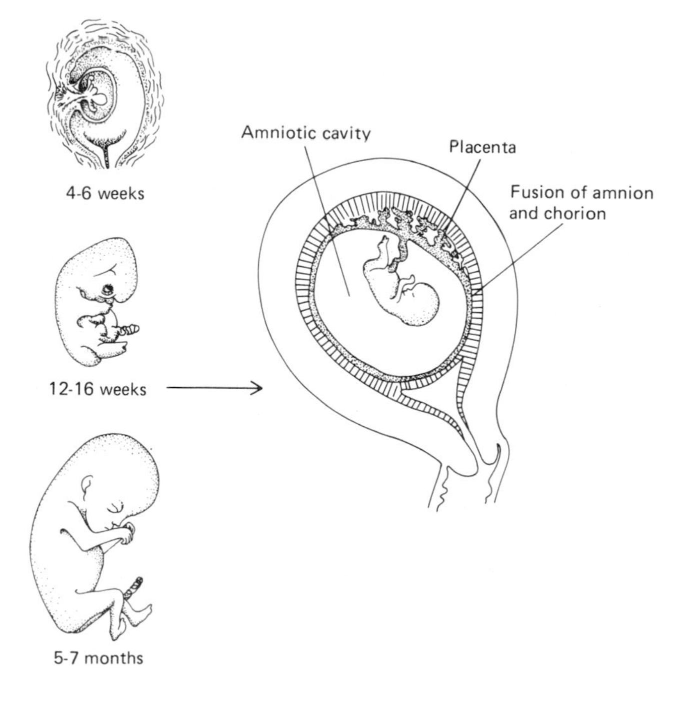

Fig. 26.11. Different stages of embryonic life (fetus), showing placenta and membranes.

Hormones in Pregnancy

The corpus luteum persists during pregnancy under the action of HCG and continues to produce estrogen and progesterone. In most mammals, removal of the ovary causes abortion but not in the human female because the placenta produces sufficient amounts of these hormones after the sixth week of pregnancy.

The secretion of HCG is lowest in early pregnancy, increases rapidly as the placenta develops, reaches a peak concentration at eight to nine weeks, and decreases greatly by 16–20 weeks when concentration levels off (Fig. 26.12). Progesterone shows a general increase during pregnancy and indicates persistent corpora lutea. Blood levels may vary from 2 μg/ml at two weeks' pregnancy to 17 μg/ml at 40 weeks. Estriol increases significantly during the last weeks of pregnancy.

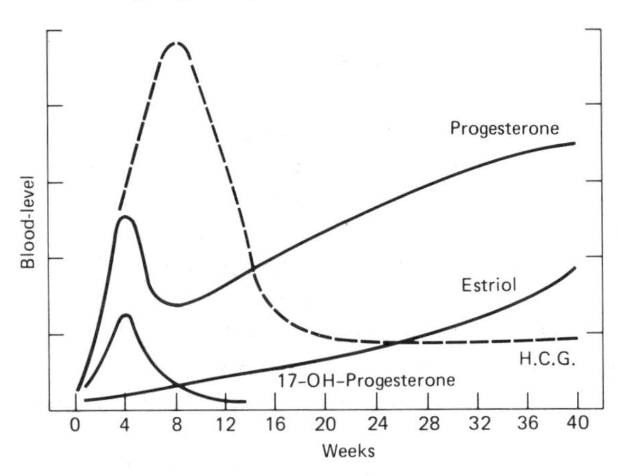

Fig. 26.12. Plasma levels of hormones during pregnancy (0–40 weeks) (see text). (After Catt KJ (1971) An ABC of endocrinology. Little, Brown and Co., Boston)

Gestation and Labor

The period of human pregnancy, or gestation, is about nine months or 270 days. Gestational periods for certain other mammals are: cattle (280 days), sheep (148 days), horse (337 days), dog (63 days), cat (63 days), and rat (22 days).

The mechanisms that initiate labor are not well understood. Certainly the uterus is more sensitive to various stimuli in the latter stages. At *onset of labor,* uterine contractions occur at a rate of 1 every 30 minutes but then increase to 1–3 per minute. The posterior pituitary hormone, oxytocin, may be involved, but birth can proceed normally if the hormone is absent or deficient.

It is likely that fetal growth first and then uterine distention are the ultimate initiators of labor. Bearing down and contraction of abdominal muscles and certain reflexes also aid in the expulsion of the fetus. In most births the head appears first, but in others the buttocks appear first.

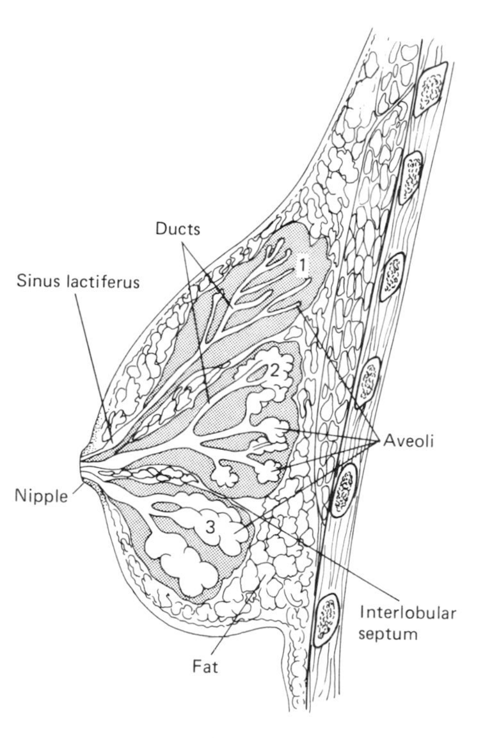

Fig. 26.13. Diagram of adult mammary gland showing development of alveolar system under different conditions: (1) nonpregnant woman, (2) middle pregnancy, and (3) during lactation.

Lactation

Lactation involves growth and development of mammary glands, milk secretion, and milk letdown, or release from the gland.

The human mammary glands, two in number, are in the upper chest lateral to the mid-line of the body. Adult female breasts consist of lobules of tissue that are firm, rounded, and the nipples are centrally located (Fig. 26.13). These glands are small and much alike in prepubertal males and females. At puberty, however, the mammary glands in females begin to grow and develop, mainly under the influence of estrogen and progesterone. Estrogen stimulates growth and development of lobules and alveoli, which secrete milk. The mammary ducts connect the alveoli with the nipples.

After the breasts are fully formed with ducts and alveoli, *milk secretion* is stimulated by the pituitary hormone, prolactin. This assumes that certain other hormones in addition to estrogen and progestrone which affect the growth and integrity of tissues are adequate, such as GH, thyroxine, insulin, and corticoids. Severe hormonal deficiencies can affect milk secretion even when prolactin is adequate.

At about the fifth month of gestation, breasts become fuller and larger because of an increase in fat and in the size and number of alveoli and lobules (Fig. 26.13). After cessation of ovarian activity (menopause), the size of the breasts decreases because of a decrease in size and numbers of alveoli and fatty tissue.

After parturition, milk secretion increases greatly and continues for several months until weaning. This coincides with the drop in secretion of estrogen and progesterone and the increase in secretion of prolactin and a decrease in FSH and LH secretion of the pituitary gland.

Secretion of prolactin is highly variable even during one day and is episodic at intervals of 20–30 minutes. Prolactin secretion in nonpregnant females average about 8–10 ng/ml blood; during pregnancy, however, it increases to levels of 100–400 ng/ml. Prolactin secretion in plasma tends to increase with the amount of milk secreted and decreases as duration of lactation increases.

Prolactin secretion increases with the onset of *nursing* as much as 10-fold. Nursing tends to inhibit the release of FSH and LH and prevents or delays the onset of ovulation and menstruation. Mothers who do not nurse their young resume their menstrual cycles within six weeks after childbirth.

Milk letdown or release. It was previously stated that oxytocin from the posterior lobe of the pituitary causes the release of milk from the ducts and milk cistern to the nipples (no effect on secretion). Nursing or reflexes associated with nursing or stimulation of breasts cause the release of oxytocin (see Chap. 25).

Coitus

The female clitoris contains erectile tissue similar to that of the glans penis, which is most sensitive to physical and psychic stimuli. Like the penis, it is also controlled by the parasympathetic motor nerves (nervi erigentes) innervating the clitoris. Stimulation of this nerve causes dilation of arteries and constriction of veins, which in turn causes the clitoris to enlarge and become erectile. Following erection, there is lubrication of the vagina; this is caused mainly by transudation of material directly from the vaginal mucosa rather than from Bartholin's glands as was formerly believed.

Physical stimuli of clitoris and vulva are reinforced by touch of the breasts and other erogenous zones. Psychic stimuli include olfactory, auditory, and visual stimuli, which involve the central nervous system. When physical and psychic stimuli reach a peak, orgasm occurs. The nervous reaction and response are similar to those in the male except that there is no secretions similar to the seminal emission of the male. Moreover, the female is capable of a return to orgasm within a short time.

Orgasm is produced by sympathetic motor impulses from the hypogastric nerves. There are marked changes in heart rate, respiratory rate, and blood pressure following sexual stimulation and the orgasm, as in the male. The heart rate which before sexual stimulation was 70–80 beats per minute increases quickly on mild stimulation to 125 and at orgasm approaches 180 beats per minute. Blood pressure is elevated 30–80 mm Hg during orgasm. Sexual behavior as influenced by the hypothalamius is discussed in Chap. 10.

Contraception

Avoidance of intercourse within three days on either side of the expected time of ovulation (period of six days) is a natural means of contraception because the ovum is capable of fertilization for only 24 hours after ovulation, and sperms in the female tract are viable for no longer than 72 hours (rhythm method). This method is extremely unreliable because of the variation in menstrual and ovarian cycle length and other factors.

Hormonal Suppression

Fertility may be suppressed or inhibited in the female by preventing ovulation. This is accomplished by ingestion of an oral contraceptive (synthetic hormones). The pill may be: (1) a combination of estrogen and progesterone for 20 days, discontinued for five days; (2) progesterone alone; or (3) estrogen alone for 15 days, followed by estrogen and progesterone for five days (sequential). The first pill (number 1) suppresses the output of FSH and LH, and therefore ovarian follicles do not grow and ovulate.

Progesterone alone in large doses also suppresses LH release and ovulation. It may do so by inhibiting the ability of estrogen to evoke ovulatory surges of LH. The use of the pill as a contraceptive device is almost foolproof, although the long-term effects of the hormones may be undesirable.

Diethylstilbestrol (synthetic) taken orally after intercourse suppresses fertility by preventing normal implantation of the fertilized ovum (an interceptive agent).

Intrauterine devices (IUD), usually spiral shaped, are placed in the uterine cervix to prevent implantation in the uterus, probably by speeding the passage of ovum through the uterus. They are effective when implanted properly.

Selected Readings

Catt KJ (1971) An ABC of endocrinology. Little, Brown, Boston

Ganong WF (1977) Review of medical physiology, 8th ed. Lange Medical, Los Altos, California

Greep RO, Koblinsky MA (1977) Frontiers in reproduction and fertility control. A review of reproductive sciences and contraceptive development. MIT Press, Cambridge

Greep RO (1977) (ed): Reproductive physiology, international review of physiology, Vol. 13, Chaps. 1, 4, and 8. University Park Press, Baltimore

McCann SM (1977) (ed) Reproductive physiology, endocrine physiology, international review of physiology, Vol. 16, Chap. 2. University Park Press, Baltimore

Masters WH, Johnson VE (1966) Human sexual response. Little, Brown, Boston

Morgan HE (1973) Endocrine control systems. In: Brobeck JH (ed) Best and Taylor's physiological basis of medical practice, 9th ed. Williams and Wilkins, Baltimore

O'Dell WD, Moyer DL (1971) Physiology of reproduction. Mosby, St. Louis

Sawin CT (1969) The hormones, endocrine physiology. Little, Brown, Boston

Tepperman J (1973) Metabolic and endocrine physiology, 3rd edn. Year Book Medical, Chicago

Review Questions

1. What are the mechanisms involved in erection of the penis?
2. What nerves are involved in erection, emission, and ejaculation?
3. Diagram the steps in spermatogenesis.
4. What are the chromosome numbers of a zygote and of a gamete in man?
5. Normal ejaculate of male humans is about how many ml? How many sperms per milliliter does it contain?

6. How long does it take for sperms to travel from vagina to fallopian tubes?
7. What are the two main substances involved in the synthesis of testosterone?
8. Where is testosterone synthesized and under what influence?
9. What are the principal effects of testosterone?
10. Diagram oogenesis.
11. Outline the main events (sequence) of the menstrual cycle, taking into account the various stages and time involved. Relate the menstrual cycle to the ovulatory cycle.
12. What pituitary hormones are involved in ovulation and how are they involved? How are estrogen and progesterone involved in ovulation?
13. Where does fertilization of the ovum occur?
14. What is the role of the corpus luteum in maintaining pregnancy?
15. What is the gestation period in human beings?
16. What factors are involved in parturition?
17. What hormones are involved in milk secretion and milk letdown?
18. What is the basis of hormonal contraceptive methods?
19. What are the principal effects of estrogens?

Chapter 27 Thyroid Hormones

The human thyroid gland consists of two main lobes, which are located laterally in the neck, attached to the upper part of the trachea, and connected by a bridge or isthmus (Fig. 27.1). It weighs about 20 g in the normal adult, but its size varies considerably. Under certain conditions it may become either enlarged (hypertrophy) or smaller than normal. The gland is highly vascular and innervated by the sympathetic system.

Light microscopy shows the gland to be made up of follicles (Fig. 27.1) about 200 μ in diameter, depending on its secretory state. In an inactive state each follicle is lined with a low, flat-cuboidal epithelium and contains much colloid material, called thyroglobulin.

In an actively secreting gland the follicles are smaller, and the epithelial cells are larger and taller and may be columnar shaped. The follicles contain less colloid, and the latter shows evidence of reabsorption as revealed by lacunae (Fig. 27.1). The histologic features of the gland may or may not

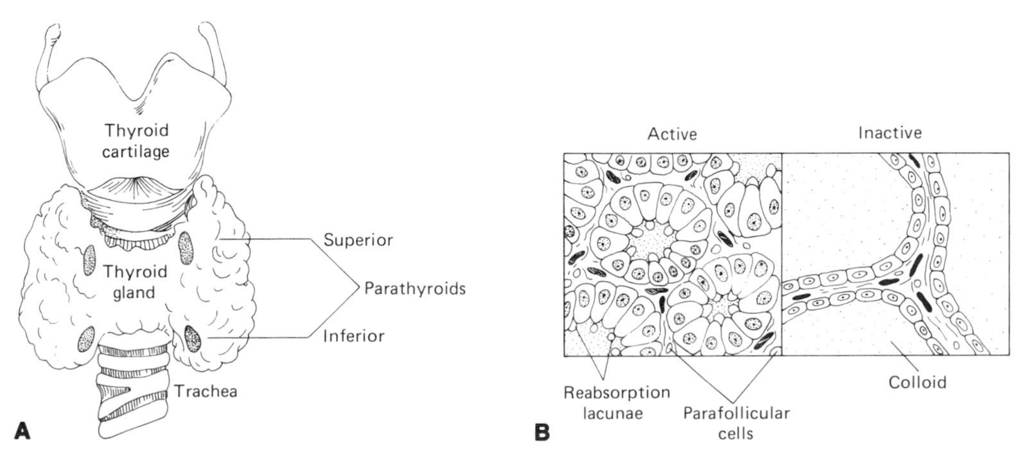

Fig. 27.1A and B. Human thyroid. **A** lateral lobes and midlobe (isthmus): **B** cross section of thyroid showing active and inactive glands. Note large colloid filled follicles with flat cuboidal cells (inactive) and larger cells and small colloid follicles (active).

reveal its true secretory state. Hypertrophy and hyperplasia of the gland (increase in size and number of cells, respectively) may reveal the effects of increased secretion of thyroid-stimulating hormone (TSH) from the pituitary gland and increased secretion of thyroid hormone.

Agents blocking the uptake or binding of iodine, or both, such as anions of chlorate, periodate, perchlorate, and nitrate and which prevent the synthesis and release of thyroxine increase the release of TSH from the pituitary. This in turn causes glandular hypertrophy and hyperplasia.

Iodinated Compounds

The steps or processes in the synthesis of thyroxine (T_4) or 3, 5, 3[1], 5[1] tetraiodothyronine are shown in Fig. 27.2 and described herein:

1. I^- is trapped by thyroid cells and actively transported across cells.
2. I^- is oxidized intracellularly to I_2 by the enzyme peroxidase and carried to the colloid of follicles where tyrosine or tyrosyl groups are attached to the thyroglobulin molecule in colloid.
3. Tyrosine is iodinated to form monoidotyrosine (MIT).
4. MIT is further iodinated to diiodotyrosine (DIT).
5. 1 MIT plus 1 DIT forms 3, 5, 3[1] triiodothyronine (T_3).
6. 1 DIT plus 1 DIT forms 1 molecule of T_4 or thyroxine.

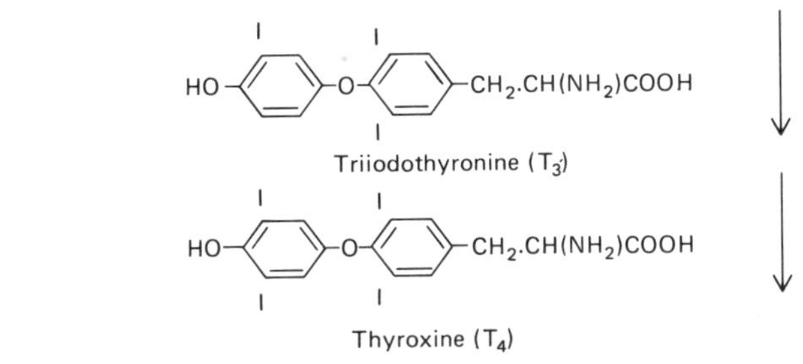

$I^- \longrightarrow I_2$ + HO–C$_6$H$_4$–CH$_2$.CH(NH$_2$)COOH Tyrosine

Monoiodotyrosine

Diiodotyrosine

Triiodothyronine (T_3)

Thyroxine (T_4)

Fig. 27.2. Steps in synthesis of thyroxine (T_4 or 3, 5, 3[1], 5[1] tetraiodothyronine). (See text.) Steps 1) I^- is trapped by thyroid; 2) I^- is oxidized to I_2; 3) tyrosine is iodinated to form monoiodotyrosine (MIT); 4) MIT is further iodinatd to diiodotyrosine (DIT); 5) 1 MIT plus 1 DIT = 3, 5, 3[1] triiodothyronine (T_3); 6) 1 DIT plus 1 DIT = 1 T_4.

Effects of Iodine on Synthesis

Dietary iodine deficiency prevents normal synthesis of thyroxine, stimulates increased TSH secretion, and produces an enlarged thyroid or goiter. This condition was once common in the "goiter belt" or iodine-deficient soils of the Great Lakes area until the practice of adding iodine to common table salt was instituted. The intake of dietary iodine must be at least 10 mg/day for normal thyroid synthesis.

Antithyroid Drugs

Thiocarbamides, among which are thiourea and thiouracil, and methimazole prevent synthesis of thyroxine and increase release of TSH by inhibiting iodination of monoiodotyrosine; they do not block iodine uptake but inhibit binding of iodine. These antithyroid agents are used clinically to treat hyperthyroid disorders.

Foods with Goitrogens

Certain vegetables of the *Brassicae* family, including cabbage and turnips, contain goitrogens or antithyroid agents. Consumers of these vegetables therefore require a higher intake of iodine than do nonconsumers.

Thyroxine Release and Secretion

Thyroglobulin found in the colloid of follicles is a substance made up of four polypeptide chains with a very high molecular weight, and it includes and binds the substances synthesized in the thyroid (MIT, DIT, T_3, and T_4). An enzyme, protease, acts on thyroglobulin and causes it to release these substances; T_3 and T_4 are available for secretion into the blood stream, but MIT and DIT are deiodinated in the thyroid.

It is generally agreed that most of the circulating T_3 is derived from T_4 by deiodination in the peripheral tissues.

A small amount of reverse T_0 (3, 3^1, 5^1 triiodothyronine) is formed by condensation of DIT and MIT. Distribution of the iodinated compounds in the human thyroid is: MIT (23.8%), DIT (33.2%), T_3 (7.4%), and T_4 (35.6%).

Circulating Thyroid Hormones

The average levels of unbound or free thyroxine (T_4) and T_3 in normal human blood are 3 ng/100 ml for T_4 and 1.5 ng/100 ml for T_3. Nearly all of T_4 (99 + %) in blood is bound by globulin (60%), prealbumin (30%), and albumin (10%). Slightly less T_3 is bound by globulin (75%) and more by albumin (25%).

Biologic activity of T_3 is several times greater than T_4, and the metabolic effects are more rapid. Only the unbound, or free, T_3 and T_4 are physiologically active. The half-lives of T_4 and T_3 are about seven and three days, respectively.

Secretion and Metabolism

The daily secretion rate of thyroxine is approximately 80 μg and of T_3, 40 to 50 μg. These rates increase or decrease depending on the degree of stimulus of TSH. Since circulating T_3 and T_4 are bound to plasma proteins, the protein-bound iodine (PBI) is a fair indicator of circulating T_3 and T_4. The normal level of PBI in man is about 6 μg/100 ml; most of this represents T_4.

Measuring Secretion Rate

Secretion and synthesis rate may be determined by techniques including (1) radioimmunoassay, (2) PBI, (3) radioiodine uptake by the thyroid, and (4) release of radioiodine from the thyroid gland.

Radioimmunoassays are very sensitive and exellent for determining small amounts of T_3 and T_4. Radioiodine uptake by the thyroid is used clinically as a measure of gross thyroid activity, and the rate of uptake and also release is a reflection of synthesis and secretion of thyroid hormones. The patient swallows the substance (tracer) containing I^{131}, and the thyroid uptake of I^{131} is determined by placing a gamma-ray counter over the thyroid in the region of the neck. The uptake is also determined in another area (nonthyroidal) and a comparison of the two is made to subtract the nonthyroidal uptake. The thyroidal uptake of I^{131} varies directly with the activity and synthesis of the gland. Ten hours after administration of I^{131}, a normal thyroid will take up 100% of the radioiodine, a hypothyroid gland 20%, and a hyperthyroid one 300%.

Thyroxine and triiodothyronine are deaminated and deiodinated in many tissues of the body; T_3 and T_4 are conjugated in the liver to form sulfates and glucuronides. The major sites for metabolism of T_4 are the liver and skeletal muscle.

Regulation and Effects of Thyroid Secretion

Factors or conditions affecting the secretion of thyroid hormones include changes in temperature, exercise, and other hormones and activities; however, the main control and regulator of thyroid secretion is the thyrotropic hormone of the anterior pituitary (TSH). The chemical nature of this hormone and the factors affecting its release from the pituitary and the releasing hormones from the hypothalamus are discussed in Chap. 25. The negative feedback mechanism (Fig.

25.6) of circulating thyroid hormones on TSH is mainly at the level of the pituitary (long loop); high levels of circulating T_3 and T_4 depress, and low levels increase, TSH release (positive feedback) from the pituitary. Part of the feedback mechanism, however, is at the level of the hypothalamus (short loop) on TSH-RH.

Metabolic Rate

The principal effect of T_3 and T_4 is the increase in O_2 consumption in most of the organs and tissues of the body, or the calorigenic effect (see also Chap. 18).

Other effects include increased absorption of nutrients and increased metabolism of proteins, carbohydrates, and fats; other effects are on reproduction, lactation and growth, and many of these effects are secondary to increased O_2 consumption.

When O_2 consumption is increased and heat is produced, protein and fat stores in the body are broken down (catabolized) and the body weight decreases; this occurs in hyperthyroidism.

Thyroid hormone action tends to increase and mobilize blood glucose, but it also catabolizes glucose so that the blood glucose level tends to fall.

Cholesterol metabolism is influenced by thyroid action, which stimulates synthesis of cholesterol as well as mechanisms that predominate in removing or decreasing blood cholesterol.

Normal growth and development of the embryo and the young depend on normal levels of thyroid hormone. Deficient levels produce a type of hypothyroidism known as cretinism (Fig. 27.3). *Metamorphosis* of amphibians from tadpoles to adult frogs is increased by thyroid hormone.

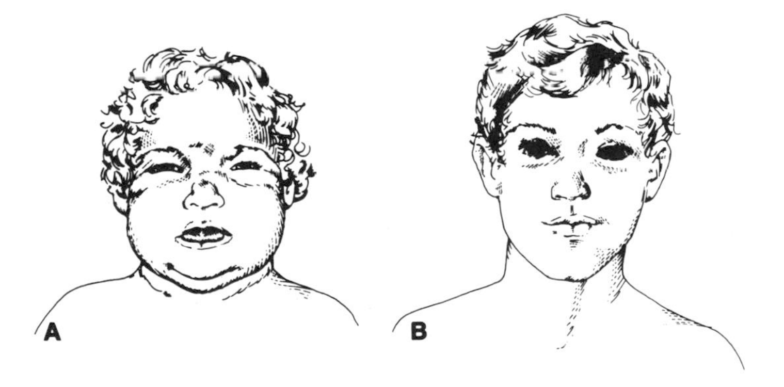

Fig. 27.3A and B. Cretinism in a child age 3½ years. **A** before treatment; **B** after prolonged treatment with thyroid hormone (age 7 years).

Thyroid Disorders

When the O_2 consumption and metabolic activity are low as a result of deficient thyroid activity, *hypothyroidism* can develop. In the young such a deficiency produces a kind of dwarfism (cretinism) that can be corrected or ameliorated by thyroxine (Fig. 27.3). Deficiencies in the hormone occurring after adulthood result in a condition known as myxedema characterized by sparse hair, dry and yellowish (puffy) skin, a slow and husky voice, and a lower-than-normal basal metabolic rate (Fig. 27.4). This condition is also improved by administration of thyroid hormone.

Hyperthyroidism

Excessive secretion of thyroid hormone increases O_2 consumption and may produce other related conditions, such as (1) nervousness, (2) tremor, (3) tachycardia, (4) sweating, (5) heat intolerance, (6) fatigue, (7) weight loss, (8) bulging eyes, and (9) goiter. Hyperthyroidism often occurs in Graves' disease (thyrotoxicosis) with associated goiter and bulging eyes (exophthalmos) (see Fig. 27.4). The secretion rate of thyroid hormones may be 3–10 times the normal rate. The hyperfunctioning thyroid of Graves' disease results from the action of two or more gamma globulins in the blood which stimulate the gland.

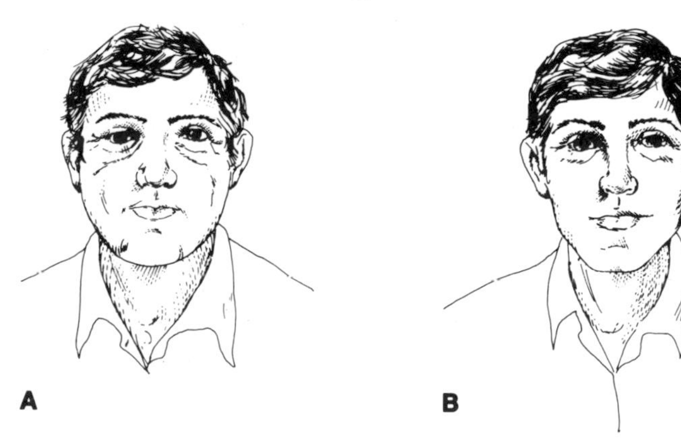

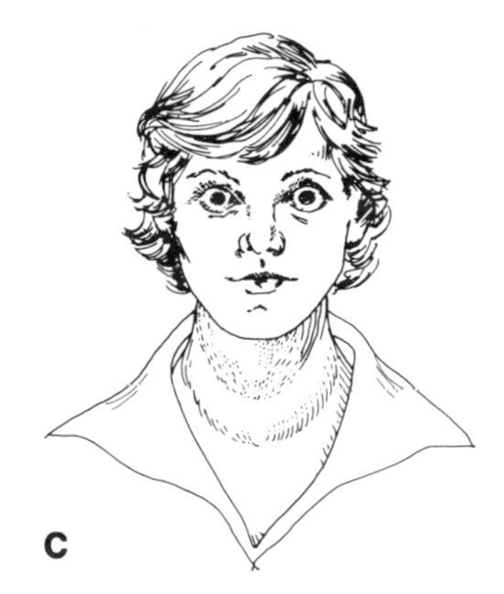

Fig. 27.4A–C. Thyroid deficiency (myxedema). **A** in an adult; **B** after treatment with thyroid hormone; **C** hyperthyroidism (Graves' disease). Note eyes (exophthalmos).

Hypothyroidism

Hypophysectomy causes decreased thyroid function because the primary stimulator, TSH, has been removed. Thyroidectomy results in depressed growth of body and changes in skin and hair. All body processes tend to be slowed because the O_2 consumption is decreased in the absence of thyroid hormone.

Selected Readings

Bentley PJ (1976) Comparative vertebrate endocrinology. Cambridge Press, New York
Catt KJ (1971) An ABC of endocrinology. Little, Brown, Boston

Ganong WF (1977) Review of medical physiology, 8th edn. Lange Medical, Los Altos, California
McCann SM (1977) (ed) Endocrine physiology II, international review of physiology, Vol. 16, Chap. 2. University Park Press, Baltimore
Morgan H (1973) Endocrine control systems. In: Brobeck Jr (ed) Best and Taylor's physiological basis of medical practice, 9th edn., Chap. 7, p. 29. Williams and Wilkins, Baltimore
Tepperman J (1973) Metabolic and endocrine physiology, 3rd edn. Year Book Medical, Chicago
Woeber KA, Braverman LE (1976) Thyroids. In: Ingbar SH (ed) The year in endocrinology, 1975–1976. Plenum, New York

Review Questions

1. What are the main steps in the synthesis of thyroxine (T_4) from tyrosine? Illustrate.
2. What effect does TSH have on secretion of thyroid hormones?
3. Explain the negative feedback of circulating T_3 and T_4.
4. What percentage of circulating T_3 and T_4 are free and bond to plasma proteins?
5. Discuss the biologic activity of T_3 and T_4.
5. What are the effects of thyroid hormones on: (a) O_2 consumption and heat production, (b) carbohydrate metabolism, (c) cholesterol metabolism, and (d) growth?
6. What is cretinism? Describe and give causes of the disorder.
7. What is myxedema? Describe and explain.
8. What is Graves' disease? Describe and explain.

Hormonal Regulation of Calcium Metabolism

Chapter 28

Calcium metabolism, particularly the levels of Ca found in blood and tissues, is regulated mainly by three hormones: (1) parathyroid (PTH) from the parathyroid glands; (2) calcitonin (CT) from the C cells of thyroid and ultimobranchial bodies, and (3) dihydroxycholecalciferol (calciferol), formed from vitamin D in liver and kidney, which increases calcium absorption from the gut.

The level of calcium in the blood and bones controls and regulates the release of calcitonin and PTH independently of the action of the pituitary gland.

Tissue Calcium Levels

The average person contains about 1000–1200 g of calcium, most of it in the skeleton, but roughly 1% is in the blood and soft tissues. Calcium in the blood exists in two forms, namely, *bound* to plasma proteins (mainly albumin) and nondiffusible through a semipermeable membrane, and *free and diffusible* (Ca^{2+}). *Total* blood Ca comprises both types. In man total blood Ca concentration is 10.0 mg/100 ml (5 mEg/liter), with Ca^{2+} making up 5.36 mg/100 ml and the bound form 4.64 mg/100 ml.

The diffusible or free form of Ca is the physiologically active ion involved in blood coagulation, irritability of muscle and nerve, and general cell activity. These activities have been discussed elsewhere. In man, levels of blood Ca well above 10 mg/100 ml are indicative of hypercalcemia; those well below, hypocalcemia.

Ablation of the parathyroid gland causes a drop in blood calcium concentration from 10 mg/100 ml to 6–7 mg/100 ml in a short time; an injection of PTH quickly restores the calcium level. In some species (birds), the laying female has blood Ca levels much higher than those in males (about 20 mg/100 ml), which is caused by the female sex hormone, estrogen, in addition to PTH.

Bone is made up of a tough organic matrix on which is deposited calcium salts (see Chap. 1). Salts are composed of calcium and phosphates called hydroxyapatites, $Ca_{10}(PO_4)_6(OH)_2$. Most of the bone calcium is in a stable form, although a small amount is in a readily exchangeable form between bone and blood. Bone calcium is continuously being deposited and released or dissolved (turnover). The annual turnover rate is about 100% in infants, 16% to 20% in adults.

The organic matrix of bone is made up of three main cell types; they are: osteoblasts, osteocytes, and osteoclasts. Osteoblasts form new bone cells (matrix), which develop into osteocytes (a later stage in the life of the osteoblast). Osteoclasts are multinuclear cells that erode and then resorb bone. They play a prominent role in making available bone Ca that is absorbed into the bloodstream.

A potent stimulator of osteoclastic activity has recently been discovered in the leukocytes, but its physiologic importance remains to be determined.

Parathyroid Glands

Normally there are two pairs of parathyroid glands, located on the superior and inferior poles of each lobe of the thyroid (Fig. 27.1a), but their location may vary considerably. They are relatively small, and their size, which is also variable, is about 6 mm long and 3 mm wide. A cross section of the adult gland reveals two main types of cells—*chief* cells and *oxyphil* cells. The former are small and arranged in cords and have a clear cytoplasm. They secrete PTH. The oxyphil cells, absent in many species, are present in man but are less numerous than chief cells; their function is not known.

Parathyroid Hormone

Parathyroid hormone is synthesized in the parathyroid glands and is a linear polypeptide containing 84 amino acids with a definite sequence for different species. Bovine and porcine PTH, although similar, differ in several amino acids. Chicken PTH has been partially purified and resembles the bovine type in molecular weight and biologic activity. Bovine PTH has a molecular weight of 9500. The first 34 amino acids nearest to the terminal amino N of bovine PTH account for nearly all of its biologic activity. Human PTH also contains 84 amino acids and although the complete sequence of amino acids has not yet been determined, it appears to be similar to the bovine hormone.

Biologic assays based on the hypercalcemic response of the intact chick or quail to PTH is more sensitive and rapid than the rat assay.

Radioimmunoassays are available, but the need for more specific antibodies is obvious because of several different immunoreactive PTH types and fragments.

Action and Release

The release of PTH is regulated by the calcium levels in the blood and soft tissues. When the level is normal (about 9–10 mg/100 ml), PTH secretion is moderate; at high levels, it is minimal or absent; and at low levels of blood calcium, PTH secretion is maximum (Fig. 28.1). The low Ca level stimulates the chief cells of the parathyroids to release PTH into the circulation; the hormone's principal action is to mobilize calcium from bone (Fig. 28.2). Parathyroid hormone activates osteoclasts, which break down and resorb stable bone, and the free calcium is absorbed into the bloodstream. The organic matrix of bone is also dissolved and forms hydroxyproline, which is then excreted in the urine.

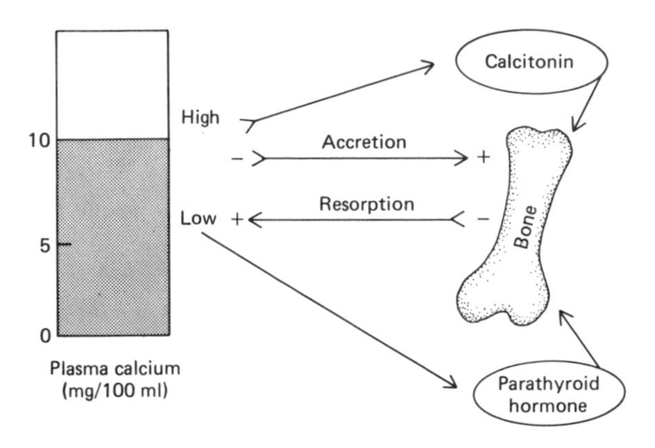

Fig. 28.1. Factors in calcium regulation. High levels of blood calcium (10+) cause release of calcitonin, which decreases blood calcium (−) and increases bone calcium (accretion). Low Ca levels release PTH, which increases blood Ca^{2+} and decreases bound calcium (resorption).

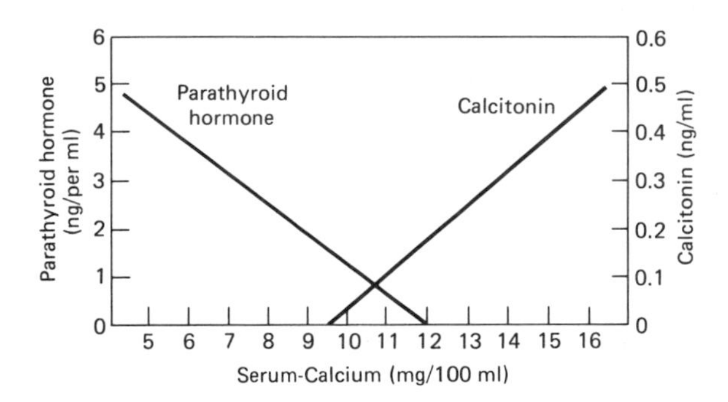

Fig. 28.2. Effects of changes in serum calcium levels on levels of calcitonin and PTH in blood. (Modified after Catt KJ (1971) An ABC of endocrinology. Little, Brown and Co., Boston.)

Parathyroid hormone activates adenyl cyclase of the receptor membrane, resulting in increased formation of cAMP (see Chap. 25 for further details). The exact manner in which cAMP is involved is unclear.

As bone is broken down by the osteoclasts, both calcium and phosphorus (phosphates) are freed. Most of the calcium is absorbed into the bloodstream, but some of it, and most of the phosphates, are excreted in the urine. Thus PTH increases calcium reabsorption and decreases phosphorus reabsorption in the kidney tubules.

The PTH also increases active transport and absorption of intestinal calcium in the presence of normal amounts of vitamin D in the diet. In the young, a deficiency of this vitamim may lead to hypocalcemia and a failure of the bones to mineralize (soft bones), producing a disorder known as *rickets*.

Hypo- and Hyperparathyroid Secretions

The surgical removal of the parathroids (hypopathyroid secretion) results in a gradual decrease in plasma calcium and an elevation of phosphates and may cause tetany, a condition characterized by muscle spasms or rapid and irregular muscle contractions. The symptoms of tetany are relieved by administration of PTH or (temporarily) by injections of calcium salts and vitamin D.
Tumors of the parathyroid gland may cause hypersecretion of PTH (hyperparathyroidism) which results in hypercalcemia, hypophosphatemia, and increased calcium in the bones and kidneys and the formation of calcium-containing kidney stones. Injections of large doses of PTH produce the same symptoms.

Calcitonin

Calcitonin (CT) is a hormone secreted by the parafollicular cells of the thyroid (C cells) of mammals, which originally came from the embryonic tissues of the ultimobranchial bodies. These cells are present mainly in the ultimobranchial bodies of reptiles, birds, and fishes. The C cells are large and distinctive and can be identified by immunofluorescence.

Biochemical Activity

Calcitonin has been isolated and purified in several species. A linear polypeptide containing 32 amino acids, it is similar in man, swine, cattle, birds, and fishes. All 32 amino acids are necessary for complete activity or potency. Fish CT (salmon) is considerably more active than the mammalian hormone and has a higher affinity for the CT receptor.
Sensitive biologic assays in vitro, capable of detecting as little as 0.5 ng/ml CT, have been developed. Radioimmunoassays have also been developed for the different types of CT. Calcitonin may be present in circulating blood at concentrations of 50 pg/ml.
Release of CT is also regulated by blood calcium level (Figs. 28.1 and 28.2). When this level is high its release (CT) is stimulated, and it acts both to increase calcium absorption

into bones (accretion), and to lower blood calcium. It does not compete directly with PTH but has an opposite effect. Although CT also lowers blood phosphorus, it has no definite effect on kidney or gut. Calcitonin acts directly on bone cells and appears to inhibit the active transport of calcium from the bone cells to the circulation. Its effect is probably mediated by cAMP. Daily secretion rate of CT in man is about 0.5 mg. The exact physiologic role of calcitonin in man is uncertain, and its importance as an essential modulator of skeletal calcium homeostasis remains to be established.

Selected Readings

Catt KJ (1971) An ABC of endocrinology. Little, Brown, Boston

Ganong WF (1977) Review of medical physiology, 8th edn. Lange Medical, Los Altos, California

Morgan HE (1973) Endocrine control systems. In: Brobeck R (ed) Best and Taylor's physiological basis of medical practice, 9th edn. Williams and Wilkins, Baltimore

Queener SF, Bell NH (1975) Calcitonin, a general survey. Metabolism 24: 555

Raisz LG, Mundy GR, Dietrich JW, Canalis EM (1977) Hormonal regulation of mineral metabolism. In: McCann SM (ed) International review of physiology; endocrine physiology II, Vol. 16. University Park Press, Baltimore

Tepperman J (1973) Metabolic and endocrine physiology, 2nd edn. Year Book Medical, Chicago

Review Questions

1. How much calcium is in the human body, and what part of it is in blood and soft tissues?
2. What are the levels and types of calcium in the blood? Name and give amounts.
3. What hormones are involved in the regulation of blood and bone calcium? Name them.
4. Explain by diagram the relationship of blood and bone calcium to parathyroid hormone (PTH) and calcitonin.
5. What bone cells are responsible for bone resorption, and what activates these cells?
6. Where is calcitonin synthesized in mammals? In reptiles and birds?
7. What is the chemical nature of calcitonin? Of PTH?

Chapter 29 Adrenal Hormones

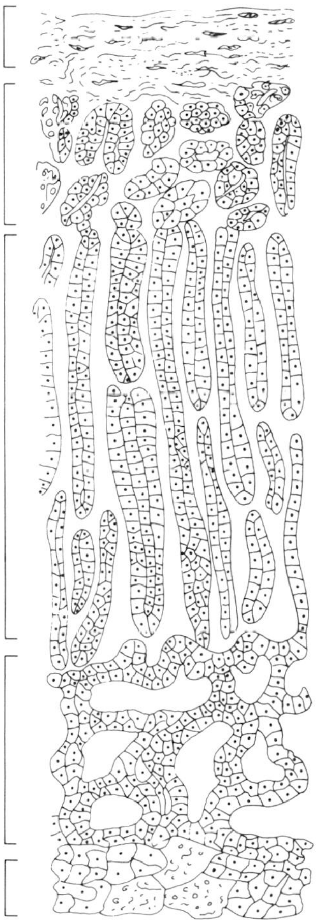

Fig. 29.1. Cross section of adrenal showing the zones of the cortex and the adjacent medulla.

Macro- and Microstructure

The adrenal glands in man are paired, pyramid-shaped and situated at the superior pole or cephalic ends of each kidney. Their lengths range from 2.5 to 5 cm, but their size may vary considerably, because of stimuli that release ACTH, which has a tropic effect on the adrenals.

The glands are made up of an outer portion—the cortex (bark)—which is of mesodermal origin and an inner medulla derived from nervous tissue and homologous to sympathetic ganglia. Preganglionic sympathetic nerve fibers innervate the medulla and when stimulated cause the release of the medullary hormones, epinephrine and norepinephrine.

The distribution of corticomedullary tissue varies widely with species, but in most mammals there is far more cortical than medullary tissue. The cortex synthesizes and releases the cortical hormones.

The adrenals of most mammalian species are composed of different cell types arranged in three different strata, or zones, from outermost to innermost areas (Fig. 29.1). They are the: zona glomerulosa (1), zona fasciculata (2), and zona reticularis (3).

The zone glomerulosa is rich in mitochondria and poor in lipids, the zona fasciculata has columnar cells rich in lipids, and the zona reticularis has flattened cells arranged in a network (reticulum). In man, aldosterone is produced in zone 1, cortisol in zones 2 and 3. Zone 1 is capable of regenerating cells to add to its own layer as well as to the other zones. Zones 2 and 3 are under the tropic influence of ACTH, but the glomerulosa is relatively independent of it.

Both cortex and medulla are highly vascular and contain venous sinusoids that absorb the hormones released.

Cortical Hormones

Many different steroid hormones have been isolated from the adrenals, but the principal ones, secreted in physiologic quantities, are the mineralocorticoid, aldosterone, which is concerned mainly with the metabolism of sodium and potassium, and the glucocorticoids, cortisol and corticosterone, which are concerned mainly with the metabolism of glucose. The structure and synthesis of these hormones are shown in Fig. 29.2.

The adrenal steroids are of two main structural types, including: (1) the C-21 steroids with 21 carbon atoms and a 2-carbon side chain attached to position 17 (D ring), and (2) the c-19 steroids with 19 carbon atoms and a keto or hydroxyl group at position 17 (ketosteroids). Two c-21 and c-19 compounds are progesterone and dehydroepiandrosterone respectively; the latter has androgenic potency. Synthesis of adrenal and sex steroids begins with cholesterol, which is converted first to pregnenolone by the mitochondria of the adrenal cortex cells and then to progesterone.

Progesterone occupies a key position in the synthesis of both cortical steroids and also androgens and estrogens (see Chap. 25). It is acted on by the enzymes 17 α-hydroxylase, 21-hydroxylase, and 11 β-hydroxylase to form 17 α-hydroxyprogesterone, 17 α-hydroxycorticosterone, and *cortisol.* The 11- and 21-hydroxylases occur in all three zones of the adrenal, but 17 α-hydroxylase is only in the fasciculata and reticularis zones.

Aldosterone, which is synthesized only in the zona glomerulosa, is derived from both progesterone and corticosterone directly by the enzyme 18-aldolase (located only in this zone; see Fig. 29.2).

Corticosterone is the principal steroid secreted by the adrenals of rats, mice, and birds; cortisone is the principle adrenocorical hormone of cows, sheep, monkeys, and man. Only 12% of the corticoids secreted in man is corticosterone.

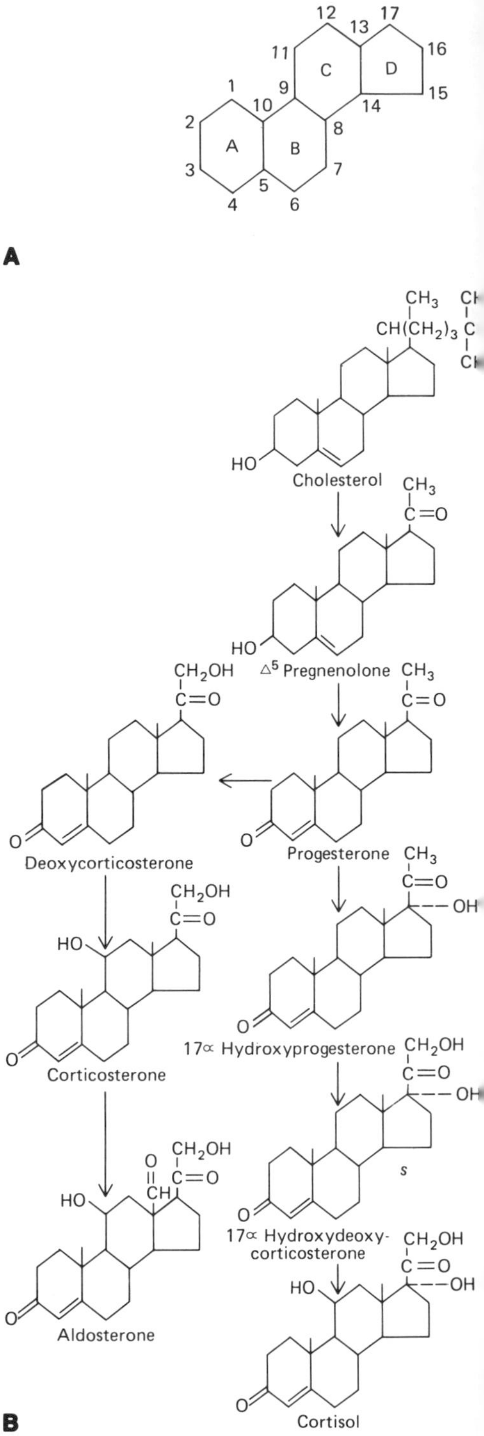

Fig. 29.2A and B. Cortical hormones. **A** basic steroid nucleus showing ring designation and numbering of carbon atoms; **B** major synthetic pathways of aldosterone, cortisol, and corticosterone.

Release of ACTH and Cortical Hormones

In Chap. 25 it was indicated that there is a releasing factor from the hypothalmus—ACTH-RH or CRH-CRF—that is involved in releasing ACTH from the pituitary. The latter in turn releases cortical hormones from the adrenal with the exception of aldosterone, the secretion of which is regulated mainly by the renin-angiotensin system, which stimulates the zona glomerulosa of the adrenal. Substances in the posterior lobe, similar to vasopressin, also may stimulate the adrenal to release aldosterone.

Table 29.1. Secretion and approximate concentrations of cortisol and ACTH.

Time of Day (h)	Cortisol (μg/100 ml)	ACTH (pg/ml)
Dark-2100	5.0	30.0
Dark-2400	5.2	33.0
Light-0600	15.0	60.0
Light-1200	9.4	44.0
Light-1800	5.3	28.0

Glucocorticoids

Cortisol (also corticosterone) and ACTH are secreted in a diurnal or circadian fashion, with the greatest flow in the early morning (6 AM) and the least in the evening. With the advent and use of sensitive radioimmunoassays, which allow frequent sampling (every 10 minutes), secretion of ACTH and cortisol occurs intermittently and in short bursts or episodes lasting only a few minutes (Table 29.1). This diurnal secretory rhythm of ACTH is attributable to the suprachiasmatic nucleus of the hypothalamus, the site of release of ACTH-RH.

Stress and ACTH

Various types of stress, including physical and emotional, cause the release of ACTH and stimulation of adrenal secretion of cortisol and may increase the size of the adrenals (hypertrophy). Stress and ACTH also lessen the ascorbic acid and cholesterol content of the adrenal gland.

Feedback Mechanism

The relationship between the plasma level of cortisol and plasma ACTH is not a close or sensitive one as indicated by the fact that: (1) plasma cortisol levels may fall to zero without stimulating ACTH release, (2) ACTH secretion may occur at a time when cortisol levels are either high or in the normal range, and (3) there is a considerable time lag between fall in plasma corticosterone (main glucocorticoid in rats) and rise in ACTH in adrenalectomized rats.
This evidence suggests that cortisol and corticosterone levels in the blood may not be directly and intimately involved (feedback mechanism) in regulating ACTH release. This is true within limits, but when the blood level of cortisol is high it inhibits ACTH secretion. At lower levels of cortisol, ACTH secretion occurs. These threshold levels of cortisol in the blood are believed to influence the level of the hormone in the central nervous system (CNS) or hypothalamus, and this finding is thought to be the key regulating factor in the ultimate release of ACTH, by affecting directly the release of ACTH from the pituitary or affecting indirectly the release of ACTH-RH from hypothalamus and median eminence. The exact mechanisms are in doubt (see also Chap. 25 for further information on ACTH-RH).

Secretion of Aldosterone

The secretion of aldosterone is affected less by ACTH release than are the glucocorticoids. Larger doses of ACTH are required to release aldosterone, and the secretion is transient even when high levels of ACTH are maintained. Stimuli other than ACTH that affect aldosterone secretion are: (1) high potassium intake, (2) low sodium intake, (3) water depletion, (4) aortic constriction, and (5) maintaining erect position; the latter causes an increased aldosterone secretion because it increases renin secretion, as does sodium restriction and an increase in plasma potassium. Stress and hemorrhage increase secretion of aldosterone and glucocorticoids.

Renin-Angiotensin System

The principal regulator of the synthesis and secretion of aldosterone is the renin-angiotensin system (Fig. 29.3). Renin is a proteolytic enzyme secreted by the cells of the glomerular apparatus of the kidney. Renin from the kidney acts on alpha gobulin to form angiotensin I, which is the precursor of angiotensin II (an octapeptide), which in man stimulates aldosterone but not corticosterone secretion. Aldosterone secretion is regulated by a feedback mechanism (Fig. 29.3). Small changes in plasma levels of K^+ (1 mEq/liter) and much larger changes in Na^+ induce aldosterone secretion.

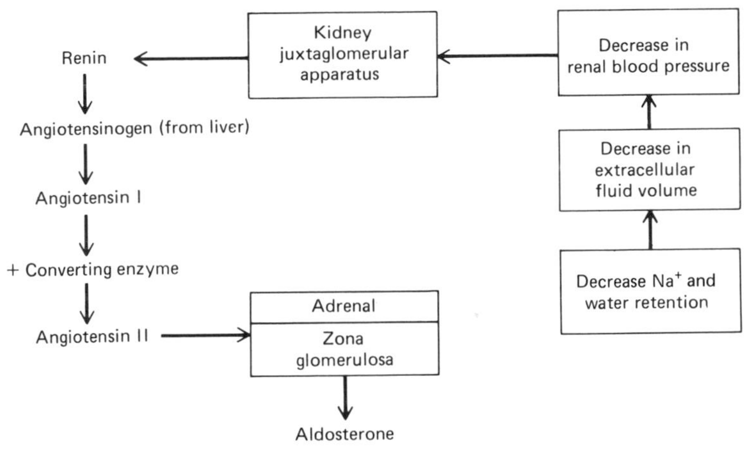

Fig. 29.3. Regulation of aldosterone secretion by renin-angiotensin system (see text). Aldosterone secretion is increased when there is decrease in Na^+ and water retention, extracellular fluid volume, and blood pressure. Changes in opposite direction (increase) decreases renin output and aldosterone secretion (---→).

The basal secretion of aldosterone under normal conditions ranges from 50–200 μg/24 h; the excretion rate is about 2–16 μg/24 h. Aldosterone in normal human blood plasma ranges from 3–15 ng/100 ml, most of it bound to plasma proteins although some is free. The free parts of aldosterone and cortisol are physiologically active.

Action of Glucocorticoids

In man, glucocorticoids and cortisol stimulate glycogen synthesis and storage (carbohydrate metabolism) and decrease

utilization, resulting in a tendency to hyperglycemia. Glucocorticoids increase or mobilize amino acids from skeletal muscle protein and prepares them for incorporation into carbohydrates (gluconeogenesis). Thus proteins are converted into carbohydrates and stored mainly in the form of glycogen. In lipid metabolism, glucocorticoids promote the mobilization of fatty acids from peripheral adipose tissue stores but inhibit fatty acid synthesis in the liver. Cortisol tends to aid other hormones like thyroxine and catecholamines in the breakdown of fats (see also Chap. 22).
Cortisol also has some effect on retention of sodium but considerably less than aldosterone. It acts to increase blood flow to the kidneys and glomerular filtration. Cortisol may cause atropy of lymphoid tissue, lymphopenia, and decrease in eosinophils. Infections and inflammations are increased in patients who lack glucocorticoids, and their response to stress is impaired.

Mineralocorticoids

The principal effect of *aldosterone* is on the kidney; it causes retention and absorption of Na^+ in the distal tubules and excretion of K^+. Reabsorption of Na^+ also occurs in sweat, saliva, and gastric juice where the ions diffuse out into the surrounding epithelial cells and are actively transported to the interstitial fluid. The *mechanism of action* of aldosterone, similar to that of cortisol, involves a specific receptor protein complex that migrates to the nucleus and stimulates DNA-dependent RNA and protein synthesis (see Chap. 22).

Adrenal Cortex Dysfunction

Atropy of the adrenal cortex or other pathologic impairment causes *Addison's disease* (primary adrenocortical insufficiency) with a decreased output of aldosterone and cortisol. This deficiency (mainly of aldosterone) leads to Na^+ deficiency in blood and tissues (hyponatremia) and increased K^+ (hyperkalemia). The salt depletion causes a loss of tissue fluids and dehydration. The decreased cortisol secretion causes weakness, drowsiness, lack of resistance to stress, and lowered blood sugar (hypoglycemia). Because of the decreased cortisol secretion, the negative feedback on the pituitary output of ACTH is impaired, and ACTH secretion increases (secondary adrenocortical insufficiency) as does beta-MSH secretion; the latter causes deposition of melanic pigment in the skin, particularly in exposed areas like the elbows. Sufferers of Addison's disease require lifelong therapeutic administration of both aldosterone and cortisol.

Cushing's disease or *syndrome* is an example of chronic excess production of cortisol caused by increased secretion of ACTH. The exact nature of the pituitary defect is not understood. Symptoms include a moon face, wasting of the arms and legs, and obesity with accumulations of fat around the middle and on the face, neck, and trunk (Fig. 29.4). Purple striations of skin are conspicuous around the abdomen and on the thighs. The patient tolerates carbohydrates poorly, may have diabetes mellitus if a familial tendency exists, and is susceptible to infections. The hypersecretion of cortisol results in Na^+ retention and hypertension.

Excess secretion of aldosterone caused by a tumor of the zona glomerulosa results in K^+ depletion and Na^+ retention, usually without edema, and is known as *Conn's syndrome*. Symptoms include: weakness, hypertension, tetany, and polyuria.

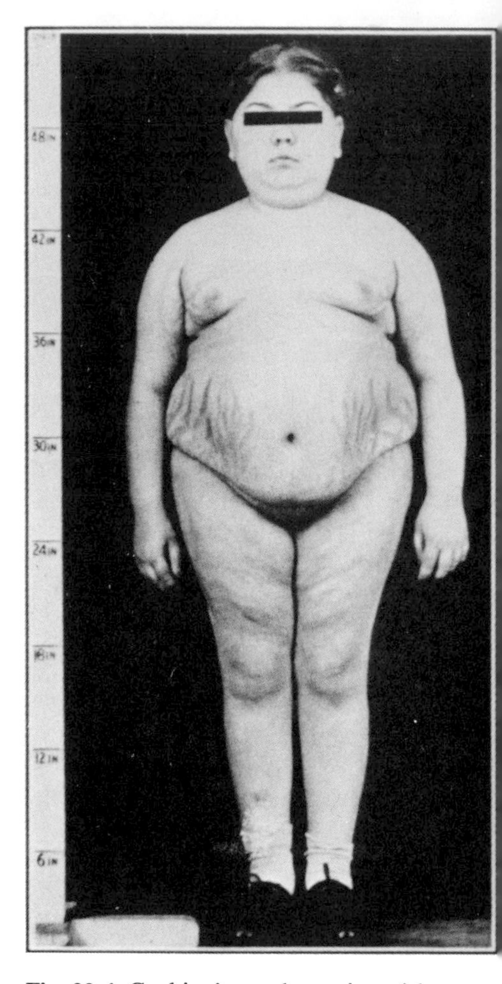

Fig. 29.4. Cushing's syndrome in a girl age 12 years.

Adrenal Medulla and Catecholamines

At birth, the medulla, the inner area of the adrenals, is made up of primitive sympathetic nerve cells, which early on differentiate into mature chromaffin cells. Portal venous blood from the cortex passes through the medulla, carrying cortical hormones (catecholamines) with it.

Synthesis and Release

The principal hormone of the medulla is epinephrine (norepinephrine is secreted in lesser amounts). The syntheses of these hormone are shown in Fig. 29.5.

The steps can be summarized as follows (substance plus enzyme): 1) TYROSINE + T. hydroxylase → 2) DOPA + L-aromatic amino acid decarboxylase → 3) DOPAMINE + dopamine beta oxidase → 4) NOREPINEPHRINE + phenylethanolamine N-methyl transferase (PNMT) → 5) EPINEPHRINE.

Catecholamines when synthesized are stored as granules in the adrenal medullary cells (in vesicles). The hormones contained in these vesicles can be released by appropriate stimuli, including: excitement, sympathetic nerve stimulation, exercise, cold, histamine, and glucagon.

Metabolism and Secretion

Epinephrine (E) and norepinephrine (NE) are metabolized by two enzymes, monamine oxidase (MAO) and catechol-o-methyl transferase (COMT). They are o-methylated in the blood and some derivatives are excreted in urine. Those not

Fig. 29.5. Pathways of synthesis of norepinephrine and epinephrine (see text).

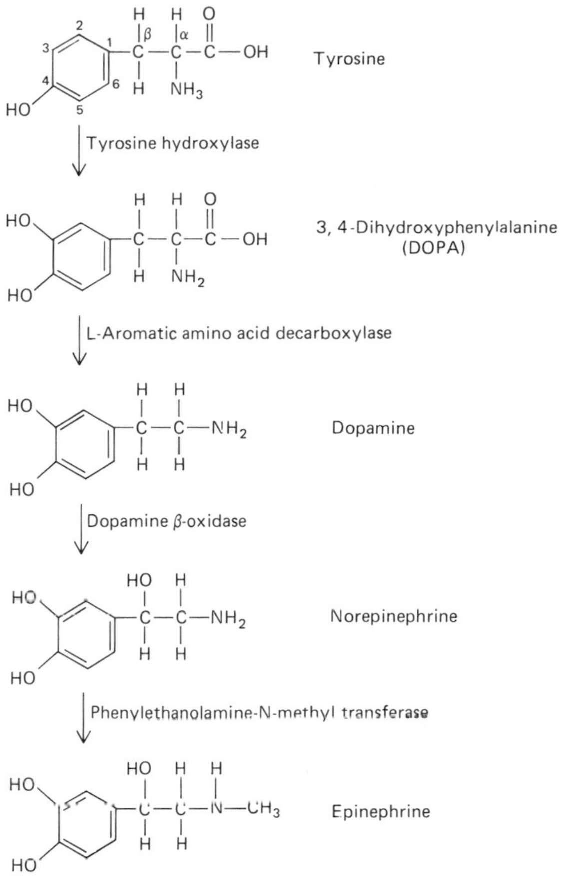

Tyrosine

↓ Tyrosine hydroxylase

3, 4-Dihydroxyphenylalanine (DOPA)

↓ L-Aromatic amino acid decarboxylase

Dopamine

↓ Dopamine β-oxidase

Norepinephrine

↓ Phenylethanolamine-N-methyl transferase

Epinephrine

excreted are then oxidized to 3-methoxy-4 hydroxymandelic acid (vanilmandelic acid). Both NE and E are rapidly metabolized, the compounds lasting only a few seconds in circulating blood.

About 80% of the catecholamines secreted by the human adrenal is E. Most of the NE in the adrenals is converted to E by the enzyme PNMT. The level of E in human blood is about 30 pg/ml and is derived mainly from the adrenal, but the level of NE is much higher (5–7 times)—about 200 pg/ml. Most of the NE is derived from sympathetic nerves discharging NE into the blood stream on stimulation (see also Chap. 10).

Many factors stimulate the release of both NE and E from the adrenal, but certain stimuli like asphyxia and hypoxia increase disproportionately the output of NE. Familiar emotional stresses also increase the output of NE, whereas facing unexpected situations increases the output of E.

Increased secretion of adrenal E is part of the emergency response system of the body, preparing the individual for sudden physical exertion (flight-or-fight reaction). Exposure to cold, shivering, and heat induces the secretion of catecholamines, which have a calorigenic or warming effect.

The effects of E and N have been considered and discussed in several sections of this book, but mainly in Chaps. 14 and 15. Some of these effects are summarized in Table 29.2.

Table 29.2. Physiologic effects of epinephrine and norepinephrine.

Effect	Norepinephrine	Epinephrine
Blood pressure	++++	++
Heart rate	+++	++
Peripheral resistance	++++	++
Heat production	+++	++++
Mobilization of fatty acids	+++	++
Glycogenolysis and hyperglycemia	+	+++
Constriction of smooth muscle of blood vessels	+++	+ or −
Motility of stomach	−−	−−
Motility of intestine	−−	−−
Sweat gland	Sweating	

Key: +, slight increase; ++, moderate; +++, heavy; ++++, severe; −, slight decrease; −−, moderate.

Selected Readings

Bentley PJ (1976) Comparative vertebrate endocrinology. Cambridge University Press, New York

Biglieri EG (1976) Aldosterone. In: Ingbar SH (ed) The year in endocrinology, 1975–1976. Plenum, New York

Brodish A, Lymangrover JR (1977) Hypothalamic pituitary adrenocortical system. In: McCann SM (ed) International review of physiology, endocrine, physiology II. University Park Press, Baltimore

Catt KJ (1971) An ABC of Endocrinology. Little, Brown, Boston

Ganong WF (1977) Review of medical physiology, 8th edn. Lange Medical, Los Altos, California

Morgan HE (1973) Endocrine control systems. In: Brobeck JR (ed) Best & Taylors, physiological bases of medical practice, 9th edn. Williams and Wilkins, Baltimore

Review Questions

1. What are the principal hormones secreted by the adrenal cortex and medulla in man?
2. Which of the hormones are mainly under the control of ACTH from the pituitary?
3. What is the relationship between ACTH secretion and blood level of cortical hormones? Discuss.
4. What is the renin-angiotensin system or complex and how is it related to aldosterone secretion?
5. In what ways are cholesterol and progesterone involved in the synthesis of adrenal cortical hormones?
6. What is the immediate precursor of aldosterone?
7. What zone of the adrenal is involved in synthesis of aldosterone?
8. What are the principal effects or action of: (1) glucocorticoids and; (2) mineralocorticoids?
9. Name a disease caused by cortical insufficiency and another caused by excess of cortical hormones.
10. Where are epinephrine and norepinephrine produced?
11. List some of the actions of epinephrine and norepinephrine.

Chapter 30 Pancreatic Hormones

The islet cells of Langerhans are the endocrine cells of the pancreas that secrete insulin and glucagon, two hormones that are important in the regulation of carbohydrate metabolism.

Diabetes mellitus, a disorder caused by lack of insulin from the pancreas, was known as early as 1500 BC. In more modern times (1674) the disease was described in detail by Thomas Willis, English anatomist and physician, who indicated that the urine had a sweet taste. Later sugar was discovered in the blood (hyperglycemia) as well as in the urine (glycosuria).

In 1889 Von Mering and Minkowski proved that the removal of the pancreas of the dog caused diabetes mellitus, and in 1900 Eugene Opie, American pathologist, indicated that the islet cells were probably involved. In 1921 Frederick Banting and Charles Best isolated a substance from pancreases that prevented diabetes mellitus in pancreatectomized dogs; their work later won the Nobel Prize. Administration of such pancreatic extracts in 1922 saved the life of a child who had diabetes mellitus. This substance, insulin, was isolated and crystallized in 1926. Its amino acid sequence was established in 1955, and it was synthesized both in 1963 and later. The discovery of the cause and successful treatment of diabetes mellitus remains one of the great landmarks of medical history.

Anatomy

The exocrine portion of the pancreas produces the pancreatic juices and enzymes and comprises most of the pancreatic tissue (see Chap. 21). The pancreas is located in the duodenal loop of the intestine (Fig. 20.1). The islet tissue

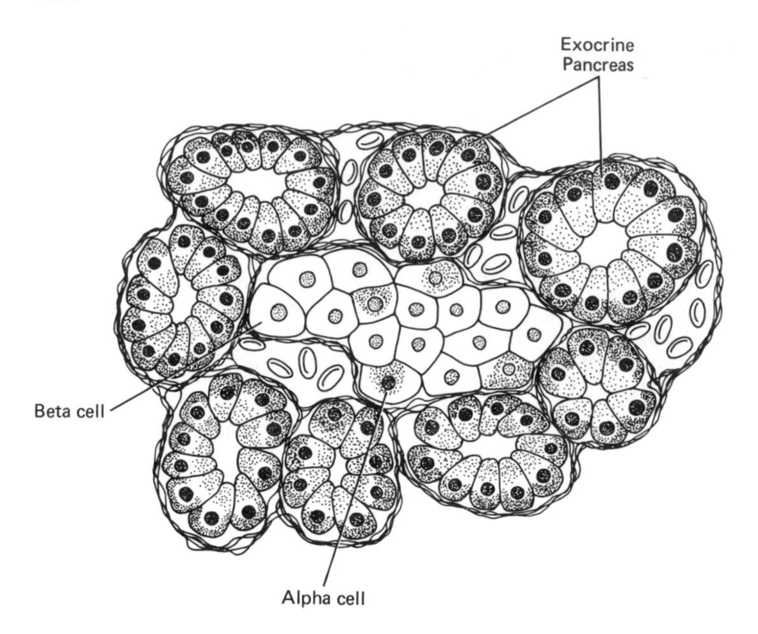

Fig. 30.1. Section of human pancreas showing the alpha and beta cells of the islet of Langerhans, surrounded by the acinar cells of the exocrine pancreas. Note blood vessels between the acinar cells.

(endocrine part) makes up about 1%–2% of the pancreas and is characterized by three types of cells: alpha, beta, and delta. They are surrounded by small acinar cells (exocrine) (Fig. 30.1). The alpha cells, which secrete glucagon, make up about 20% of the islet cells and the beta cells, which produce insulin, make up about 65%–75%. The delta cells contain other hormones, such as gastrin and somatostatin. The cytoplasm of the alpha cells stain reddish-pink with acid stains, and the beta-cell cytoplasmic granules stain bluish-purple with basic stains. Beta-cell granules contain packets of insulin, and the alpha granules contain glucagon.

Insulin

Insulin is a polypeptide containing two separate chains (A and B) of amino acids linked by disulfide bridges. The type and sequence of amino acids vary with the species, but the insulin structure in most species is similar. The A chain contains 21 amino acids, the B chain 30 (Fig. 30.2). The molecular weight of human insulin is 5734. Insulin in man differs from other mammalian insulins in the type and sequence of amino acids at positions 8, 9, and 10 of chain A and position 30 of the B chain.

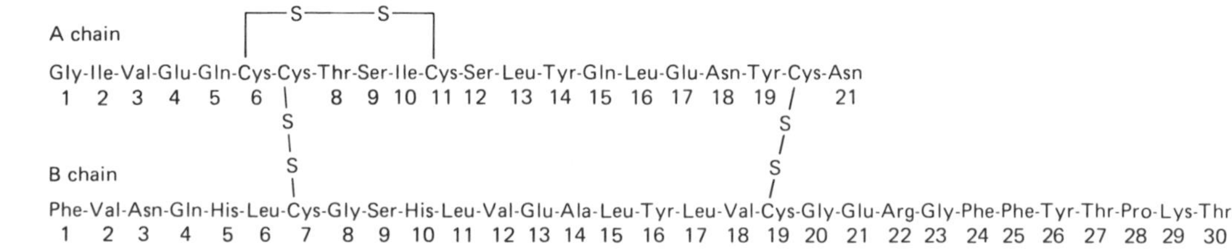

Fig. 30.2. Structure of human insulin. Two polypeptide chains A and B linked by disulfide bonds (S—S).

Insulin is synthesized in the endoplasmic reticulum of beta cells and packaged into granules that move to the cell membrane and are expelled to the outside of the cell mainly by exocytosis (see Chap. 2). It is first synthesized as a large single-chain structure known as proinsulin, which later forms two chains.
Almost all the tissues can metabolize insulin, but much of it (more than 80%) is metabolized in the liver or kidneys. Insulin has a half-life of 10 to 25 minutes in man, in smaller animals it is much shorter.

Secretion

Insulin is usually prepared commercially and dispensed as 24 Units/mg, or IU = 40 μg of insulin. Daily human secretion is roughly 25 to 50 Units, or 1 to 2 mg, into the pancreatic vein, about one-half of which is destroyed in the liver and never reaches the peripheral circulation. The secretion is in response to various stimuli. The rate of secretion is influenced in part by the rate of metabolism. Insulin exists in the beta cell in two forms, namely, a storage or preformed pool, which responds to stimuli by acute insulin release and a pool synthesized and secreted to maintain the storage pool and provide insulin in basal and prolonged stimulation.

Stimuli-Affecting Release

The principal stimuli-affecting release of insulin are: (1) glucose level in blood, particularly high-level (hyperglycemia); (2) amino acid levels; when high, increases insulin secretion; (3) alpha adrenergic receptor; stimulated by release of epinephrine and norepinephrine; decreases release of insulin by inhibiting formation of cAMP; (4) beta adrenergic receptor; stimulation by epinephrine causes increased insulin release and cAMP formation; (5) glucagon; stimulates increased release of insulin by increasing blood glucose, which then causes release of insulin, and by acting directly on beta cells; (6) other hormones (secretin and cholecystokinin); and (7) vagus nerve (some species). Some of the factors affecting insulin release are shown in Fig. 30.3.

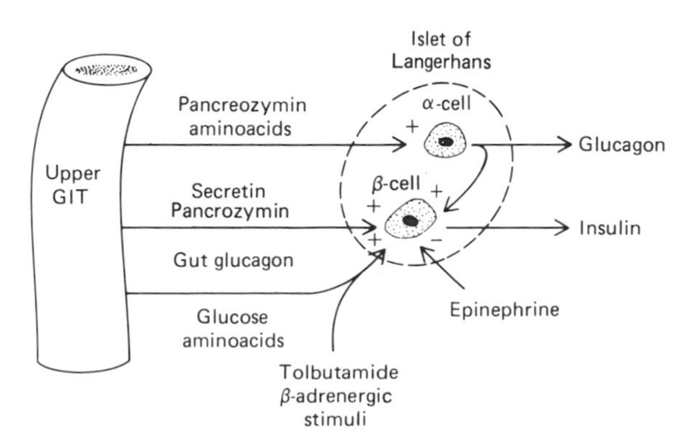

Fig. 30.3. Factors affecting secretion of insulin and glucagon. (Modified after Catt KJ (1971) An ABC of endocrinology. Little, Brown and Co, Boston.)

Glucose

The most potent releaser of insulin, glucose when ingested causes a rapid release of preformed insulin within one to two minutes; the level then drops and later rises (30–60 minutes) and is sustained at this level. This phase is probably derived from newly synthesized insulin. Following ingestion of glucose, the insulin level in plasma may rise from 20 to 50–150 μunits/ml.

Actions

The major actions of insulin are: reduction of blood sugar level by increasing glucose uptake and utilization in tissues, and stimulation of synthesis of (1) glycogen in muscle and liver, (2) lipids in adipose tissue and liver, and (3) protein, RNA and DNA in cells.

Glucose Uptake and Utilization

Blood glucose level is a reflection of the regulating action of insulin. After fasting for eight to 24 hours the normal blood glucose level in man is 70–120 mgs/100 ml. Levels appreciably above the upper limit indicate diabetic symptoms. After a meal, particularly one high in carbohydrates, there is a rapid increase in blood glucose level, which then returns to normal within two hours. This happens because insulin is released after glucose ingestion, which lowers blood sugar. The *glucose tolerance test* determines normal insulin activity or diabetic symptoms after ingestion of a known amount of glucose. The comparative responses of a normal subject and a diabetic are shown in Fig. 30.4. The test shows that the diabetic does not produce enough insulin to reduce the blood sugar level appreciably within $1^1/_2$–2 hours as the normal subject does.

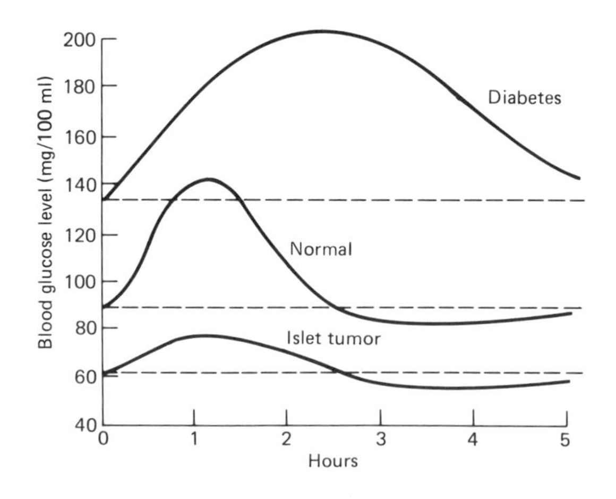

Fig. 30.4. Glucose tolerance curve in normal subject, in a diabetic, and in patient with tumor of pancreas (hyper insulinism) receiving 50 g of glucose orally. (Modified after Guyton AC (1971) Textbook of medical physiology. Saunders, Philadelphia.)

Insulin increases glucose uptake in: muscle (smooth, skeletal, and cardiac), adipose tissue, hypophysis, mammary glands, and leukocytes. Some other tissues may take up glucose, but its uptake is not facilitated by insulin in brain, kidney tubules, red blood cells, and gastrointestinal tract.
Although insulin increases uptake and utilization of glucose in the peripheral tissues, the liver takes up blood sugar and converts and stores it as glycogen (glycogenesis); insulin, however, does not affect directly the movement of glucose across the liver cells. Insulin breaks down glycogen to glucose (glycogenolysis) and releases it into the bloodstream as it is needed by the peripheral tissues.
Gluconeogenesis is another important pathway by which glucose is formed by the liver. It represents the formation of glucose from proteins and fats and occurs when body stores of carbohydrates are low, as in starvation. As much as 60% of the amino acids in the body can be converted into carbohydrates.
Glucocorticoids and thyroxine mobilize amino acids which may be deaminated in the liver to form substrates for glucose conversion. For further details on pathways of metabolism of proteins, carbohydrates, and fats, see Chap. 22.

Diabetes Mellitus

Diabetes mellitus is a disorder characterized by excess urine flow (diabetes) which has a sweet taste (mellitus). It results from decreased production of insulin by the islet cells and the tendency appears to be inherited. It affects both the young and adults. The juvenile type occurs early in life and is usually more severe, requiring immediate treatment with insulin replacement. The adult type may require treatment with either insulin or a substance that stimulates release of insulin, such as the sulfonylurea tolbutamide. Mild diabetes may be controlled by diet.
The principal symptoms of diabetes mellitus, or lack of insulin activity, in man, are: (1) increased blood sugar (hyperglycemia), (2) increased urinary sugar (glycosuria), (3) increased urination, (4) increased water consumption (polydipsia), (5) weight loss and muscle weakness, (6) increased urination, (4) increased water consumption (polydipsia), (5) weight loss and muscle weakness, (6) increased catabolism of proteins and lipids, and (7) increased ketosis and acidosis.
The defect prevents normal tissue uptake and utilization of carbohydrates, which are the principal and immediate sources of energy in the body. This defect in carbohydrate metabolism interferes with the normal metabolism of lipids and proteins, which are then catabolized at a higher rate to provide alternative sources of energy.
There is also an increased release of glucose from the liver, which increases the tendency to hyperglycemia, glycosuria,

dehydration, and osmotic diuresis; this in turn increases water consumption (polydipsia).

All of the just-cited abnormalities are correctable by insulin administration.

Pancreatectomy has been performed in many species, including man. It produces the symptoms of severe diabetes mellitus in most species except in certain avian ones. In them the procedure produces mild diabetes or no symptoms, suggesting that there may be non pancreatic sources of insulin secretion, or that other hormones or agents may be involved. Further evidence for this view derives from the failure of certain chemicals like streptozotocin to destroy avian pancreatic beta cells as it does in mammals and produce symptoms of diabetes mellitus.

Insulin Excess

Too much insulin produces hypoglycemia (blood sugar as low as 20–35 mg/100 ml); this may be accompanied by increased secretion of adrenocortical hormones, which tends to mobilize glucose and compensate for the hypoglycemia. Low blood sugar may lead to mental confusion, weakness, dizziness, hunger, and finally convulsions and coma. Insulin excess occurs in patients with tumors of the endocrine pancreas in whom the blood sugar level may drop as low as 20 mg/100 ml.

Glucagon: Secretion and Metabolism

Glucagon, secreted by the alpha cells of the pancreas, is a linear polypeptide, containing 29 amino acids arranged in sequence (Fig. 30.5); the structures of human and porcine hormones are the same. Glucagon has a molecular weight of 3485. Avian (chicken) glucagon also has 29 amino acids but differs from the human type by the substitution of asparagine (ASN) for serine at position 28.

The rate of pancreatic glucagon secretion in man is very small when compared to the doses of glucagon previously employed to study the metabolic effects of the hormone. Basal secretory rates based on radioimmunoassay in normal subjects are about 100–150 μg/day; these values are probably high because the immunoreactivity of the hormone is greater than its biologic activity. Factors that increase secretion are: amino acids, glucocorticoids, stress and infections, exercise, gastrin, cholecystokinin, and insulin. Factors that decrease secretion are: glucose, secretin, free fatty acids, and ketones.

Fig. 30.5. Structure of glucagon in man.

His-Ser-Gln-Gly-Thr-Phe-Thr-Ser-Asp-Tyr-Ser-Lys-Tyr-Leu-Asp-Ser-Arg-Arg-Ala-Gln-Asp-Phe-Val-Gln-Trp-Leu-Met-Asn-Thr
1 2 3 4 5 6 7 8 9 10 11 12 13 14 15 16 17 18 19 20 21 22 23 24 25 26 27 28 29

Glucagon has a half-life of 5–10 minutes and is degraded (catabolized) by a number of tissues, primarily by the kidney in contrast to insulin, which is catabolized mainly by the liver.

Action

The principal action of glucagon is elevation of blood glucose, which it does by stimulating adenyl cyclase in the liver which increases synthesis of cAMP. This results in the breakdown of glycogen (glycogenolysis) and release of glucose in the liver, but not in muscle. Glucagon increases gluconeogenesis from amino acids in the liver.

Many of the experiments on glucagon have dealt with very large, unphysiologic doses (pharmacologic doses), which have definite effects on carbohydrate metabolism. Recent experiments, however, (see review of Sherwin and Felig) raise questions about the significance of such observations and whether or not glucagon in physiologic amounts plays an important role in glucose regulation (homeostasis). Physiologic doses in man produce a small and transient rise in plasma glucose.

Recent results also suggest that the hyperglycemia of diabetes mellitus is caused by insulin deficiency and is not influenced significantly by excess glucagon secretion, but this view is debatable.

Selected Readings

Catt KJ (1971) An ABC of endocrinology. Little, Brown, Boston

Ganong WF (1977) Review of medical physiology, 8th edn. Lange Medical, Los Altos, California

Guyton AC (1977) Basic human physiology, 2nd edn. Chap. 25. Saunders, Philadelphia

Hazelwood RL (1976) Carbohydrate metabolism (Chap. 11) and Pancreas (Chap. 21). In: Sturkie PD (ed) Avian physiology, 3rd edn. Springer Verlag, New York

Sherwin R, Felig P (1977) Glucagon physiology in health and disease. In: McCann SM (ed) International review of physiology-endocrine physiology II, Vol. 16. University Park Press, Baltimore

Tepperman J (1973) Metabolic and endocrine physiology, 3rd edn. Year Book Medical, Chicago

Turner CD, Bagnara JT (1976) General endocrinology, 6th edn. Saunders, Philadelphia

Review Questions

1. Who were Frederick Banting and Charles Best and what did they discover?

2. What cells of pancreas secrete insulin? Glucagon?
3. Insulin is a polypeptide made up of two chains of amino acids containing how many acids?
4. What is the principal stimulus for secretion of insulin?
5. What is the principal action of insulin?
6. What are the symptoms of diabetes mellitus? Discuss.
7. What are: (a) glycogenolysis, (b) glycogenesis, and (c) gluconeogenesis.
8. What disorder is caused by an excess of insulin? Explain.
9. Glucagon is a linear polypeptide containing how many amino acids?
10. What is the principal action of glucagon?
11. Is glucagon as important as insulin in carbohydrate regulation? Discuss.

Index

A

B

C

D

E

F

I

K

L

M

N

O

P

R

S

T

U

V

W